Emerging Protozoan Pathogens

Emerging Protozoan Pathogens

Edited by Naveed Ahmed Khan

School of Biological and Chemical Sciences
Birkbeck College
University of London
UK

Taylor & Francis
Taylor & Francis Group

Published by:

Taylor & Francis Group

In US: 270 Madison Avenue
New York, N Y 10016
In UK: 2 Park Square, Milton Park
Abingdon, OX14 4RN

© 2008 by Taylor & Francis Group

ISBN: 978-0-415-42864-4

Library of Congress Cataloging-in-Publication Data

Emerging protozoan pathogens / edited by Naveed Ahmed Khan.
p. ; cm.
Includes bibliographical references and index.
ISBN 978-0-415-42864-4 (alk. paper)
1. Protozoa, Pathogenic. 2. Protozoan diseases. 3. Emerging infectious diseases.
I. Khan, Naveed Ahmed.
[DNLM: 1. Protozoan Infections--physiopathology. 2. Host-Parasite Relations.
3. Opportunistic Infections. 4. Protozoa--pathogenicity. 5. Protozoan
Infections--therapy. WC 700 E53 2008]
QR251.E44 2008
571.9′94--dc22

2007043118

Editor: Elizabeth Owen
Editorial Assistant: Kirsty Lyons
Production Editor: Simon Hill
Typeset by: Techset Composition
Printed by: Cromwell Press

Printed on acid-free paper

10 9 8 7 6 5 4 3 2 1

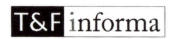

Taylor & Francis Group, an informa business   Visit our Web site at http://www.garlandscience.com

Contents

Contributors

Aurélien Dumètre   Parasitologie-Mycologie, Faculté de Médecine, CNR Toxoplasmosis, Limoges, France

Guy A. Cabral   Virginia Commonwealth University School of Medicine, Department of Microbiology and Immunology, Richmond, VA, USA

Dominique Aubert   Parasitologie-Mycologie, Faculté de Médecine, CNR Toxoplasmosis, Reims, France

Raina E. Fichorova   Department of Obstetrics and Gynecology, Brigham and Women's Hospital, Harvard Medical School, Boston, MA, USA

Francis Derouin   Parasitologie-Mycologie, Faculté de Médecine Denis Diderot, Université Paris and Hôpital Saint-Louis, Assistance Publique-Hôpitaux de Paris, Paris, France

Lynne S. Garcia   LSG & Associates, Santa Monica, CA, USA

Jeremy Gray   Agriculture and Food Science Centre, School of Biology and Environmental Science, University College Dublin, Belfield, Dublin, Republic of Ireland

Hervé Pelloux   Parasitologie-Mycologie, Département des Agents Infectieux, CHU de Grenoble and Adaptation et Pathogénie des Microorganismes, Institut Jean Roget, Grenoble, France

Isabelle Villena   Laboratoire de Parasitologie-Mycologie, Faculté de Médecine, CNR Toxoplasmosis, Reims, France

Edward L. Jarroll   Department of Biology, Northeastern University, Boston, MA, USA

Naveed Ahmed Khan   School of Biological and Chemical Sciences, Birkbeck College, University of London, UK

Marie-Laure Dardé   Parasitologie-Mycologie, Faculté de Médecine, CNR Toxoplasmosis, Limoges, France

John. J. Lucas   Department of Biochemistry and Molecular Biology, SUNY Upstate Medical University, Syracuse, NY, USA

Francine Marciano-Cabral   Virginia Commonwealth University School of Medicine, Department of Microbiology and Immunology, Richmond, VA, USA

Stanley Dean Rider Jr. Department of Veterinary Pathobiology, College of Veterinary Medicine and Biomedical Sciences, Texas A & M University, College Station, TX, USA

Frederick L. Schuster California Department of Health Services, Viral and Rickettsial Disease Laboratory, California, USA

Bibhuti N. Singh Department of Biochemistry and Molecular Biology, SUNY Upstate Medical University, Syracuse, NY, USA

Kevin S. W. Tan Laboratory of Molecular and Cellular Parasitology, Department of Microbiology, Yong Loo Lin School of Medicine, National University of Singapore, Singapore

Govinda S. Visvesvara Centers for Disease Control and Prevention, Division of Parasitic Diseases, Atlanta, GA, USA

Louis M. Weiss Division of Infectious Diseases, Albert Einstein College of Medicine, Bronx, New York, NY, USA

Guan Zhu Department of Veterinary Pathobiology, College of Veterinary Medicine and Biomedical Sciences, Texas A & M University, College Station, TX, USA

Introduction

This book is intended as a guide for microbiologists, molecular and cell biologists and health professionals interested in the study of 'emerging protozoan pathogens'. It provides the most current understanding of the biology, genomics, pathways of the cell's life cycle, differentiation, adherence and intracellular trafficking, host-parasite cross-talk leading to infection, molecular mechanisms, risk factors, epidemiology, clinical diagnosis, current therapies, virulence factors, possible vaccine and therapeutic targets, and future research needs in this important group of pathogens.

With the rapid advances in this field, we attempted to collect and organize the major and most up-to-date advancements in this discipline. This volume provides a reference that is useful to newcomers to this field, especially young scientists, experienced researchers and the public health professionals. It could be adopted for parasitology, medical microbiology, and infectious disease courses or used to develop an independent course. This view is strengthened further with the addition of sections on general properties of protozoa and human immune responses. The contributors to this book have kindly shared the best possible updates in their own research. This book, therefore, represents a platform to which further developments in the fascinating biology of these organisms can be added.

Naveed Ahmed Khan
December 2007

Cover Illustration:

The front cover shows *Acanthamoeba* phagocytosing human cells. *Acanthamoeba* incubated with human corneal epithelial cells exhibited the presence of amoebastomes (mouth-like openings) within 30 minutes of incubation. These structures are known be involved in the pathogenesis of *Acanthamoeba*.

A

Amoebae

A1 *Acanthamoeba* spp.

Naveed Ahmed Khan

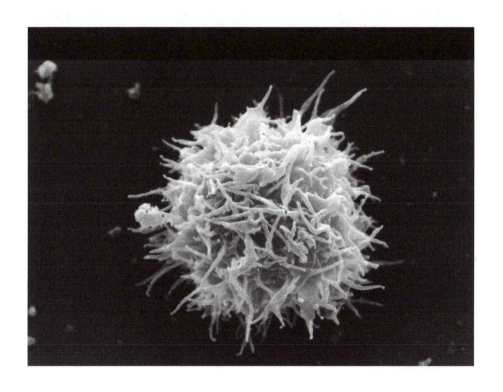

1 Introduction

Acanthamoeba is an opportunistic protozoan pathogen, a member of free-living amoebae that are widely distributed in the environment. During the last few decades, *Acanthamoeba* has become an increasingly important microbe. It is now well-recognized as a human pathogen causing serious as well as life-threatening infections, has a potential role in ecosystems, and acts as a carrier and a reservoir for prokaryotes and viruses.

2 Discovery of pathogenic free-living amoebae

The term 'amoebae' encompasses the largest diverse group of protist organisms that have been studied since the discovery of the early microscope, the largest being *Amoeba proteus* (Figures 1 and 2). Although these

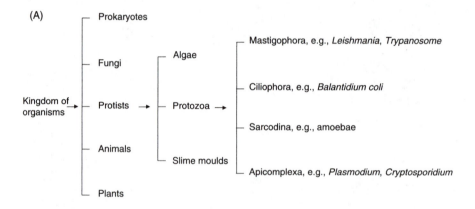

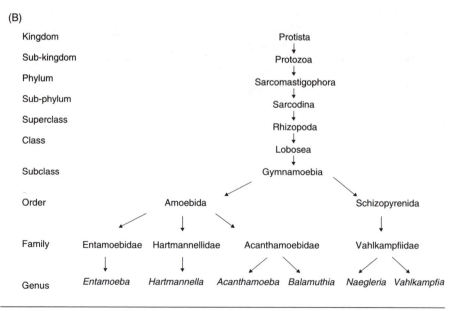

Figure 1 The traditional classification scheme of (A) protozoan pathogens, and (B) free-living amoebae, based largely on morphological characteristics.

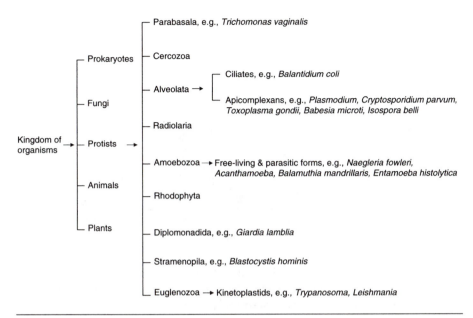

Figure 2 The present classification scheme of protozoan pathogens, based largely on their genetic relatedness.

organisms have a common amoeboid motion, that is a crawling-like movement, they have been classified into several different groups. These include potent parasitic organisms such as *Entamoeba* spp. that were discovered in 1873 from a patient suffering from bloody dysentery and named *E. histolytica* in 1903. Among free-living amoebae, *Naegleria* was first discovered by Schardinger in 1899, who named it *Amoeba gruberi*. In 1912, Alexeieff suggested its genus name *Naegleria*, and much later, in 1970, Carter identified *Naegleria fowleri* as the causative agent of fatal human infections (reviewed in De Jonckheere, 2002). In 1913, Puschkarew isolated an amoeba from dust, and later in 1931, the genus *Acanthamoeba* was created (Castellani, 1930; Douglas, 1930; Volkonsky, 1931). *Balamuthia mandrillaris* has only been discovered relatively recently in 1986 from the brain of a baboon that died of meningoencephalitis and was described as a new genus, *Balamuthia* (Visvesvara *et al.*, 1990, 1993). Over the years, these free-living amoebae have gained increasing attention from the scientific community due to their diverse roles, in particular, in causing serious and sometimes fatal human infections (Figure 3).

3 *Acanthamoeba* spp.

In 1913, Puschkarew isolated an amoeba from dust and named it *Amoeba polyphagus*. Later in 1930, Castellani isolated an amoeba that occurred as a contaminant in a culture of the fungus *Cryptococcus pararoseus* (Castellani, 1930). These amoebae were round or oval in shape with a diameter of

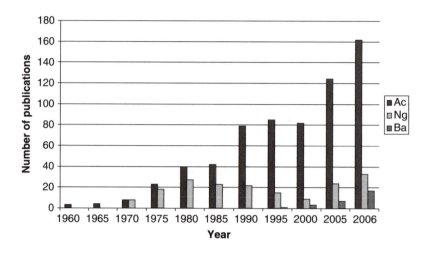

Figure 3 The number of published articles on free-living amoebae. Data for *Acanthamoeba* (Ac), *Naegleria* (Ng) and *Balamuthia* (Ba) were collected from PubMed (http://www.ncbi.nlm.nih.gov/entrez/query.fcgi).

13.5–22.5 μm and exhibited the presence of pseudopodia (now known as acanthopodia). In addition, the encysted form of these amoebae exhibited double walls with an average diameter of between 9 and 12 μm. This amoeba was placed in the genus *Hartmannella*, and named *Hartmannella castellanii*. A year later, Volkonsky subdivided the *Hartmannella* genus into three genera (Volkonsky, 1931), based on the following characteristics:

1. *Hartmannella*: amoebae characterized by round, smooth-walled cysts were placed in this genus.
2. *Glaeseria*: this genus included amoebae characterized by nuclear division in the cysts.
3. *Acanthamoeba*: amoebae characterized by the appearance of pointed spindles at mitosis, and double-walled cysts and an irregular outer layer were placed in this genus.

Singh (1950) and Singh and Das (1970) argued that the classification of amoebae by morphology, locomotion and appearance of cysts was of limited phylogenetic value and that these characteristics were not diagnostic. They concluded that the shape of the mitotic spindle was inadequate as a generic character and discarded the genus *Acanthamoeba*. In 1966, Pussard agreed with Singh (1950) that the spindle shape was an unsatisfactory feature for species differentiation but considered the distinctive morphology of the cyst to be a decisive character at the generic level and recognized the genus *Acanthamoeba*. After studying several strains of *Hartmannella* and *Acanthamoeba*, Page (1967a, 1967b) also concluded that the shape of the spindle was a doubtful criterion for species differentiation. He considered the presence of acanthopodia and the structure of the cyst to be sufficiently distinctive to justify the generic designations of *Hartmannella* and *Acanthamoeba*. He also stated that the genus *Hartmannella* had nothing in common with *Acanthamoeba* except for a general mitotic pattern, which is a property shared with many other amoebae.

In 1975, Sawyer and Griffin created the family Acanthamoebidae and Page (1988) placed *Hartmannella* in the family Hartmannellidae. The current position of *Acanthamoeba* in relation to *Hartmannella*, *Naegleria* and other free-living amoebae is shown in Figures 1 and 2.

The prefix 'acanth' (Greek meaning 'spikes') was added to the term amoeba to indicate the presence of spine-like structures (now known as acanthopodia) on the surface of this organism. After the initial discovery, this organism was largely ignored for several decades. However, in the late 1950s, it was discovered as a tissue culture contaminant (Culbertson *et al.*, 1958; Jahnes *et al.*, 1957). Later, Culbertson and colleagues (Culbertson *et al.*, 1958, 1959), for the first time, demonstrated its pathogenic potential by exhibiting its ability to produce cytopathic effects on monkey kidney cells *in vitro*, and to kill laboratory animals *in vivo*. The first clearly identified *Acanthamoeba* granulomatous encephalitis (AGE) in humans was observed in 1972 (Jager and Stamm, 1972). *Acanthamoeba* keratitis cases were reported for the first time in the early 1970s (Jones *et al.*, 1975; Nagington *et al.*, 1974). *Acanthamoeba* was first shown to be infected with bacteria in 1954 (Drozanski, 1956), demonstrated to harbour bacteria as endosymbionts in 1975 (Proca-Ciobanu *et al.*, 1975) and later identified as a reservoir for pathogenic facultative mycobacteria (Krishna-Prasad and Gupta, 1978). *Acanthamoeba* was first linked with Legionnaires' disease by Rowbotham (1980). Since then the worldwide research interest in the field of *Acanthamoeba* has increased dramatically and continues to do so (Figure 3).

3.1 Ecological distribution

Acanthamoeba has the ability to survive in diverse environments and has been isolated from public water supplies, swimming pools, bottled water, sea water, pond water, stagnant water, fresh-water lakes, salt-water lakes, river water, distilled water bottles, ventilation ducts, the water-air interface, air-conditioning units, sewage, compost, sediments, soil, beaches, vegetables, surgical instruments, contact lenses and their cases, and from the atmosphere with the recent demonstration of *Acanthamoeba* isolation even by air sampling, indicating the ubiquitous nature of this organism. In addition, *Acanthamoeba* has been recovered from hospitals, dialysis units, eye-wash stations, human nasal cavities, pharyngeal swabs, lungs tissues, skin lesions, from bone graft of the mandible suffering from osteomyelitis, corneal biopsies, cerebrospinal fluids (CSF) and brain necropsies (reviewed in Khan, 2003; Marciano-Cabral and Cabral, 2003; Schuster and Visvesvara, 2004; Visvesvara *et al.*, 2007). It is not surprising that the majority of healthy individuals have been shown to possess anti-*Acanthamoeba* antibodies indicating our common exposure to this organism (Cursons *et al.*, 1980).

3.2 Biology

The vegetative state of *Acanthamoeba* is known as trophozoites. Trophozoites are normally in the range of 12–35 μm in diameter; however the size varies significantly between isolates belonging to different species/genotypes. In general, *Acanthamoeba* does not differ greatly at the ultrastructural level from a mammalian cell, and thus presents an excellent model for cell biology studies. The trophozoites exhibit spine-like structures on their surface known as acanthopodia. The acanthopodia are most likely important in adhesion to surfaces (biological or inert), cellular movements or capturing prey. The acanthopodia are composed of hyaline (transparent) cytoplasm, which excludes various vacuoles and particles that are normally present on the interior of the cell (Bowers and Korn, 1968). The trophozoites normally possess a single nucleus that is approximately one-sixth the size of the trophozoite. The nucleus possesses a prominent nucleolus (approximately 2.4 μm in diameter, occupying one-eighth of the nuclear volume, while two nucleoli per nucleus are not uncommon), but multinucleate amoebae have been observed. The membranes of the nucleus are separated by a distance of about 350 Å (Bowers and Korn, 1968). There are numerous nuclear pores that are about 1040 Å in diameter. RNA synthesis in *Acanthamoeba* is a nuclear function. As for other eukaryotic cells, *Acanthamoeba* possess an extensive network of both smooth and rough endoplasmic reticulum (ER), with ribosomes stubbed on the cytoplasmic surface of the rough ER (Bowers and Korn, 1968). These ribosomes translate messenger RNA and actively synthesize proteins. However, not all proteins are synthesized on the ER-bound ribosomes. Protein synthesis also occurs on the unbound ribosomes that are free in the cytosol. Generally speaking, secretory proteins and membrane proteins are made by ER-bound ribosomes, while proteins intended for use within the cytosol are made on free ribosomes. The pattern of ribosomal proteins analyzed by 2-dimensional gel electrophoresis revealed that the small subunit of cytoplasmic ribosome contains 25 proteins, while the large subunit contains 40 proteins. Furthermore, exponentially growing *Acanthamoeba* contains a 45-kDa phosphorylated small subunit ribosomal protein that is absent in the cyst ribosomes. In contrast, smooth ER does not possess ribosomes and thus has no protein synthesis. Smooth ER is involved in the synthesis of lipids and may play a role in the inactivation or detoxification of drugs that might otherwise be toxic or harmful to the cell. Proteins from the ER-bound ribosomes bud off the ER in vesicles and arrive at the Golgi complex (named after its Italian discoverer, Camillo Golgi), which plays an important role in the synthesis of complex polysaccharides. Here, proteins (and other substances) undergo posttranslational modifications (most notably glycosylation). Carbohydrates are shown to be either linked to the nitrogen atom of an amino group (*N*-linked glycosylation, that is shown to be limited to the asparagine amino acid) or to the oxygen atom of a hydroxyl group (*O*-linked glycosylation, i.e., limited to serine or threonine amino acids but may also be attached to hydroxylysine or hydroxyproline, which are derivatives of the amino acids lysine and proline, respectively). The modified contents are then passed on to other compartments of the cell by means of vesicles that arise by budding off the Golgi complex. The process of glycosylation initiates within the lumen of the ER and is completed within the Golgi complex and its products are finally destined for the cell membrane or for export.

Under the microscope, an actively feeding trophozoite exhibits one or more prominent contractile vacuoles (contractility meaning shortening), whose function is to expel water (i.e., water expulsion vacuoles), and which are involved in osmotic regulation. Usually, these do not contain any precipitates but exhibit alkaline phosphatase activity in the membrane; this activity may also be present in other vesicles (Bowers and Korn, 1974). During discharge, a contractile vacuole becomes associated with the plasmalemma and may account for the alkaline phosphatase activity in the plasma membrane. In addition, it is shown that alkaline phosphatase activity in the plasma membrane of stationary-phase cells (high density populations) is remarkably reduced as compared to the exponentially growing cells. The contents of a contractile vacuole are hypotonic compared with the cytoplasm. The ionic content of the contractile vacuole in other amoeba, for example *Pelomyxa carolinensis*, is high in sodium and low in potassium, with these ions accounting for essentially all of the osmotically active content of the vacuole. *Acanthamoeba* appears to lack a 'sodium pump' similar to that found in many cells since no Na-K ATPases activity was found in whole homogenates or isolated membranes. Other types of vacuoles present in the cytoplasm are digestive vacuoles containing precipitates and amorphous substances that are bounded by membranes similar to a plasma membrane in size and staining characteristics (Bowers and Korn, 1968). The size of digestive vacuoles may range from 0.1 μm to larger than the nucleus. Trophozoites possess large numbers of mitochondria, generating the energy required for metabolic activities involved in feeding, as well as movement, reproduction and other cellular functions. The mitochondria are in either of two forms: elongate or dumbbell-shaped, or spherical (Bowers and Korn, 1968). The cristae are formed by branching tubular extensions of the inner mitochondrial membrane and are about 600 Å in diameter (Bowers and Korn, 1968). The plasma membranes are approximately 100 Å in thickness and consist of phospholipids (25%), proteins (33%), sterols (13%), and lipophosphonoglycan (29%), with sugars exposed on both sides of the membrane (Bowers and Korn, 1974). The presence of lipophosphonoglycan is unusual as it is absent from mammalian cells (Korn *et al.*, 1974).

The trophozoite possesses large numbers of mitochondria, generating the energy required for metabolic activities involved in feeding, as well as movement, reproduction and other cellular functions. The cytoplasm possesses large numbers of fibrils, and glycogen and lipid droplets. Both lysosomes and peroxisomes have been isolated from the Neff's strain of *Acanthamoeba castellanii*. A variety of lysosomal enzymes have been identified including alpha- and beta-glycosidases, amylase, beta-galactosidase, beta-*N*-acetylglucosaminidase, beta-glucuronidase, protease, phosphatase, acid hydrolase, RNAse, and DNAse (reviewed in Weisman, 1976). However, it is reasonable to suppose that some of the autolyzed material is used as a source of precursors for cell-wall synthesis or as a source of energy for the cell that has no external food source.

3.3 Motility

Amoeboid movement is accompanied by protrusions of the cytoplasm called acanthopodia (pseudopodia). These cells have an outer thick, gelatinous cytoplasm called the ectoplasm and an inner layer of more fluid cytoplasm called the endoplasm. As acanthopodia extend from the cell, fluid endoplasm streams forward in the direction of extension and coagulates into the ectoplasm at the tip of acanthopodia. At the same time, in the rear of the cell, the ectoplasm changes into more fluid endoplasm and streams towards the acanthopodia. These changes in the actin cytoskeleton are referred to as gelation-solation. The gelation and solation serve two functions: cell motility and cytoplasmic structure, that is they generate the force for cellular movement and also serve as cytoskeletal elements by forming a solid gel. More than 20 cytoskeletal proteins have been isolated from trophozoites, including actin (constituting 20% of the total protein), myosin, myosin cofactor, gelation factors and a Ca^{2+}-ATPase that interact with actin. Of these, actin is the major component of the gel that forms filaments, and these filaments must bind to each other/crosslink to form the solid gel. In *Acanthamoeba*, ATP, $MgCl_2$ and temperature are important regulators of gel formation (gel forms at 25°C and can be liquified by cooling). Other proteins such as myosin, actin-binding protein and other unidentified proteins, can accelerate the polymerization as well as support the gel.

3.4 Life cycle

Acanthamoeba undergoes two stages during its life cycle: a vegetative trophozoite stage and a resistant cyst stage (Figure 4). The trophozoites are normally in the range of 12–35 μm in diameter, but the size varies

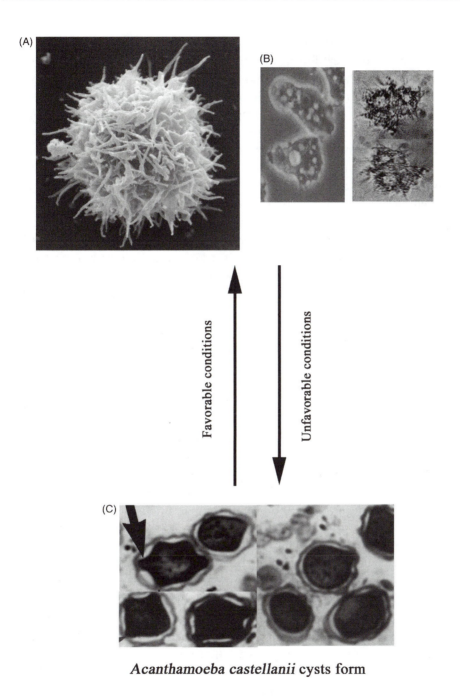

Acanthamoeba castellanii cysts form

Figure 4 The life cycle of *Acanthamoeba castellanii*. The infective form of *A. castellanii*, also known as trophozoites, as observed under (A) scanning electron microscope, and (B) phase-contrast microscope. Under unfavorable conditions, trophozoites differentiate into cysts. (C) Cysts form of *A. castellanii*, characterized by a double wall as indicated by arrows, (published with permission from Elsevier).

significantly between isolates belonging to different species/genotypes. The trophozoites exhibit acanthopodia that are most likely important in adhesion to surfaces, cellular movements or capturing prey. The trophozoites normally possess a single nucleus that is approximately one-sixth the size of the trophozoite. During the trophozoite stage, *Acanthamoeba* actively feeds on bacteria, algae, yeasts or small organic particles and many food vacuoles can be seen in the cytoplasm of the cell. Under optimal growth conditions, *Acanthamoeba* reproduces by binary fission. In axenic cultures, *Acanthamoeba* shows a typical exponential growth phase, followed by a period of reduced growth rate called the growth deceleration phase, and finally a stationary phase during which no further increase in cell density occurs. The generation time differs between isolates belonging to different species/genotypes from 8–24 h, during which time the cell passes through a series of discreet stages, collectively known as the cell cycle. Cell division to produce two genetically identical daughter cells is called mitosis and involves two overlapping events in which the nucleus divides first, followed by the division of the cytoplasm. Nuclear division is called mitosis and the division of the cytoplasm to produce two daughter cells is termed cytokinesis.

For exponentially growing cells, cell division is largely occupied with the G2 phase (up to 90%) and a limited G1 phase (5–10%), 2–3% M phase (mitosis) and 2–3% S phase (synthesis; Band and Mohrlok, 1973; Byers *et al.*, 1990, 1991). *Acanthamoeba* can be maintained in the trophozoite stage with an abundant food supply, neutral pH, appropriate temperature (i.e., 30°C) and osmolarity between 50 and 80 mosmol. However, harsh conditions (i.e., lack of food, increased osmolarity or hypo-osmolarity, extremes in temperatures and pH, and high cell densities) induce the transformation of trophozoites into the cyst stage, a process known as encystment (Neff and Neff, 1969). Encystment occurs both under harsh conditions as well as after the growth phase. Encystment is accompanied by morphological changes, termination of cell growth, and biochemical differences. Under the scanning electron microscope, the morphological changes include thickening or shortening of acanthopodia until the surface is covered with short stubby processes, followed by the development of the thick, interconnecting ridges that characterize the surface of the mature cyst. In simple terms, the trophozoite becomes metabolically inactive (minimal metabolic activity) and encloses itself within a resistant shell. More precisely, during the encystment stage, excess food, water and particulate matter is expelled with a decrease in cytoplasmic mass and the trophozoite condenses itself into a rounded structure (i.e., pre-cyst or immature cyst), accompanied by the synthesis of a chemically and structurally complex cell wall, which matures into a double-walled cyst with the wall serving only as a shell to help the parasite survive hostile conditions. The mature cyst wall consists of a laminar, fibrous outer layer, (i.e., ectocyst), and an inner layer of fine fibrils, (i.e., endocyst). Both layers are normally separated by a space, except at certain points where they form opercula in the center of ostioles (exits for excysting amoebae). The ectocyst is the first to appear in the encysting cells, even before the cell rounds up completely, and consists of several layers. It appears as an amorphous, discontinuous layer, about 10 μm outside the plasma membrane. By the time sufficient material accumulates to form a layer covering the entire surface, the cell has become nearly spherical. In mature cysts, the ectocyst terminates in a loose fibrous layer and the entire ectocyst is 0.3–0.5 μm thick, and fibrils appear to be less than 50 Å in diameter. The endocyst is shown to possess cellulose (not present in the trophozoite stage) that accounts for nearly one-third of the dry weight of the cyst wall. The ineffectiveness of cellulase against intact cysts indicates that cellulose exists as an inner and inaccessible layer of the wall (Neff and Neff, 1969). It is shown that the ectocyst consists of proteins and polysaccharide.

Overall, the cyst wall composition for *A. castellanii* (T4 genotype) has been shown to be 33% protein, 4–6% lipid, 35% carbohydrates (mostly cellulose), 8% ash, and 20% unidentified materials (Neff and Neff, 1969), although the cyst wall composition varies between isolates belonging to different species/genotypes. The precursor of cellulose is glucose, a source of which is glycogen. Encystment is aerobic, but the respiration rates of whole cells and isolated mitochondria, as well as the intracellular ATP levels of cells, diminish throughout the course of encystment. Once *Acanthamoeba* cells enter stationary phase, phagocytic activity ceases, while pinocytic activity is halved. The cellular levels of RNA, proteins, triacylglycerides and glycogen declines substantially during the encystment process resulting in decreased cellular volume and dry weight (Weisman, 1976). The cyst stage is 5–20 μm in diameter but again this varies between isolates belonging to different species/genotypes.

Cysts are airborne, which may help spread *Acanthamoeba* in the environment and/or carry these pathogens to the susceptible hosts. Recent studies report that cysts can remain viable for several years while maintaining their pathogenicity, thus presenting a role in the transmission of *Acanthamoeba* infections (Mazur *et al.*, 1995). As indicated above, cysts possess pores known as ostioles, which are used to monitor environmental changes. The trophozoites emerge from the cysts under favorable conditions leaving behind the outer shell. In the pre-emergent stage, a contractile vacuole becomes evident and moves towards the wall. *Acanthamoeba* then pulls away from the endocyst and move freely within the cyst wall. The first indication of emergence of trophozoites, visible by scanning electron microscopy is the appearance of a cytoplasmic bud pushing through the ostioles from which the operculum had been removed. The emerging cytoplasmic buds do not possess the long acanthopodia, which are characteristics of the trophozoites. These acanthopodia do not appear until the trophozoites have completely excysted. The holes through which the trophozoites emerge are apparent on the surfaces of the empty cyst walls, otherwise surfaces of the empty cyst walls are indistinguishable from those of mature cysts and exhibit sculptured interconnecting ridges surrounded by shallow craters. Thus *Acanthamoeba* trophozoites emerge without the complete digestion of the cyst wall. The emerged trophozoites actively reproduce as described above, thus completing the cycle.

One reason that *Acanthamoeba* infections are notoriously difficult to treat is the rapid propensity of the trophozoite to transform into a highly resistant cyst, thus an understanding of the molecular pathways associated with these events should provide possible targets for treatment. Both the encystment and the excystment processes require active macromolecule synthesis and can be blocked by cycloheximide, a protein synthesis inhibitor. In addition, recent studies have shown that *Acanthamoeba* serine proteases are crucial in the differentiation of *A. castellanii* (Dudley *et al.*, unpublished findings). It was observed that the inhibition of serine proteases attenuates *Acanthamoeba* metamorphosis, as demonstrated by arrest of both encystment and excystment. This suggested that *Acanthamoeba* serine proteases are not simply allied to trophozoite host invasion and cellular cytolysis, but are a requisite for trophozoite metamorphosis and re-emergence during excystment and may prove a potential target for future therapies.

3.5 Feeding

Acanthamoeba feeds on microorganisms present on surfaces, in diverse environments and even at the air-water interface (Preston *et al.*, 2001). The spiny structures or acanthopodia that arise from the surface of *Acanthamoeba* trophozoites may be used to capture food particles, which usually are bacteria, but algae, yeast, and other protists are also grazed upon. *Acanthamoeba* does not actively transport solutes. As a result, it depends totally upon pinocytosis and phagocytosis for the uptake of nutrients and these endocytic processes directly involve the plasma membrane. Both processes are energy dependent and can be inhibited by metabolic inhibitors, such as NaN_3, NaCN, NaF, iodoacetate, and 2,4-dinitrophenol and by incubation below 12°C but were not affected by inhibitors of glycolysis (Bowers and Olszewski, 1972). Phagocytosis is a receptor-dependent process (discussed in section 7.1), while pinocytosis is a nonspecific process (continuous in growing cells) that is used to take up large volumes of solutes/medium, through membrane invaginations and thus requires turnover of the cell surface (Bowers and Olszewski, 1972). It is shown that *Acanthamoeba* uses both specific phagocytosis and nonspecific pinocytosis for the uptake of food particles and large volumes of solutes with optimal uptake at 30–35°C (Bowers and Olszewski, 1972). Solutes of varying molecular weights, including albumin (M_r 65 000), inulin (M_r 5000), glucose

(M_r 180) and leucine (M_r 131), enter amoebae at a similar rate. The calculated volume of fluid taken in during pinocytosis in culture medium is about $2\,\mu l\,h^{-1}$ per 10^6 amoebae at 30°C (Bowers and Olszewski, 1972). There is no uptake at 0°C indicating that the uptake process is not simply a physical binding of solutes to the amoebae surfaces. But how *Acanthamoeba* discriminate between pinocytosis and phagocytosis; why they use one or the other, and whether there are any differences in this respect between pathogenic and nonpathogenic *Acanthamoeba*, remain incompletely understood (Alsam *et al.*, 2005a). However, there is an optimal size for the formation of the phagocytic vesicle that can be satisfied by one bead with diameter of 1.3, 1.9 or 2.7 μm. For example, smaller beads are accumulated at the surface of the cell until they reach the critical volume, at which point they are ingested collectively within one vesicle. In ingesting particular matter greater than 1 μm, the amoeba closely surrounds each particle with its plasma membrane. Subsequent to particle uptake, the membrane-enclosed particles bud off into the cytoplasm giving rise to the newly formed phagosome, which is subsequently (within 30 min) fused with other vacuoles to acquire hydrolytic enzymes (acid hydrolases) such as acid phosphatase, acid beta-glucosidase, *N*-acetylglucosaminidase, alpha-glucosidase, galactosidase, and bacteriolytic *N*-acetylmuramidase. The acid hydrolases exhibit distinct mitochondrial and peroximal patterns indicating their lysosomal origin and thus form phagolysosomes. Furthermore, it is shown that metabolic inhibitors, for example KCN and 2,4-dinitrophenol (DNP), have no effects on the fusion of phagosomes with lysosomes suggesting that the fusion reaction can proceed in the absence of oxidative metabolism. Overall, these data indicate that the newly formed phagosomes rapidly enter into *Acanthamoeba*'s lysosomal system with many phagosomes becoming phagolysosomes soon after ingestion. The rate of phagolysosomes fusion may be regulated by cyclic nucleotides with enhancement of the fusion rate by cAMP and inhibition of the rate by cGMP. Moreover, it is shown that *Acanthamoeba* exhibit the ability to distinguish vacuoles containing digestible and indigestible particles. For example, Bowers and Olszewski (1983) have shown that the fate of vacuoles within *Acanthamoeba* is dependent on the nature of particles: latex beads *versus* food particle. Vacuoles containing food particles are retained and digested, while latex beads are exocytosed upon presentation of new particles. It is estimated that during active uptake, the rate of turnover of surface membrane by *Acanthamoeba* is 5–50 times an hour, which is remarkably high compared to the other cell types (Bowers and Olszewski, 1972). Of interest, the macrophage interiorizes 50% of its surface in 2–5 h during active pinocytosis. Similarly, up to 50% of the surface of *Amoeba proteus* may be ingested during one pinocytotic cycle, but then a rest period is required before a new cycle can be initiated. Overall, particle uptake in *Acanthamoeba* is a complex process that may play a significant role, both in food uptake as well as in the pathogenesis of *Acanthamoeba.*

3.6 Acanthamoeba *genome*

The trophozoites contain cellular, nuclear and mitochondrial DNA with nuclear DNA comprising 80–85% of the total DNA. In addition, cytoplasmic nonmitochondrial DNA has been reported (Ito *et al.*, 1969), but its origin is not known. Total cellular DNA ranges between 1 and 2 pg for single-cell uninucleate amoebae during the log phase (Byers *et al.*, 1990). The genome size of mitochondrial DNA was originally calculated to be about 3.4×10^7 Da, from measurements of renaturation kinetics, sedimentation coefficient and electron micrographs. In comparison, it was shown that the genome size of mitochondrial DNA in other organisms is $3.0–3.5 \times 10^7$ Da in *Tetrahymena pyriformis*, 5×10^7 Da for yeast, and 7×10^7 Da for higher plants. The number of nuclear chromosomes is uncertain, but may be numerous. The measurements of nuclear DNA content (*A. castellanii* Neff strain, belonging to the T4 genotype) showed a total DNA content of 10^9 bp. Measurements of kinetic complexity suggest a haploid genome size of about $4–5 \times 10^7$ bp (Byers *et al.*, 1990). Pulse field gel electrophoresis suggests a genome of around $2.3–3.5 \times 10^7$ bp, which express more than 5000 transcripts. For comparison, the haploid genome size of *Saccharomyces* is around 2×10^7 bp, and *Dictyostelium* is around 5×10^7 bp (reviewed in Byers *et al.*, 1990).

3.7 Methods of isolation

In natural environments, *Acanthamoeba* feeds on yeasts, other protozoa, bacteria, small organisms and organic particles. Any of the aforementioned can be used as growth substrates for *Acanthamoeba* in the laboratory but there are some technical problems. For example, the use of yeast and protozoa as growth substrates is problematic due to complexity in their preparations, their possible overwhelming growth and the difficulty in eradicating yeast to obtain pure axenic *Acanthamoeba* cultures. Organic matter such as glucose, proteose peptone or other substrates provide rich nutrients for unwanted organisms, for example yeasts, fungi, other protozoa and bacteria. To overcome these technical problems and to maximize the likelihood of *Acanthamoeba* isolation from the environmental as well as clinical samples, protocols have been developed using simple plating assays as described below. Both of the following methods can be used to obtain a large number of *Acanthamoeba* trophozoites for biochemical studies.

Isolation of *Acanthamoeba* using non-nutrient agar plates seeded with Gram-negative bacteria

This method has been used extensively in the isolation of *Acanthamoeba* from both environmental and clinical samples, worldwide. The basis of this method is the use of Gram-negative bacteria (*Escherichia coli* or *Enterobacter aerogenes*, are most commonly used) that are seeded on the non-nutrient agar plate as a food source for *Acanthamoeba*. The non-nutrient agar contains minimal nutrients and thus inhibits the growth of unwanted organisms. Briefly, non-nutrient agar plates containing 1% (w/v) Oxoid no. 1 agar in Page's amoeba saline (PAS; 2.5 mM NaCl, 1 mM KH_2PO_4, 0.5 mM Na_2HPO_4, 40 μM $CaCl_2.6H_2O$ and 20 μM $MgSO_4.7H_2O$) supplemented with 4% (w/v) malt extract and 4% (w/v) yeast extract are prepared, and the pH adjusted to 6.9 with KOH. Approximately 5 ml of late-log-phase cultures of Gram-negative bacteria (*E. coli* or *E. aerogenes*) are poured onto non-nutrient agar plates and left for 5 min, after which excess culture fluid is removed and plates are left to dry before their inoculation with an environmental sample or a clinical specimen. Once inoculated, plates are incubated at 30°C and observed daily for the presence of *Acanthamoeba* trophozoites (Khan and Paget, 2002; Khan *et al.*, 2001). Depending on the number of *Acanthamoeba* in the sample, trophozoites can be observed within a few hours (up to 12 h). However in the absence of amoebae, plates should be monitored for up to 7 days. Once bacteria are consumed, *Acanthamoeba* differentiates into the characteristic cyst (Figure 4). The precise understanding of bacterial preference by *Acanthamoeba*, that is Gram-negative *versus* Gram-positive bacteria, or why *E. coli* or *E. aerogenes* are used most commonly as food substrate, and whether bacterial preferences vary between *Acanthamoeba* isolates belonging to different species/genotypes are questions for future studies.

'Axenic' cultivation of *Acanthamoeba*

Acanthamoeba can be grown 'axenically' in the absence of external live food organisms. This is typically referred to as 'axenic culture' to indicate that no other living organisms are present. However, *Acanthamoeba* cultures may never be truly axenic as they may contain live bacteria or viruses surviving internally as endosymbionts. Under laboratory conditions, axenic growth is achieved using liquid PYG medium proteose peptone 0.75% (w/v), yeast extract 0.75% (w/v) and glucose 1.5% (w/v). Briefly, non-nutrient agar plates overlaid with bacteria are placed under ultraviolet light (UV) for 15–30 min to kill the bacterial lawn. A small piece of non-nutrient agar (stamp-sized) containing amoebic cysts is placed on plates containing these UV-killed bacteria. When amoebae begin to grow, a stamp-sized piece of the agar containing trophozoites or cysts is transferred into 10 ml sterile PYG medium containing antibiotics (penicillin and streptomycin). The *Acanthamoeba* switch to the PYG medium as a food source, and their multiplication can be observed within several days. Once multiplying in PYG medium, *Acanthamoeba* are typically grown aerobically in tissue-culture flasks with filter caps at 30°C in static conditions. The trophozoites adhere to the flask walls and are collected by chilling the flask for 15–30 min, followed by centrifugation of the medium containing the cells.

3.8 Methods of encystment

Both xenic and axenic methods have been developed to obtain *Acanthamoeba* cysts. For xenic cultures, *Acanthamoeba* are inoculated onto non-nutrient agar plates seeded with bacteria as described above. Plates are incubated at 30°C until the bacteria are cleared and trophozoites have transformed into cysts. Cysts can be scraped off the agar surface using phosphate-buffered saline (PBS) and used for assays. This resembles the most likely natural mode of encystment and can be very effective, achieving up to 100% encystment. However, one major limitation may be the presence of bacterial contaminants that could hamper molecular and biochemical studies. Alternatively, at least for pathogenic *Acanthamoeba* (T4 genotype), parasites are inoculated onto non-nutrient agar plates without bacterial lawns. This will allow rapid transformation of trophozoites into cysts and can be considered as an axenic method of encystment. Alternatively, for axenic encystment, *Acanthamoeba* are grown in PYG medium for 17–20 h. After this, encystment can be induced by incubating amoebae in Neff's encystment medium [100 mM KCl, 8 mM $MgSO_4$, 0.4 mM $CaCl_2$, 20 mM 2-amino-2-methyl-1,3-propanediol (AMPL), pH 9.0; Neff *et al.*, 1964); Tris-buffered encystment medium (95 mM NaCl, 5 mM KCl, 8 mM $MgSO_4$, 0.4 mM $CaCl_2$, 1 mM $NaHCO_3$, 20 mM Tris-HCl, pH 9.0; Hirukawa *et al.*, 1998); or by adding 8% glucose (as an osmolarity trigger) in RPMI 1640 (Invitrogen). Plates are incubated at 30°C for up to 48 h. To confirm transformation of trophozoites into cysts, sodium dodecyl sulfate (SDS, 0.5% final concentration) is added: trophozoites are SDS-sensitive and any remaining are lysed immediately upon addition of SDS, while cysts remain intact (Cordingley *et al.*, 1996; Dudley *et al.*, 2005). This method allows the simple counting of cysts using a hemocytometer and is useful in studying the process of encystment.

3.9 Storage of Acanthamoeba

For short-term storage, *Acanthamoeba* are maintained on non-nutrient agar plates. Plates inoculated with *Acanthamoeba* can be kept at 4°C under moist conditions for several months or as long as plates are protected from drying out. Cysts can be re-inoculated into PYG medium in the presence of antibiotics to obtain axenic cultures as described above. Alternatively, *Acanthamoeba* trophozoites can be stored as axenic cultures, long-term. Briefly, log-phase amoebae (actively dividing) are re-suspended at a density of $3–5 \times 10^6$ parasites ml^{-1} in freezing medium (PYG containing 10% dimethylsulfoxide, DMSO; John and John, 1994). Finally cultures are transferred to -20°C for around 2 h, followed by their storage at -70°C or in liquid nitrogen indefinitely. *Acanthamoeba* cultures can be revived by thawing at 37°C, followed by immediate transfer to PYG medium in a T-75 flask at 30°C. However, the inclusion of 20% fetal bovine serum in freezing medium has been shown to improve the revival viability of *Acanthamoeba* in long-term storage (John and John, 1994).

3.10 Classification of Acanthamoeba

Following the discovery of *Acanthamoeba*, several isolates belonging to the genus *Acanthamoeba* with distinct morphology were isolated and given different names based on the isolator, source, or other criteria. In an attempt to organize the increasing number of isolates belonging to this genus, Pussard and Pons (1977) classified this genus based on morphological characteristics of the cysts, which were the most appropriate criteria at the time. The genus *Acanthamoeba* was classified into three groups based only on two obvious characters: cyst size and number of arms within a single cyst (Figure 4). Based on this scheme, Pussard and Pons (1977) divided the *Acanthamoeba* genus (18 species at the time) into three groups. Subsequently, the classification of Pussard and Pons gained acceptance (De Jonckheere, 1987; Page, 1988).

Group 1: Four species were placed in this group: *A. astronyxis, A. comandoni, A. echinulata and A. tubiashi*. These species exhibit large trophozoites, while in the cyst, ectocyst and endocyst are widely separated and exhibit the following properties:

1. Less than six arms with the average diameter of cysts at ≥18 μm: *A. astronyxis*
2. 6–10 arms and the average diameter of cysts at ≥25.6 μm: *A. comandoni*
3. 12–14 arms and the average diameter of cysts at ≥25 μm: *A. echinulata*
4. The average diameter of cysts at ≥22.6 μm: *A. tubiashi*

Group 2: This group included 11 species, which are the most widespread and commonly isolated *Acanthamoeba*. The ectocyst and endocyst may be close together or widely separated. The ectocyst may be thick or thin and the endocyst may be polygonal, triangular or round with the mean diameter of 18 μm or less. The species included in this group were *A. mauritaniensis, A. castellanii, A. polyphaga, A. quina, A. divionensis, A. triangularis, A. lugdunensis, A. griffini, A. rhysodes, A. paradivionensis* and *A. hatchetti*.

Group 3: Five species were included in this group: *A. palestinensis, A. culbertsoni, A. royreba, A. lenticulata* and *A. pustulosa*. The ectocyst in this group is thin and the endocyst may have 3–5 gentle corners with the mean cyst diameter at <18 μm.

Later, *A. tubiashi* in group 1 and *A. hatchetti* in group 2 were added by Visvesvara (1991).

From the above, it is obvious that the identification of the various species of *Acanthamoeba* by morphological features alone is problematic. In addition, several studies have demonstrated inconsistencies and/or variations in the cyst morphology belonging to the same isolate/strain. For example, Sawyer discovered that the ionic strength of the growth medium could alter the shape of the cyst walls (Sawyer, 1971), thus substantially reducing the reliability of cyst morphology as a taxonomic characteristic. Furthermore, this scheme had limited value in associating pathogenesis with a named species. For example, several studies demonstrated that strains/isolates within *A. castellanii* can be virulent, weakly virulent or avirulent. This discrepancy in assigning an unambiguous role to a given species presented a clear but urgent need to reclassify this genus. The discovery of advanced molecular techniques led to the pioneering work of the late Dr. T. Byers (Ohio University, USA) in the classification of the genus *Acanthamoeba* based on ribosomal RNA (rRNA) gene sequences. Because life evolved in the sea, most likely through self-replicating RNA as the genetic material or as common ancestor, and evolved into diverse forms, it is reasonable to study the evolutionary relationships through such molecules, that is rRNA. In addition, this is a highly precise, reliable, and informative scheme. Each base presents a single character providing an accurate and diverse systematic. Based on rRNA sequences, the genus *Acanthamoeba* is divided into 15 different genotypes (T1–T15, Table 1; Schuster and Visvesvara, 2004). Each genotype exhibits 5% or more sequence divergence between different genotypes. Note that in a recent study, Maghsood *et al.* (2005) proposed to subdivide T2 into a further two groups, T2a and T2b. This is due to the sequence dissimilarity of 4.9% between these two groups, which is very close to the current cut-off limit of 5% between different genotypes. This should help differentiate pathogenic and nonpathogenic isolates within this genotype. With the clear advantage of the rRNA sequences over morphology-based classification, an attempt is made to refer to the genotype rather than species name wherever possible in this chapter. Based on this classification scheme, the majority of human infections due to *Acanthamoeba* have been associated with the T4 genotype. For example, nearly 90% of *Acanthamoeba* keratitis cases have been linked with this genotype. Similarly, T4 has been the major genotype associated with non-keratitis infections such as *Acanthamoeba* granulomatous encephalitis (AGE) and cutaneous infections. Moreover, recent findings suggest that the abundance of T4 isolates in human infections is most likely due to their greater virulence and/or properties that enhance their transmissibility, as well as their decreased susceptibility to chemotherapeutic agents (Maghsood *et al.*, 2005). Future studies will identify virulence traits and genetic markers limited only to certain genotypes, which may help clarify these issues. A current list of genotypes and their association with human infections is presented in Table 1. With increasing research interest in the field of *Acanthamoeba* and the worldwide availability of advanced molecular techniques, undoubtedly additional genotypes will be identified. These studies will help clarify the

Table 1 Known *Acanthamoeba* genotypes and their associations with the human diseases, keratitis and granulomatous encephalitis

Acanthamoeba genotypes	Human disease association
T1	Encephalitis
T2a[a]	Keratitis
T2b[a] – ccap1501/3c-alike sequences	NA
T3	Keratitis
T4[b]	Encephalitis, keratitis
T5	Keratitis
T6	Keratitis
T7	NA
T8	NA
T9	NA
T10	Encephalitis
T11	Keratitis
T12	Encephalitis
T13	NA
T14	NA
T15	NA

[a] Basis of T2 division into T2a and T2b has been proposed by Maghsood *et al.* (2005).
[b] This genotype has been most associated with both diseases.
NA – no disease association has been found yet.

role of *Acanthamoeba* within the ecosystem, in bacterial symbiosis, as well as in causing primary and secondary human infections.

3.11 Human infections

Acanthamoeba causes two well-recognized diseases that are major problems in human health: a rare AGE involving the central nervous system (CNS) that is limited typically to immunocompromised patients and almost always results in death, and a painful keratitis that can result in blindness.

4 *Acanthamoeba* keratitis

First discovered in the early 1970s, *Acanthamoeba* keratitis has become a significant ocular infection due to microbes. A key predisposing factor in *Acanthamoeba* keratitis is the use of contact lenses exposed to contaminated water (Figure 5), but the precise mechanisms associated with this process are not fully understood. Overall this is a multifactorial process

Figure 5 *Acanthamoeba* keratitis has become a significant problem in recent years, especially in contact-lens wearers exposed to contaminated water.

that involves: (i) contact lens wear for extended periods of time, (ii) lack of personal hygiene, (iii) inappropriate cleaning of contact lenses, (iv) biofilm formation on contact lenses, and (v) exposure to contaminated water. For example, Beattie *et al.* (2003) have shown that *Acanthamoeba* exhibits higher binding to used contact lenses as compared with unworn contact lenses. Tests on used contact lenses showed the presence of saccharides, including mannose, glucose, galactose, fucose, *N*-acetyl-D-glucosamine, *N*-acetyl-D-galactosamine and *N*-acetyl neuraminic acid (sialic acid), and proteins, glycoproteins, lipids, mucins, polysaccharides, calcium, iron, silica, magnesium, sodium, lactoferrin, lysozyme and immunoglobin (Ig) molecules (Gudmundsson *et al.*, 1985; Klotz *et al.*, 1987; Tripathi and Tripathi, 1984) on the surface of contact lenses after only 30 min of contact lenses wear. These may act as receptors for *Acanthamoeba* trophozoites and/or enhance parasite ability to bind to contact lenses. For example, *Acanthamoeba* expresses a mannose-binding protein (MBP) on its surface, which specifically binds to mannose residues. This may explain the ability of *Acanthamoeba* to exhibit higher binding to used rather than unworn contact lenses (Beattie *et al.*, 2003).

Alternatively, biofilm formation on contact lenses may provide increased affinity for *Acanthamoeba*. This is shown by increased *Acanthamoeba* binding to biofilm-coated lenses as opposed to contact lenses without biofilms (Beattie *et al.*, 2003; Tomlinson *et al.*, 2000). Moreover, biofilms may enhance *Acanthamoeba* persistence during contact lens storage/cleaning as well as providing nutrients for *Acanthamoeba*.

Once an *Acanthamoeba*-contaminated lens is placed over the cornea, parasites transmit to the cornea. *Acanthamoeba* transmission onto the cornea is dependent on the virulence of *Acanthamoeba* (discussed later) and the physiological status of the cornea. For example, several studies showed that corneal trauma is a prerequisite in *Acanthamoeba* keratitis *in vivo*, and animals with intact corneas (i.e., epithelial cells) do not develop this infection (Niederkorn *et al.*, 1999). Corneal trauma followed by exposure to contaminated water, soil, or vector (inert objects or biological) is sufficient to contract amoebae resulting in *Acanthamoeba* keratitis and is the most likely cause of *Acanthamoeba* keratitis in contact lens non-wearers (Sharma *et al.*, 1990). The requirement of corneal trauma can be explained by the fact that the expression of *Acanthamoeba*-reactive glycoprotein(s) on damaged corneas is 1.8 times higher than on the healthy ones, suggesting that corneal injury contributes to *Acanthamoeba* infection (Jaison *et al.*, 1998). Future studies should determine whether corneal injury simply exposes mannose-containing glycoprotein(s), providing additional binding sites for *Acanthamoeba*, or whether the expression of mannose glycoprotein(s) is generally higher on the healing corneal epithelial cells. It is important to note that *Acanthamoeba* must be present in the trophozoite stage to bind to human corneal epithelial cells. Recent studies have shown that *Acanthamoeba* cysts do not bind to human corneal epithelial cells, indicating that cysts are a noninfective stage (Dudley *et al.*, 2005; Garate *et al.*, 2006).

4.1 Epidemiology

Originally thought to be a rare infection, *Acanthamoeba* keratitis has become increasingly important in human health. This is due to increased awareness and the availability of diagnostic methods. Over the last few decades, it has become clear that users of contact lenses are at increased risk of corneal infections. For example, contact-lens wearers are 80-fold more likely to contract corneal infection than those who do not (Alvord *et al.*, 1998; Dart *et al.*, 1991). The incidence rate of microbial keratitis in users of extended-wear contact lenses is determined at 20.9 per 10 000 wearers per annum in the USA (Poggio *et al.*, 1989). Similar findings have been reported from Sweden (Nilsson and Montan, 1994), Scotland (Seal *et al.*, 1999) and the Netherlands (Cheng *et al.*, 1999). By the late 1990s there were approximately 70 million people throughout the world wearing contact lenses (Barr, 1998) and, with their wider potential application beyond vision correction, such as UV protection and cosmetic purposes, this number will undoubtedly rise. With an increasing number of people wearing contact lenses, it is important to assess any associated risks, and to make both existing and new users aware. Among other microbial agents, bacteria including *Pseudomonas* and *Staphylococcus*, fungi including *Fusarium*, and protozoa including *Acanthamoeba*, are the major causes of corneal infections in the users of contact lenses. The incidence rate of *Acanthamoeba* keratitis varies between different geographical locations. For example, in Hong Kong, an incidence rate of 0.33 per 10 000 contact-lens wearers is reported, 0.05 per 10 000 in Holland, 0.01 per 10 000 in the USA (Stehr-Green *et al.*, 1989), 0.19 per 10 000 in England (Radford *et al.*, 2002), and 1.49 per 10 000 in Scotland (Lam *et al.*, 2002; Seal *et al.*, 1999). However, these variations do not reflect the geographical distribution of *Acanthamoeba*, but perhaps due to extended wear of soft contact lenses, lack of awareness of the potential risks associated with wearing contact lenses, enhanced detection, and/or local conditions that promote the growth of pathogenic *Acanthamoeba* only, for example, water hardness or salinity.

4.2 Pathophysiology

The onset of symptoms can take from a few days to several weeks, depending on the inoculum size of *Acanthamoeba* and/or the extent of corneal trauma. During the course of infection, symptoms may vary depending on the clinical management of the disease. Most commonly, *Acanthamoeba* keratitis is associated with considerable production of tears, epithelial defects, photophobia which leads to inflammation with redness, stromal infiltration, edema, stromal opacity together with excruciating pain due to radial neuritis (with suicidal pain), epithelial loss and stromal abscess formation with vision-threatening consequences (Figure 6). Other symptoms may involve scattered subepithelial infiltrates, anterior uveitis, stromal perforation, and the presence of scleral inflammation. Secondary infection due to bacteria may additionally complicate the clinical management of the disease. Glaucoma is commonly reported, and occasionally posterior segment signs such as nerve edema, optic atrophy and retinal detachment are observed. In untreated eyes, blindness may eventually result as the necrotic region spreads inwards (Niederkorn *et al.*, 1999).

4.3 Clinical diagnosis

The clinical diagnosis of *Acanthamoeba* keratitis includes both clinical syndromes and/or demonstration of the presence of amoebae (Martinez and Visvesvara, 1991). In the majority of cases, this infection is misdiagnosed as Herpes simplex virus or adenovirus infection. The clinical symptoms are indicated above, but the use of contact lenses by the patient, together with excruciating pain, is strongly indicative of *Acanthamoeba* keratitis. The confirmatory evidence comes from the isolation of *Acanthamoeba* from the corneal biopsy. To this end, several methods are available. For example, light microscopy has been used for rapid identification of *Acanthamoeba* on contact lenses, in lens-case solution, or in corneal biopsy specimens (Epstein *et al.*, 1986). Winchester *et al.* (1995) demonstrated the use of noninvasive confocal microscopy to aid in the diagnosis of *Acanthamoeba*

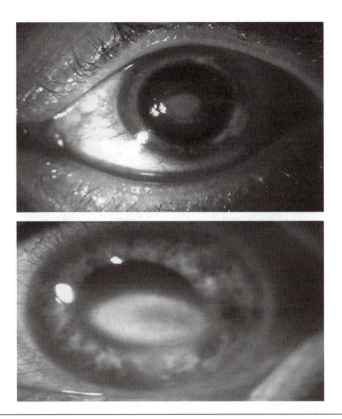

Figure 6 *Acanthamoeba*-infected eye. Note the ulcerated epithelium and stromal infiltration exhibiting corneal opacity in acute *Acanthamoeba* keratitis (published with permission from Elsevier).

keratitis. Confocal microscopy has the advantage over conventional optical microscopy in that it can image layers within the substance of a specimen of substantial thickness, so it is effective in imaging the cornea. Such microscopic identification based on morphological characteristics requires skill and the use of robust keys for identification. The examiners must have familiarity with the morphological characteristics of *Acanthamoeba* species otherwise diagnosis may require histological examination of material obtained by corneal biopsy or keratoplasty.

PCR-based methods using the 18S rRNA gene have also been developed for the rapid detection of *Acanthamoeba*. This method is highly specific and can detect fewer than five cells (Khan *et al.*, 2001, 2002; Lehmann *et al.*, 1998; Schroeder *et al.*, 2001). Despite the development of microscopic and molecular-based approaches, cultivation of *Acanthamoeba* from corneal biopsy specimens or from contact lenses or lens cases remains the most widely used assay in clinical settings because it is simple, inexpensive, and there is no loss of cells during centrifugation and/or washing steps. This method provides large numbers of *Acanthamoeba*, which could be used for typing, sequencing, epidemiological studies or pathogenicity assays. Briefly, specimens (contact lenses or corneal biopsy specimens) are inoculated onto non-nutrient agar plates seeded with Gram-negative bacteria. Plates are incubated at 30°C and observed daily for the presence of *Acanthamoeba* as described in section 3.7. *Acanthamoeba* can be identified at the genus level, based on the morphological characteristics of trophozoites and cysts using a phase-contrast microscope (Figure 7) or PCR-based assays as described above.

In addition, recent studies have suggested that matrix-assisted laser desorption-ionization time of flight MS (MALDI-TOF-MS) may be of potential value in the rapid identification of *Acanthamoeba* in clinical specimens (Visvesvara *et al.*, 2007). This method has been used to identify and characterize protists at the strain level within 15 min, based on protein profiles (Moura *et al.*, 2003).

4.4 Host susceptibility

Previous studies have demonstrated clearly the host specificity in *Acanthamoeba* keratitis. For example, successful *Acanthamoeba* keratitis models that mimic the human form of the disease were only produced in pigs and Chinese hamsters, not in rats, mice or rabbits, suggesting that the expression of specific molecular determinants may be limited to certain mammalian species (reviewed in Niederkorn *et al.*, 1999). Even in susceptible species, corneal injury is a prerequisite for *Acanthamoeba* keratitis, and animals that have intact epithelial layers do not develop *Acanthamoeba* keratitis (Niederkorn *et al.*, 1999). The importance of corneal injury is demonstrated further by reports that injury to the surface of the cornea, even with only a splash of *Acanthamoeba*-contaminated water, can lead to *Acanthamoeba* keratitis in individuals who do not wear contact lenses (Sharma *et al.*, 1990). It has been shown that the expression of *Acanthamoeba*-reactive glycoprotein(s) on surface-damaged corneal epithelial cells is significantly higher than on the surface of normal corneal epithelial cells, suggesting that corneal injury contributes markedly to *Acanthamoeba* keratitis (Jaison *et al.*, 1998). In addition, it may be that some individuals lack antiacanthamoebic defense determinants in tear film (discussed in section 9.1) or exhibit corneal properties both at the surface and molecular level which could render the cornea more susceptible to *Acanthamoeba* keratitis.

4.5 Risk factors

As indicated earlier, the major risk factor for *Acanthamoeba* keratitis is poor hygiene in the use of contact lenses (Figure 8). In support of this statement, more than 85% of *Acanthamoeba* keratitis cases occur in wearers of contact lenses. However, this may be associated with individual behavior. For example, *Acanthamoeba* keratitis has been associated frequently with young males (Niederkorn *et al.*, 1999), which could be due to their poor personal hygiene, poor handling and care of their lenses or lens storage

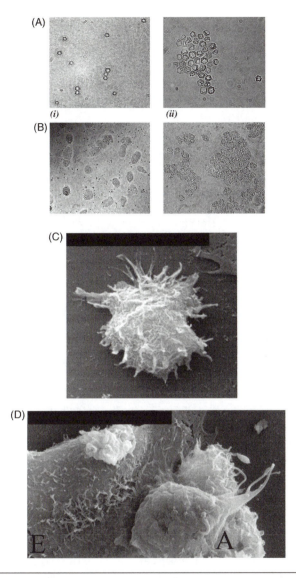

Figure 7 (A) *Acanthamoeba* cysts under phase-contrast microscope. *(i)* Non-nutrient agar plates exhibiting *Acanthamoeba* cysts. *(ii) Acanthamoeba* cysts were collected from non-nutrient agar plates using PBS and observed under the phase-contrast microscope. Note cysts formed clusters in PBS ×250. (B) *Acanthamoeba* trophozoites on non-nutrient agar plates observed under phase-contrast microscope. Note the characteristic contractile vacuole in *Acanthamoeba* trophozoites ×250. (C) *Acanthamoeba* trophozoite binding to glass cover slips observed under scanning electron microscope. Note the large number of acanthopodia on the surface of *A. castellanii* trophozoites belonging to the T4 genotype. (D) *Acanthamoeba* binding to corneal epithelial cells. *A. castellanii* (T4 isolate) were incubated with corneal epithelial cells, followed by several washes, and observed under scanning electron microscope. Note that parasites were able to bind to the host cells and binding was mediated by the acanthopodia. A, amoeba; E, corneal epithelial cell.

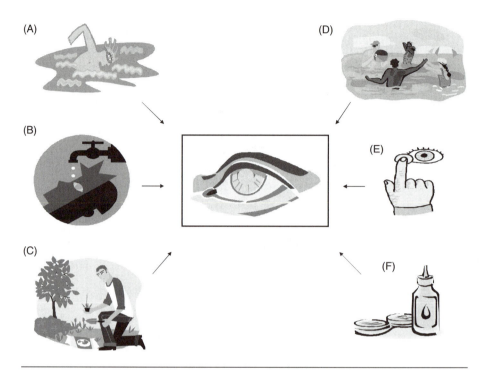

Figure 8 The risk factors contributing to *Acanthamoeba* keratitis: (A) swimming, especially while wearing contact lenses; (B) washing eyes during or immediately after contact-lens wear; (C) working with soil and rubbing eyes; (D) water-related activities (splashing water), especially during or immediately after contact lens wear; (E) handling contact lenses without proper hand washing; (F) use of homemade saline (or even chlorine-based disinfectants) for contact-lens cleaning.

cases, and noncompliance with disinfection procedures such as using homemade saline. Contact lenses that have been scratched or fragmented through mishandling should not be used. Additional factors are swimming or washing the eyes while wearing contact lenses, and the use of chlorine-based disinfectants for contact lens cleaning, because *Acanthamoeba* are highly resistant to chlorine. In addition, *Acanthamoeba* exhibits significantly higher binding to silicone hydrogel contact lenses than to the conventional hydrogel contact lenses (Beattie *et al.*, 2003), suggesting that polymer characteristics of the lens or surface treatment procedures may increase the risk of *Acanthamoeba* keratitis. Thus, the extended wear of lenses without proper maintenance and recommended replacement, together with the lens type, can be important risk factors for *Acanthamoeba* keratitis.

Overall, these characteristics suggest that although the intact cornea is highly resistant to *Acanthamoeba* infection, corneal trauma (microscopic defects) followed by exposure to contaminated water (during swimming, eye washing, water splash), dust, vegetable matter or any foreign particle are important risk factors associated with *Acanthamoeba* keratitis.

Because *Acanthamoeba* is ubiquitously present in water, air, the water-air interface and soil, susceptible hosts should be warned of the risks associated with the wearing of contact lenses while swimming or bathing/washing, and cleaning lenses with homemade saline, and so on. Proper cleaning of the contact lenses is crucial to prevent this devastating infection. Chlorine-based cleaning solutions should not be used, because *Acanthamoeba* are highly resistant to chlorine but a two-step hydrogen peroxide system at the concentration of 3% is highly effective against cysts and trophozoites (Beattie *et al.*, 2002). In addition, strategies and/or control measures to reduce the formation of biofilms and the build-up of carbohydrate moieties on contact lenses should help to prevent this serious infection (Table 2).

Table 2 Risk factors associated with *Acanthamoeba* infections

No.	Risk factors associated with *Acanthamoeba* keratitis
1	Handling of contact lenses (CL) with unclean hands
2	Washing CL with homemade saline/tap water
3	CL wear for more than recommended times
4	CL wear during swimming, performing water sport activities or relaxing in a hot tub
5	Washing eyes and/or swimming with corneal trauma – splashing eyes with contaminated water
6	Re-using CL without proper cleaning
7	Incubating CL in disinfectants for less than recommended times
8	Chlorine-based disinfectants are less effective in killing *Acanthamoeba*

No.	Risk factors associated with *Acanthamoeba* encephalitis
1	HIV/AIDS patients are particularly at risk
2	Individuals with lymphoproliferative disorders, hematologic disorders, diabetes mellitus, pneumonitis, renal failure, liver cirrhosis, rhinitis, pharyngitis, gammalobulinemia, pregnancy, systemic lupus erythematosus, glucose-6-phosphate deficiency, and tuberculosis are at risk
3	Alcohol abuse
4	Organ/tissue transplantation with immunosuppressive therapy
5	Excessive use of steroids or antibiotics
6	For all of the above, exposure to contaminated soil/water are potential risk factors
7	For all of the above, activities which may result in skin cuts/bruises followed by exposure to contaminated soil/water are potential risk factors

4.6 Treatment

Acanthamoeba keratitis is a difficult infection to treat. Early diagnosis followed by aggressive treatment is essential for the successful prognosis of the infection (Perez-Santonja *et al.*, 2003). The recommended treatment regimen includes a biguanide (0.02% polyhexamethylene biguanide, PHMB, or 0.02% chlorhexidine digluconate, CHX) together with a diamidine (0.1% propamidine isethionate, also known as Brolene, or 0.1% hexamidine, also known as Desomedine). If bacteria are also associated and/or suspected with the infection, the addition of antibiotics, for example neomycin or chloramphenicol, is recommended. The initial treatment involves hourly topical application of drugs, day and night for 2–3 days, followed by hourly topical application during the day for a further 3–4 days. Subsequently, application is reduced to 2-hourly application during the day for up to a month. This is followed by application six times a day, for the next few months for up to a year. This clearly presents an appreciable social and economic burden due to this infection. Persistent inflammation and severe pain may be managed by topical application of steroids, such as 0.1% dexamethasone, together with pain killers. However, it should be noted that dexamethasone causes a significant increase in the proliferation of amoebae numbers and increases severity in *Acanthamoeba* keratitis in an *in vivo* model (McClellan *et al.*, 2001). Thus prolonged application of steroids should be carried out with care. This aggressive, complicated and prolonged management is required because of the ability of *Acanthamoeba* to adapt rapidly to harsh conditions and to switch to the resistant cyst form, and because of the lack of available methods for the targeted killing of both trophozoites and cysts. The presence of antibiotics (neomycin or chloramphenicol) limits possible bacterial infection, as well as eliminating a food source for *Acanthamoeba*. As a last resort a keratoplasty may be indicated, especially in drug-resistant cases. In the case of penetrating keratoplasty, to obtain rehabilitation due to corneal scarring, topical treatment with the above is essential as a first measure. Rejection can occur, but is rare. Recurrence of infection does occur even though it is recommended that topical treatment continues for up to a year post operatively, as cysts may survive in the acceptor cornea.

5 *Acanthamoeba* granulomatous encephalitis

Acanthamoeba granulomatous encephalitis (AGE) is a rare infection but it almost always proves fatal. The mechanisms associated with the pathogenesis are unclear; however the pathophysiological complications involving the CNS most likely include induction of the pro-inflammatory responses, invasion of the blood-brain barrier and the connective tissue and neuronal damage leading to brain dysfunction. The routes of entry include the lower respiratory tract, leading to amoebae invasion of the intravascular space, followed by the hematogenous spread. Skin lesions may provide direct amoebae entry into the bloodstream, thus bypassing the lower respiratory tract (Figure 9). Amoebae entry into the CNS most likely occurs at the sites of the blood-brain barrier (Martinez, 1985, 1991). The cutaneous and respiratory infections can last for months but the involvement of the CNS can result in fatal consequences within days or weeks. In addition, the olfactory neuroepithelium provides another route of entry into the CNS and has been studied in experimental models (Martinez, 1991; Martinez and Visvesvara, 1997; Figure 9).

5.1 Epidemiology

The epidemiology of AGE is rather confusing. The fact that normally this infection is secondary, makes diagnosis difficult and thus contributes to our inability to assess the actual number of AGE infections. Perhaps the number of AGE cases in HIV patients, although not completely accurate, may indicate the real burden of this infection. This has only been made possible by the pioneering work of G. S. Visvesvara (Center for Disease Control, USA) and the late Dr. A. J. Martinez (University of Pittsburgh School of Medicine, USA). In the USA, there were approximately 350 000

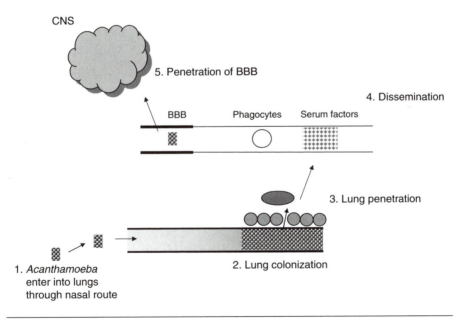

Figure 9 The model of *Acanthamoeba* granulomatous encephalitis. *Acanthamoeba* are thought to enter lungs via the nasal route. Next, amoebae traverse the lungs into the bloodstream, followed by hematogenous spread. Finally, *Acanthamoeba* crosses the blood-brain barrier (BBB) and enters into the central nervous system (CNS) to produce disease. It is noteworthy that *Acanthamoeba* may bypass the lower respiratory tract and directly enter into the bloodstream via skin lesions. The olfactory neuroepithelium may provide an alternative route of entry into the CNS.

deaths due to HIV/AIDS during 1981–1996 with the highest mortality during the mid 1990s: 49 000 in 1994 and 50 000 in 1995, which declined to 39 000 in 1996 (Heath *et al.*, 1998; Center for Disease Control, www.cdc.gov). Over a similar period, the number of AGE deaths in HIV/AIDS patients was approximately 55 (Martinez and Visvesvara, 1997). Thus, the approximate rate can be calculated as 1.57 AGE deaths per 10 000 HIV/AIDS deaths in the USA, even though the number of AGE infections may be much higher in countries with warmer climates due to increased ubiquity and/or increased outdoor activities. At present, the estimated worldwide number of HIV/AIDS patients is a massive 40–45 million (as of 2005) and continues to rise sharply.

Hypothetically, this figure represents the number of AGE-susceptible hosts. If this is so, then why are there not a large number of AGE infections? There could be several explanations for this. At least for the USA, the number of deaths due to HIV/AIDS has been declining since the late 1990s: 22 000 deaths in 1997 and 18 000 deaths in 2003 (Heath *et al.*, 1998; CDC), thus reducing the number of AGE-susceptible hosts. This decline in HIV/AIDS deaths in the USA is attributed to early diagnosis followed by the introduction of novel antiretroviral therapies, that is, highly active antiretroviral therapy (HAART), which was first introduced in 1996. As well as improving AIDS symptoms, HAART has protective effects against *Acanthamoeba* and other opportunistic pathogens (Carter *et al.*, 2004; Pozio and Morales, 2005; Seijo Martinez *et al.*, 2000). However, these therapies are not available to the majority of HIV/AIDS patients in the less-developed or developing countries in other parts of the world. Thus the approximate rate of 1.57 AGE infections per 10 000 HIV/AIDS deaths in such countries may provide only a minimum estimate of the burden of AGE infections. The fact that AGE cases are not being reported in developing countries (especially in Africa) is due to the lack of expertise, reporting problems, lack of proper monitoring and the lack of proper health care systems.

Of interest, there were 5 million new reported cases of HIV/AIDS in 2003 alone (approximately 14 000 infections per day), while 3 million deaths occurred due to HIV/AIDS-related diseases (approximately 8 500 deaths per day), mostly in Africa (even though there hasn't been a single reported case of AGE in Africa). And applying 1.57 AGE deaths per 10 000 HIV/AIDS deaths, the total number of AGE infections in 2003 can be estimated at approximately 471. Although this number is significantly less than the 3 million deaths in total, AGE is certainly a contributing factor in AIDS-related deaths and needs continued attention. Moreover, other conditions such as diabetes, malignancies, malnutrition, alcoholism, or having a debilitated immune system due to immunosuppressive therapy or other complications may all contribute to AGE infections.

5.2 Pathophysiology

The clinical symptoms may resemble viral, bacterial or tuberculosis meningitis: headache, fever, behavioral changes, hemiparesis, lethargy, stiff neck, aphasia, ataxia, vomiting, nausea, cranial nerve palsies, increased intracranial pressure, seizures, and death. These are due to hemorrhagic necrotizing lesions with severe meningeal irritation and encephalitis (Figure 10; Martinez, 1985, 1991). Patients with respiratory infections, skin ulcerations or brain abscesses should be strongly suspected of infections due to free-living amoebae. Post mortem examination often shows severe edema and hemorrhagic necrosis. It is not known whether this necrotic phase is caused by actively feeding trophozoites or inflammatory processes such as the release of the cytokines. The lesions due to AGE are most numerous in the basal ganglia, midbrain, brainstem, and cerebral hemiparesis with characteristic lesions in the CNS parenchyma resulting in chronic granulomatous encephalitis. A granulomatous response may be absent or minimal in patients with a severely impaired immune system that is interpreted as impairment of the cellular immune response (Martinez *et al.*, 2000). The affected tissues other than the CNS may include subcutaneous tissue, skin, liver, lungs, kidneys, adrenals, pancreas, prostate, lymph nodes, and bone marrow.

5.3 Diagnosis

Due to the rarity of the disease and complicated symptoms, which are common to other pathogens causing CNS infections, the diagnosis of AGE is problematical. The symptoms may be similar to other CNS pathogens

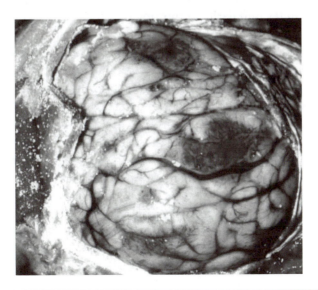

Figure 10 *Acanthamoeba*-infected brain exhibiting the severity of the disease.

including viruses, bacteria and fungi. This makes AGE diagnosis difficult and requires very high suspicion and expertise. Brain image analyses using computed tomography (CT) or magnetic resonance imaging (MRI) scan may show multifocal areas of signal intensities or lesions indicating brain abscess or tumors suggestive of the CNS defects. The cerebrospinal fluid (CSF) findings, although not confirmatory of AGE, are suggestive of CNS involvement. Pleocytosis with lymphocytic predominance is an important feature with elevated polymorphonuclear leukocytes, increased protein concentrations, decreased glucose concentrations and minimal cloudiness. The absence of viral and bacterial pathogens should be strongly suggestive of AGE. Due to the low density of parasites, the detection of host immune response should be attempted primarily. The demonstration of high levels of *Acanthamoeba*-specific antibodies in the patient's serum may provide a useful and straightforward method to further suspect AGE infection. This is performed by indirect immunofluorescence (IIF) assays. The serial dilutions of the patient's serum are incubated with fixed *Acanthamoeba*-coated slides (preferably T1, T4, T12 isolates as they have been shown to cause AGE infections), followed by incubation with fluorescein isothiocyanate (FITC)-labeled antihuman antibody and visualized under a fluorescent microscope. It is important to remember that the levels of anti-*Acanthamoeba* antibodies in normal populations may be in the range of 1:20 to 1:60 (Cerva, 1989; Cursons *et al.*, 1980). However, patients with a severely impaired immune system may not develop a high titer, thus other clinical findings should be taken into account for the correct diagnosis.

The confirmatory evidence comes from direct microscopic observation of *Acanthamoeba* in the CSF (after centrifugation at low speed) or in the brain biopsy, but requires familiarity of morphological characters. Giemsa-Wright, acridine orange or Calcofluor White staining may facilitate morphological-based positive identification of these amoebae. The lack of familiarity with *Acanthamoeba* morphological characteristics may require immunohisto-chemical studies using antisera made against *Acanthamoeba* to identify the etiologic agent, which should aid in the clinical diagnosis of AGE. It is helpful to use a few drops of the CSF and/or a portion of brain biopsy for *Acanthamoeba* culturing as described in section 3.7 (using non-nutrient agar plates). *Acanthamoeba* uses bacteria as a food source, and depending on the number of amoebae in the specimen, trophozoites can be observed within a few hours (up to 12 h). However in the absence of *Acanthamoeba*, plates should be monitored for up to 7 days. This method is particularly useful if problems are encountered in differentiating *Acanthamoeba* from monocytes, polymorphonuclear leukocytes and macrophages. As indicated in section 4.3, PCR-based methods have been developed but microscopy and plating-based analysis remain methods of choice. As indicated earlier MALDI-TOF-MS may be of potential value in the rapid identification of *Acanthamoeba* in clinical specimens (Moura *et al.*, 2003; Visvesvara *et al.*, 2007).

5.4 Host susceptibility

AGE is a rare disease that occurs mostly in immunocompromised or debilitated patients due to HIV infection, diabetes, immunosuppressive therapy, malignancies, malnutrition, and alcoholism, usually as a secondary infection. This is due to the inability of *Acanthamoeba* to evade the immune system of immunocompetent individuals. Indeed, sera from healthy individuals exhibit amoebicidal activities by activating the alternative complement pathway. Of interest, protozoan parasites with the ability to evade the host immune system possess sialic acid on the surface of their plasma membranes, which blocks alternative pathway convertase, or have a special coat or a capsule. For example, variable surface glycoproteins (VSG) on the surface of African *Trypanosoma* cover underlying components of the plasma membrane, thus preventing activation of the alternative complement pathway. However, the plasma membrane of *Acanthamoeba* lacks sialic acid (Korn and Olivecrona, 1971) or any protective coat or capsule (Bowers and Korn, 1968) and thus the amoebae are exposed to complement-mediated attack in an antibody-independent pathway (Ferrante and Rowan-Kelly, 1983). The presence of anti-*Acanthamoeba* antibodies in normal populations provides additional protection against these opportunistic pathogens (Cursons *et al.*, 1980).

Overall, complement pathways and their antibodies together with neutrophils and macrophages show potent amoebalytic activities thus suppressing infection. The conclusion from these findings is clear, in that a debilitated immune status of the host is usually a prerequisite for AGE, but the core basis of host susceptibility in contracting AGE requires further study as it may involve other factors such as host ethnicity (i.e., genetic basis of the host) or the inability of the host to induce a specific immune response against these pathogens. Interestingly, in a study by Chappell *et al.* (2001), Hispanic subjects were 14.5 times less likely to be seropositive against a T4 isolate than Caucasians. But whether Hispanics may be more susceptible to *Acanthamoeba* (T4 genotype) infections or they are simply exposed frequently to *Acanthamoeba* (Hispanics form a major workforce in agriculture in the USA), remains to be determined. Future studies will identify the precise host factors that play an important role in controlling this fatal infection, and may help develop therapeutic interventions for susceptible hosts.

5.5 Risk factors

Acanthamoeba granulomatous encephalitis is normally a secondary infection to other primary diseases. Almost all reported cases have occurred in immunocompromised patients due to HIV (AIDS patients), and/or in individuals with lymphoproliferative disorders, hematologic disorders, diabetes mellitus, pneumonitis, renal failure, liver cirrhosis, rhinitis, pharyngitis, gammaglobulinaemia, pregnancy, systemic lupus erythematosus, glucose-6-phosphate deficiency, tuberculosis, chronic alcoholism, malnourishment, chronic illness or debilitation, or undergoing radiotherapy. Patients undergoing organ/tissue transplantation with immunosuppressive therapy, steroids and excessive antibiotics are also at risk (Table 2). The risk factors for patients suffering from the above diseases include exposure to contaminated water such as swimming pools, on beaches, or working with garden soil (Fig. 8).

5.6 Treatment

For AGE, there are no recommended treatments and the majority of cases due to this are identified, post-mortem. This is due to low sensitivity of *Acanthamoeba* to many antiamoebic agents but, more importantly, the inability of these compounds to cross the blood-brain barrier into the CNS. Current therapeutic agents include a combination of ketoconazole, fluconazole, sulfadiazine, pentamidine isethionate, amphotericin B, azithromycin, itraconazole, or rifampin that may be effective against CNS infections due to free-living amoebae, but have severe side effects. Recent studies have suggested that alkylphosphocholine compounds, such as hexadecylphosphocholine, exhibit anti-*Acanthamoeba* properties as well as the ability to cross the blood-brain barrier and thus may have value in the treatment of AGE (Kotting *et al.*, 1992; Walochnik *et al.*, 2002). Further studies are needed to determine their precise mode of action on *Acanthamoeba* and to develop methods of application and, more importantly, to assess the success of these compounds *in vivo*. Even with treatment, survivors may develop disability such as hearing loss, vision impairment, and so on.

6 Cutaneous acanthamebiasis

Other infections due to *Acanthamoeba* involve nasopharyngeal and, more commonly, cutaneous infections. The cutaneous infections are characterized by nodules and skin ulcerations and demonstrate *Acanthamoeba* trophozoites and cysts. In the healthy individuals, these infections are very rare and are self-limiting. However, in immunocompromised patients, this may provide a route of entry into the bloodstream, followed by the hematogenous spread to various organs/tissues, which may lead to fatal consequences. The involvement of the CNS warrants death within weeks. Both AGE and cutaneous infections can occur in combination or independent of each other. The direct demonstration of amoebae in biopsy using Calcofluor staining, IIF or PCR-based assays, or isolation of amoebae from the clinical specimen using plating assays, provides positive diagnosis as described above. There is no recommended treatment, but topical application of itraconazole, 5-fluorocytosine, ketoconazole and chlorohexidine may be of value.

7 Pathogenesis

As indicated above, the ability of *Acanthamoeba* to produce keratitis is not from mere exposure to the eye, but due to its virulent nature and tissue specificities. For example, Gray *et al.* (1995) tested 101 contact-lens storage cases of asymptomatic daily wearers for the presence of microbes and found that 81% were contaminated with bacteria, fungi and protozoa including *Acanthamoeba*. The occurrence of fungi was higher than protozoa: 24 *versus* 20% respectively, and significantly higher than *Acanthamoeba*, 24 *versus* 8% respectively. Even though fungi rarely cause corneal infections in wearers of contact lenses, this provides further supporting evidence for the proposition that keratitis-causing microbes possess specific virulence properties enabling them to become potential ocular pathogens. Of interest, 75% of individuals in this study used hydrogen peroxide as a disinfectant for their lenses indicating the ability of microbes to resist and survive the many available disinfection methods for cleaning contact lenses. Indeed, these organisms identified as contaminants in the cases exhibit catalase activity, an enzyme that breaks down hydrogen peroxide to oxygen and water (Gray *et al.*, 1995). The pathogenesis of *Acanthamoeba* is highly complex and involves several determinants working in concert to produce disease. The sequence of events, at least for *Acanthamoeba* keratitis, involves breaching of the surface epithelium, keratocyte depletion by *Acanthamoeba*, stromal necrosis, and induction of an intense inflammatory response. For simplicity, in the following account these factors are described separately as contact-dependent and contact-independent mechanisms (Figure 11).

7.1 Contact-dependent factors

The ability of amoebae to bind to host cells is the first crucial step in the pathogenesis of *Acanthamoeba* infections. This leads to secondary events such as interference with host intracellular signaling pathways, toxin secretions, and ability to phagocytose host cells, ultimately leading to cell death.

Mannose-binding protein

Morton *et al.* (1991) showed that the binding of *Acanthamoeba* to corneal epithelial cells of rabbit is mediated by amoeba adhesin, an MBP expressed on the surface of the parasite. The role of the MBP was subsequently established with the discovery that it is important in parasite binding to various cell types including rabbit corneal epithelial cells (Morton *et al.*, 1991; Yang *et al.*, 1997), pig corneal epithelium (van Klink *et al.*, 1992), Chinese hamster corneal epithelium (van Klink *et al.*, 1993), human corneal fibroblasts (Badenoch *et al.*, 1994), rat microglial cells (Shin *et al.*, 2001), and human corneal epithelial cells (Sissons *et al.*, 2004a). Later, Alsam *et al.* (2003) extended these to include human brain microvascular endothelial cells,

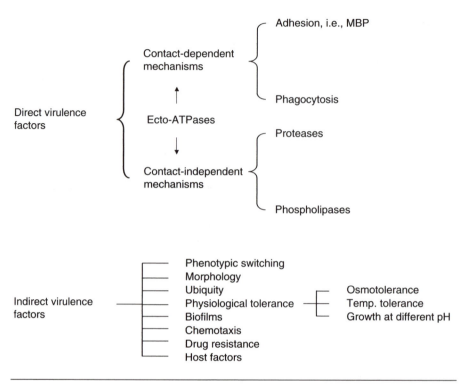

Figure 11   Direct and indirect virulence factors that contribute to *Acanthamoeba* infections.

suggesting a possible role of MBP in AGE. It is not clear whether there are other more specific mechanisms of amoebic binding to host cells, but at least the initial binding seems to be dependent on the expression of MBP and its binding to mannose-containing glycoproteins on the surface of the host cell. Indeed, amoebae will bind even to mannose-coated tissue culture plates (Yang *et al.*, 1997). The significance of MBP is further shown by the observation that it is expressed only during the infective trophozoite stage of *Acanthamoeba*; cysts lack MBP and therefore cannot bind to the host cells (Dudley *et al.*, 2005; Garate *et al.*, 2006). Garate *et al.* (2004) have identified a *mbp* gene in *Acanthamoeba* containing six exons and five introns that spans 3.6 kb. The 2.5 kb cDNA codes for an 833 amino-acid precursor protein with a signal sequence (residues 1–21 aa), an *N*-terminal extracellular domain (residues 22–733 aa) with five *N*- and three *O*-glycosylation sites, a transmembrane domain (residues 734–755 aa), and a *C*-terminal intracellular domain (residues 756–833 aa; Garate *et al.*, 2004).

Besides these studies, an understanding of the precise events in MBP binding to mannose-containing glycoprotein and/or additional *Acanthamoeba* determinants secondary to MBP should be a focus for future studies. Of interest in this context, is a recently identified laminin-binding protein from *Acanthamoeba* (Hong *et al.*, 2004); laminin is a major mannosylated glycoprotein constituent of the extracellular matrix (ECM) and the basement membrane of the host cells.

Host intracellular signaling in response to *Acanthamoeba*

The initial binding of *Acanthamoeba* to the surface of host cells interferes with the host intracellular signaling pathways. Several studies have shown that *Acanthamoeba* induces apoptosis in the host cells (Alizadeh *et al.*, 1994; Sissons *et al.*, 2005). Apoptosis, or programmed cell death, is known to be dependent on the host cell's own signaling pathways, involving Ca^{2+} responses. Previous studies have shown that increase in cytosolic levels of Ca^{2+} in response to *Acanthamoeba* metabolites are dependent on trans-membrane influx of extracellular Ca^{2+} (Mattana *et al.*, 1997). Among other roles, the changes in the levels of intracellular Ca^{2+} exert effects on cytoskeletal structure, induce morphological changes, or alter the permeability of the plasma membrane, finally leading to target cell death within minutes.

An understanding of the complex intracellular signaling pathways is crucial to identify targets for therapeutic interventions. There are more than 10 000 signaling molecules in a single host cell at any one time, so the identification of key molecules and how they interact in response to *Acanthamoeba*, leading to a functional outcome, is clearly a challenge. However, it most likely involves events both at the transcriptional and post-translational level. It is well-established that proteins that regulate cell fate require tyrosine as well as serine/threonine phosphorylations for intracellular signaling. To this end, recent studies have shown that *Acanthamoeba* up-regulates or down-regulates the expression of a number of genes important for regulating the cell cycle (Sissons *et al.*, 2004a). Overall, it is shown that *Acanthamoeba* up-regulates the expression of genes such as GADD45A and p130 Rb, associated with cell-cycle arrest, as well as inhibiting the expression of other genes, such as those for cyclins F, G1 and cyclin-dependent kinase-6 that encode proteins important for cell-cycle progression. The overall response of these events is shown to be arrest of the host cell cycle (Figure 12). This is further supported by the dephosphorylation of retinoblastoma protein (pRb). In the unphosphorylated form, pRb remains bound to E2F transcription factors (required for DNA synthesis) in the cytoplasm and inhibits E2F translocation into the nucleus. However, when phosphorylated by cyclin-dependent kinases (CDKs), pRb undergoes a conformational change resulting in E2F-pRb complex dissociation. The released E2F translocates to the nucleus and initiates DNA synthesis for the S phase (Figure 12). Thus pRb is a potent inhibitor of G1/S cell-cycle progression. Recent studies showed that *Acanthamoeba* inhibits pRb phosphorylations in human corneal epithelial cells as well as in human brain microvascular endothelial cells, indicating that amoebae induce cell-cycle arrest in host cells. Other studies have shown that

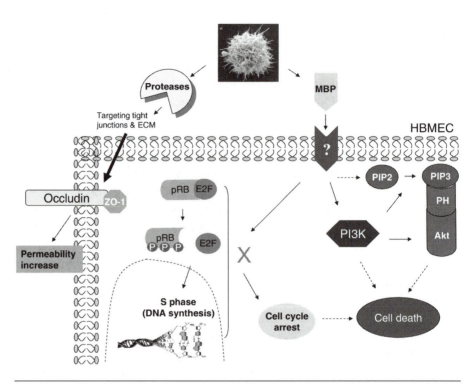

Figure 12 Host intracellular signaling in response to *Acanthamoeba*. Note that *Acanthamoeba* induces cell-cycle arrest in the host cells by altering expression of genes as well as by modulating protein retinoblastoma (pRb) phosphorylations. In addition, *Acanthamoeba* have also been shown to induce host cell death via phosphatidylinositol 3-kinase (PI3K). By secreting proteases, amoebae disrupt tight junctions by targeting zonula-1 and occludin proteins. MBP, mannose-binding protein; E2F, a transcription factor that controls cell proliferation through regulating the expression of essential genes required for cell-cycle progression; PIP2, phosphatidylinositol-4,5-bisphosphate; PIP3, phosphatidylinositol-3,4,5-trisphosphate; Akt (protein kinase B)-PH domain, a serine/threonine kinase – a critical enzyme in signal transduction pathways involved in cell proliferation, apoptosis, angiogenesis, and diabetes.

Acanthamoeba induces apoptosis in host cells, but whether these events are independent of each other, or whether cell-cycle arrest is a primary event that leads subsequently to apoptosis, is not clear. However, recent studies have shown that *Acanthamoeba*-mediated host cell death is dependent on the activation of phosphatidylinositol 3-kinase (PI3K; Figure 12). This was confirmed using LY294002, a specific PI3K inhibitor, as well as using host cells expressing mutant p85, a regulatory subunit of PI3K (dominant negative PI3K). PI3K has been known traditionally to be important for cell survival pathways, so these results are not surprising. For example, Thyrell *et al.* (2004), have shown that IFNα induces PI3K-mediated apoptosis in myeloma cells without Akt phosphorylations. It was further shown that downstream effectors of PI3K-mediated apoptosis involve activation of the proapoptotic molecules Bak and Bax, loss of mitochondrial membrane potential and release of cytochrome c: all well-known mediators of apoptosis. Similar mechanisms may exist in *Acanthamoeba*-mediated host cell death.

Phagocytosis

Adhesion of *Acanthamoeba* to the host cells leads to secondary processes such as phagocytosis or secretion of toxins (discussed later). Phagocytosis is an actin-dependent process involving polymerization of monomeric G-actin into filamentous F-actin. The ecological significance of *Acanthamoeba* phagocytosis is in the uptake of food particles such as bacteria, plasmids, or fungal cells. The ability of *Acanthamoeba* to form food cups, or amoebastomes, during incubations with host cells suggests they have a role in *Acanthamoeba* pathogenesis (Khan, 2001; Pettit *et al.*, 1996). In support of this, it is shown that cytochalasin D (a toxin that blocks actin polymerization) inhibits *Acanthamoeba*-mediated host cell death, confirming that actin-mediated cytoskeletal rearrangements play an important role in *Acanthamoeba* phagocytosis (Niederkorn *et al.*, 1999). The ability of *Acanthamoeba* to phagocytose is an intracellular signaling-dependent process. Genistein (a protein tyrosine kinase inhibitor) inhibits while sodium orthovanadate (protein tyrosine phosphatase inhibitor) enhances *Acanthamoeba* phagocytosis, indicating that a tyrosine-kinase-induced actin polymerization signal is important in *Acanthamoeba* phagocytosis (Alsam *et al.*, 2005a). Rho GTPases are the major regulators of the actin cytoskeleton, which link external signals to the cytoskeleton (Mackay and Hall, 1998). As their name indicates, Rho GTPases bind and hydrolyze GTP and, in the process, stimulate pathways that induce specific cytoskeletal rearrangements resulting in distinct phenotypes. There are three well-studied pathways: the RhoA pathway, leading to stress fibre formation; Rac1 activation, triggering lamellipodia formation; and Cdc42 activation, promoting filopodia formation. Recent studies have shown that the Rho kinase inhibitor, Y27632, partially blocks *Acanthamoeba* phagocytosis. Y27632 blocks stress fiber formation by inhibiting myosin light chain phosphorylation and cofilin phosphorylations but independent of the profilin pathway. Overall, these findings suggested that, in addition to the RhoA pathway, Rac1 and Cdc42 pathways may also be involved in *Acanthamoeba* phagocytosis. This is further supported with the finding that LY294002, a specific inhibitor of PI3K, inhibits *Acanthamoeba* phagocytosis. The PI3K is shown to be involved in Rac1-dependent lamellipodia formation (Wennström *et al.*, 1994) and Cdc42-dependent cytoskeletal changes (Jimenez *et al.*, 2000). Future research into the identification of additional molecules/pathways, how various intracellular signaling pathways interact, and/or whether they are independent of each other, will enhance our understanding of *Acanthamoeba* phagocytosis, and should be of value in the development of therapeutic interventions. Of interest, mannose, but not other saccharides, inhibits *Acanthamoeba* phagocytosis suggesting that *Acanthamoeba* phagocytosis is a receptor-dependent process, and *Acanthamoeba* adhesin (or MBP), is involved in its ability to phagocytose (Allen and Dawidowicz, 1990; Alsam *et al.*, 2005a). Overall, these findings suggest that MBP is crucial in *Acanthamoeba* binding to the target cells, as well as it's phagocytotic processes.

7.2 Ecto-ATPases

Ecto-ATPases are glycoproteins present in plasma membranes that have their active sites facing the external medium rather than the cytoplasm. Ecto-ATPases hydrolyze extracellular ATP and other nucleoside triphosphates. The resultant ADP can have toxic effects on the host cells. For example, Mattana *et al.* (2002) have shown that ADP released by *Acanthamoeba* binds to $P2y_2$ purinergic receptors on the host cells, causing an increase in the intracellular Ca^{2+}, inducing caspase-3 activation and finally resulting in apoptosis. A P2 receptor antagonist, suramin, inhibits *Acanthamoeba*-mediated host cell death (Mattana *et al.*, 2002; Sissons *et al.*, 2004b), suggesting that ecto-ATPases play an important role in *Acanthamoeba* pathogenesis, in this case by a contact-independent mechanism. Furthermore, clinical isolates exhibit higher ecto-ATPase activities compared with environmental isolates. The ecto-ATPase activity is significantly increased in the presence of mannose but not other sugars. Interestingly, weakly and/or nonpathogenic *Acanthamoeba* shows no differences in ecto-ATPase activities in the presence of mannose. Taken together, these findings suggest that the engagement of *Acanthamoeba* adhesin (i.e., MBP) enhances ecto-ATPase activities and thus ecto-ATPases may also play a role in *Acanthamoeba* pathogenesis but this time by a contact-dependent mechanism. How MBP is associated with ecto-ATPase activities is not clear. Of interest, several ecto-ATPases with approximate molecular weights of 62, 100, 218, 272 and >300 kDa have been described in *Acanthamoeba* (Sissons *et al.*, 2004b). The differences in ecto-ATPases have been attributed to strain differences. Future research will elucidate their function in *Acanthamoeba* biology, investigate their precise role in contact-dependent and contact-independent mechanisms of *Acanthamoeba* pathogenesis, and their usefulness as diagnostic targets for genotype differentiation.

7.3 Contact-independent factors

To produce damage to the host cell and/or tissue migration, the majority of pathogens rely upon their ability to produce hydrolytic enzymes. These enzymes may be constitutive, that is required for routine cellular functions, or inducible, produced under specific conditions, for example upon contact with the target cells. These enzymes can have devastating effects on host cells by causing membrane dysfunction or physical disruptions. Cell membranes are made of proteins and lipids, and *Acanthamoeba* is known to produce two types of hydrolytic enzymes: proteases, which hydrolyze peptide bonds, and phospholipases, which hydrolyze phospholipids.

Proteases

Proteases are well-known virulence factors in the majority of viral, bacterial, protozoan and multicellular pathogens. These enzymes hydrolyze peptide bonds and thus exhibit the ability to degrade various substrates. *Acanthamoeba* secretes large amounts of proteases. Interestingly, both clinical and nonclinical isolates of *Acanthamoeba* exhibit protease activities, but larger amounts are observed with the former. This suggests that the principal physiological role of proteases is to degrade the substrate for feeding purposes. This is consistent with the idea that *Acanthamoeba* is primarily a free-living environmental organism and its role in human/animal infections is secondary or opportunistic. There are four major types of proteases: aspartic, cysteine, serine, and metalloproteases; so far *Acanthamoeba* are known to produce all but aspartic protease activities, as described below.

All *Acanthamoeba* isolates tested to date exhibit proteolytic activities and serine proteases seem to be the most abundant in almost all genotypes. For example, Hadas and Mazur (1993) demonstrated the presence of a 35 kDa serine protease in *Acanthamoeba* isolates (some most likely belonging to the T4 genotype). Other studies using T4 isolates have identified extracellular serine proteases with approximate molecular weights of 36, 49 and 66 kDa (Mitro *et al.*, 1994), 107 kDa (Khan *et al.*, 2000), and 55, 97 and 230 kDa (Cao *et al.*, 1998), but other studies show the presence of 27, 47, 60, 75, 100, >110 kDa serine proteases (Alfieri *et al.*, 2000). Furthermore, Mitra *et al.* (1995), demonstrated a 40 kDa serine protease in *Acanthamoeba* belonging to the T4 genotype. This protease was shown to activate plasminogen, whose physiological function is to degrade ECM components. This diversity of serine proteases within a single genotype of *Acanthamoeba* may be due to strain differences, differences in their virulence, culture under diverse conditions or differences in assay methods. Kong *et al.* (2000) demonstrated a 33 kDa serine protease from *A. healyi* (T12 isolate), which demonstrated degradation of types I and IV collagen and fibronectin, which are main components of the ECM, as well as fibrinogen, IgG, IgA, albumin, and hemoglobin. These properties of serine proteases most likely facilitate *Acanthamoeba* invasion of corneal stroma and lead to secondary reactions such as edema, necrosis and inflammatory responses. Other studies have identified serine proteases with approximate molecular weights of 42 kDa (Cho *et al.*, 2000) and 12 kDa that degrade immunoglobulins, protease inhibitors and interleukin-1 (Na *et al.*, 2001, 2002), but their genotypes are not known. A direct functional role of serine proteases in *Acanthamoeba* infection is indicated by the observations that intrastromal injections of *Acanthamoeba*-conditioned medium produces corneal lesions *in vivo* similar to those observed in *Acanthamoeba* keratitis patients, and this effect is inhibited by PMSF, a serine protease inhibitor (He *et al.*, 1990; Na *et al.*, 2001).

Among cysteine proteases, although data are limited, a 65 kDa extracellular cysteine protease is reported in *Acanthamoeba* isolates belonging to the T4 genotype as well as 43, 70 and 130 kDa cysteine proteases (Alfieri *et al.*, 2000; Hadas and Mazur, 1993). It is worth noting that in these studies 65 and 70 kDa may refer to the same protease as the difference may have been due to variation in protease mobility in the substrate gels. Recently, Hong *et al.*, (2002) have identified a 24 kDa cysteine protease (most likely intracellular) from an *Acanthamoeba* isolate belonging to the T12 genotype. The physiological roles of cysteine proteases remain to be identified.

In addition to serine and cysteine proteases, there is evidence for metalloprotease activity in *Acanthamoeba* (Alfieri *et al.*, 2000; Mitro *et al.*, 1994). Cao *et al.* (1998) showed evidence of an 80 kDa metalloprotease, but its origin (whether *Acanthamoeba* or the host cells) is not known. Recent studies have identified a 150 kDa extracellular metalloprotease from *Acanthamoeba* isolated from an AGE patient (belonging to the T1 genotype; Alsam *et al.*, 2005b). This metalloprotease exhibited properties of ECM degradation as evidenced by its activity against collagen I and III (major components of collagenous ECM), elastin (elastic fibrils of ECM), plasminogen (involved in proteolytic degradation of ECM), as well as degradation of casein, gelatin, and hemoglobin, suggesting a role both in AGE and *Acanthamoeba* keratitis infections (Sissons *et al.*, 2006a).

Overall, these studies have shown that amoebae exhibit diverse proteases and elastases (Ferrante and Bates, 1988), which could play important roles in *Acanthamoeba* infections. However, their precise modes of action at the molecular level are only beginning to emerge. That some of the above proteases are secreted only by clinical isolates may indicate their role as potent virulence factors and/or diagnostic targets. Future studies in the role of proteases as vaccine targets, search for novel inhibitors by screening of chemical libraries, or rational development of drugs based on structural studies should enhance our ability to target these pathogens.

Phospholipases

Phospholipases are a diverse group of enzymes that hydrolyze the ester linkage in glycerophospholipids and can cause membrane dysfunction. The five major phospholipases are A1, A2, B, C, and D, and each has the ability to cleave a specific ester bond in the substrate of the target membrane. All phospholipases are present in multiple forms. Our knowledge of phospholipases in the virulence of *Acanthamoeba* is fragmented, however several studies have shown the presence of phospholipase activities in *Acanthamoeba* (Cursons *et al.*, 1978; Victoria and Korn, 1975a, 1975b). Cursons *et al.* (1978) were the first to demonstrate phospholipase A in *Acanthamoeba* and suggested a role in AGE infection *in vivo*. Because phospholipases cleave phospholipids, their possible role in membrane disruptions, penetration of the host cells and cell lysis should be the subject for future studies. Other actions of phospholipases may involve interference with intracellular signaling pathways. For example, phospholipases generate lipids and lipid-derived products that act as secondary messengers. Oishi *et al.* (1988) showed that lysophospholipids, by/end products of phospholipase B, induce activation of protein kinase C, which has diverse functions in host cell signaling pathways. Phospholipase C of *Clostridium perfringens* induces expression of interleukin-8 synthesis in endothelial cells (Bryant and Stevens, 1996). Overall, these studies suggest that *Acanthamoeba* phospholipases and/or lysophospholipases may play a direct role in causing host cell damage or affect other cellular functions, such as the induction of inflammatory responses, thus facilitating the virulence of *Acanthamoeba*; however, this remains to be fully established. More studies are needed to identify and characterize *Acanthamoeba* phospholipases and to determine their potential role for therapeutic intervention. This is not a novel concept: earlier studies have shown that phospholipase C from *C. perfringens* induces protection against *C. perfringens*-mediated gas gangrene (Kameyama *et al.*, 1975). In addition, the targeting of phospholipases using synthetic inhibitor compounds has been shown to have the potential to prevent *Candida* infections (Hanel *et al.*, 1995). Antibodies produced against *Acanthamoeba* phospholipases may also be of potential value in the development of sensitive and specific diagnostic assays.

8 Indirect virulence factors

The ability of *Acanthamoeba* to produce human diseases is a multifactorial process and is, amongst other factors, dependent on its ability to survive outside its mammalian host for various times and under diverse environmental conditions (high osmolarity, varying temperatures, food deprivation, and resistance to chemotherapeutic drugs). The ability of *Acanthamoeba* to overcome such conditions can be considered as contributory factors towards disease and are indicated as indirect virulence factors (Figure 11).

8.1 Morphology

The infective forms of *Acanthamoeba* or trophozoites do not have a distinct morphology. However, they do possess spine-like structures known as acanthopodia on their surface, which may play a key role in the pathogenesis of *Acanthamoeba* infections by modulating binding of pathogenic

Acanthamoeba to corneal epithelial cells (Khan, 2001; Figure 4). It would not be surprising if it was found that MBP, which is involved in the binding of *Acanthamoeba* to host cells, is localized on the acanthopodia. In addition, their amoeboid motion resembles that of macrophages/neutrophils. From this it can be speculated that *Acanthamoeba* may use similar mechanisms to traverse the biological barriers such as the blood-brain barrier.

8.2 Temperature tolerance and osmotolerance

Upon contact with tear film and corneal epithelial cells, *Acanthamoeba* is exposed to high osmolarity (due to salinity in tears) as well as high temperatures. For successful transmission, *Acanthamoeba* must withstand these burdens and exhibit growth. Growth at high temperature and high osmolarity are the hallmarks of pathogenic *Acanthamoeba* (De Jonckheere, 1983; Khan *et al.*, 2001, 2002; Walochnik *et al.*, 2000). These studies have shown that the ability of *Acanthamoeba* to grow at high temperature and osmolarity correlates with the pathogenicity of *Acanthamoeba* isolates. However, the precise mechanisms by which pathogenic *Acanthamoeba* adapt to higher temperatures and maintain their metabolic activities remaining, entirely unknown. Interestingly, temperature tolerance studies in *Candida neoformans* have identified the Ca^{2+}-dependent protein phosphatase calcineurin as a requirement for its growth at 37°C (Odom *et al.*, 1997a, 1997b). Furthermore, strains of *C. neoformans* in which the calcineurin gene has been disrupted are avirulent in a model of cryptococcal meningitis *in vivo*. These studies might serve as a basis for research into determining the physiological properties of *Acanthamoeba*.

8.3 Growth at different pH

Pathogenic *Acanthamoeba* can grow at pH ranging from 4 to 12 (Khan *et al.*, unpublished data), which gives it the potential to colonize several niches. For example, the ability of *Candida albicans* to grow at diverse pH is crucial for its virulence (Davis *et al.*, 2000) and two pH-regulating genes, *PHR1* (expressed at neutral and basic pH) and *PHR2* (expressed at acid pH), have been identified. Deletion of *PHR2* results in the loss of virulence, while deletion of *PHR1* results in reduced virulence in a systemic model (De Bernardis *et al.*, 1998). The clinical significance of the ability of *Acanthamoeba* to exhibit growth at different pH remains to be determined.

8.4 Phenotypic switching

Phenotypic switching in *Acanthamoeba* is the ability to differentiate into a morphologically distinct dormant cyst form or a vegetative trophozoite form. This is a reversible change dependent on environmental conditions (Figure 4). Cysts are resistant to various antimicrobial agents and adverse conditions such as extreme temperatures, pH, osmolarity and desiccation, and they can be airborne: all of which present a major problem in chemotherapy because their persistence may lead to recurrence of the disease. Furthermore, *Acanthamoeba* cysts can survive for several years while maintaining pathogenicity (Mazur *et al.*, 1995). These characteristics suggest that the primary functions of cysts lie in withstanding adverse conditions and in the spread of amoebae throughout the environment. This may represent the ability of *Acanthamoeba* to alternate expression of surface proteins/glycoproteins in response to changing environments and/or immune surveillance. Overall, phenotypic switching represents a major factor in the transmission of *Acanthamoeba* infections: however, the underlying molecular mechanisms in these processes remain to be eluci-dated. At present it is not clear whether *Acanthamoeba* shows antigenic variations, and their possible involvement in phenotypic switching should be investigated in future studies.

8.5 Drug resistance

Current treatment for *Acanthamoeba* keratitis involves topical application of mixtures of drugs including CHX, PHMB, neomycin and propamidine isethionate as indicated in section 4.6. These drugs have been shown to be most effective in killing *Acanthamoeba* trophozoites. Both chlorhexidine diacetate and PHMB are 'membrane-acting' cationic biocides. At alkaline pH, surface proteins of *Acanthamoeba* are negatively charged, interacting rapidly with these cationic biocides inducing structural and permeability changes in cell membrane leading to leakage of ions, water and other cyto-plasmic components resulting in cellular damage (Perrine *et al.*, 1995). Drugs such as propamidine isethionate belong to the diamidine family and are effective inhibitors of DNA synthesis (Duguid *et al.*, 1997). Of concern, several studies have recently shown the increasing resistance of *Acanthamoeba* to antimicrobial chemotherapy, but the mechanisms of such drug resistance in *Acanthamoeba* remain incompletely understood (Ficker *et al.*, 1990; Larkin *et al.*, 1992; Lim *et al.*, 2000; Lloyd *et al.*, 2001; Murdoch *et al.*, 1998). One intriguing report was made by Ficker *et al.* (1990), who observed the development of propamidine resistance during the course of therapy for *Acanthamoeba* keratitis, which led to recurrence of the infection. This may be due to the fact that although propamidine isethionate inhibits DNA synthesis, at presently recommended concentrations it may not be active against the cyst forms of *Acanthamoeba* due to their reproductive inactivity and/or very limited metabolic activity. Also, it is likely that the double-walled structure of *Acanthamoeba* cysts comprising an inner endo-cyst and an outer ectocyst (composed of 33% protein, 4–6% lipids and 35% carbohydrates, mostly cellulose), provides a physical barrier against chemotherapeutic agents (Neff and Neff, 1969; Turner *et al.*, 2000).

Additionally, the dangers of increased selection pressure induced by con-tinuous drug exposure should not be ignored. The precise understanding of these mechanisms is crucial for the rationale development of much needed drugs against this serious infection. As indicated above, the current treatment against *Acanthamoeba* infections involves the use of a mixture of drugs for up to a year. The aggressive and prolonged management is due to the ability of *Acanthamoeba* to rapidly adapt to harsh conditions and switch phenotypes into a resistant cyst form. One possibility of improving the treatment of *Acanthamoeba* infections is to inhibit the ability of these parasites to switch into the cyst form. The cyst wall is made partially of cellulose. In support, recent studies have shown that a cellulose synthesis inhibitor, 2,6-dichlorobenzonitrile (DCB) can enhance the effects of the antiamoebic drug, pentamidine isethionate (PMD; Dudley *et al.*, 2007). These studies suggested that the inclusion of a cellulose synthesis inhibitor in an 'associative' therapy should act in consort, and thus may enhance the potency of other compounds resulting in improved treatment, which can be achieved in a limited time. Of interest, cellulose biosynthesis is limited to bacteria, some protists and higher plants. Consequently, a specific cellulose biosynthesis inhibitor should have no and/or minimal effects on nontarget cells, that is human cells. Future studies should test the use of a nontoxic cellulose synthesis inhibitor *in vivo*, which may aid in the improved therapy against *Acanthamoeba* infections.

8.6 Ubiquity

Acanthamoeba have been found in diverse environments, from drinking water to distilled water wash-bottles, so it is not surprising that humans regularly encounter these organisms, as is evidenced by the finding that, in some areas, the majority of the population possess *Acanthamoeba* antibodies. This clearly suggests that these are one of the most ubiquitous protozoa and often come in contact with humans. This provides amoebae with a wider access to approach the limited susceptible hosts.

8.7 Biofilms

Biofilms are known to play an important role in the pathogenesis of *Acanthamoeba* keratitis. Biofilms are microbially derived sessile communities, which can be formed in aqueous environments as well as on any materials and medical devices, including intravenous catheters, contact lenses, scleral buckles, suture material, and intraocular lenses. In the instance of contact lenses, biofilms are formed through contamination of the storage case. Once established, biofilms provide attractive niches for *Acanthamoeba*, by fulfilling their nutritional requirements as well as providing resistance to disinfectants. For example, Beattie *et al.* (2003) have shown recently that *Acanthamoeba* exhibits significantly higher binding to used and *Pseudomonas* biofilm-coated hydrogel lenses compared with unworn contact lenses. In addition, the abundant nutrients provided by the biofilm encourages transformation of *Acanthamoeba* into the vegetative, infective trophozoite form, and it is important to remember that binding of *Acanthamoeba* to human corneal epithelial cells most likely occurs during the trophozoite stage, as cysts exhibit no and/or minimal binding (Dudley *et al.*, 2005; Garate *et al.*, 2006). Overall, these findings suggest that biofilms play an important role in *Acanthamoeba* keratitis in wearers of contact lenses and preventing their formation is an important preventative strategy.

8.8 Host factors

The factors that enable *Acanthamoeba* to produce disease are not limited solely to the parasite, but most likely involve host determinants. Evidence for this comes from recent studies in the UK, Japan and New Zealand, which suggest that storage cases of contact lenses of 400–800 per 10 000 asymptomatic wearers were contaminated with *Acanthamoeba* (Devonshire *et al.*, 1993; Gray *et al.*, 1995; Larkin *et al.*, 1990; Watanabe *et al.*, 1994). This number is remarkably high compared with the incidence of *Acanthamoeba* keratitis in wearers of contact lenses, which is 0.01–1.49 per 10 000. These findings suggest that factors such as host susceptibility, tissue specificity, tear factors, secretory immunoglobulin A (sIgA), corneal trauma, as well as environmental factors such as osmolarity, may be important in initiating *Acanthamoeba* infections. In addition, malnutrition, mental stress, age, metabolic factors, and other primary diseases may play a role in the pathogenesis of *Acanthamoeba* infections.

However, the extent to which such host factors contribute to the outcome of *Acanthamoeba* infections is unclear because host factors are more complex and difficult to study than those of the parasite. For example, in bacterial infections such as *Salmonella*, the genetic constitution of the host determines susceptibility (Fleiszig *et al.*, 1994; Harrington and Hormaeche, 1986). These studies with *Salmonella* were possible only because transgenic animals were available.

9 Immune response

Despite the ubiquitous presence of these organisms in diverse environments, the number of infections due to *Acanthamoeba* has remained very low. This is due to the fact that *Acanthamoeba* are opportunistic pathogens and their ability to produce disease is dependent on host susceptibility (e.g., immunocompromised patients or contact-lens wearers), environmental conditions (personal hygiene, exposure to contaminated water/soil), and their own virulence. For AGE, in normal circumstances, a competent immune system is sufficient to control these pathogens. But due to the complexity of the immune system, the precise factors which contribute to host resistance and the associated mechanisms remain unclear. By contrast, *Acanthamoeba* keratitis can occur in normal individuals, although oral immunization can prevent this infection. This suggests that the immune system still plays an important role in this condition. Although there are obvious similarities in the immune response to AGE and to *Acanthamoeba* keratitis, they are described separately for simplicity.

9.1 Acanthamoeba *keratitis and the immune response*

Because the normal cornea is avascular, many primary host defenses are provided by the eyelids and secreted tear film. The tear fluid produced by the lacrimal system, together with constant eye-lid movement, provides the first line of defense in *Acanthamoeba* keratitis. The tear film contains lysozyme, lactoferrin, β-lysins, sIgA, prostoglandins, and other compounds with antimicrobial and immunological properties (Nassif, 1996; Qu and Leher, 1998). Of the three distinct layers in the tear film (the oil layer, the aqueous layer and the mucous layer), the aqueous layer is the source of compounds with antimicrobial properties such as nonspecific antimicrobial lysozyme, lactoferrin and specific sIgA. A role for sIgA in *Acanthamoeba* keratitis is indicated by the findings that *Acanthamoeba* keratitis patients show decreased levels of sIgA, and specific antiamoebic sIgA inhibits *Acanthamoeba* binding to corneal epithelial cells, suggesting that sIgA normally plays an important role in alleviating this infection (Alizadeh *et al.*, 2001; Leher *et al.*, 1999; Walochnik *et al.*, 2001).

The mechanisms of how binding is inhibited are unclear, but could involve interference with *Acanthamoeba* adhesins and the associated pathways. Overall, the tear film in conjunction with the blink reflex is highly effective in blocking microbial access to the corneal epithelial cells, expelling amoebae from the surface of the eye and carrying it to the conjunctiva. Although the cornea is protected by only limited immune mechanisms, the conjunctiva is highly vascular with lymphoid tissue and contains mainly IgA-producing plasma cells, T-lymphocytes, natural killer cells and macrophages, which are highly effective in clearing *Acanthamoeba* and stimulating humoral and T-cell responses. For example, conjunctival macrophage depletion exacerbates *Acanthamoeba* keratitis symptoms *in vivo*, and increases the rate of infection to 100% (van Klink *et al.*, 1996). Thus in order to proceed with infection, *Acanthamoeba* must remain at the corneal surface and invade into the stromal tissue. Of interest, the tear film in wearers of contact lenses differs in terms of volume and make up, and it seems that use of extended-wear contact lenses alters the levels of inflammatory mediators in tears and contributes to increased inflammation (Thakur and Willcox, 2000).

Other major changes are exclusion of atmospheric oxygen and thinning of the basal tear film. It has been proposed that hypoxia is responsible for the metabolic compromise of the cornea and also leads to less resistance to microbial infection (Weisman and Mondino, 2002). This indicates that the cornea is a stressed environment in the presence of a contact lens, and this affects components of the tear film making invasion by amoebae more likely. In addition, the highly virulent strains of *Acanthamoeba* evade these primary defenses to traverse the cornea and invade the stroma. Once in the stroma, *Acanthamoeba* secrete proteolytic enzymes causing stromal degradation leading to macrophage/neutrophil infiltration, which modulate B- and T-lymphocyte activity to clear *Acanthamoeba*. The macrophages provide defense against *Acanthamoeba* keratitis directly by clearing amoebae and by inducing an inflammatory response, in particular secretion of macrophage inflammatory protein 2 (MIP-2). This results in the recruitment of other immune cells, such as neutrophils, which are highly potent in destroying *Acanthamoeba* in a myeloperoxidase-dependent manner (Hurt *et al.*, 2001, 2003). Overall, these studies suggest that upon entry into the eye, *Acanthamoeba* has to deal with normal tear film containing nonspecific antimicrobial compounds together with the sweeping action of the eye lids, as well as specific immunoglobulins and cell-mediated immunity.

9.2 AGE and the immune response

As AGE is a rare infection, much of our understanding of its immunopathogenesis comes from studies using animals. The following account highlights some of the findings arising during the last decade. Complement is the first line of powerful defense against invading pathogens. Complement activation is achieved by the classical pathway (activated by specific antibodies attached to *Acanthamoeba* surface), alternative pathways (activated by opsonization) or mannose-binding lectin pathways (activated directly by components of the surface composition of the pathogen). The ultimate effects are to induce the deposition of complement proteins, leading to opsonization of *Acanthamoeba* followed by their uptake by phagocytes, or the formation of the membrane attack complex (MAC) resulting in target cell death.

In support of MAC attack, previous studies have shown that normal human serum exhibits complement-mediated lysis in *Acanthamoeba*, resulting in up to 100% death of amoebae (Ferrante, 1991; Ferrante and Rowan-Kelly, 1983; Stewart *et al.*, 1992). This may be due to the fact that the plasma membranes of *Acanthamoeba* lack sialic acid (Korn and Olivecrona, 1971) or any protective coat or capsule (Bowers and Korn, 1968) and thus *Acanthamoeba* are exposed to complement-mediated killing (Ferrante and Rowan-Kelly, 1983). However recent studies have shown that a subpopulation of the virulent strains of *Acanthamoeba* are resistant to complement-mediated lysis (Sissons *et al.*, 2006b; Toney and Marciano-Cabral, 1998). But the complement pathway in the presence of phagocytes (macrophages/neutrophils) is highly effective in clearing *Acanthamoeba* (Stewart *et al.*, 1994). Overall, complement pathways and their antibodies, together with neutrophils and macrophages, show potent amoebalytic activities thus suppressing the infection by clearing the amoebae (Marciano-Cabral and Toney, 1998; Toney and Marciano-Cabral, 1998).

It is important to indicate that although macrophages/neutrophils from naive animals are able to destroy *Acanthamoeba*, cells from immune animals exhibit significantly increased amoebalytic activities. The macrophage-mediated killing is contact-dependent and is inhibited with cytochalasin D, that is, an actin polymerization inhibitor (van Klink *et al.*, 1997). This is further confirmed with the findings that conditioned medium obtained from macrophages after treatment with lipopolysaccharide and interferon-gamma had no effects on *Acanthamoeba* (Marciano-Cabral and Toney, 1998). Similarly, the neutrophils exhibit potent amoebicidal effects. The neutrophil-mediated killing is significantly increased in the presence of antiacanthamoebic antibodies (Stewart *et al.*, 1994). These interactions also stimulate secretion of pro-inflammatory cytokines including IL-1β, IL-6 and TNF-α (Marciano-Cabral and Toney, 1998; Toney and Marciano-Cabral, 1998). Other studies in mice have shown significant increased natural killer (NK) cell activities in *Acanthamoeba*-infected animals suggesting that NK cells may also play a role in protective immunity (Kim *et al.*, 1993). The fact that the majority of individuals possess anti-*Acanthamoeba* antibodies suggests that immunocompetent individuals exhibit *Acanthamoeba*-specific T-cell responses (in particular CD4 T-cells; Cursons *et al.*, 1980; Tanaka *et al.*, 1994). Based on these findings, it is clear that a debilitated immune status of the host is usually a prerequisite in AGE, but the core basis of host susceptibility in contracting AGE is not fully understood and may also involve the host's ethnic origin (i.e., genetic basis of the host) or the inability of the host to induce a specific immune response against these pathogens. Of interest, both *Acanthamoeba* cysts and trophozoites are immunogenic, and immune sera from animals infected with *Acanthamoeba* trophozoites cross-react with cysts suggesting that some of the antigenic epitopes are retained during the encystment process (McClellan *et al.*, 2002). Future studies will identify the precise host factors which play an important role in controlling this fatal infection and may help develop therapeutic interventions in susceptible hosts.

10 *Acanthamoeba* and bacteria interactions

Acanthamoeba were first shown to be infected and lysed by bacteria in 1954 (Drozanski, 1956) and to harbor bacteria as endosymbionts in 1975 (Proca-Ciobanu *et al.*, 1975). Later studies revealed that *Acanthamoeba* act as a reservoir for pathogenic facultative mycobacteria (Krishna-Prasad and Gupta, 1978). *Acanthamoeba* have also been shown to harbor virulent *Legionella* spp. associated with Legionnaires' disease (Rowbotham, 1980). At the same time, it is well-established that *Acanthamoeba* consumes bacteria in the environment, so the interactions of *Acanthamoeba* and bacteria are highly complex and dependent on the virulence of amoebae, the virulence of bacteria, and the environmental conditions. The outcome of these convoluted interactions may be beneficial to *Acanthamoeba* or to bacteria or may result in the development of a symbiotic relationship. Adding to this complexity, *Acanthamoeba* are known to interact with various Gram-positive and Gram-negative bacteria resulting in a range of outcomes. A complete understanding of *Acanthamoeba*-bacteria interactions is beyond the scope of this chapter and for further information readers are referred to Greub and Raoult (2004). For simplicity and for researchers new to this area, these interactions are discussed in three sections below.

10.1 Acanthamoeba *as bacterial predators*

This group includes bacteria that are used as a food source for *Acanthamoeba*. Bacteria are taken up by phagocytosis, followed by their lysis in phagolysosomes. Although *Acanthamoeba* consumes both Gram-positive and Gram-negative bacteria, they preferentially graze upon Gram-negative bacteria, which are used widely as a food source in the isolation of *Acanthamoeba* (Bottone *et al.*, 1994). However, the ability of *Acanthamoeba* to consume bacteria is dependent on the virulence properties of the bacteria and environmental conditions. For example, recent studies have shown that in the absence of nutrients, virulent strains of *Escherichia coli* K1 invade *Acanthamoeba* and remain viable intracellularly. Upon the availability of nutrients, K1 escapes *Acanthamoeba*, grows exponentially and lyses the host *Acanthamoeba* (Alsam *et al.*, 2006). By contrast, in the absence of nutrients, the avirulent strains of *E. coli* K12 are phagocytosed (instead of bacterial invasion) by *Acanthamoeba* and are killed (Alsam *et al.*, 2006). This demonstrates clear differences in the ability of *Acanthamoeba* to interact with *E. coli* that are dependent on the virulence properties of the *E. coli* and the environmental conditions. Future studies should examine the basis of these differences as well as how bacteria are taken up (invasion *versus* phagocytosis) and the molecular mechanisms involved.

10.2 Acanthamoeba *as bacterial reservoirs*

The ability of *Acanthamoeba* to act as a bacterial reservoir has gained much attention (Greub and Raoult, 2004) because many of these bacteria are human pathogens. These include: *Legionella pneumophila* (causative agent of Legionnaires' disease; Rowbotham, 1980), *E. coli* O157 (causative agent of diarrhea; Barker *et al.*, 1999), *Coxiella burnetii* (causative agent of Q fever; La Scola and Raoult, 2001), *Pseudomonas aeruginosa* (causative agent of keratitis; Michel *et al.*, 1995), *Vibrio cholerae* (causative agent of cholera; Thom *et al.*, 1992), *Helicobacter pylori* (causative agent of gastric ulcers; Winiecka-Krusnell *et al.*, 2002), *Simkania negevensis* (causative agent of pneumonia; Kahane *et al.*, 2001), *Listeria monocytogenes* (causative agent of listeriosis; Ly and Muller, 1990), and *Mycobacterium avium* (causative agent of respiratory diseases; Krishna-Prasad and Gupta, 1978; Steinert *et al.*, 1998). This property of *Acanthamoeba* is particularly important as these bacterial pathogens not only survive intracellularly but also multiply. This allows bacteria to transmit throughout the environment, evade the host defenses and/or chemotherapeutic drugs, and reproduce in sufficient numbers to produce disease. Upon favorable conditions, the increasing bacterial densities lyse their host amoebae and infect new amoebae and/or produce disease. In the long term, these *Acanthamoeba*-bacteria interactions can be considered as 'parasitic'. It is important to recognize that the virulence determinants responsible for bacterial invasion of *Acanthamoeba*, their intracellular survival by inhibition of phagolysosomes formation or growth at acidic pH, and their escape from *Acanthamoeba* vary between bacteria.

10.3 Acanthamoeba *as bacterial Trojan horse*

One of the key requirements for many bacterial pathogens is to survive harsh environments during transmission from one host to another. Once bacteria invade host tissue such as nasal mucosa, lung epithelial cells, or gut mucosa to produce disease, they must resist the innate defenses as well as cross the biological barriers. To this end, *Acanthamoeba* may act as a 'Trojan horse' for bacteria. The term 'Trojan horse' is used to describe bacterial presence inside *Acanthamoeba* as opposed to 'carrier' which may be mere attachment/adsorption to the surface. Recent studies have shown that *Burkholderia cepacia* (a causative agent of lung infection) remains viable within *Acanthamoeba* but does not multiply (Landers *et al.*, 2000; Marolda *et al.*, 1999). Essig *et al.* (1997) have shown similar findings using *Chlamydophila pneumoniae* (causative agent of respiratory disease). This property has also been observed with other bacterial pathogens as indicated in section 10.1. Overall, these findings suggest that *Acanthamoeba* facilitate bacterial transmission and/or provide protection against the human immune system. In support of this, Cirillo *et al.* (1997) showed increased *M. avium* colonization of mice, when inoculated in the presence of amoebae. The ability of *Acanthamoeba* to resist harsh conditions (such as extreme temperatures, pH, and osmolarity), especially during their cyst stage, suggest their usefulness as bacterial vectors. In particular, *Acanthamoeba* cysts are notoriously resistant to chlorine (a key and sometimes the only compound used in cleaning water systems). This poses clear challenges in eradicating bacterial pathogens from public water supplies, especially in developing countries. In addition, *Acanthamoeba*-bacteria interactions also affect bacterial virulence. For example, *L. pneumophila* grown within *Acanthamoeba* exhibited increased motility, virulence and drug resistance compared with axenically grown *Legionella* (Greub and Raoult, 2002).

11 Conclusions

Studies of *Acanthamoeba* have grown exponentially. These organisms have gained attention from the broad scientific community studying cellular microbiology, environmental biology, physiology, cellular interactions, molecular biology and biochemistry. This is due to their versatile roles in ecosystems and their ability to capture prey by phagocytosis (similar to macrophages), act as vectors, reservoirs and Trojan horses for bacterial pathogens, and to produce serious human infections such as blinding keratitis and fatal encephalitis. In addition, this unicellular organism has been used extensively to understand the molecular biology of motility. The ability of *Acanthamoeba* to switch phenotypes makes it an attractive model for the study of cellular differentiation processes. Moreover, the increasing numbers of HIV/AIDS patients, contact–lens wearers, and warmer climates, will add to the increasing burden of *Acanthamoeba* infections. Our understanding of *Acanthamoeba* pathogenesis both at the molecular and cellular level, as well as its ability to transmit, adapt to diverse conditions, overcome host barriers and emerge as infective trophozoite will provide targets for therapeutic interventions. The availability of the *Acanthamoeba* genome, together with the recently developed transfection assays (Peng *et al.*, 2005; Yin and Henney, 1997) and RNA interference methods (Lorenzo-Morales *et al.*, 2005), will undoubtedly increase the pace of our understanding of this complex but fascinating organism.

12 Acknowledgments

The author is grateful to Selwa Alsam, James Sissons, Samantha Jayasekera, Ricky Dudley, Abdul Matin, and Ruqaiyyah Siddiqui, School of Biological and Chemical Sciences, Birkbeck, University of London, London, UK; Kwang Sik Kim, Division of Infectious Diseases, Johns Hopkins University School of Medicine, Baltimore, MD, USA; Ed Jarroll, Northeastern University, Boston, MA, USA; David Warhurst, London School of Hygiene and Tropical Medicine, University of London, London, UK; Sim Webb, University of East Anglia, Norwich, UK; Amir Maghsood, Department of Medical Parasitology and Mycology, School of Public Health and Institute of Health Research, Tehran University of Medical Sciences, Tehran, Iran for their assistance; Joe Harrison, Kemnal Technology College, Kent, UK for technical support and Graham Goldsworthy, School of Biological and Chemical Sciences, Birkbeck College, University of London, London, UK for critical review. This work was supported by grants from the Faculty Research Fund, Central Research Fund, University of London, the Nuffield

Foundation, the Royal Society and the British Council for Prevention of Blindness.

13 References

Alfieri, S.C., Correia, C.E., Motegi, S.A. and Pral, E.M. (2000) Proteinase activities in total extracts and in medium conditioned by *Acanthamoeba polyphaga* trophozoites. *J. Parasitol.* **86**(2): 220–227.

Alizadeh, H., Pidherney, M.S., McCulley, J.P. and Niederkorn, J.Y. (1994) Apoptosis as a mechanism of cytolysis of tumor cells by a pathogenic free-living amoeba. *Infect. Immun.* **62**(4): 1298–1303.

Alizadeh, H., Apte, S., El-Agha, M.S., Li, L., Hurt, M., Howard, K., Cavanagh, H.D., McCulley, J.P. and Niederkorn, J.Y. (2001) Tear IgA and serum IgG antibodies against *Acanthamoeba* in patients with *Acanthamoeba* keratitis. *Cornea* **20**(6): 622–627.

Allen, P.G. and Dawidowicz, E.A. (1990) Phagocytosis in *Acanthamoeba*. 1. A mannose receptor is responsible for the binding and phagocytosis of yeast. *J. Cell Physiol.* **145**: 508–513.

Alsam, S., Kim, K.S., Stins, M., Rivas, A.O., Sissons, J. and Khan, N.A. (2003) *Acanthamoeba* interactions with human brain microvascular endothelial cells. *Microb. Pathogen.* **35**(6): 235–241.

Alsam, S., Sissons, J., Dudley, R. and Khan, N.A. (2005a) Mechanisms associated with *Acanthamoeba castellanii* (T4) phagocytosis. *Parasitol. Res.* **96**(6): 402–409.

Alsam, S., Sissons, J., Jayasekera, S. and Khan, N.A. (2005b) Extracellular proteases of *Acanthamoeba castellanii* (encephalitis isolate belonging to T1 genotype) contribute to increased permeability in an *in vitro* model of the human blood-brain barrier. *J. Infect.* **51**(2): 150–156.

Alsam, S., Jeong, S.R., Sissons, J., Dudley, R., Kim, K.S. and Khan, N.A. (2006) *Escherichia coli* interactions with *Acanthamoeba*: a symbiosis with environmental and clinical implications. *J. Med. Microbiol.* **55**(6): 689–694.

Alvord, L., Court, J., Davis, T., Morgan, C.F., Schindhelm, K., Vogt, J. and Winterton, L. (1998) Oxygen permeability of a new type of high risk soft contact lens material. *Optom. Vis. Sci.* **75**: 30–36.

Badenoch, P.R., Adams, M. and Coster, D.J. (1994) Corneal virulence, cytopathic effect on human keratocytes and genetic characterization of *Acanthamoeba*. *Int. J. Parasitol.* **25**: 229–239.

Band, R.N. and Mohrlok, S. (1973) The cell cycle and induced amitosis in *Acanthamoeba*. *J. Protozool.* **20**(5): 654–657.

Barker, J., Humphrey, T.J. and Brown, M.W. (1999) Survival of *Escherichia coli* O157 in a soil protozoan: implications for disease. *FEMS Microbiol. Lett.* **173**(2): 291–295.

Barr, J. (1998) The 1997 report on contact lenses. *Contact Lens Spectrum* **1**: 23–33.

Beattie, T.K., Tomlinson, A. and Seal, D.V. (2002) Anti-*Acanthamoeba* efficacy in contact-lens disinfecting systems. *Br. J. Ophthalmol.* **86**: 1319–1320.

Beattie, T.K., Tomlinson, A., McFadyen, A.K., Seal, D.V. and Grimason, A.M. (2003) Enhanced attachment of *Acanthamoeba* to extended-wear silicone hydrogel contact lenses: a new risk factor for infection? *Ophthalmology* **110**(4): 765–771.

Bottone, E.J., Perez, A.A., Gordon, R.E. and Qureshi, M.N. (1994) Differential binding capacity and internalisation of bacterial substrates as factors in growth rate of *Acanthamoeba* spp. *J. Med. Microbiol.* **40**(2): 148–154.

Bowers, B. and Korn, E.D. (1968) The fine structure of *Acanthamoeba castellanii*. I. The trophozoite. *J. Cell Biol.* **39**(1): 95–111.

Bowers, B. and Korn, E.D. (1974) Localization of lipophosphonoglycan on both sides of *Acanthamoeba* plasma membrane. *J. Cell Biol.* **62**(2): 533–540.

Bowers, B. and Olszewski, T.E. (1972) Pinocytosis in *Acanthamoeba castellanii*, kinetics and morphology. *J. Cell Biol.* **53**(3): 681–694.

Bowers, B. and Olszewski, T.E. (1983) *Acanthamoeba* discriminates internally between digestible and indigestible particles. *J. Cell Biol.* **97**: 317.

Bryant, A.E. and Stevens, D.L. (1996) Phospholipase C and perfringolysin O from *Clostridium perfringens* upregulate endothelial cell-leukocyte adherence molecule 1 and intercellular leukocyte adherence molecule 1 expression and induce interleukin-8 synthesis in cultured human umbilical vein endothelial cells. *Infect. Immun.* **64**(1): 358–362.

Byers, T.J., Hugo, E.R. and Stewart, V.J. (1990) Genes of *Acanthamoeba*: DNA, RNA and protein sequences (a review). *J. Protozool.* **37**(4): 17S–25S.

Byers, T.J., Kim, B.G., King, L.E. and Hugo, E.R. (1991) Molecular aspects of the cell cycle and encystment of *Acanthamoeba*. *Rev. Infect. Dis.* **13**: S373–S384.

Cao, Z., Jefferson, M.D. and Panjwani, N. (1998) Role of carbohydrate-mediated adherence in cytopathogenic mechanisms of *Acanthamoeba*. *J. Biol. Chem.* **273**: 15838–15845.

Carter, W.W., Gompf, S.G., Toney, J.F., Greene, J.N. and Cutolo, E.P. (2004) Disseminated *Acanthamoeba* sinusitis in a patient with AIDS: a possible role for early antiretroviral therapy. *AIDS Read.* **14**(1): 41–49.

Castellani, A. (1930) An amoeba found in culture of yeast: preliminary note. *J. Trop. Med. Hyg.* **33**: 160.

Cerva, L. (1989) *Acanthamoeba culbertoni* and *Naegleria fowleri*: occurrence of antibodies in man. *J. Epidemiol. Microbiol. Immunol.* **33**: 99–103.

Chappell, C.L., Wright, J.A., Coletta, M. and Newsome, A.L. (2001) Standardized method of measuring *Acanthamoeba* antibodies in sera from healthy human subjects. *Clin. Diagn. Lab. Immunol.* **8**(4): 724–730.

Cheng, K.H., Leung, S.L., Hoekman, H.W., Beekhuis, W.H., Mulder, P.G.H., Geerards, A.J.M. and Kijlstra, A. (1999) Incidence of contact-lens-associated microbial keratitis and its related morbidity. *Lancet* **354**(9174): 181–185.

Cho, J.H., Na, B.K., Kim, T.S. and Song, C.Y. (2000) Purification and characterization of an extracellular serine proteinase from *Acanthamoeba castellanii*. *IUBMB Life* **50**: 209–214.

Cirillo, J.D., Falkow, S., Tompkins, L.S. and Bermudez, L.E. (1997) Interaction of *Mycobacterium avium* with environmental amoebae enhances virulence. *Infect. Immun.* **65**(9): 3759–3767.

Cordingley, J.S., Wills, R.A. and Villemez, C.L. (1996) Osmolarity is an independent trigger of *Acanthamoeba castellanii* differentiation. *J. Cell Biochem.* **61**: 167–171.

Culbertson, C.G., Smith, J.W. and Miner, J.R. (1958) *Acanthamoeba*: observation on animal pathogenicity. *Science* **127**: 1506.

Culbertson, C.G., Smith, J.W., Cohen, H.K. and Miner, J.R. (1959) Experimental infection of mice and monkeys by *Acanthamoeba*. *Am. J. Pathol.* **35**: 185–197.

Cursons, R.T.M., Brown, T.J. and Keys, E.A. (1978) Virulence of pathogenic free-living amoebae. *J. Parasitol.* **64**: 744–745.

Cursons, R.T., Brown, T.J., Keys, E.A., Moriarty, K.M. and Till, D. (1980) Immunity to pathogenic free-living amoebae: role of humoral antibody. *Infect. Immun.* **29**(2): 401–407.

Dart, J.K.G., Stapleton, F. and Minassian, D. (1991) Contact lenses and other risk factors in microbial keratitis. *Lancet* **338**: 650–653.

Davis, D., Edwards, J.E., Mitchell, A.P. and Ibrahim, A.S. (2000) *Candida albicans RIM101* pH response pathway is required for host pathogen interactions. *Infect. Immun.* **68**: 5953–5959.

De Bernardis, F., Muhlschlegel, F.A., Cassone, A. and Fonzi, W.A. (1998) The pH of the host niche controls gene expression in and virulence of *Candida albicans*. *Infect. Immun.* **66**: 3317–3325.

De Jonckheere, J.F. (1983) Growth characteristics, cytopathic effect in cell culture and virulence in mice of 36 type strains belonging to 19 different *Acanthamoeba* spp. *Appl. Environ. Microbiol.* **39**: 681–685.

De Jonckheere, J. (1987) Taxonomy. In: *Amphizoic Amoebae Human Pathology* (ed. E.G. Rondanelli). Piccin Nuova Libraria, Padua, Italy, pp. 25–48.

De Jonckheere, J.F. (2002) A century of research on the amoeboflagellate genus *Naegleria*. *Acta Protozool.* **41**: 309–342.

Devonshire, P., Munro, F.A., Abernethy, C. and Clark, B.J. (1993) Microbial contamination of contact lens cases in the west of Scotland. *Br. J. Ophthalmol.* **77**(1): 41–45.

Douglas, M. (1930) Notes on the classification of the amoeba found by Castellani on culture of a yeast-like fungus. *J. Trop. Med. Hyg.* **33**: 258–259.

Drozanski, W. (1956) Fatal bacterial infection in soil amoebae. *Acta Microbiol. Pol.* **5**: 315–317.

Dudley, R., Matin, A., Alsam, S., Sissons, J., Mahsood, A.H. and Khan, N.A. (2005) *Acanthamoeba* isolates belonging to T1, T2, T3, T4 but not T7 encyst in response to increased osmolarity and cysts do not bind to human corneal epithelial cells. *Acta Trop.* **95**: 100–108.

Dudley, R., Alsam, S., Khan, N.A. (2007) Cellulose biosynthesis pathway is a potential target in the improved treatment of *Acanthamoeba* keratitis. *Appl. Microbiol. Biotech.* **75**(1):133–140.

Duguid, I.G., Dart, J.K., Morlet, N., Allan, B.D., Matheson, M., Ficker, L. and Tuft, S. (1997) Outcome of *Acanthamoeba* keratitis treated with polyhexamethyl biguanide and propamidine. *Ophthalmology* **104**: 1587–1592.

Epstein, R.J., Wilson, L.A., Visvesvara, G.S. and Plourde Jr, E.G. (1986) Rapid diagnosis of *Acanthamoeba* keratitis from corneal scrapings using indirect fluorescent antibody staining. *Arch. Ophthalmol.* **104**(9): 1318–1321.

Essig, A., Heinemann, M., Simnacher, U. and Marre, R. (1997) Infection of *Acanthamoeba castellanii* by *Chlamydia pneumoniae. Appl. Environ. Microbiol.* **63**(4): 1396–1399.

Ferrante, A. (1991) Immunity to *Acanthamoeba. Rev. Infect. Dis.* **13**: S403–S409.

Ferrante, A. and Bates, E.J. (1988) Elastase in the pathogenic free-living amoebae, *Naegleria* and *Acanthamoeba* spp. *Infect. Immun.* **56**: 3320–3321.

Ferrante, A. and Rowan-Kelly, B. (1983) Activation of the alternative pathway of complement by *Acanthamoeba culbertsoni. Clin. Exp. Immunol.* **54**(2): 477–485.

Ficker, L., Seal, D., Warhurst, D. and Wright, P. (1990) *Acanthamoeba* keratitis: resistance to medical therapy. *Eye* **4**: 835–838.

Fleiszig, S.M., Zaidi, T.S., Ramphal, R. and Pier, G.B. (1994) Modulation of *Pseudomonas aeruginosa* adherence to the corneal surface by mucus. *Infect. Immun.* **62**(5): 1799–1804.

Garate, M., Cao, Z., Bateman, E. and Panjwani, N. (2004) Cloning and characterization of a novel mannose-binding protein of *Acanthamoeba. J. Biol. Chem.* **279**(28): 29849–29856.

Garate, M., Marchant, J., Cubillos, I., Cao, Z., Khan, N.A. and Panjwani, N. (2006) Pathogenicity of *Acanthamoeba* is associated with the expression of the mannose-binding protein. *Invest. Ophthalmol. Vis. Sci.* **47**(3): 1056–1062.

Gray, T.B., Cursons, R.T., Sherwan, J.F. and Rose, P.R. (1995) *Acanthamoeba*, bacterial, and fungal contamination of contact lens storage cases. *Br. J. Ophthalmol.* **79**(6): 601–605.

Greub, G. and Raoult, D. (2002) *Parachlamydiaceae*: potential emerging pathogens. *Emerg. Infect. Dis.* **8**(6): 625–630.

Greub, G. and Raoult, D. (2004) Microorganisms resistant to free-living amoebae. *Clin. Microbiol. Rev.* **17**(2): 413–433.

Gudmundsson, O.G., Woodward, D.F., Fowler, S.A. and Allansmith, M.R. (1985) Identification of proteins in contact lens surface deposits by immunofluorescence microscopy. *Arch. Ophthalmol.* **103**(2): 196–197.

Hadas, E. and Mazur, T. (1993) Biochemical markers of pathogenicity and virulence of *Acanthamoeba* sp. strains. *Trop. Med. Parasitol.* **44**: 197–200.

Hanel, H., Kirsch, R., Schmidts, H.L. and Kottmann, H. (1995) New systematically active antimycotics from the beta-blocker category. *Mycoses* **38**(7–8): 251–264.

Harrington, K.A. and Hormaeche, C.E. (1986) Expression of the innate resistance gene *Ity* in mouse Kupffer cells infected with *Salmonella typhimurium in vitro*. *Microb. Pathogen.* **1**: 269–274.

He, Y.G., Neiderkorn, J.Y., McCulley, J.P., Stewart, G.L., Meyer, D.R., Silvany, R. and Doughtery, J. (1990) *In vivo* and *in vitro* collagenolytic activity of *Acanthamoeba castellanii*. *Invest. Ophthalmol. Vis. Sci.* **31**: 2235–2240.

Heath, K.V., Frank, O., Montaner, J.S., O'Shaughnessy, M.V., Schechter, M.T. and Hogg, R.S. (1998) Human immunodeficiency virus (HIV)/acquired immunodeficiency syndrome (AIDS) mortality in industrialized nations, 1987–1991. *Int. J. Epidemiol.* **27**(4): 685–690.

Hirukawa, Y., Nakato, H., Izumi, S., Tsuruhara, T. and Tomino, S. (1998) Structure and expression of a cyst specific protein of *Acanthamoeba castellanii*. *Biochim. Biophys. Acta* **1398**: 47–56.

Hong, Y.C., Hwang, M.Y., Yun, H.C., Yu, H.S., Kong, H.H., Yong, T.S. and Chung, D.I. (2002) Isolation and characterization of a cDNA encoding a mammalian cathepsin L-like cysteine proteinase from *Acanthamoeba healyi*. *Korean J. Parasitol.* **40**(1): 17–24.

Hong, Y.C., Lee, W.M., Kong, H.H., Jeong, H.J. and Chung, D.I. (2004) Molecular cloning and characterization of a cDNA encoding a laminin-binding protein (AhLBP) from *Acanthamoeba healyi*. *Exp. Parasitol.* **106**(3–4): 95–102.

Hurt, M., Apte, S., Leher, H., Howard, K., Niederkorn, J. and Alizadeh, H. (2001) Exacerbation of *Acanthamoeba* keratitis in animals treated with anti-macrophage inflammatory protein 2 or antineutrophil antibodies. *Infect. Immunol.* **69**(5): 2988–2995.

Hurt, M., Proy, V., Niederkorn, J.Y. and Alizadeh, H. (2003) The interaction of *Acanthamoeba castellanii* cysts with macrophages and neutrophils. *J. Parasitol.* **89**(3): 565–572.

Ito, S., Chang, R.S. and Pollard, T.D. (1969) Cytoplasmic distribution of DNA in a strain of hartmannellid amoeba. *J. Protozool.* **16**(4): 638–645.

Jager, B.V. and Stamm, W.P. (1972) Brain abscesses caused by free-living amoeba probably of the genus *Hartmannella* in a patient with Hodgkin's disease. *Lancet* **2**: 1343–1345.

Jahnes, W.G., Fullmer, H.M. and Li, C.P. (1957) Free-living amoebae as contaminants in monkey kidney tissue culture. *Proc. Soc. Exp. Biol. Med.* **96**: 484–488.

Jaison, P.L., Cao, Z. and Panjwani, N. (1998) Binding of *Acanthamoeba* to mannose-glycoproteins of corneal epithelium: effect of injury. *Curr. Eye Res.* **17**(8): 770–776.

Jimenez, C., Portela, R.A., Mellado, M., Rodriguez-Frade, J.M., Collard, J., Serrano, A., Martinez, A.C., Avila, J. and Carrera, A.C. (2000) Role of the PI3K regulatory subunit in the control of actin organization and cell migration. *J. Cell Biol.* **151**(2): 249–262.

John, D.T. and John, R.A. (1994) Cryopreservation of pathogenic free-living amoebae. *Folia Parasitol.* **41**: 110–114.

Jones, D.B., Visvesvara, G.S. and Robinson, N.M. (1975). *Acanthamoeba polyphaga* keratitis and *Acanthamoeba* uveitis associated with fatal meningoencephalitis. *Trans Ophthalmol. Soc. UK* **95**: 221–232.

Kahane, S., Dvoskin, B., Mathias, M. and Friedman, M.G. (2001) Infection of *Acanthamoeba polyphaga* with *Simkania negevensis* and *S. negevensis* survival within amoebal cysts. *Appl. Environ. Microbiol.* **67**(10): 4789–4795.

Kameyama, S., Sato, H. and Murata, R. (1975) The role of alpha-toxin of *Clostridium perfringens* in experimental gas gangrene in guinea pigs. *Jpn J. Med. Sci. Biol.* **25**: 200.

Khan, N.A. (2001) Pathogenicity, morphology and differentiation of *Acanthamoeba*. *Curr. Microbiol.* **43**: 391–395.

Khan, N.A. (2003) Pathogenesis of *Acanthamoeba* infections. *Microb. Pathogen.* **34**(6): 277–285.

Khan, N.A. and Paget, T.A. (2002) Molecular tools for speciation and epidemiological studies of *Acanthamoeba*. *Curr. Microbiol.* **44**(6): 444–449.

Khan, N.A., Jarroll, E.L., Panjwani, N., Cao, Z. and Paget, T.A. (2000) Proteases as markers of differentiation of pathogenic and non-pathogenic *Acanthamoeba*. *J. Clin. Microbiol.* **38**: 2858–2861.

Khan, N.A., Jarroll, E.L. and Paget, T.A. (2001) *Acanthamoeba* can be differentiated by the polymerase chain reaction and simple plating assays. *Curr. Microbiol.* **43**(3): 204–208.

Khan, N.A., Jarroll, E.L. and Paget, T.A. (2002) Molecular and physiological differentiation between pathogenic and non-pathogenic *Acanthamoeba*. *Curr. Microbiol.* **45**(3): 197–202.

Kim, K.H., Shin, C.O. and Im, K. (1993) Natural killer cell activity in mice infected with free-living amoeba with reference to their pathogenicity. *Korean J. Parasitol.* **31**(3): 239–248.

Klotz, S.I., Misra, R.P. and Butrus, S.I. (1987) Carbohydrate deposits on the surfaces of worn extended-wear soft contact lenses. *Arch. Ophthalmol.* **105**: 974–977.

Kong, H.H., Kim, T.H. and Chung, D.I. (2000) Purification and characterization of a secretory proteinase of *Acanthamoeba healyi* isolates from GAE. *J. Parasitol.* **86**(1): 12–17.

Korn, E.D. and Olivecrona, T. (1971) Composition of an amoeba plasma membrane. *Biochem. Biophys. Res. Commun.* **45**(1): 90–97.

Korn, E.D., Dearborn, D.G. and Wright, P.L. (1974) Lipophosphonoglycan of the plasma membrane of *Acanthamoeba castellanii*. Isolation from whole amoebae and identification of the water-soluble products of acid hydrolysis. *J. Biol. Chem.* **249**(11): 3335–3341.

Kotting, J., Berger, M.R., Unger, C. and Eibl, H. (1992) Alkylphosphocholines: influence of structural variations on biodistribution of antineoplastically active concentrations. *Cancer Chemother. Pharmacol.* **30**: 105–112.

Krishna-Prasad, B.N. and Gupta, S.K. (1978) Preliminary report on engulfment and retention of mycobacteria by trophozoites of axenically grown *Acanthamoeba castellanii* Douglas. *Curr. Sci.* **47**: 245–247.

La Scola, B. and Raoult, D. (2001) Survival of *Coxiella burnetii* within free-living amoeba *Acanthamoeba castellanii*. *Clin. Microbiol. Infect.* **7**(2): 75–79.

Lam, D.S.C., Houang, E., Fan, S.P., Lyon, D., Seal, D., Wong, E. and the Hong Kong microbial keratitis study group (2002) Incidence and risk factors for microbial keratitis in Hong Kong: comparison with Europe and North America. *Eye* **16**: 608–618.

Landers, P., Kerr, K.G., Rowbotham, T.J., Tipper, J.L., Keig, P.M., Ingham, E. and Denton, M. (2000) Survival and growth of *Burkholderia cepacia* within the free-living amoeba *Acanthamoeba polyphaga*. *Eur. J. Clin. Microbiol. Infect. Dis.* **19**(2): 121–123.

Larkin, D.F., Kilvington, S. and Easty, D.L. (1990) Contamination of contact lens storage cases by *Acanthamoeba* and bacteria. *Br. J. Ophthalmol.* **74**(3): 133–135.

Larkin, D.F.P., Kilvington, S. and Dart, J.K.G. (1992) Treatment of *Acanthamoeba* keratitis with polyhexamethylene biguanide. *Ophthalmology* **99**: 185–191.

Leher, H., Zaragoza, F., Taherzadeh, S., Alizadeh, H. and Niederkorn, J.Y. (1999) Monoclonal IgA antibodies protect against *Acanthamoeba* keratitis. *Exp. Eye Res.* **69**(1): 75–84.

Lehmann, O.J., Green, S.M., Morlet, N., Kilvington, S., Keys, M.F., Matheson, M.M., Dart, J.K., McGill, J.I. and Watt, P.J. (1998) Polymerase chain reaction analysis of corneal epithelial and tear samples in the diagnosis of *Acanthamoeba* keratitis. *Invest. Ophthalmol. Vis. Sci.* **39**(7): 1261–1265.

Lim, L., Coster, D.J. and Badenoch, P.R. (2000) Antimicrobial susceptibility of 19 Australian corneal isolates of *Acanthamoeba*. *Clin. Exp. Ophthalmol.* **28**: 119–124.

Lloyd, D., Turner, N.A., Khunkitti, W., Hann, A.C., Furr, J.R. and Russell, A.D. (2001) Encystation in Acanthamoeba castellanii: development of biocide resistance. *J. Eukaryot. Microbiol.* **48**(1):11–6.

Lorenzo-Morales, J., Ortega-Rivas, A., Foronda, P., Abreu-Acosta, N., Ballart, D., Martinez, E. and Valladares, B. (2005) RNA interference (RNAi) for the silencing of extracellular serine proteases genes in *Acanthamoeba*: molecular analysis and effect on pathogenecity. *Mol. Biochem. Parasitol.* **144**(1): 10–15.

Ly, T.M. and Muller, H.E. (1990) Ingested *Listeria monocytogenes* survive and multiply in protozoa. *J. Med. Microbiol.* **33**(1): 51–54.

Mackay, D.J. and Hall, A. (1998) Rho GTPases. *J. Biol. Chem.* **273**(33): 20685–20688.

Maghsood, A.H., Sissons, J., Rezaian, M., Nolder, D., Warhurst, D. and Khan, N.A. (2005) *Acanthamoeba* genotype T4 from the UK and Iran and isolation of the T2 genotype from clinical isolates. *J. Med. Microbiol.* **54**(8): 755–759.

Marciano-Cabral, F. and Cabral, G. (2003) *Acanthamoeba* spp. as agents of disease in humans. *Clin. Microbiol. Rev.* **16**(2): 273–307.

Marciano-Cabral, F. and Toney, D.M. (1998) The interaction of *Acanthamoeba* spp. with activated macrophages and with macrophage cell lines. *J. Eukaryot. Microbiol.* **45**(4): 452–458.

Marolda, C.L., Hauroder, B., John, M.A., Michel, R. and Valvano, M.A. (1999) Intracellular survival and saprophytic growth of isolates from the *Burkholderia cepacia* complex in free-living amoebae. *Microbiology* **145**(7): 1509–1517.

Martinez, A.J. (1985) *Free-living Amebas: Natural History, Prevention, Diagnosis, Pathology and Treatment of Disease.* CRC Press, Boca Raton, FL, p. 156.

Martinez, A.J. (1991) Infections of the central nervous system due to *Acanthamoeba. Rev. Infect. Dis.* **13**: S399–S402.

Martinez, A.J. and Visvesvara, G.S. (1991) Laboratory diagnosis of pathogenic free-living amoebas: *Naegleria, Acanthamoeba,* and *Leptomyxid. Clin. Lab. Med.* **11**(4): 861–872.

Martinez, A.J. and Visvesvara, G.S. (1997) Free-living, amphizoic and opportunistic amebas. *Brain Pathol.* **7**(1): 583–598.

Martinez, M.S., Gonzalez-Mediero, G., Santiago, P., Rodriguez de Lope, A., Diz, J., Conde, C. and Visvesvara, G.S. (2000) Granulomatous amebic encephalitis in a patient with AIDS: isolation of *Acanthamoeba* sp. group II from brain tissue and successful treatment with sulfadiazine and fluconazole. *J. Clin. Microbiol.* **38**: 3892–3895.

Mattana, A., Bennardini, F., Usai, S., Fiori, P.L., Franconi, F. and Cappuccinelli, P. (1997) *Acanthamoeba castellanii* metabolites increase the intracellular calcium level and cause cytotoxicity in wish cells. *Microb. Pathogen.* **23**(2): 85–93.

Mattana, A., Cappai, V., Alberti, L., Serra, C., Fiori, P.L. and Cappuccinelli, P. (2002) ADP and other metabolites released from *Acanthamoeba castellanii* lead to human monocytic cell death through apoptosis and stimulate the secretion of proinflammatory cytokines. *Infect. Immun.* **70**(8): 4424–4432.

Mazur, T., Hadas, E. and Iwanicka, I. (1995) The duration of the cyst stage and the viability and virulence of *Acanthamoeba* isolates. *Trop. Med. Parasitol.* **46**: 106–108.

McClellan, K., Howard, K., Niederkorn, J.Y. and Alizadeh, H. (2001) Effect of steroids on *Acanthamoeba* cysts and trophozoites. *Invest. Ophthalmol. Vis. Sci.* **42**(12): 2885–2893.

McClellan, K., Howard, K., Mayhew, E., Niederkorn, J. and Alizadeh, H. (2002) Adaptive immune responses to *Acanthamoeba* cysts. *Exp. Eye Res.* **75**(3): 285–293.

Michel, R., Burghardt, H. and Bergmann, H. (1995) *Acanthamoeba*, naturally intracellularly infected with *Pseudomonas aeruginosa*, after their isolation from a microbiologically contaminated drinking water system in a hospital. *Zentralbl. Hyg. Umweltmed.* **196**(6): 532–544.

Mitra, M.M., Alizadeh, H., Gerard, R.D. and Niederkorn, J.Y. (1995) Characterization of a plasminogen activator produced by *Acanthamoeba castellanii. Mol. Biochem. Parasitol.* **73**(1–2): 157–164.

Mitro, K., Bhagavathiammai, A., Zhou, O.M., Bobbett, G., McKerrow, J.H., Chokshi, R., Chokshi, B. and James, E.R. (1994) Partial characterization of the proteolytic secretions of *Acanthamoeba polyphaga. Exp. Parasitol.* **78**(4): 377–385.

Morton, L.D., McLaughlin, G.L. and Whiteley, H.E. (1991) Effect of temperature, amebic strain and carbohydrates on *Acanthamoeba* adherence to corneal epithelium *in vitro. Infect. Immun.* **59**: 3819–3822.

Moura, H., Ospina, M., Woolfit, A.R., Barr, J.R. and Visvesvara, G.S. (2003) Analysis of four human microsporidian isolates by MALDI-TOF mass spectrometry. *J. Eukaryot. Microbiol.* **50**(3):156–163.

Murdoch, D., Gray, T.B., Cursons, R. and Parr, D. (1998) *Acanthamoeba* keratitis in New Zealand, including two cases with *in vivo* resistance to polyhexamethylene biguanide. *Aust. NZ J. Ophthalmol.* **26**: 231–236.

Na, B.K., Kim, J.C. and Song, C.Y. (2001) Characterization and pathogenetic role of proteinase from *Acanthamoeba castellanii. Microb. Pathogen.* **30**(1): 39–48.

Na, B.K., Cho, J.H., Song, C.Y. and Kim, T.S. (2002) Degradation of immunoglobulins, protease inhibitors and interleukin-1 by a secretory proteinase of *Acanthamoeba castellanii. Korean J. Parasitol.* **40**(2): 93–99.

Nagington, J., Watson, P.G., Playfair, T.J., McGill, J., Jones, B.R. and Steele, A.D. (1974) Amoebic infections of the eye. *Lancet* **2**: 1537–1540.

Nassif, K.F. (1996) Ocular surface defense mechanisms. *Infections of the eye*. Little, Brown and Company, New York, NY, USA.

Neff, R.J. and Neff, R.H. (1969) The biochemistry of amoebic encystment. *Symp. Soc. Exp. Biol.* **23**: 51–81.

Neff, R.J., Ray, S.A., Benton, W.F. and Wilborn, M. (1964) Induction of synchronous encystment (differentiation) in *Acanthamoeba* sp. In: *Methods in Cell Physiology*, Vol. 1 (ed. D.M. Prescott), Academic Press, New York, NY, pp. 55–83.

Niederkorn, J.Y., Alizadeh, H., Leher, H. and McCulley, J.P. (1999) The pathogenesis of *Acanthamoeba* keratitis. *Microbes Infect.* **1**: 437–443.

Nilsson, S.E.G. and Montan, P.G. (1994) The annualized incidence of contact lens induced keratitis in Sweden and its relation to lens type and wear schedule: results of a three-month prospective study. *CLAO J.* **20**: 225–230.

Odom, A., Del Poeta, M., Perfect, J. and Heitman, J. (1997a) The immunosuppressant FK506 and its nonimmunosuppressive analog L-685,818 are toxic to *Cryptococcus neoformans* by inhibition of a common target protein. *Antimicrob. Agents Chemother.* **41**: 156–161.

Odom, A., Muir, S., Lim, E., Toffaletti, D.L., Perfect, J. and Heitman, J. (1997b) Calcineurin is required for virulence of *Cryptococcus neoformans*. *EMBO J.* **16**: 2576–2589.

Oishi, K., Raynor, R.L., Charp, P.A. and Kuo, J.F. (1988) Regulation of protein kinase C by lysophospholipids. Potential role in signal transduction. *J. Biol. Chem.* **263**(14): 6865–6871.

Page, F.C. (1967a) Taxonomic criteria for limax amoebae, with descriptions of 3 new species of *Hartmannella* and 3 of *Vahlkampfia*. *J. Protozool.* **14**: 499–521.

Page, F.C. (1967b) Re-definition of the genus *Acanthamoeba* with descriptions of three species. *J. Protozool.* **14**: 709–724.

Page, F.C. (1988) *A new key to fresh water and soil amoebae.* Freshwater Biological Association Scientific Publications, Cumbria, UK, pp. 1–122.

Peng, Z., Omaruddin, R. and Bateman, E. (2005) Stable transfection of *Acanthamoeba castellanii*. *Biochim. Biophys. Acta* **1743**(1–2): 93–100.

Perez-Santonja, J.J., Kilvington, S., Hughes, R., Tufail, A., Metheson, M. and Dart, J.K.G. (2003) Persistently culture positive *Acanthamoeba* keratitis: *in vivo* resistance and *in vitro* sensitivity. *Ophthalmology* **110**(8): 1593–1600.

Perrine, D., Chenu, J.P., Georges, P., Lancelot, J.C., Saturnino, C. and Robba, M. (1995) Amoebicidal efficiencies of various diamidines against two strains of *Acanthamoeba polyphaga*. *Antimicrob. Agents Chemother.* **39**(2): 339–342.

Pettit, D.A., Williamson, J., Cabral, G.A. and Marciano-Cabral, F. (1996) *In vitro* destruction of nerve cell cultures by *Acanthamoeba* spp.: a transmission and scanning electron microscopy study. *J. Parasitol.* **82**(5): 769–777.

Poggio, E.C., Glynn, R.J., Schein, O.D., Seddon, J.M., Shannon, M.J., Scardino, V.A. and Kenyon, K.R. (1989) The incidence of ulcerative keratitis among users of daily-wear and extended-wear soft contact lenses. *N. Engl. J. Med.* **321**: 779–783.

Pozio, E. and Morales, M.A. (2005) The impact of HIV-protease inhibitors on opportunistic parasites. *Trends Parasitol.* **21**(2): 58–63.

Preston, T.M., Richard, H. and Wotton, R.S. (2001) Locomotion and feeding of *Acanthamoeba* at the water-air interface of ponds. *FEMS Lett.* **194**: 143–147.

Proca-Ciobanu, M., Lupascu, G.H., Petrovici, A. and Ionescu, M.D. (1975) Electron microscopic study of a pathogenic *Acanthamoeba castellanii* strain: the presence of bacterial endosymbionts. *Int. J. Parasitol.* **5**(1): 49–56.

Pussard, M. and Pons, R. (1977) Morphologies de la paroi kystique et taxonomie du genre *Acanthamoeba* (Protozoa, Amoebida). *Protistologica* **13**: 557–610.

Qu, X.D. and Leher, R.L. (1998) Secretory phospholipase A2 is the principal bactericide for *Staphylococci* and other Gram positive bacteria in human tears. *Infect. Immun.* **6**(66): 2791–2797.

Radford, C.F., Minassian, D.C. and Dart, J.K. (2002) *Acanthamoeba* keratitis in England and Wales: incidence, outcome, and risk factors. *Br. J. Ophthalmol.* **86**(5): 536–542.

Rowbotham, T.J. (1980) Preliminary report on the pathogenicity of *Legionella pneumophila* for freshwater and soil amoebae. *J. Clin. Pathol.* **33**(12): 1179–1183.

Sawyer, T. (1971) *Acanthamoeba griffini*, a new species of marine amoeba. *J. Protozool.* **18**: 650–654.

Sawyer, T.K. and Griffin, J.L. (1975) A proposed new family, Acanthamoebidae (order Amoebida), for certain cyst-forming filose amoebae. *Trans Am. Microsc. Soc.* **94**: 93–98.

Schroeder, J.M., Booton, G.C., Hay, J., Niszl, I.A., Seal, D.V., Markus, M.B., Fuerst, P.A. and Byers, T.J. (2001) Use of subgenic 18S ribosomal DNA PCR and sequencing for genus and genotype identification of acanthamoebae from humans with keratitis and from sewage sludge. *J. Clin. Microbiol.* **39**(5): 1903–1911.

Schuster, F.L. and Visvesvara, G.S. (2004) Free-living amoebae as opportunistic and non-opportunistic pathogens of humans and animals. *Int. J. Parasitol.* **34**(9): 1–27.

Seal, D.V., Kirkness, C.M. and Bennett, H.G.B. (1999) Population-based cohort study of microbial keratitis in Scotland: incidence and features. *Contact Lens Ant. Eye* **22**: 49–57.

Seijo Martinez, M., Gonzalez-Mediero, G., Santiago, P., De Lope Rodriguez, A., Diz, J., Conde, C. and Visvesvara, G.S. (2000) Granulomatous amebic encephalitis in a patient with AIDS: isolation of *Acanthamoeba* sp. Group II from brain tissue and successful treatment with sulfadiazine and fluconazole. *J. Clin. Microbiol.* **38**(10): 3892–3895.

Sharma, S., Srinivasan, M. and George, C. (1990) *Acanthamoeba* keratitis in non-contact lens wearers. *Arch. Ophthalmol.* **108**(5): 676–678.

Shin, H.J., Cho, M.S., Jung, S.Y., Kim, H.I., Park, S., Seo, J.H., Yoo, J.C. and Im, K.I. (2001) Cytopathic changes in rat microglial cells induced by pathogenic *Acanthamoeba culbertsoni*: morphology and cytokine release. *Clin. Diagn. Lab. Immunol.* **8**(4): 837–840.

Singh, B.N. (1950) Nuclear division in nine species of small free-living amoebae and its bearing on the classification of the order Amoebida. *Philos. Trans R. Soc. Lond. [Biol.]* **236**: 405–461.

Singh, B.N. and Das, S.R. (1970) Studies on pathogenic and non-pathogenic small free-living amoebae and the bearing of nuclear division on the classification of the order Amoebida. *Philos. Trans R. Soc. Lond. [Biol.]* **259**: 435–476.

Sissons, J., Alsam, S., Jayasekera, S., Kim, K.S., Stins, M. and Khan, N.A. (2004a) *Acanthamoeba* induces cell-cycle arrest in the host cells. *J. Med. Microbiol.* **53**(8): 711–717.

Sissons, J., Alsam, S., Jayasekera, S. and Khan, N.A. (2004b) Ecto-ATPases of clinical and non-clinical isolates of *Acanthamoeba. Microb. Pathogen.* **37**(5): 231–239.

Sissons, J., Kim, K.S., Stins, M., Jayasekera, S., Alsam, S. and Khan, N.A. (2005) *Acanthamoeba castellanii* induces host cell death via a phosphatidylinositol 3-kinase-dependent mechanism. *Infect. Immun.* **73**(5): 2704–2708.

Sissons, J., Alsam, S., Goldsworthy, G., Lightfoot, M., Jarroll, E.L. and Khan, N.A. (2006a) Identification and properties from an *Acanthamoeba* isolate capable of producing granulomatous encephalitis. *BMC Microbiol.* **6**: 42.

Sissons, J., Alsam, S., Stins, M., Rivas, A.O., Morales, J.L., Faull, J. and Khan, N.A. (2006b) Use of *in vitro* assays to determine effects of human serum on biological characteristics of *Acanthamoeba castellanii. J. Clin. Microbiol.* **44**(7):2595–2600.

Stehr-Green, J.K., Baily, T.M. and Visvesvara, G.S. (1989) The epidemiology of *Acanthamoeba* keratitis in the United States. *Am. J. Ophthalmol.* **107**: 331–336.

Steinert, M., Birkness, K.K., White, E., Fields, B. and Quinn, F. (1998) *Mycobacterium avium* bacilli grow saprozoically in coculture with *Acanthamoeba polyphaga* and survive within cyst walls. *Appl. Environ. Microbiol.* **64**(6): 2256–2261.

Stewart, G.L., Kim, I., Shupe, K., Alizadeh, H., Silvany, R., McCulley, J.P. and Niederkorn, J.Y. (1992) Chemotactic response of macrophages to *Acanthamoeba castellanii* antigen and antibody-dependent macrophage-mediated killing of the parasite. *J. Parasitol.* **78**(5): 849–855.

Stewart, G.L., Shupe, K.K., Kim, I., Silvany, R.E., Alizadeh, H., McCulley, J.P. and Niederkorn, J.Y. (1994) Antibody-dependent neutrophil-mediated killing of *Acanthamoeba castellanii. Int. J. Parasitol.* **24**(5): 739–742.

Tanaka, Y., Suguri, S., Harada, M., Hayabara, T., Suzumori, K. and Ohta, N. (1994) *Acanthamoeba*-specific human T-cell clones isolated from healthy individuals. *Parasitol. Res.* **80**(7): 549–553.

Thakur, A. and Willcox, D.P. (2000) Contact lens wear alters the production of certain inflammatory mediators in tears. *Exp. Eye Res.* **70**: 255–259.

Thom, S., Warhurst, D. and Drasar, B.S. (1992) Association of *Vibrio cholerae* with fresh water amoebae. *J. Med. Microbiol.* **36**(5): 303–306.

Thyrell, L., Hjortsberg, L., Arulampalam, V., Panaretakis, T., Uhles, S., Zhivotovsky, B., Leibiger, I., Grander, D. and Pokrovskaja, K. (2004) Interferon-alpha induced apoptosis in tumor cells is mediated through PI3K/mTOR signaling pathway. *J. Biol. Chem.* **279**(23): 24152–24162.

Tomlinson, A., Simmons, P.A., Seal, D.V. and McFadyen, A.K. (2000) Salicylate inhibition of *Acanthamoeba* attachment to contact lenses: a model to reduce risk of infection. *Ophthalmology* **107**(1): 112–117.

Toney, D.M. and Marciano-Cabral, F. (1998) Resistance of *Acanthamoeba* species to complement lysis. *J. Parasitol.* **84**(2): 338–344.

Tripathi, R.C. and Tripathi, B.J. (1984) Lens spoilage. In: Contact Lenses: The Contact Lens Assoc. Ophth. Guide to Basic Science and Clinical Practice (ed. O. Diabezies). Grune and Stratton, Orlando, FL, pp. 45.1–45.33.

Turner, N.A., Russell, A.D., Furr, J.R. and Lloyd, D. (2000) Emergence of resistance to biocides during differentiation of *Acanthamoeba castellanii*. *J. Antimicrob. Chem.* **46**: 27–34.

van Klink, F., Alizadeh, H., Stewart, G.L., Pidherney, M.S., Silvany, R.E., He, Y., McCulley, J.P. and Niederkorn, J.Y. (1992) Characterization and pathogenic potential of a soil isolate and an ocular isolate of *Acanthamoeba castellanii* in relation to *Acanthamoeba* keratitis. *Curr. Eye Res.* **11**(12): 1207–2012.

van Klink, F., Alizadeh, H., He, Y., Mellon, J.A., Silvany, R.E., McCulley, J.P. and Niederkorn, J.Y. (1993) The role of contact lenses, trauma, and Langerhans cells in a Chinese hamster model of *Acanthamoeba* keratitis. *Invest. Ophthalmol. Vis. Sci.* **34**(6): 1937–1944.

van Klink, F., Taylor, W.M., Alizadeh, H., Jager, M.J., van Rooijen, N. and Niederkorn, J.Y. (1996) The role of macrophages in *Acanthamoeba* keratitis. *Invest. Ophthalmol. Vis. Sci.* **37**(7): 1271–1281.

van Klink, F., Leher, H., Jager, M.J., Alizadeh, H., Taylor, W. and Niederkorn, J.Y. (1997) Systemic immune response to *Acanthamoeba* keratitis in the Chinese hamster. *Ocul. Immunol. Inflamm.* **5**(4): 235–244.

Victoria, E.J. and Korn, E.D. (1975a) Plasma membrane and soluble lysophospholipases of *Acanthamoeba castellanii*. *Arch. Biochem. Biophys.* **171**(1): 255–258.

Victoria, E.J. and Korn, E.D. (1975b) Enzymes of phospholipid metabolism in the plasma membrane of *Acanthamoeba castellanii*. *J. Lipid Res.* **16**(1): 54–60.

Visvesvara, G.S. (1991) Classification of *Acanthamoeba*. *Rev. Infect. Dis.* **13**(Suppl 5): 369–372.

Visvesvara, G.S., Martinez, A.J., Schuster, F.L., Leitch, G.J., Wallace, S.V., Sawyer, T.K. and Anderson, M. (1990) Leptomyxid ameba, a new agent of amebic meningoencephalitis in humans and animals. *J. Clin. Microbiol.* **28**(12): 2750–2756.

Visvesvara, G.S., Schuster, F.L. and Martinez, A.J. (1993) *Balamuthia mandrillaris*, N. G., N. sp., agent of amebic meningoencephalitis in humans and other animals. *J. Eukaryot. Microbiol.* **40**(4): 504–514.

Visvesvara, G.S., Moura, H. and Schuster, F.L. (2007) Pathogenic and opportunistic free-living amoebae: *Acanthamoeba* spp., *Balamuthia mandrillaris*, *Naegleria fowleri*, and *Sappinia diploidea*. *FEMS Immunol. Med. Microbiol.* **50**(1):1–26.

Volkonsky, M. (1931) *Hartmanella castellanii* Douglas, et classification des hartmannelles. *Arch. Zool. Exp. Gen.* **72**: 317–339.

Walochnik, J., Obwaller, A. and Aspock, H. (2000) Correlations between morphological, molecular biological, and physiological characteristics in clinical and nonclinical isolates of *Acanthamoeba* spp. *Appl. Environ. Microbiol.* **66**(10): 4408–4413.

Walochnik, J., Obwaller, A., Haller-Schober, E.M. and Aspock, H. (2001) Anti-*Acanthamoeba* IgG, IgM, and IgA immunoreactivities in correlation to strain pathogenicity. *Parasitol. Res.* **87**(8): 651–658.

Walochnik, J., Duchene, M., Seifert, K., Obwaller, A., Hottkowitz, T., Wiedermann, G., Eibl, H. and Aspock, H. (2002) Cytotoxic activities of alkylphosphocholines against clinical isolates of *Acanthamoeba* spp. *Antimicrob. Agents Chemother.* **46**(3): 695–701.

Watanabe, R., Ishibashi, Y., Hommura, S. and Ishii, K. (1994) *Acanthamoeba* isolation from contact lens solution of contact lens wearers without keratitis. *J. Jpn Ophthalmol. Soc.* **98**(5): 477–480.

Weisman, R.A. (1976) Differentiation in *Acanthamoeba castellanii. Annu. Rev. Microbiol.* **30**: 189–219.

Weisman, B. and Mondino, B. (2002) Risk factors for contact lens associated microbial keratitis. *Contact Lens Ant. Eye* **25**: 3–9.

Wennström, S., Hawkins, P., Cooke, F., Hara, K., Yonezawa, K., Kasuga, M., Jackson, T., Claesson-Welsh, L. and Stephens, L. (1994) Activation of phospho-inositide 3-kinase is required for PDGF-stimulated membrane ruffling. *Curr. Biol.* **4**: 385–396.

Winchester, K., Mathers, W.D., Sutphin, J.E. and Daley, T.E. (1995) Diagnosis of *Acanthamoeba* keratitis *in vivo* with confocal microscopy. *Cornea* **14**(1): 10–17.

Winiecka-Krusnell, J., Wreiber, K., von Euler, A., Engstrand, L. and Linder, E. (2002) Free-living amoebae promote growth and survival of *Helicobacter pylori. Scand. J. Infect. Dis.* **34**(4): 253–256.

Yang, Z., Cao, Z. and Panjwani, N. (1997) Pathogenesis of *Acanthamoeba* keratitis: carbohydrate mediated host-parasite interactions. *Infect. Immun.* **65**: 439–445.

Yin, J. and Henney Jr, H.R. (1997) Stable transfection of *Acanthamoeba. Can. J. Microbiol.* **43**(3): 239–244.

A2 *Balamuthia mandrillaris*

Frederick L. Schuster
and Govinda S. Visvesvara

1 Introduction

Granulomatous amoebic encephalitis is caused by two different genera of amoebae: *Balamuthia mandrillaris* and *Acanthamoeba* spp. Although they are morphologically different during *in vitro* culture, they can be readily confused with one another in clinical specimens such as formalin-fixed brain and skin tissue sections. *Balamuthia* is an opportunist pathogen causing infections in immunocompromised hosts as well as seemingly immunocompetent hosts. Cases have been reported worldwide: Mexico and a number of South American countries (Argentina, Brazil, Chile, Peru, Venezuela), United States, Australia, Japan, India, Thailand, England, Portugal, Canada and the Czech Republic.

Prior to the identification of *Balamuthia* as a 'new' pathogen, amoebae and cysts in brain and other tissue sections were often identified as *Acanthamoeba*, even though immunofluorescent staining of sections did not support the conclusion. Once anti-*Balamuthia* antibodies were produced in rabbits, it became possible to distinguish between the two amoebae using indirect immunofluorescent staining (see below). With this diagnostic tool *Balamuthia* infections could be identified in hospitalized patients, as well as retrospectively in archived slides and paraffin-embedded tissues.

Similarities between *Balamuthia* and *Acanthamoeba* are numerous: (i) they have similar life cycles and look alike in formalin-fixed clinical specimens, (ii) both amoebae cause granulomatous amoebic encephalitis (GAE), (iii) GAE caused by the two amoebae runs a similar course, and (iv) both are free-living amoebae and are found in soil and water. There are also differences: (i) *Balamuthia* trophozoites and cysts are larger than those of *Acanthamoeba*, (ii) *Acanthamoeba* but not *Balamuthia*, in addition to causing encephalitis, can cause amoebic keratitis mostly in contact-lens users, (iii) *Acanthamoeba* encephalitis occurs almost exclusively in immunocompromised persons, and, (iv) *Acanthamoeba* is the hardier organism with virtually universal distribution in soil and water, and tolerance of a broad range of environmental stresses.

Several reviews have been published on *Balamuthia* and balamuthiasis (Glaser *et al.*, 2006b; Schuster and Visvesvara, 2004a; Visvesvara *et al.*, 2007). In this review, we focus on more recent findings about the organism and the disease. Because of the similarities between balamuthiasis and acanthamoebiasis, we include comparisons between the two organisms and diseases where such information is informative. A comprehensive review of the *Acanthamoeba* literature has recently been published (Khan, 2006). The medical aspects of earlier pediatric cases of balamuthiasis have been reviewed by Rowen *et al.* (1995) and the paper is recommended to readers in search of background data pertaining to infections in children.

2 Taxonomy

When first isolated into culture, *Balamuthia* amoebae were thought to resemble members of a group of soil amoebae, the Leptomyxidae (Visvesvara *et al.*,

1990). With time, however, the amoeba was recognized as a new genus and species, and was named *Balamuthia mandrillaris*: the genus was named for William Balamuth, a protozoologist who promoted the study of free-living amoebae as research organisms, and the species designation denotes the source of the first isolate, the brain of a mandrill baboon (Visvesvara *et al.*, 1993). With the development of sophisticated techniques for comparing protistan nuclear (18S) and mitochondrial (16S) DNAs, it became clear that the organism was closely related to *Acanthamoeba* spp. (Booton *et al.*, 2003a; Stothard *et al.*, 1998), and was not a leptomyxid (Amaral Zettler *et al.*, 2000). In accordance with recent taxonomic treatment of the protists, *Balamuthia* and *Acanthamoeba* are members of the Acanthamoebidae, a subgroup of the Amoebozoa (Adl *et al.*, 2005). Genomic studies indicate that all isolates, from clinical specimens and environmental samples, belong to a single species, *B. mandrillaris* (Booton *et al.*, 2003a). Much of the literature from the early 1990s refers to *Balamuthia* as *Leptomyxa* or as a leptomyxid amoeba.

3 Life Cycle

The amoeba is found in soil with a life cycle that consists of a trophic amoeba and a dormant thick-walled cyst. In soil it is most likely present in the cyst stage, enabling it to survive lack of food, periods of desiccation, and other environmental hardships. The amoeba measures from 12 to 60 μm in length (Figure 1), and the rounded cyst 6 to 30 μm in diameter. The trophic amoeba is pleomorphic and shows variation in shape in cultures. Locomotion in cultures is typically amoeboid by means of lobose pseudopodia, although the amoeba exhibits a spider- or crab-like locomotion on extended leg-like pseudopodia when feeding on tissue culture cells in fluid cultures (Visvesvara *et al.*, 1993). In young cultures or on agar, the amoeba assumes a dendriform shape (Figure 2).

The most distinctive morphological feature of both the trophic and the cystic amoebae is the vesicular nucleus (Figure 1), measuring around 5 μm in diameter and containing typically a large centrally placed nucleolus, the entire unit resembling a 'fried egg'. Multiple nucleoli are sometimes seen, a feature that is helpful in distinguishing balamuthiae from acanthamoebae in formalin-fixed tissue biopsy/autopsy sections (Figures 3 and 4). The nuclear profile also helps to distinguish the amoebae from leukocytes and other host cells with diffuse nucleolar material in stained tissue samples. At mitosis the nuclear envelope remains intact with an intranuclear spindle until about metaphase, at which time the envelope breaks down and reforms at interphase (Figure 5). Centrioles are not seen at the poles, nor are they seen during interphase (Visvesvara *et al.*, 1993). A microtubule-organizing center (MTOC) is present (Figure 6), as in *Acanthamoeba* (Bowers and Korn, 1968).

The cyst, when it forms, has a three-layered wall: the outer wall or ectocyst, the inner wall or endocyst and, between these two, a fibrous mesocyst. The cyst wall structure is best seen in sectioned material with the transmission electron microscope and may be indistinct when viewed by light microscopy (Figures 7 and 8). Pores are not seen in the wall, and at

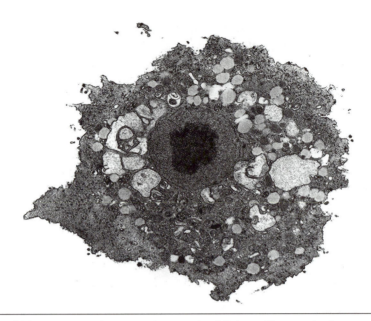

Figure 1 Transmission electron micrograph of *Balamuthia* amoeba from axenic culture (BM-3) medium. The vesicular nucleus, the large structure in the center of the amoeba, contains a central nucleolus. Numerous pinocytotic vacuoles and/or vesicles are seen in the cytoplasm, as well as mitochondria. A thin rim of ectoplasm is evident at the periphery of the amoeba, inside the cell membrane.

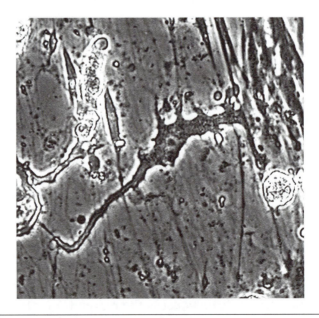

Figure 2 Phase-contrast micrograph of *Balamuthia* amoebae in tissue culture flask feeding on tissue culture cells, photographed *in situ*. The amoeba exhibits typical dendriform morphology with pseudopodia extending in multiple directions. The vesicular nucleus can be clearly seen.

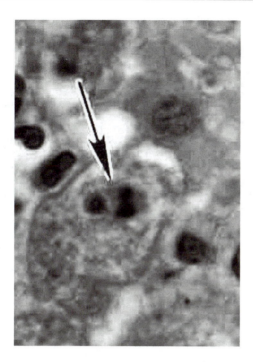

Figure 3 Hematoxylin-eosin stain of trophic amoeba with a nucleus that contains two nucleoli (*arrow*) This feature is useful in initial identification of *Balamuthia*, and is of help in distinguishing *Balamuthia* from *Acanthamoeba*.

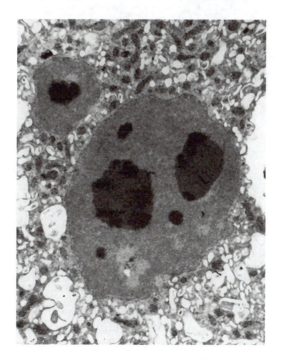

Figure 4 Transmission electron micrograph of nucleus with multiple nucleoli. (cf. Figure 2).

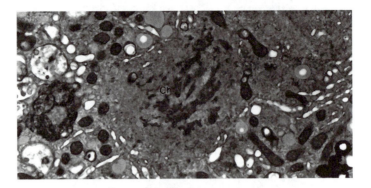

Figure 5 A portion of a late mitotic stage in a *Balamuthia* trophozoite. The nuclear envelope has broken down and chromosomes are seen in the nucleoplasm. Traces of the old nuclear envelope surround the mitotic figure as small elongate vesicles. Ch, chromosomes (Visvesvara *et al.*, 1993; reprinted from the J. Eukaryot. Microbiol.)

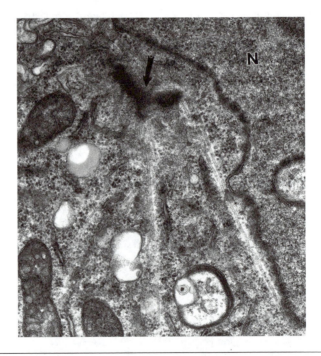

Figure 6 The microtubule-organizing center (MTOC) seen alongside of the nucleus. Two dense bodies (*arrow*) are seen in the cytoplasm with microtubules radiating outward. A portion of the nucleus (N) is seen in the upper right corner of the micrograph. Centrioles, however, are not seen in the amoebae. (Visvesvara *et al.*, 1993; reprinted from the J. Eukaryot. Microbiol.)

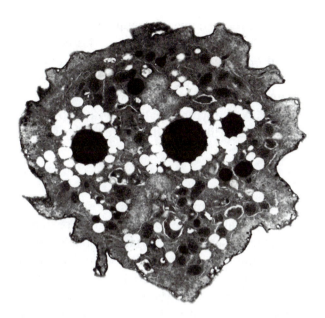

Figure 7 Transmission electron micrograph of an early stage in encystation of a *Balamuthia* amoeba from an axenic culture (BM-3). The density of the cytoplasm has increased and numerous electron-transparent vesicles are seen, some of which surround electron-dense bodies of unknown composition. The cyst wall appears in an early stage of formation as indicated by the deposit of electron-dense material on the outer surface of the cell membrane.

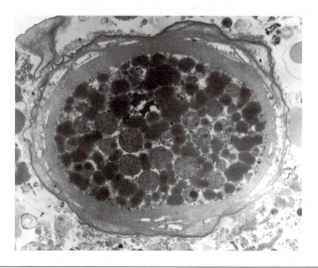

Figure 8 Transmission electron micrograph of a *Balamuthia* mature cyst from mouse brain. The three-layered wall is made up of a thin, wavy ectocyst, a fibrous mesocyst, and a thick endocyst. The cytoplasm is filled with vesicles and rounded mitochondria.

excystation the wall breaks down allowing the amoeba to revert to its trophic existence. The ability to encyst in tissue is shared by both *Acanthamoeba* and *Balamuthia* and, before specific fluorescent-antibody staining was developed to identify *Balamuthia*, several cases of balamuthiasis were attributed to *Acanthamoeba* (Visvesvara *et al.*, 1993). This was due partly to similar disease symptomatology, but also because of the presence of the *Acanthamoeba*-like cysts in affected tissues.

4 Cultivation

4.1 Growth on tissue culture cells

Balamuthiae, whether from clinical specimens or soil, cannot be cultured on bacteria *in vitro*, the usual food source for isolation of free-living amoebae (Schuster, 2002). In the presence of bacteria, amoebae *in vitro* may be able to utilize nutrients that diffuse outward from bacteria, and balamuthiae have been shown to ingest fluorescently labeled, heat-killed bacteria, but the bacteria do not support active growth (Matin *et al.*, 2006a). In place of bacteria, the amoebae have been cultured on monolayers of tissue culture cells. Different laboratories have used different lines of tissue culture cells: African green monkey kidney, rat glioma, mouse macrophage, murine mastocytoma, human lung fibroblasts and human brain microvascular endothelial cells. The amoebae *in vitro* will ingest other free-living amoebae such as *Acanthamoeba* and *Naegleria*, as it would tissue culture cells.

The technique for isolation from brain tissue or cerebrospinal fluid (CSF) is to add macerated brain tissue or CSF to the tissue culture (Jayasekara *et al.*, 2004; Visvesvara *et al.*, 1990). After a period ranging from days to several weeks, amoebae will start feeding on the monolayer, initially producing plaques like those produced by a virus and soon consuming virtually all the tissue culture cells (Figure 9a, b). All isolates in culture grow at 37°C.

4.2 Axenic growth

Isolates of *Balamuthia*, whether from humans or animals, can be grown in an enriched cell-free axenic medium (BM-3), supplemented with fetal calf serum, vitamins, nonessential amino acids, liver extract, yeast RNA, and hemin (see Table 1; Schuster and Visvesvara, 1996). At 37°C, such cultures yield around 1.5 to 3×10^5 amoebae ml^{-1} depending on the isolate. In the axenic medium, the amoeba has a relatively long generation time of 20 to 30 h and still longer, up to 50 h, for several isolates. At least one isolate, obtained from soil, will not grow in BM-3 but requires maintenance on tissue culture cells (Dunnebacke *et al.*, 2004). Michel and Janitschke (1996) have used a variant of Chang's growth medium for *Naegleria* (Chang, 1971) to culture *Balamuthia*, with generation times of 32 to 36 h, as have Kiderlen *et al.* (2006a).

Although optimal growth is at 37°C, balamuthiae will also grow at lower temperatures but at a slower rate. Slower growth is sometimes helpful in

(A)

(B)

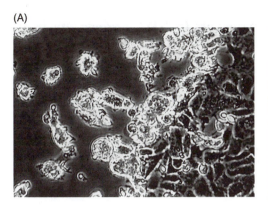
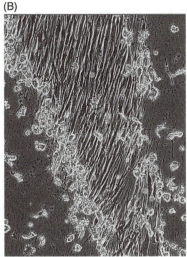

Figure 9 *Balamuthia* strain, isolated from a horse, feeding on tissue culture (TC) cells. (A) The dense TC monolayer is seen on the right of the micrograph with numerous trophic amoebae making up the edge of a plaque in the TC layer. (B) A raft of TC cells are being fed upon by *Balamuthia* amoebae.

increasing the time intervals needed for transferring cultures. For growth in BM-3 medium, recommended transfer interval is every 3 weeks at around 20°C, every 2 to 3 weeks at room temperature and 30°C, and around 2 weeks at 37°C. Weekly transfer is recommended for amoebae growing on tissue culture cells at 37°C. Weekly transfer is also recommended for maintaining actively growing/log phase cultures for research purposes. It is possible to increase yield and maintain continuously active growth of amoebae by aspirating 'old' medium from a flask of a growing culture and replacing it with fresh medium. Remarkably, 2-month-old cultures will contain trophic amoebae, which will revive after several days with addition of fresh medium. The ability of the amoebae to tolerate temperatures above 37°C has not been critically tested. However, preliminary studies indicate that the index strain (from the mandrill baboon; CDC:V039), upon inoculation into human lung fibroblast cell culture, can grow well at 41.5°C and destroy the tissue culture monolayer in 1 week.

Prolonged cultivation in axenic medium led to a reduction of amoeba cytopathogenicity when transferred onto tissue culture monolayers. Amoebae maintained on tissue culture cells, however, retained their cytopathogenicity (Kiderlen *et al.*, 2006b). Loss of virulence with extended axenic cultivation has also been described for pathogenic *Naegleria fowleri* and *Acanthamoeba* strains.

Most of the strains of *Balamuthia* now in culture came from human or animal brain tissue and one from human CSF that was obtained at autopsy.

Table I Formulation of a cell-free medium (BM-3) for axenic growth of Balamuthia mandrillaris (Schuster and Visvesvara, 1996)

Basal medium	
Proteose peptone[a]	2.0 gm
Yeast extract	2.0 gm
Torula yeast RNA (Sigma)	0.5 gm
Double-distilled water	345 ml
Autoclaved separately or filter sterilized	
Supplements: added to the sterile basal medium	
Hanks' balanced salt solution (10×)	34.0 ml
Liver digest (5%) in Hanks' salts (Sigma)[a]	100.0 ml
MEM vitamin mixture (100×)	5.0 ml
Lipid mixture (1000×) (Gibco)	0.5 ml
MEM nonessential amino acids (100×)	5.0 ml
Glucose (10%)	5.0 ml
Hemin (2 mg/ml)	0.5 ml
Taurine (0.5%)	5.0 ml
Fetal calf serum	to 10%
Penicillin-streptomycin (20 000 U/ml)	5.0 ml (optional)
pH adjusted to 7.2 with 1 N NaOH	

---

[a]The original formulation used Biosate peptone and Panmede ox liver extract, which are no longer available.

Two strains have been isolated from soil samples, one from a flowerpot in the home of a child who died of balamuthiasis and the other from an outdoor potted tree (Dunnebacke *et al.*, 2004; Schuster *et al.*, 2003). The original strain of *Balamuthia* isolated from the brain of the mandrill baboon is available from the American Type Culture Collection as ATCC 50209. It is classified as a Biosafety Level 2 organism (presenting moderate risk, and requiring BSL 2 laboratory guidelines).

5 Ecology

Prior to its isolation from soil, it was unclear as to whether *Balamuthia* was a free-living amoeba or a parasite. Its recognition as a soil amoeba placed it among the other free-living pathogenic amoebae.

5.1 Isolation from soil

Although *Balamuthia* strains have only been isolated from clinical and soil samples, evidence points to their being present in water (Finnin *et al.*, 2007;

Foreman *et al.*, 2004). During the isolation of *Balamuthia* from soil on non-nutrient agar plates with bacteria, successive growth waves of ciliates, soil amoebae (*Acanthamoeba, Naegleria, Hartmannella*, etc.), fungi and nematodes preceded the appearance of *Balamuthia*. The sealed plates, kept for several months at 17°C, yielded amoebae in sparse numbers in or upon the agar surface (Dunnebacke *et al.*, 2004; Schuster *et al.*, 2003). The delay in growth was likely due to the prolonged generation times of *Balamuthia*, the inability of the amoebae to compete successfully against other, more robust and numerous soil organisms, and lack of a still-to-be-determined optimal food source. Since *Balamuthia* amoebae can be maintained *in vitro* on smaller free-living amoebae, the assumption is that they are a likely food source in nature. Thus, in its soil ecosystem, *Balamuthia* is a predator on smaller bacterivorous soil amoebae.

5.2 Are all strains of Balamuthia pathogenic?
The presence of *Balamuthia* in clinical specimens is necessary and sufficient proof of the pathogenicity of the strain. Additionally, these isolates have been tested for pathogenicity in mice (Visvesvara *et al.*, 1990). The environmental strains have yet to be tested for pathogenicity in an animal model of disease (see below). *Acanthamoeba* strains isolated from soil may be intrinsically pathogenic or nonpathogenic as determined by mouse infection studies, and the same may be true for *Balamuthia*. With other free-living amoebae, ability to survive at 37°C or higher is an important property for infecting mammals, though not all amoebae that survive at 37°C are pathogenic (Schuster and Visvesvara, 2004a). The two *Balamuthia* environmental isolates in culture grow at 37°C.

5.3 Amoebae as bacterial hosts
Considerable information, based on *in vivo* and *in vitro* studies, has been gathered about the presence of endosymbiotic bacteria in acanthamoebae, the implication being that amoebae can serve as repositories and vectors of pathogenic bacteria (Schuster and Visvesvara, 2004a). Although none of the strains of *Balamuthia* now in culture has been found to harbor endosymbionts, at least one strain has been shown to be capable of serving as host for *Legionella pneumophila*, the causal agent of legionellosis (Shadrach *et al.*, 2005).

6 Physiology
The organism does best in iso-osmotic culture media. In aging cell-free and tissue cultures, though synchronous encystment does not occur, cyst numbers increase slowly over time. In old cultures, the average size of the amoebae decreases but, upon refeeding with fresh medium, amoeba size returns to 'normal' parameters. No data are available on the longevity of *Balamuthia* cysts *in vitro*. Encystment cannot be induced, as for *Acanthamoeba*, by suspension of trophic amoebae in a solution of 50 mM $MgCl_2$.

7 Molecular biology of *Balamuthia*

The amoeba has been the focus of several genomic studies. Booton *et al.* (2003a, 2003b, 2003c) have developed PCR protocols for genomic, phylogenetic, and diagnostic testing of *Balamuthia*. They sequenced the nuclear 18S (nDNA) and mitochondrial 16S (mDNA) DNAs of *Balamuthia* isolates and found that the nDNA of four isolates was identical and the mDNA of seven isolates varied only to a maximum of 1.8%. They concluded, therefore, that all isolates of *Balamuthia* were of a single species with a worldwide distribution. Based on these results, they developed a primer set that produced a 1075-bp product that successfully distinguished between *Balamuthia* and *Acanthamoeba* in clinical specimens (Booton *et al.*, 2003b, 2003c):

> 5′Balspec16S (5′-CGCATGTATGAAGAAGACCA-3′)
> 3′Balspec16S (5′-TTACCTATATAATTGTCGATACCA-3′)

given that balamuthiasis was only recently described, there is interest in retrospective studies of brain tissue sections from persons who have died recently or even years ago from encephalitis. A problem that exists with formalin-fixed archival material is the cross-linking and fragmentation of DNA that occurs as a consequence of fixation. These small fragmented sequences do not readily react with the above primer set. To overcome this difficulty, Foreman *et al.* (2004) developed a primer set producing an amplicon of only 230 bp:

> 5′Balspec16S (5′-CGCATGTATGAAGAAGACCA-3′)
> Bal16Sr610 (5′-CCCCTTTTTAACTCTAGTCATATAGT-3′),

that was successful in identifying *Balamuthia* mDNA in 50% of sectioned archival slide material from less than 1 to more than 30 years (Yagi *et al.*, 2006, unpublished observation). Optimal corroboration of immunofluorescent staining results for *Balamuthia* in archival specimens was less than or equal to 5 years. Although the study examined tissues up to 30 years of age, identifications of these older tissues were less likely to agree with the immunofluorescent diagnoses, although a 1982 balamuthiasis archival case (~25 years old) was correctly identified by PCR. Thus, even with the shorter primer set, there was difficulty in reliably detecting *Balamuthia* mDNA in sectioned tissues older than about 5 years. Amplification of mDNA in older specimens (>5 years) could result in misidentification (as *Acanthamoeba*) and/or false positives and false negatives.

8 Mechanisms of Pathogenesis

8.1 Cytopathology

Balamuthia amoebae feeding on tissue culture cells rapidly destroy the monolayer (Figure 9a, b). The basic mechanism involved appears to be phagocytosis by the amoebae. Feeding cups (amoebastomes), present in *Naegleria* and *Acanthamoeba*, have also been reported in *Balamuthia* trophozoites (Rocha-Azevado *et al.*, 2007). Other factors are also involved since, in mouse macrophage cultures being fed upon by amoebae, the target cells swell and apparently rupture (Kiderlen *et al.*, 2006b). Using murine mastocytoma cells (transfected with a bacterial β-galactosidase reporter gene) as targets for amoebae, Kiderlen *et al.* (2006b) found that destruction of feeder cells (as detected by the release of β-galactosidase) was dependent upon contact between amoebae and the target cells. Separation of amoebae from target cells by a semipermeable membrane reduced or prevented destruction, as did lysates of the amoebae and inhibitors of actin polymerization (cytochalasin B, latrunculin and an algal toxin). Transmission electron microscopy of amoeba-target cell contacts indicated that amoeba pseudopodia penetrate target cells.

8.2 Enzymatic potential

Amoebae produce a number of enzymes which function both intra- and extracellularly. Protease activity, specifically metalloprotease, has been detected in *Balamuthia* amoebae (Matin *et al.*, 2006b). This group of enzymes is postulated to have the ability to weaken or destroy the extracellular matrix that provides the cement to hold brain cells together. By attacking components of the extracellular matrix such as collagen, elastin and plasminogen, the metalloproteases can facilitate amoeba penetration into the brain parenchyma. Balamuthiae may also bind to the extracellular matrix as shown *in vitro* on surfaces coated with laminin, collagen and fibronectin (Rocha-Azevado *et al.*, 2007). Binding was inhibited by a coat of sialic acid. The sugars galactose and mannose reduced or increased ability to bind, respectively, and divalent cations were needed for binding to laminin.

Lipolytic enzymes have also been detected: lipase, phospholipase A and lysophospholipase A (Shadrach *et al.*, 2004). The authors have conjectured that balamuthiae cannot utilize bacteria as a food source because they require phosphatidylcholine, present in mammalian cells but not in bacteria that, instead, contain phosphatidylgylcerol. In the infected host, the combined action of protease and phospholipase activity may be a significant virulence factor, aiding in necrosis of brain infrastructure.

9 Infection and Disease

There have been between 100 and 150 published and unpublished cases of balamuthiasis. Most of these cases were diagnosed at the Centers for Disease Control and Prevention and a few at the California Department of Health Services. It is assumed that many more cases have gone undiagnosed, since autopsies are less likely to be performed because of the cost or reluctance of family members to consent to the procedure based on personal or religious beliefs. There have been several cases in California that, were it not for a proactive program to identify causal agents of encephalitis, would have been classified as 'encephalitis of unknown origin' (Glaser *et al.*, 2006a). In Africa, the Indian subcontinent and Southeast Asia where the HIV/AIDS epidemic is largely uncontrolled, it is probable that *Balamuthia* is one of the many opportunistic pathogens that infect HIV/AIDS patients but, for lack of diagnostic facilities, are not identified and certainly do not appear in the literature.

Among the diseases caused by free-living amoebae, a distinction is made between meningoencephalitis (caused by *Naegleria fowleri*) and granulomatous encephalitis (caused by *Balamuthia* and *Acanthamoeba*). Meningoencephalitis involves the meninges surrounding the CNS, while encephalitis affects the brain itself, giving rise to classic symptoms such as seizures, cranial nerve palsies, and ataxia. In much of the literature on *Balamuthia* encephalitis, the disease is referred to either as encephalitis or meningoencephalitis, depending on the degree of meningeal involvement.

9.1 Animal model of disease

A number of studies have focused on testing pathogenicity and modes of infection by amoebic agents of encephalitis. The mouse has been the preferred animal for testing pathogenicity of strains of *Naegleria*, *Acanthamoeba* and, more recently, *Balamuthia*. Diseased mice show hunched posture, ruffled fur, partial paralysis, aimless wandering, coma and, ultimately, death. The baboon isolate was tested on 6- to 8-week-old immunocompetent mice with inocula of 1000 trophic and cystic amoebae harvested from tissue cultures (Visvesvara *et al.*, 1990). Animals received both intranasal and intraperitoneal inoculations. Within a week's time, some of the mice died and others that showed symptoms of the disease as described above were sacrificed. The CNS infections were confirmed histopathologically in all animals.

Another study used severe combined immunodeficient mice (SCID) to replicate what then appeared to be the major at-risk group for balamuthiasis, the immunocompromised human (Janitschke *et al.*, 1996). SCID mice are unable to mobilize a humoral or cell-mediated immune response. Intranasal inoculation of mice with suspensions of 1000 trophic and cystic amoebae caused death of 70% of the animals within a period of 2–4 weeks. Histopathologic study of brains of the dead mice found amoebae in the parenchyma without evidence of inflammatory response.

Intranasal inoculation was used in another study in which amoebae were tracked histopathologically from the point of instillation to invasion of the brain over a period of 3–22 days (Kiderlen and Laube, 2004). The amoebae were traced from the nasal epithelium to the submucosal surface, through the cribriform plate of the ethmoid bone along olfactory nerve filaments, and then into the brain parenchyma.

It was also possible to demonstrate infection *via* the gastrointestinal tract using a tube passed into the stomach (gavage) to introduce a dose of

10 000 trophic and cystic amoebae (Kiderlen *et al.*, 2006a). The amoebae killed 20% and 40% of the immunocompetent and immunodeficient mice, respectively. Evidence of *Balamuthia* antigen and/or amoebae was detected by immunofluorescent staining in the CNS, as well as stomach lining, intestinal tract, lungs, liver and gall bladder of the dead mice. Amoebae were not recovered from fecal pellets though antigen was detected. Animal-to-animal transmission did not occur when infected and noninfected mice were housed in the same cage; however, *Balamuthia* antigen appeared in the intestinal tract of the latter group of animals. It is therefore possible that *Balamuthia* may gain entry into the host via the oral route. Experimentally, mice have also been infected via intravascular and intracoelomic injections of amoebae (Kiderlen *et al.*, 2006a).

9.2 Portals of entry

Contact with soil appears as a common denominator in a number of balamuthiasis cases, but because of an extended incubation period of the disease, the source and mode of infection are often indeterminable. Evidence points to the following as portals of entry: (i) cutaneous lesions, (ii) the nasal passages and/or the sinuses, and (iii) inhalation of soil particles containing cysts carried on air currents into the respiratory tract. These pathways are surmised from the nature and site of the initial infection and the organs or tissues involved.

Once gaining entrance to a skin lesion, lung tissue, or nasal epithelium, the amoebae are transported hematogenously to the CNS. They may also bypass the circulatory system by going directly from nasal epithelium to the CNS via olfactory nerves (Kiderlen and Laube, 2004). In the case of skin lesions, the presence of amoebae may induce a granulomatous response from the host allowing proliferation of amoebae prior to being transported to the CNS (Bravo *et al.*, 2006). The blood-brain barrier must be breached before amoebae can gain entrance to the brain. Jayasekera *et al.* (2004) have postulated that the vascularized choroid plexus is the site of invasion of the brain, with further spread in the CNS facilitated by the CSF produced there.

The oral route of infection is also feasible, as Kiderlen *et al.* (2006a) have shown in mouse-model experiments. This is supported by a report of a fatal case of *Acanthamoeba* encephalitis in a patient with ulcerative colitis (Kidney and Kim, 1998). The authors postulated that the portal of entry was the inflamed bowel. The oral route of infection was also hypothesized in a sheep that died of amoebic meningoencephalitis caused by a leptomyxid amoeba (*Balamuthia*; Fuentealba *et al.*, 1992). They based their conclusion on the presence of amoebae in the enlarged submandibular lymph node resulting from an oral lesion.

As examples of varying modes of infection, a 64-year-old man who had been digging in his backyard and, after being pricked by a rose thorn, developed a cutaneous lesion that most likely provided ingress for amoebae in soil (Deetz *et al.*, 2003). Histopathology of the skin lesion, diagnosed as foreign body granuloma, provided retrospective evidence for the presence of balamuthiae. Another case involved a 72-year-old woman who was working with composted garden soil (Jung *et al.*, 2004). Although no obvious skin lesions were detected, even a soil-contaminated microabrasion can suffice for amoeba entry. Other cases have developed as infections associated with the nasal passages or the lungs, suggesting inhalation of cysts. A case of balamuthiasis in a five-year-old Australian boy originated as a nasal infection with secondary spread to the CNS (Reed *et al.*, 1997). Two California balamuthiasis cases may have resulted from riding in open vehicles across terrain with exposure to wind-blown soil (Schuster *et al.*, 2006b). Most infections in North America are first recognized as CNS disease; in contrast, cases in South America present as cutaneous lesions, often facial, with secondary development of neurologic disease (Bravo *et al.*, 2006; Valdeverde *et al.*, 2006).

The circumstances leading to the death of two dogs are themselves instructive. One animal had been receiving prednisone to suppress an inflammatory disease (Foreman *et al.*, 2004); the second had CNS lymphoma (Finnin *et al.*, 2007). Both canines had a history of swimming in stagnant pond water, suggesting that the source of infection was balamuthiae in the water.

In several other animal and human cases, there has been suspicion raised about water as a source of infection. In a human case in Thailand, the patient sustained an injury to his nose following a fall from a motorcycle into a swamp (Intalapaporn *et al.*, 2004). In Peru, individuals who developed cutaneous and/or CNS balamuthiasis often bathed in ponds or creeks (Bravo *et al.*, 2006). Clogged drains and exposure to accumulated water 3 months before a gorilla in a zoological park developed signs of balamuthiasis, may have been a source of infection (Anderson *et al.*, 1986). Attempts, however, to isolate *Balamuthia* from water and soil samples in the gorilla's enclosure were unsuccessful.

Both *Naegleria* and *Acanthamoeba*, even pathogenic species and strains, have been isolated from the nasal passages of healthy persons (Schuster and Visvesvara, 2004a). Once gaining a foothold they are able to colonize the human (or animal) nasal epithelium, perhaps using it as a staging area for the start of the disease process. No attempts at screening humans have been made to detect and/or isolate *Balamuthia* from the nasal epithelium, a difficult undertaking given that the organism is not readily cultured. But it is conceivable that such colonization might occur.

9.3 The Balamuthia paradox

Balamuthia does not compete well against other soil protists, has complex nutritional requirements with a prolonged generation time, is not likely to occur in soil in large numbers, is not known to produce toxins (though it produces a variety of enzymes), and appears to be a pathogen of low virulence with a less than or equal to 2-year subacute prodromal period. It is therefore puzzling that amoebae can invade across effective host barriers, evade the humoral and cellular immunologic defenses of the host, and ultimately find their way to the CNS causing the death of the host. Compare balamuthiasis to infections caused by other free-living amoebae: (i) in primary amoebic meningoencephalitis due to *Naegleria fowleri*, the individual is likely to have been exposed to numerous amoebae in a water source, and (ii) in *Acanthamoeba* infections, the person is already predisposed to infection by immunodepression. Neither of these appears to be the case in many *Balamuthia* infections. It is not known how many balamuthiae are required to establish infection. In theory, it takes only a single amoeba, but given its vulnerabilities, multiple amoebae are probably required to cause disease.

9.4 Groups or individuals at risk

Table 2 provides an overview of individuals who developed balamuthiasis. In assembling the table, we have been selective for cases that are representative of: (i) the disease itself, (ii) the age and health (predisposing factors) of the patient, (iii) the CSF data that is helpful in diagnosis, (iv) symptoms at presentation as the basis for the initial diagnosis, and (v) the duration of the acute phase of disease. There are many more cases described in the literature that are not included because of space limitations.

9.5 Immunocompromised versus immunocompetent hosts

When the earliest cases of balamuthiasis were reported in humans, it appeared to be a disease of the immunocompromised host: HIV/AIDS patients, alcoholics, intravenous drug users and persons of advanced age and/or in poor health (Anzil *et al.*, 1991; Gordon *et al.*, 1992; Visvesvara *et al.*, 1990). With AIDS in abatement in the developed world, children are now among those most frequently reported with the infection. In virtually all cases of childhood balamuthiasis, the patient was in good health with no overt evidence of immunodeficiency.

9.6 Hispanic Americans

In North America, Hispanic Americans figure significantly among those with balamuthiasis (see Table 2). In the United States, about half the number of reported cases has been in Hispanic Americans, though they comprise little more than about 13% of the population. In a recent survey of encephalitis patients in California, Hispanics made up eight of a total of 11 (73%) of the known cases of balamuthiasis that have occurred in the state since the disease was first described in 1990, although the proportion of Hispanics in the state is only 32% (Schuster *et al.*, 2004). The disproportionate incidence of

Table 2 Recent human cases of Balamuthia amoebic granulomatous encephalitis, in order of increasing age

Age (Years)	Sex	Ethnicity (Country or State)	CSF Data[b,c]	Symptoms at Presentation	Duration of Illness[d]	Predisposing Factors	Initial Diagnosis	Reference
<7 months	M	H[a] Chile	P: 890 WBC: 800 G: normal	Febrile, convulsions, bradycardia, personality change	~5 w	None apparent	Toxoplasmosis	Oddó et al., 2006
2	F	H California	P: 84 WBC: 124 G: 58	Hemiparesis, facial palsy, lethargy, headache, febril, hydrocephalus	3–4 w	None apparent	Tuberculous meningitis	Bakardjiev et al., 2003
2.5	M	H Texas	P: 116 WBC: 14 G: 39	Intermittently febrile, emesis, ataxia	14 w	Possible play in a backyard wading pool	Amoebic infection	Bakardjiev et al., 2003
3	F	C-H California	P: 1247 WBC: 354 G: 6	Febrile, comatose, emesis, tonic-clonic seizures, hydrocephalus	4 w	Handling soil from an indoor flowerpot; otitis media	Tuberculous meningitis	Bakardjiev et al., 2003
3	M	C Czech Republic	ND	Lethargy, tiredness, seizures	<3 m	Rhinitis, otitis media	ND	Kodet et al., 1998
5	M	C Australia	P: 0.5 gl^{-1} WBC: 34 × 10^6 lymphocytes l^{-1} G: 4.5 mmol l^{-1}	Facial cutaneous lesion, CNS involvement, unsteady gait, febrile, diplopia	~15 m	Midfacial lesion on bridge of nose; cause unknown	Mycosis, Wegener's granulomatosis, tuberculous disease	Reed et al., 1997
5	F	H California	P: 41 WBC: 162 G: 73	Febrile, generalized seizures, headache	Recovered after treatment; minor neurologic sequelae	None apparent	Herpes simplex, neuro-cysticercosis	Deetz et al., 2003

5	F	H Chile	P: 200/128 WBC: 423/245 G: 29/35	Facial cutaneous lesion	~1 y	None apparent	Myco-bacteriosis	Cuevas et al., 2006
7	M	H California	P: 308 WBC: 230 G: <20	Febrile, personality change, focal seizures, wide gait	6 w	None apparent	ND	Bakardjiev et al., 2003
7	M	H California	ND	Headache, neck pain, tonic-clonic seizures, febrile, nystagmus	~2 w	History of pneumonia	Tuberculosis, mycosis, toxoplasmosis nephrotic syndrome, proteinuria, unknown kidney disease	Unpublished data from CEP
7 to 42	4 M 2 F	H Mexico	ND	Headache, seizures, emesis, febrile, coma	~1 w to ~7 m	Alcoholism, malnutrition, SLE, respiratory tract infection	Retrospective study	Uribe-Uribe et al., 2001
8	M	C Portugal	ND	Intracranial hypertension, esotropia, diplopia, emesis, headache, lethargy	~4 w	History of swimming in pools	Tuberculosis	Tavares et al., 2006
12	M	H California	P: 422–1230 WBC: 78–733 G: 40–75	Headache, febrile, emesis, altered mental status, mental confusion	>4 m	Exposure to blowing soil while motorcycling	Acute disseminated encephalo-myelitis (ADEM)	D. Michelson, unpublished data
23	M	A-P Thailand	P: 100 WBC: 262 G: 48	Severe headache, cutaneous lesion (nose), febrile	6 m	Cutaneous lesion (nose)	Inflamed keloid	Intalapaporn et al., 2004

(Continued)

Table 2 (Continued)

Age (Years)	Sex	Ethnicity (Country or State)	CSF Data[b,c]	Symptoms at Presentation	Duration of Illness[d]	Predisposing Factors	Initial Diagnosis	Reference
32	M	H England	ND	Cutaneous lesion, edema, hydrocephalus, febrile	10 w	Cutaneous lesion (elbow)	Cutaneous lymphoma, cutaneous tuberculosis, mycosis	White et al., 2004
38	M	H Texas	ND	Ulceration of thigh lesion, tonic-clonic seizures	~4 m	Drug user, male prostitute; HIV negative	Tumor	Deol et al., 2000
43	M	H New York	P: 50; WBC: 200 μl^{-1} G: 145	Seizure, dysarthria, afebrile, headache	3–4 w	Diabetes, hypertension	Toxoplasmosis, pyogenic abscess	Rahimian et al., 2005
52	F	ND Massachusetts	P: 2.6; 10.8 gl^{-1} WBC: 0.4; 0.002×10^9 l^{-1} G: 28 mg dl^{-1}	Speech difficulty, hemiparesis, lethargy	>6 w	Chronic neutropenia, ulcerated lesion on ankle	ND	Katz et al., 2000
64	M	California	ND	Hemiparesis, focal seizures, lethargy, emesis	Recovered after treatment; neurologic sequelae	Cutaneous lesion with likely soil contamination	Treated for tuberculous disease	Deetz et al., 2003

Age	Sex	Ethnicity/Location	CSF data	Symptoms	Duration	Underlying condition	Initial diagnosis	Reference
64	M	H California	P: Very high WBC:~250 G:ND	Headache, lethargy, emesis, disorientation, speech difficulty	~2 w	Reportedly alcoholic, worked in construction	Pyogenic brain abscess	Unpublished data from CEP
72	F	ND New York	ND	Focal seizures, visual loss, aphasia	Recovered after treatment; no neurologic sequelae	Gardening	ND	Jung et al., 2004
78	F	A-P Japan	P: 580; WBC: 1888 G: 56	Febrile, emaciated, anemic, nuchal rigidity, coma	~2 w	Chronic interstitial nephritis	Sjogren's syndrome	Shirabe et al., 2002
84	M	H California	ND	Headache, lethargy, diplopia, speech and hearing difficulties	~1 m	Diabetes, leukemia	ND	Unpublished data from CEP
89	M	C Texas	P: slightly elevated WBC: 40 microscope field^{-1} G: borderline	Erythematous plaque on nose, ulceration, CNS involvement (MRI)	~9 m	Advanced age; cattle rancher	Rosacea	Pritzker et al, 2004

Abbreviations: F, female; M, male; H, Hispanic ethnicity; C, Caucasian; A-P, Asian-Pacific ethnicity; ND, no data; P, protein; G, glucose; WBC, white blood cells; w, week; m, month; CEP, California Encephalitis Project.

[a] Ethnicity was based on the patient's medical history or the family name. Since the designation 'Hispanic' includes a wide variety of nationals (Mexicans, Central and South Americans) we have indicated as best we can the original countries of origin of the patients.
[b] CSF data in mg dl^{-1}, unless otherwise specified in table.
[c] Normal values for CSF components: protein (15–45 mg dl^{-1}); white blood cells (0–5 mm^{-3}); glucose (40–80 mg dl^{-1}).
[d] Where the referenced authors do not provide specific information about duration of illness, we have estimated the duration from information given in the text. These dates are usually from the first day of hospitalization.

the disease in Hispanics may be due to genetic predisposition or other factors such as lifestyle, socioeconomic status, access to clean water and availability of medical services. In California, Hispanics comprise a large part of the workforce in agriculture and construction and/or may reside in rural settings with exposure to soil, either directly or indirectly. Hispanics in the United States are more likely to be diagnosed with neurocyticercosis, typhoid fever and amoebiasis than non-Hispanics, most of the diseases having been acquired before entering the country. Of possible relevance, severe forms of coccidioidomycosis surveyed in California are more likely to occur in persons of Hispanic ethnicity than in other racial groups (Louie *et al.*, 1999). This susceptibility was found to be associated with HLA class II genotypes, as well as ABO blood groups.

9.7 Disease in animals

In addition to humans, animals have been diagnosed with balamuthiasis. The first description and isolation of the amoeba was from a pregnant female mandrill baboon that died of encephalitis (Visvesvara *et al.*, 1990). Since that index case, balamuthiasis has been described in other nonhuman primates (a gibbon, an ape, an orangutan and a gorilla), as well as a horse, a sheep and dogs (see Table 3; Schuster and Visvesvara, 2004a). Some of the animals that developed balamuthiasis, like their human counterparts, were immunocompromised and/or ailing due to other diseases. The disease in animals follows a course similar to that in humans. Many of these infected animals have died in zoological parks, hence the interest in determining cause of death. Other animals, either in the wild or on farms or ranches, may develop balamuthiasis but are never diagnosed.

10 Balamuthiasis: Nature of the Disease

The disease expresses itself in two different modes.

10.1 Central nervous system disease

The prodromal period, during which the disease is incubating asymptomatically, can vary from weeks to as long as 2 years. Once it passes from indolence to a fulminant stage, the acute disease progresses rapidly causing death within weeks (see Table 2). The variation in length of the prodromal period may be due to the virulence of the infecting strain of *Balamuthia*, the size of the initial infectious inoculum, portal of entry and the intrinsic health and immunologic status of the individual. In addition to the CNS, various organs may also show evidence of *Balamuthia* amoebae: thyroid gland, kidneys, pancreas, liver, myometrium, prostate, lymph nodes, adrenal glands and lungs (Martinez and Visvesvara, 1997).

Table 3 Data from known animal cases of *Balamuthia* amoebic granulomatous encephalitis

Animal	Location	Gender	Age (Years)	Symptoms at Presentation	Duration of Illness[a]	Predisposing Factors	Reference
Gorilla[b]	California	F	1	Muscle weakness, drowsiness, posterior paresis, coma, CNS involvement	19 d	No prior history of illness	Anderson et al., 1986
Orangutan	Australia	M	20	Depression, lethargy, headache, loss of appetite, ataxia	9 d	History of aspiration pneumonia	Canfield et al., 1997
Canine (Golden retriever)	Australia	M	6	Neurologic signs (hearing loss, circling, head tilt), respiratory distress	8 m	History of seizures, CNS lymphoma, prednisolone treatment	Finnin et al., 2007
Canine (Great Dane)	California	M	2	Weight loss, emesis, diarrhea, hematuria, lethargy, seizures, nystagmus, coma	~2 m	Prednisone treatment for inflammatory bowel disease	Foreman et al., 2004
Sheep	Texas	F	1.5	Blindness, incoordination	<2 w	Self-inflicted cutaneous lesions	Fuentealba et al., 1992
Horse	California	Gelding	20	Excessive salivation, staggering gait	~3 d	No prior history of illness, sudden onset; negative: rabies, herpes virus I, WEE, EEE	Kinde et al., 1998
Four Old World primates[b]	California (1944–1965)	3 F 1 M	4–13	Paresis, paralysis, weakness, nystagmus, depression	2 d– >1 y	Retrospective study; no prior conditions noted	Rideout et al., 1997

Abbreviations: M, male; f, female; d, day; m, month; EEE, eastern equine encephalitis; WEE, western equine encephalitis

[a] As in Table 2, where the referenced authors do not provide specific information about duration of illness, we have estimated the duration from information given in the text.

[b] These are primates that died at the San Diego Zoological Park or the Wildlife Preserve: Kikuyu colobus monkey (~6 years); mandrill baboon (~4 years); gibbon (5 years); gorilla (13 years). A fifth primate (gorilla, 1 year old) was also noted in this study but is included separately in the table above (Anderson et al., 1986).

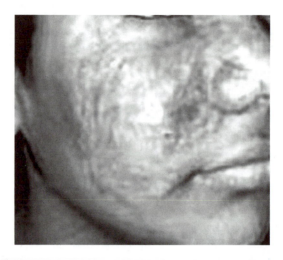

Figure 10 Patient (Peru) with a facial lesion typical of cutaneous balamuthiasis. Note involvement of the nose. Photograph courtesy of Dr. Francisco Bravo.

10.2 Cutaneous disease

Overshadowed by the neurologic form of balamuthiasis found in North America and Europe, the disease in South America is more likely initially to follow a cutaneous route before secondarily attacking the CNS. A erythematous plaque (Figure 10) of rubbery-to-hard consistency may appear on the face, the trunk, or the limbs (Bravo *et al.*, 2006). Plaques appearing on the face often include the nose, resulting in its enlargement. Unusual for North America, a case of facial cutaneous balamuthiasis causing ulceration of the nose, and a second case involving the thigh, face and neck, developed secondarily into CNS disease in a non-Hispanic 89-year-old Texas rancher and a 38-year-old Hispanic male, respectively (see Table 2; Deol *et al.*, 2000; Pritzker *et al.*, 2004). Neither case was associated with prior breaks in the skin surface. Lesions appearing on other parts of the body may appear as single or multiple plaques or as nodules (Gordon *et al.*, 1992). A cutaneous lesion may also appear at the site of an abrasion of the skin surface (Deetz *et al.*, 2003; White *et al.*, 2004). If recognized and treated successfully, spread to the CNS can be prevented (Bravo *et al.*, 2006; personal communication). The time period for transition from the cutaneous to the neurologic phase can range from 30 days to 2 years, with an average of 5–8 months (Bravo *et al.*, 2006).

11 Balamuthiasis at Presentation

11.1 CNS disease

Initial signs of balamuthiasis include headache, tonic-clonic seizures, vomiting, personality change and neck stiffness, features that are shared with other kinds of infectious and noninfectious encephalitidies. The patient may be febrile or afebrile and may be comatose or soon become so. Other signs at presentation may include diplopia, papilledema from increased intracranial pressure, hemiparesis, aphasia, ataxia, and cranial nerve palsies (mainly of the third and sixth cranial nerves). The complete blood count (CBC) is uninformative. The CSF, obtained by lumbar puncture, is clear, though CSF samples will show elevated protein (increasing from normal [15–45 mg dl^{-1}] to >1000 mg dl^{-1} with progression of the disease), lymphocytic pleocytosis with counts greater than 1000 cells mm^{-3} (normal 0–5 cells mm^{-3}), and normal or slightly lowered glucose level (normal 40–80 mg dl^{-1}; Table 2). Cultured and Gram-stained preparations of CSF are negative for bacteria and fungi. Neuroimaging (MRI or CT) may show uni- or multifocal ring-enhancing lesions of 2–3 cm in diameter or a space-occupying mass (Figure 11a, b). Hydrocephalus and edema may be present. Elevated intracranial pressure can lead to herniation of the uncal and cerebellar tonsils and death (Martinez and Visvesvara, 2001). Histopathology may reveal a granulomatous process, though this may be absent in immuno-compromised individuals. Areas of the brain more likely to be affected are the cerebral and cerebellar hemispheres, the pons, the brain stem and cervical spinal cord, and areas of gray-white junction. Differential diagnoses include neurotuberculosis, neurocysticercosis, bacterial abscess, viral or bacterial encephalitis, acute disseminated encephalomyelitis (ADEM), leptomeningitis, and brain tumor. Diagnosis of balamuthiasis based on clinical presentation alone is not possible and requires confirmatory evidence from immunohistochemistry or PCR of biopsied tissue.

(A)   (B)

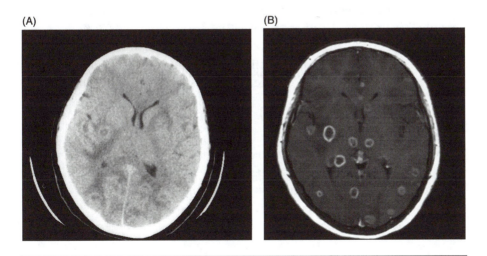

Figure 11 Neuroimages of the brain of a 12-year-old balamuthiasis patient. The two scans are at approximately the same level of the brain. (A) Computerized tomography (CT) image with visible ring-enhancing lesions but at low contrast. (B) Magnetic resonance image (MRI) with more readily apparent lesions. Neuroimages courtesy of Dr. David Michelson.

11.2 Cutaneous disease

A plaque, nodule or abscess appearing on the face or trunk should alert the clinician to the possibility of cutaneous balamuthiasis. A persistent or chronic sore that does not heal following general or topical application of antimicrobials is also suspect (Reed *et al.*, 1997; White *et al.*, 2004). Confirmation is made by a biopsy of the affected area with examination of the sections for amoebae with vesicular nuclei. The cutaneous form of disease can mimic acanthamoebiasis, which also may present as a cutaneous infection. Typically, however, the latter form of cutaneous lesion is more likely to occur in an immunocompromised patient. Differential diagnoses for cutaneous balamuthiasis are fungal infection (sporotrichosis), leishmaniasis, lupus vulgaris (cutaneous tuberculosis) and Wegener's granulomatosis (necrotizing granulomatous vasculitis with or without CNS involvement; rarely in children; Bravo *et al.*, 2006).

12 Pathophysiology

Once having passed from a subacute to acute status, death usually occurs in a matter of weeks. In most cases described in the literature, the patients were hospitalized following the shift to a fulminant process. Delayed diagnosis and treatment with inappropriate antibacterial/antiviral therapy are important factors responsible for the known high mortality (>95%) of balamuthiasis. In immunocompromised patients, death may be the result of cumulative opportunistic infections and not solely due to balamuthiasis. In immunocompetent individuals damage to the brain, particularly herniation resulting from elevated intracranial pressure, is a major cause of death.

Middle ear infections (otitis media) have preceded *Balamuthia* encephalitis in several pediatric cases (Bakardjiev *et al.*, 2003; Kodet *et al.*, 1998; Rowen *et al.*, 1995). It is not known if the ear infection may have predisposed the children to infection or whether otitis media may be a subtle indication of immune deficiency in these children (Mogi and Maeda, 1982).

13 Epidemiology

Anyone who comes in contact with soil is potentially at risk for *Balamuthia* infection, but not necessarily disease. Many individuals have been shown to have antibodies to *Balamuthia*, presumably as a result of contact with the amoebae in soil. There is no evidence, however, that these antibodies are protective in any way.

13.1 Balamuthia *antibodies in humans*

Huang *et al.* (1999), in a study of healthy individuals in Australia tested for anti-*Balamuthia* antibodies with a fluorescent activated cell sorter (FACS), found levels of IgM and IgG antibodies ranging from 1:64 to 1:256. A positive serum control from a balamuthiasis patient had an IgG titer of 1:1024.

They also found antibodies in umbilical-cord blood indicating transplacental transfer at a much lower titer (1:4 or lower).

In a study that screened 290 immunocompetent patients with encephalitis for *Balamuthia* neurologic disease, seven cases (2%) were detected using immunofluorescent antibody staining (IFA; Schuster *et al.*, 2006b). The titers were generally lower than those detected by FACS, with greater than or equal to 1:128 taken as an IFA positive. PCR of tissue and/or CSF confirmed IFA and H&E staining. The difference in titers detected in the two studies may be due to the greater sensitivity of FACS, or the subjectivity of the IFA determinations (i.e., estimating barely perceptible levels of fluorescence as seen with the fluorescence microscope). A positive serum sample (1:128 to 1:256) from this study, when tested by FACS, gave a titer of 1:400 to 1:800 (A. Kiderlen, personal communication). All patients tested had significantly higher CSF protein levels, elevated leukocyte levels and normal or near-normal glucose (see Table 2, CSF data). What these two studies show is that *Balamuthia* antibodies are not rare in either healthy individuals or encephalitis patients. It is not certain that all the titers detected by FACS or IFA were actually anti-*Balamuthia* or were products of cross-reactivity, unless confirmed by some other testing technique (e.g., IIF or PCR). Similar studies have found anti-*Naegleria* and anti-*Acanthamoeba* antibodies in otherwise normal humans and feral and domesticated animals (Schuster and Visvesvara, 2004a). Are antibody levels present in healthy humans the result of contact with nonpathogenic strains of these free-living amoebae? What factor or factors have made the handful of balamuthiasis patients vulnerable to the disease? These are questions that have yet to be answered.

13.2 Balamuthiasis in populations

The occurence of balamuthiasis in the population is unknown. In a study of 1100 cerebral tumors operated on at the National Pediatric Hospital in Buenos Aires, Argentina (1988 to 2000), three balamuthiasis cases (0.27%) were retrospectively diagnosed following histopathologic review (Galarza *et al.*, 2002). The figure for the California survey of 290 encephalitis patients for balamuthiasis was 2% (Schuster *et al.*, 2006b). Immunostaining is likely to be more effective in detecting amoebae because a fluorescing amoeba, even a single one, will stand out, more so than by conventional histopathology. There have been 45 presumptive balamuthiasis cases reported from the Cayetano Heredia General Hospital in Lima, Peru. Twenty of those cases were confirmed as balamuthiasis by immunofluorescence staining (Bravo *et al.*, 2006).

Balamuthia cases are discrete occurrences with no evidence of a single source of infections. All indications point to soil or water as a source of infection, and there is no evidence for 'epidemics' associated with a particular site, such as a lake as there has been for some *Naegleria fowleri* infections. Most cases occur in the young or the old. In the cases noted in Table 2, one individual was a reported alcoholic (64 years of age) and another was of advanced age (89 years of age). An unpublished fatal case involved an elderly (84-year-old) immunodepressed male Californian with diabetes and leukemia who was being treated with steroids. Among children included in Table 2, ages range from under one to about 12 years of age.

In considering all known human balamuthiasis cases in California from 1990 to the present time, nine of the 12 human cases (75%) were from southern California, an area with relatively mild-to-warm climate year-round and little rainfall. Three of the 12 human cases (25%) were from northern California (Schuster *et al.*, 2006b) where temperatures are cooler and rainfall is more abundant. Seven animal cases also occurred in southern California (vs. two in northern California), but this is more a reflection of animal deaths in zoological parks or other facilities caring for animals, in particular the San Diego Zoo and the San Diego Wild Life Park (see Table 3). Balamuthiasis cases in Peru have been in individuals from semi-desert, rural areas – mostly farmers – of the country, with few cases from urban settings (Bravo *et al.*, 2006).

Some infections, such as *Naegleria fowleri* meningoencephalitis, may be traced to a disturbed environment (e.g., a manmade lake or thermally polluted stream or river) that favors growth of the pathogen, providing warm water and bacteria as a food source. Thus, infections are more likely to be encountered during the summer months of the year. There is, however, no suggestion of seasonality in the occurrence of CNS balamuthiasis cases.

13.3 Prevention

Given the information available now about how the organism infects humans (respiratory tract, skin, or via the nasal epithelium) it is difficult to propose realistic measures to lessen the chance of infection: avoid occupations involving agriculture or construction work that maximize exposure to soil or dust or wear protective gear; give up gardening or, if you must garden, wear gloves to prevent contamination of wounds or microabrasions with soil; and avoid swimming in stagnant waters or again, if you must, keep your head above water. It is, of course, all but impossible to prevent contact with soil, whether blowing soil, or stirred up mud from the bottom of a lake. But balamuthiasis cases are few and the circumstances that predispose healthy, immunocompetent individuals to the disease have yet to be discovered.

14 Diagnosis

14.1 Gross pathology

At the macroscopic level, purulent exudate may be seen at the leptomeninges imparting a clouded or opaque appearance to an otherwise translucent covering. Hemorrhagic necrosis in the brain parenchyma is evident as are uni- or multifocal areas of brain softening (encephalomalacia). The olfactory lobes often appear normal, arguing against invasion *via* the olfactory nerve tracts and for hematogenous spread from skin or respiratory tract.

14.2 Basic pathology

In CNS disease, histopathologic slide preparations (e.g., H&E stains) may be expected to show large numbers of trophic amoebae and cysts, typically around blood vessels (Figures 12 and 13). Amoebae may contain evidence of ingested erythrocytes and damaged parenchymal components (Figure 14). Blood vessels show perivascular cuffing mainly by lymphocytes, other types of leukocytes and giant multinucleate cells. Thrombi may be seen within the blood vessels. Angiitis (vasculitis) may be evident, leading to hemorrhagic necrosis. Amoebae may also invade the parenchyma, though neurons are generally not affected. Granulomatous histopathology may also be evident, though granuloma formation may depend upon the immune status of the patient.

Depending on the part of the brain or cutaneous lesion that is biopsied, few amoebae may be present and may require an intensive search of the sectioned specimen. The clinician should alert the pathologist about suspicions of balamuthiasis because amoebae may otherwise pass unrecognized or be dismissed as macrophages. It is also important, in arriving at a diagnosis, to keep in mind that balamuthiasis is not solely a disease of an immunocompromised host, and that most of the recent cases have occurred in healthy children.

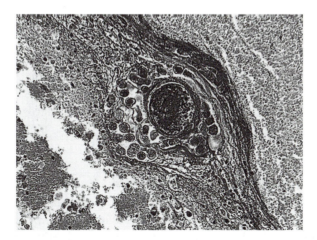

Figure 12 Hematoxylin-eosin-stained cross-section of a blood vessel in brain. Numerous rounded amoebae are clustered in the perivascular space (*arrow*).

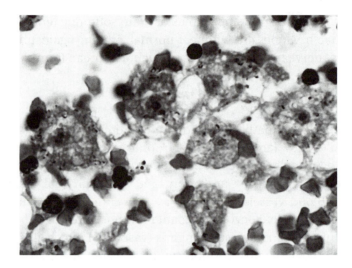

Figure 13 Hematoxylin-eosin stain of necrotic area of brain from balamuthiasis patient. Several trophic amoebae are seen, one of which shows the typical vesicular nucleus, distinguishing the amoeba from host tissue nuclei.

14.3 Antigen detection: indirect immunofluorescent staining of tissue sections

If biopsied or necropsied brain or other types of tissues (e.g., kidney, lung, skin etc.) are available, indirect immunofluorescent staining (IIF) can be employed for specific identification of *Balamuthia* amoeba *in situ*. Formalin-fixed, paraffin-embedded tissue sections are suitable. The tissue sections are deparaffinized and exposed to rabbit anti-*Balamuthia* serum (30 min at 37°C), washed thrice in PBS buffer, followed by FITC-conjugated goat anti-rabbit IgG (30 min at 37°C), and then washed. The slides are mounted with a coverslip using a suitable antiquenching mounting medium (polyvinyl alcohol/DABCO) and examined under a microscope equipped with epifluorescence optics. *Balamuthia* amoebae, if present, will fluoresce apple green (Figures 15 and 16). Typically, clusters of amoebae will be found in the perivascular compartment of blood vessels. Currently, IIF staining for *Balamuthia* antigen is the gold standard for laboratory diagnosis of balamuthiasis.

14.4 Clinical recognition

Balamuthiasis cannot be diagnosed based entirely upon signs and symptoms exhibited by the patient. The symptomatology is not unique and may mimic other types of encephalitidies (Glaser *et al.*, 2006a). Patients typically present with headache, seizures, nuchal rigidity and/or may be comatose. Although they have been isolated from CSF that was obtained at autopsy (Jayasakara *et al.*, 2004), balamuthiae have not been reported in wet mounts or stained smears. If, however, CSF is available for microscopic

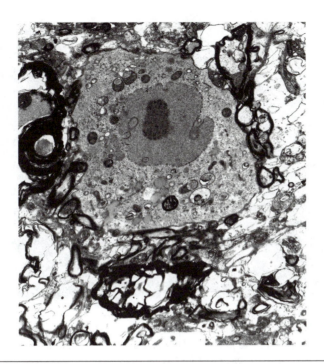

Figure 14 Transmission electron micrograph of *Balamuthia* trophozoites in the brain of an experimentally infected mouse following nasal instillation of amoebae. The nuclear membrane shows several invaginations, which is typical of *Balamuthia* trophozoites. The electron-dense material around the amoeba represents degenerating myelin sheaths, some of which appear in food vacuoles inside the amoeba.

examination, it should be kept at ambient temperature and not frozen. The sample should be centrifuged to concentrate particulate material and cells, and examined as soon as possible, preferably with a phase-contrast microscope. *Balamuthia* amoebae move slowly (0.24 μm s^{-1}) as compared to *Acanthamoeba* ($\leq$0.42 μm s^{-1}) and *Naegleria* (~1 μm s^{-1}). Because of the sluggish movement of the amoebae, medical and laboratory personnel anxious for a diagnosis have identified leukocytes present in CSF wet-mount preparations as balamuthiae.

Once having developed as a fulminant process, the disease progresses insidiously to death in about 4 weeks or less (see Table 2, Duration of illness). Cases of *Acanthamoeba* encephalitis that have occurred in immunocompetent children could pass for balamuthiasis. IFA or IIF staining can resolve the difference.

14.5 Neuroimaging

Both magnetic resonance imaging (MRI) and computerized tomography (CT) have been used during the course of CNS balamuthiasis. Disease may present as ring-enhancing (Healy, 2002; Michelson, personal

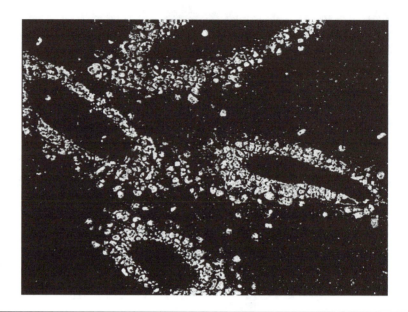

Figure 15 Indirect immunofluorescent stain of brain tissue of a balamuthiasis patient. The amoebae, in the perivascular compartments of the blood vessels, outline the vessels and are visible from having been stained with goat anti-human fluorescein isothiocyanate. In a color image, the amoebae would appear apple green.

communication) or linear (Zagardo *et al.*, 1997) lesions or as a space-occupying mass (Martinez and Visvesvara, 1997). Single lesions seen initially may develop as multifocal lesions as the disease progresses (Schumacher *et al.*, 1995). The lesions are most frequently hypodense, though hyperdense and isodense types are seen in a minority of cases (Figure 11a, b).

MRIs from two surviving patients have shown changes in cranial lesions paralleling recovery. Ring-enhancing lesions were seen in early MRIs of a five-year-old child (Deetz *et al.*, 2003). The lesions came to be enclosed by calcified walls that, in turn, were surrounded by edematous areas (Deetz *et al.*, 2003; Healy, 2002). With time, the lesions regressed in size and the edematous areas resolved. The child recovered with no gross neurologic sequelae after surgery and treatment with anti-*Balamuthia* medications. A similar regression of lesions, but without calcification, was seen in a 64-year-old man after several years (Figure 17a, b). In this case, however, there were significant neurologic deficits (Deetz *et al.*, 2003).

14.6 Molecular techniques: the polymerase chain reaction
PCR of tissue

Preliminary efforts have been made to use the polymerase chain reaction (PCR) for confirming diagnoses. DNA is extracted from samples of patient brain or other tissues or centrifuged CSF samples and amplified using

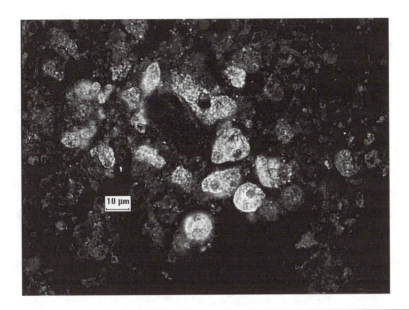

Figure 16 An indirect immunofluorescent image at a higher magnification than in Figure 15, showing a cluster of amoebae from human brain tissue surrounding a blood vessel.

primer sets producing PCR products 1075-bp and/or 230-bp in length (Booton *et al.*, 2003b; Foreman *et al.*, 2004; Yagi *et al.*, 2005). These have successfully demonstrated the presence of *Balamuthia* DNA in fresh and formalin-fixed brain tissue and CSF, even from archival brain tissue scraped from slides (Yagi *et al.*, unpublished observations).

Booton *et al.* (2003c) developed a genus-specific primer set that produced PCR products of 1075 bp from two isolates of *Balamuthia* amoebae, one from a child who died of balamuthiasis and the second from the soil of a potted plant in the child's home (Booton *et al.*, 2003c). The products were sequenced, providing evidence that the two isolates were identical. They also employed an additional, inner primer to perform seminested PCR of formalin-fixed tissue samples. The first round of amplification utilized the primers for the 1075-bp product, and was followed by a second round using the inner primer to amplify a 500-bp product. This technique gave a more distinct band than the one-step PCR, allowing sequencing of the PCR product and its identification as *Balamuthia* DNA.

Conventional PCR

Yagi *et al.* (2005) amplified *Balamuthia* 16S rRNA gene mitochondrial DNA from tissue and CSF of several individuals who died of *Balamuthia* encephalitis, confirmed by IFA and IIF. The tissue samples used included fixed and unfixed brain, kidney and lung. Sections of lung tissue from one of the patients were negative by IIF, but were positive by PCR, suggesting

(A) (B)

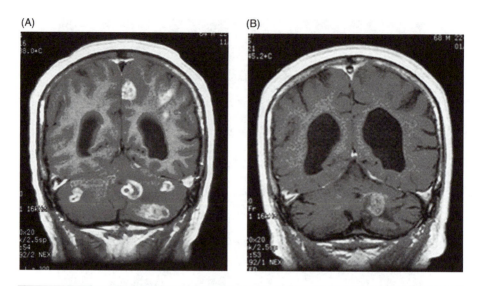

Figure 17 Two magnetic resonance images (MRI) of the same patient who survived a *Balamuthia* infection taken 5 years apart. The two scans are at approximately the same level of the brain. (A) An MRI was taken in 1996 during the acute phase of the disease showing multiple lesions. (B) An MRI taken in 2001 during the convalescent/later phase, showing regression of the lesions. Although the patient recovered with antimicrobial treatment, he sustained neural deficits. MRIs courtesy of Dr. Thomas Deetz.

that PCR might be more sensitive in detecting the amoeba than IFA, particularly when amoebae may be sparse. In assaying the sensitivity of the PCR technique, mitochondrial DNA from less than one amoeba per reaction mixture was detected because of the presence of multiple copies of the targeted sequence.

Real-time PCR

More recently, a multiplex real-time PCR assay was developed for simultaneous testing of 18S rRNA gene DNA from three different amoebae: *Naegleria fowleri*, *Acanthamoeba* spp., and *Balamuthia* (Qvarnstrom *et al.*, 2006). This technique has advantages over the conventional PCR in that: (i) it can simultaneously detect DNA from the three organisms, (ii) it can do so faster (≤ 5 h) since it does not require gel electrophoresis, (iii) risk of contamination of the test sample is less likely since there is no postamplification handling of the PCR product, and (iv) it can also provide an estimate of the parasite load in the specimen. As for sensitivity, the technique was able to detect DNA from less than one *Balamuthia* amoeba.

PCR of CSF

Balamuthia mitochondrial DNA has been demonstrated retrospectively in CSF by PCR in three cases (Yagi *et al.*, 2005). Its presence in the CSF may be a result of necrosis of brain tissue and/or lysis of infecting amoebae.

14.7 Cultivation as a diagnostic method

Amoebae can be cultured from necrotic tissues but its use for rapid diagnosis is precluded by the long time it takes for the balamuthiae to acclimate to *in vitro* growth conditions. It may take as long as a month for the amoebae to feed on the tissue culture cells and multiply. Therefore, while it may be of limited use in the diagnosis of infection, it is still of value to isolate strains of *Balamuthia* for diagnostic confirmation, assessment of virulence, drug sensitivity, genomic composition and epidemiologic study. For optimal results, fresh unfrozen brain tissue should be used as a potential source of amoebae. If the tissue is frozen, there is a greater chance that the freezing process will destroy trophic amoebae. If the frozen tissue contains cysts, which are protected from destruction by their thick walls, there is a greater likelihood of amoeba survival.

15 Host Response

15.1 Mobilization of the immune system

The presence of balamuthiae may elicit a vigorous humoral response. Because of the prolonged subacute phase of the disease, a strong IgM and IgG antibody reaction may occur. In the acute phase of disease, IgG levels are typically elevated while IgM levels are low (Schuster, unpublished observation). Despite the elevated antibody titers, the infection progresses insidiously. All patients with confirmed neurologic balamuthiasis have decidedly elevated antibody titers as shown by IFA staining. In two survivors studied (United States), the antibody titers dropped from about 1:256 to 1:64 after several years (Deetz *et al.*, 2003).

15.2 Antibody production and detection: fluorescent antibody staining

Antibodies in serum

Antibody in patients' sera is detected by indirect immunofluorescent antibody (IIF) staining. Whole *Balamuthia* amoebae fixed and dried on slides are treated with dilutions of patient serum (1:2 to 1:4096) to determine antibody titer (Visvesvara *et al.*, 1990, 1993; Schuster *et al.*, 2006b). This is followed by incubation with goat anti-human IgG conjugated with FITC, washing, and mounting with antiquenching polyvinyl alcohol/DABCO. When the slides are examined with epifluorescence optics, an apple green, fluorescent halo will be seen at the amoeba membrane. The wells are rated subjectively from +1 to +4 depending on the degree of fluorescence. The titer is that concentration of patient serum at which the fluorescence of the amoebae (the antigen) begins to fade. Most commonly, serum titers for *Balamuthia* antibodies range from no detectable titer to about 1:128 or higher (Schuster *et al.*, 2006b). Although titer alone may be misleading, a titer of 1:128 or higher is indicative of an immune response to the presence of *Balamuthia*. Ideally, acute and convalescent (or later) sera should be tested. A titer of 1:64, while not definitive, would suggest a careful appraisal of the case, with follow-up serum samples or tissue if available. No cross-reactivity between *Balamuthia* and *Acanthamoeba* antibodies, which are usually tested simultaneously, has been detected (Huang *et al.*, 1999; Schuster *et al.*, 2006b). However, definitive diagnosis of balamuthiasis based on serum antibody titers alone is not recommended since many healthy individuals show evidence of anti-*Balamuthia* antibodies in their serum and there could be cross-reactivity with an unknown antigen (Huang *et al.*, 1999; Schuster *et al.*, 2006b). It is not known if these antibodies afford protection from infection or if the absence or low levels of antiamoeba antibodies make individuals vulnerable to infection and disease. In a study of the effects of human serum from healthy individuals on *Balamuthia*, 40% of the trophic amoebae were killed (Matin *et al.*, 2007). Serum also inhibited binding and cytotoxicity of balamuthiae to human brain microvascular endothelial cells.

CSF samples from *Balamuthia*-suspect cases generally have low or no antibody titers (≤1:32). In two cases when sequential CSF samples from the patient were available, the CSF antibody titer increased from 1:32 to 1:128 shortly before death, perhaps as a result of hemorrhagic necrosis (Schuster, unpublished observations).

Cytokine production

Aside from the serologic studies that have been carried out, there is little information on the host response generated by *Balamuthia*. One such study was based on the ability of balamuthiae to stimulate cytokine production in the host (Jayasekera *et al.*, 2005). Human brain microvascular endothelial cells regulate the blood-brain barrier, which prevents penetration of the CNS by pathogens. The *in vitro* study examined how the presence of *Balamuthia* affects these cells. Amoebae induced release of the cytokine interleukin-6 (IL-6), after stimulating the cells to produce phosphatidylinositol kinase. The release of IL-6 caused alterations in the blood-brain barrier facilitating movement of leukocytes (and perhaps *Balamuthia*) across the barrier.

16 Antimicrobial Therapy

Infectious diseases of the CNS present obstacles in treatment that are encountered in few other types of infections. The blood-brain barrier prevents free movement of drug into the CSF, but not so into the brain parenchyma. In both balamuthiasis and acanthamoebiasis, the brain parenchyma is the major target, and less so the CSF; thus, hematogenous transport of drug into the parenchyma is essential for antiamoebic therapy. Azithromycin, for example, does not penetrate well into the CSF, but achieves levels in brain parenchyma of one- to twofold that in serum (Jaruratanasirikul *et al.*, 1996).

16.1 What works: clinical successes

In the United States, of the reported cases of CNS balamuthiasis, three individuals (~2%) are known to have survived, an unacceptably high mortality. In Peru, four individuals have survived cutaneous and/or CNS balamuthiasis (Seas and Bravo, 2006). One of the reasons for the high mortality is the difficulty in diagnosing the disease; the second cause is the absence of an optimal antimicrobial therapy (Schuster and Visvesvara, 2003, 2004b).

The three surviving patients with CNS balamuthiasis in the United States (Deetz *et al.*, 2003; Jung *et al.*, 2004) received a specific anti-*Balamuthia* regimen, whereas other patients who did not survive received either general antibacterial therapy or received the anti-*Balamuthia* regimen only when the disease had progressed to a late stage and the patients were in terminal condition. The regimen (see Table 4) consists of several different antimicrobials: pentamidine isethionate, flucytosine (5-fluorocytosine), fluconazole, a macrolide antibiotic (azithromycin or clarithromycin), sulfadiazine and a phenothiazine (thioridazine or trifluoperazine). Some of these drugs are antifungal compounds (fluconazole, flucytosine), and others are antibacterial (the macrolides). Phenothiazines, a group of antipsychotic drugs, have been shown to have an antiamoebic effect on *Acanthamoeba* (Schuster and Mandel, 1984) and *Balamuthia* (Schuster, unpublished; see Table 5). During the course of treatment, that spanned several years for two of the survivors, some of the drugs had to be suspended intermittently or permanently because of side effects and/or toxicity: hyperglycemia and elevated creatinine levels caused by pentamidine, elevated liver enzyme levels caused by fluconazole, rigidity or seizures caused by trifluoperazine, and elevated

---

Table 4 Antimicrobial regimen for *Balamuthia* infections of the CNS[a]

Flucytosine (5-fluorocytosine)	2 gm q6h po
Fluconazole	400 mg qd
Pentamidine isethionate	4 mg kg^{-1} qd iv
Sulfadiazine	1.5 gm q6h po
Azithromycin[b]	500 mg qd
Trifluoperazine[c]	10 mg q12h

Abbreviations: qd, every day; q6h, every 6h; q12h, every 12h; po, by mouth; iv, intravenous.
[a]Dosages used for treating a 64-year-old male patient (Deetz *et al.*, 2003).
[b]Changed to clarithromycin at the same concentration.
[c]Halted after brief use due to muscle rigidity. In treatment of five-year-old female patient, thioridazine was used but briefly halted due to seizure.

Table 5 *In vitro* sensitivity of *Balamuthia mandrillaris* isolates to antimicrobials[a]

Drug	Highest Concentration Tested	Approximate Inhibition (%)
Strong anti-*Balamuthia* activity		
Pentamidine isethionate	$\leq 10\ \mu g\ ml^{-1}$	93
Propamidine isethionate (Brolene®)	$\leq 10\ \mu g\ ml^{-1}$	93
Trifluoperazine	$\leq 10\ \mu g\ ml^{-1}$	64
Chlorpromazine	$\leq 10\ \mu g\ ml^{-1}$	71
Gramicidin S	$\leq 10\ \mu g\ ml^{-1}$	97
Polymyxin B	$\leq 10\ \mu g\ ml^{-1}$	96
Amphotericin B[a]	$\leq 10\ \mu g\ ml^{-1}$	47
Colistin	$\leq 20\ \mu g\ ml^{-1}$	90
Miltefosine (hexadecylphosphocholine)	$\leq 100\ \mu M$	90
Moderate anti-*Balamuthia* activity		
Ketoconazole	$\leq 10\ \mu g\ ml^{-1}$	80
Bifonazole	$\leq 1\ \mu g\ ml^{-1}$	67
Azithromycin (Zithromax®)	$\leq 10\ \mu g\ ml^{-1}$	29
Clarithromycin	$\leq 10\ \mu g\ ml^{-1}$	24
Clotrimazole	$\leq 1\ \mu g\ ml^{-1}$	34
Fluconazole (Diflucan®)	$\leq 10\ \mu g\ ml^{-1}$	34
Little or no anti-*Balamuthia* activity		
Amphotericin methyl ester	$\leq 10\ \mu g\ ml^{-1}$	6
Sulfadiazine	$\leq 10\ \mu g\ ml^{-1}$	10
Sulfamethoxazole	$\leq 10\ \mu g\ ml^{-1}$	<5
Trimethoprim-sulfamethoxazole (1:5)	$\leq 10\ \mu g\ ml^{-1}$	<5
Flucytosine (5-fluorocytosine)	$\leq 100\ \mu g\ ml^{-1}$	<5
Arsobal	$\leq 10\ \mu g\ ml^{-1}$	~10
Voriconazole (Vfend®)	$\leq 40\ \mu g\ ml^{-1}$	18 to 25
Erythromycin	$\leq 20\ \mu g\ ml^{-1}$	<5
Trifluraline	$\leq 100\ \mu M$	~5
Caspofungin (Cancidas®)	$\leq 20\ \mu g\ ml^{-1}$	~20
Terbinafine (Lamasil®)	$\leq 20\ \mu g\ ml^{-1}$	~60

[a]Schuster and Visvesvara, 1996, and unpublished observations.

creatinine levels and concern about interstitial nephritis caused by azithromycin (Deetz *et al.*, 2003).

Two of the Peruvian survivors – one had progressed from cutaneous to CNS disease – were successfully treated with albendazole and itraconazole. In one case, excision of the lesion on the chest wall may have reduced the amoeba load, allowing the host immune system to cope with the remaining balamuthiae in conjunction with antimicrobial therapy (Bravo *et al.*, 2006). No specific course of therapy was recommended, however, nor were drug regimens or dosages published.

The multiplicity of drugs used in empirical treatment makes it difficult to single out one or more of these that could be the basis for optimal therapy. Unfortunately, with a patient in the acute stage of the disease, there is little time for the clinician to experiment. Some of the drugs, singly or in different combinations/concentrations, when tested *in vitro* may show little or no efficacy and others may be amoebistatic rather than amoebicidal. But consideration of synergistic drug interactions, amount of free drug available (compared to serum-bound drug), drug penetration into the brain parenchyma, drug breakdown and elimination *in vivo* versus *in vitro*, dosage and continuity of treatment (multiple dosage *in vivo* vs. one dose added *in vitro*), and the overall condition of the patient are undefined or partially defined factors in the present empirical regimen.

The use of corticosteroids in the course of the disease is controversial. Steroids are used as antiinflammatory agents in both acanthamoebiasis (particularly keratitis) and balamuthiasis and have been suspected of promoting amoeba growth by suppression of the inflammatory response. The use of prednisolone in treating a 5-year-old Australian boy appeared to precede a worsening of symptoms by 1 week (Reed *et al.*, 1997), as did prednisone in the treatment of a 12-year-old Argentinian boy with facial lesions (Taratuto *et al.*, 1991).

In a study (*in vitro* and *in vivo*) that focused on the effect of steroids (dexamethasone) on keratitis-causing strains of *Acanthamoeba*, a four- to 10-fold increase of the amoeba population occurred. Additionally, dexamethosone aggravated keratitis in the hamster model of disease (McClellan *et al.*, 2001). *In vitro* studies on clinical isolates of *Balamuthia*, however, indicate that the steroids dexamethasone and methyl prednisolone (Solu-Medrol®) at concentrations less than or equal to 100 μg ml^{-1} have neither stimulatory nor inhibitory effects on growth of the amoebae (Schuster and Visvesvara, unpublished observations).

A matter of concern in the small number of recovered balamuthiasis patients is whether potentially viable *Balamuthia* cysts in the CNS may activate and release trophic amoebae with cessation of drug therapy. Evidence is lacking for or against such delayed activation but erring on the side of caution, fluconazole (Diflucan®) treatment has been extended over several years after one patient's apparent recovery (Deetz *et al.*, 2003; personal communication).

16.2 What may work: in vitro drug testing

In vitro drug testing depends on the strain of *Balamuthia* tested, the drug concentrations used, the duration of exposure to the drug and the follow-up to determine if the drug is amoebacidal or amoebastatic.

A wide variety of drugs have been tested against *Balamuthia* but all tests have been *in vitro* (see Table 5). Amphotericin B (Fungizone®), the drug of choice against *Naegleria fowleri* (Seidel *et al.*, 1982), is less effective *in vitro* against *Balamuthia*. Different strains of clinical isolates that were tested have shown varied sensitivity. But amphotericin B has been used as an antifungal drug added to tissue cultures for suppressing or eliminating fungal contamination during isolation of balamuthiae from brain tissue without adverse effect on the amoebae (Visvesvara, unpublished observations).

Tata *et al.* (2003; personal communication) using β-galactosidase release from transfected murine mastocytoma cells as an indicator of target cell lysis, found that amphotericin B nanosuspensions had the highest amoebicidal activity (1 μM or 1 μg ml^{-1}) of all drugs tested against *Balamuthia*, followed by dihydroemetine and pentamidine isethionate. Schuster and Visvesvara (1996) found that 1 and 10 μg ml^{-1} of amphotericin B to be 19 and 47% inhibitory, respectively (see Table 5). Nanosuspensions of drugs, however, may facilitate drug-uptake by the amoebae, reducing the amount of tissue culture cell lysis. Ketoconazole was only amoebastatic, but miltefosin, azithromycin, and flucytosine had little effect on amoebae at 5 μM. Other things being equal, drug use in a clinical environment is a more reliable indicator of drug efficacy than *in vitro* studies; research is needed on drug efficacy using the animal model of disease.

One drug, miltefosine, of those indicated in Table 5 has potential for treating balamuthiasis (Schuster *et al.*, 2006a). The minimal inhibitory concentration against *Balamuthia* was 65 μM, and the minimal amoebacidal concentration was 75 μM. Tata *et al.*, however, found miltefosine to be ineffective, though at a lower concentration of 5 μM (Tata *et al.*, 2003; personal communication). Miltefosine has been used successfully in treating sleeping sickness (trypanosomiasis) in Africa, and has also been shown to have *in vitro* activity against *Balamuthia* in addition to *Acanthamoeba* (Schuster *et al.*, 2006a; Walochnik *et al.*, 2002).

The basis for rating the drugs in Table 5 has been the standard growth curve of *Balamuthia* strains growing axenically (BM-3 medium) compared to a growth series containing differing concentrations of the drug (Schuster and Visvesvara, 1996). Apparent amoebacidal drug effects may, in reality, be amoebastatic when apparently 'dead' amoebae recover after transfer to drug-free medium. Studies of the protective value of drugs in an animal model have not been done. As implied in the previous paragraph, *in vitro* efficacy is not, *ipso facto*, a sign of a successful therapeutic agent. Variation in sensitivity between *Balamuthia* strains occurs among those in culture, as does variation in the health of the patient being treated, the stage of the disease and the degree of damage to the affected organs. As a general observation, it appears that drugs that are effective against *Acanthamoeba* encephalitis are also effective against balamuthiasis. A number of other drugs listed in Table 5 are not meant for parenteral use even though they are amoebicidal, may be toxic at clinically effective concentrations, or were found to have no or minimal activity against *Balamuthia* (Schuster and Visvesvara, 1996).

17 Overview, Conclusions and Questions

Balamuthia encephalitis is a deadly disease with a poor prognosis. However, it remains a rarity in humans. Lack of familiarity with the organism and the disease, and the length of time taken to make a diagnosis are factors responsible for the high mortality. Relatively few clinicians and pathologists are familiar enough with the organism or the disease to recognize it in a patient. The list of differential diagnoses adds further uncertainty to early identification. Early recognition and initiation of antimicrobial therapy are essential to survival. Many humans (and animals, too) are undoubtedly exposed in one way or another to the amoeba, as demonstrated by testing for *Balamuthia* antibodies in serum, but very few individuals progress to acute neurologic disease. While much information has been gathered about *Balamuthia* in the 25 or so years since its recognition as an infectious agent, there remains much to learn about the amoeba and the disease. We have attempted to focus on some of the issues and topics that require further research in the following list of questions:

* If many individuals carry anti-*Balamuthia* antibodies in their serum, why do some develop disease?
* Are there both pathogenic and nonpathogenic strains of *Balamuthia*?
* Can *Balamuthia* colonize the mammalian nasal epithelium?
* Do humans and animals develop disease only if infected by pathogenic *Balamuthia*?
* Do the clinical isolates of *Balamuthia* possess virulence factors that render them pathogenic?
* Are there subclinical cases of balamuthiasis among healthy individuals in the population?
* Can *Balamuthia* amoebae encysted in brain tissue of a recovered patient activate at some later time?
* Is there a difference between the *Balamuthia* amoebae that cause neurologic disease (typical of North America and Europe) and those that cause cutaneous lesions (as in South America)?
* Does human genotype (e.g., HLA complex) influence susceptibility to balamuthiasis?

18 Acknowledgments

The authors thank: Dr. Carol Glaser (Chief, Viral and Rickettsial Disease Laboratory, Richmond, CA) for her support and for allowing us to use information collected by the California Encephalitis Program; Dr. B. Joseph Guglielmo (University of California School of Pharmacy, San Francisco, CA) for making available drugs for testing against amoebae and for his pharmacologic insights; Drs. David Michelson (Loma Linda University School of Medicine, Loma Linda, CA) and Thomas Deetz (Sutter Health, Santa Cruz, CA) for permission to use data on an unpublished case of balamuthiasis and MRIs, respectively; Dr. Francisco Bravo for providing a photograph of a balamuthiasis patient; Drs. Albrecht Kiderlen and Phiroze Tata (Robert Koch Institute, Berlin) for sharing their unpublished observations; Dr. Yvonne Qvarnstrom for her valuable editorial comments; and to our late friend, colleague and unrivaled authority on histopathology of amoebic diseases, Dr. A. Julio Martinez (University of Pittsburgh School of Medicine, Pittsburgh, PA), who had generously provided us with electron micrographs of clinical and cultured specimens that we have used in this chapter. This work is supported in part by the National Institutes of Health (NIH/NICHD-HD054451) and the Emerging Infectious Disease Program of the CDC (USOCCU915548_09 to CG).

19 References

Adl, S.M., Simpson, A.G.B., Farmer, M.A., *et al.* (2005) The new higher level classification of eukaryotes with emphasis on the taxonomy of protists. *J. Eukaryot. Microbiol.* **52**: 399–451.

Amaral Zettler, L.A., Nerad, T.A., O'Kelly, C.J., Peglar, M.T., Gillevet, P.M., Silberman, J.D. and Sogin, M.L. (2000) A molecular reassessment of the leptomyxid amoebae. *Protist* **151**: 275–282.

Anderson, M.P., Oosterhuis, J.E., Kennedy, S. and Benirschke, K. (1986) Pneumonia and meningoencephalitis due to amoeba in a lowland gorilla. *J. Zool. Anim. Med.* **17**: 87–91.

Anzil, A.P., Rao, C., Wrzolek, M.A., Visvesvara, G.S., Sher, J.H. and Kozlowski, P.B. (1991) Amebic meningoencephalitis in a patient with AIDS caused by a newly recognized opportunistic pathogen. *Arch. Pathol.* **115**: 21–25.

Bakardjiev, A., Azimi, P.H., Ashouri, N., Ascher, D.P., Janner, D., Schuster, F.L., Visvesvara, G.S. and Glaser, C. (2003) Amebic encephalitis caused by *Balamuthia mandrillaris*: report of four cases. *Pediatr. Infect. Dis.* **22**: 447–452.

Booton, G.C., Carmichael, J.R., Visvesvara, G.S., Byers, T.J. and Fuerst, P.A. (2003a) Genotyping of *Balamuthia mandrillaris* based on nuclear 18S and mitochondrial 16S rRNA genes. *Am. J. Trop. Med. Hyg.* **68**: 65–69.

Booton, G.C., Carmichael, J.R., Visvesvara, G.S., Byers, T.J. and Fuerst, P.A. (2003b) Identification of *Balamuthia mandrillaris* by PCR assay using the mitochondrial 16S rRNA gene as a target. *J. Clin. Microbiol.* **41**: 453–455.

Booton, G.C., Schuster, F.L., Carmichael, J.R., Fuerst, P.A. and Byers, T.J. (2003c) *Balamuthia mandrillaris*: identification of clinical and environmental isolates using genus-specific PCR. *J. Eukaryot. Microbiol.* **50**: 508S–509S.

Bowers, B. and Korn, E.D. (1968) The fine structure of *Acanthamoeba castellanii*. The trophozoites. *J. Cell Biol.* **39**: 95–111.

Bravo, F.G., Cabrera, J., Gotuzzo, E. and Visvesvara, G.S. (2006) Cutaneous manifestations of infection by free-living amebas. In: *Tropical Dermatology* (eds. S.K. Tyring, O. Lupi, and U.R. Hengge). Elsevier, Philadelphia, PA, pp. 49–55.

Canfield, P.J., Vogelnest, L., Cunningham, M.L. and Visvesvara, G.S. (1997) Amoebic meningoencephalitis caused by *Balamuthia mandrillaris* in an orang utan. *Aust. Vet. J.* **75**: 97–100.

Chang, S.L. (1971) Small free-living amebas: cultivation, quantitation, identification, classification, pathogenesis, and resistance. *Curr. Top. Comp. Pathobiol.* **1**: 201–254.

Cuevas, M., Smoje, G., Jofré, L., Ledermann, W., Noemi, I., Berwart, F., Latorre, J.J. and González, S. (2006) Meningoencefalitis granulomatosa por *Balamuthia mandrillaris*: reporte de un caso y revisión de la literature. *Rev. Chil. Infect.* **23**: 237–242. (In Spanish with English abstract).

Deetz, T.R., Sawyer, M.H., Billman, G., Schuster, F.L. and Visvesvara, G.S. (2003) Successful treatment of *Balamuthia* amoebic encephalitis: presentation of two cases. *Clin. Infect. Dis.* **37**: 1304–1312.

Deol, I., Robledo, L., Meza, A., Visvesvara, G.S. and Andrews, R.J. (2000) Encephalitis due to a free-living amoeba (*Balamuthia mandrillaris*): case report with literature review. *Surg. Neurol.* **53**: 611–616.

Dunnebacke, T.H., Schuster, F.L., Yagi, S. and Booton, G.C. (2004) *Balamuthia mandrillaris* from soil samples. *Microbiology* **150**: 2837–2842.

Finnin, P.J., Visvesvara, G.S., Campbell, B.E., Fry, D.R. and Gasser, R.B. (2007) Multifocal *Balamuthia mandrillaris* infection in a dog in Australia. *Parasitol. Res.* **100**: 423–426.

Foreman, O., Sykes, J., Ball, L., Yang, N. and De Cock, H. (2004) Disseminated infection with *Balamuthia mandrillaris* in a dog. *Vet. Pathol.* **41**: 506–510.

Fuentealba, I.C., Wikse, S.E., Read, W.K., Edwards, J.F. and Visvesvara, G.S. (1992) Amebic meningoencephalitis in a sheep. *Am. J. Vet. Med. Assoc.* **200**: 363–365.

Galarza, M., Cuccia, V., Sosa, F.P. and Monges, J.A. (2002) Pediatric granulomatous cerebral amebiasis: a delayed diagnosis. *Pediatr. Neurol.* **26**: 153–156.

Glaser, C.A., Honarmand, S., Anderson, L.J., Schnurr, D.P., Forghani, B., Cossen, C.K., Schuster, F.L., Christie, L.J. and Tureen, J.H. (2006a) Beyond viruses: clinical profiles and etiologies associated with encephalitis. *Clin. Infect. Dis.* **43**: 1565–1577.

Glaser, C.A., Lewis, P.F. and Schuster, F.L. (2006b) Fungal, rickettsial, and parasitic diseases of the nervous system. In: *Pediatric Neurology: Principles and Practice,*

Vol. 2, 4th Edn (eds. K.F. Swaiman, S. Ashwal and D.M. Ferriero). Mosby/Elsevier, Philadelphia, PA, pp. 1631–1684.

Gordon, S.M., Steinberg, J.P., DuPuis, M.H., Kozarsky, P.E., Nickerson, J.F. and Visvesvara, G.S. (1992) Culture isolation of *Acanthamoeba* species and leptomyxid amebas from patients with amebic meningoencephalitis, including two patients with AIDS. *Clin. Infect. Dis.* **15**: 1024–1030.

Healy, J.F. (2002) *Balamuthia* amebic encephalitis: radiographic and pathologic findings. *Am. J. Neuroradiol.* **23**: 486–489.

Huang, Z.H., Ferrante, A. and Carter, R.F. (1999) Serum antibodies to *Balamuthia mandrillaris*, a free-living amoeba recently demonstrated to cause granulomatous amoebic encephalitis. *J. Infect. Dis.* **179**: 1305–1208.

Intalapaporn, P., Suankratay, C., Shuangshoti, S., Phantumchinda, L., Keelawat, S. and Wilde, H. (2004) *Balamuthia mandrillaris* meningoencephalitis: the first case in Southeast Asia. *Am. J. Trop. Med. Hyg.* **70**: 666–669.

Janitschke, K., Martinez, A.J., Visvesvara, G.S. and Schuster, F. (1996) Animal model *Balamuthia mandrillaris* CNS infection: contrast and comparison in immunodeficient and immunocompetent mice: a murine model of 'granuloma-tous' amebic encephalitis. *J. Neuropathol. Exp. Neurol.* **55**: 815–821.

Jaruratnasirikul, S., Hortiwakul, R., Tantisarasart, T., Phuenpathom, N. and Tussanasunthorwong, S. (1996) Distribution of azithromycin into brain tissue, cerebrospinal fluid, and aqueous humor of the eye. *Antimicrob. Agents Chemother.* **40**: 825–826,

Jayasekera, S., Sissons, J., Tucker, J., Rogers, C., Nolder, D., Warhurst, D., Alsam, S., White, J.M.L., Higgins, E.M. and Khan, N.A. (2004) Post-mortem culture of *Balamuthia mandrillaris* from the brain and cerebrospinal fluid of a case of gran-ulomatous amoebic meningoencephalitis, using human brain microvascular endothelial cells. *J. Med. Microbiol.* **53**: 1007–1012.

Jayasekera, S., Matin, A., Sissons, J., Maghsood, A.H. and Khan, N.A. (2005) *Balamuthia mandrillaris* stimulates interleukin-6 release in primary human brain microvascular endothelial cells via a phosphatidylinositol 3-kinase-dependent pathway. *Microbes Infect.* **7**: 1345–1351.

Jung, S., Schelper, R.L., Visvesvara, G.S. and Chang, H.T. (2004) *Balamuthia mandrillaris* meningoencephalitis in an immunocompetent patient. *Arch. Pathol. Lab. Med.* **128**: 466–468.

Katz, J.D., Ropper, A.H., Adelman, L., Worthington, M. and Wade, P. (2000) A case of *Balamuthia mandrillaris* meningoencephalitis. *Arch. Neurol.* **57**: 1210–1212.

Khan, N.A. (2006) *Acanthamoeba*: biology and increasing importance in human health. *FEMS Microbiol. Rev.* **30**: 564–595.

Kiderlen, A.F. and Laube, U. (2004) *Balamuthia mandrillaris*, an opportunistic agent of encephalitis, infects the brain via the olfactory nerve pathway. *Parasitol. Res.* **94**: 49–52.

Kiderlen, A.F., Laube, U., Radam, E. and Tata, P.S. (2006a) Oral infection of immunocompetent and immunodeficient mice with *Balamuthia mandrillaris* amebae. *Parasitol. Res.* **100**(4): 775–782.

Kiderlen, A.F., Tata, P., Özel, M., Laube, U., Radam, E. and Schäfer, H. (2006b) Cytopathogenicity of *Balamuthia mandrillaris*, an opportunistic causative agent of granulomatous amoebic encephalitis. *J. Eukrayot. Microbiol.* **53**: 456–463.

Kidney, D.D. and Kim, S.H. (1998) CNS infections with free-living amebas: neuroimaging findings. *Am. J. Neuroradiol.* **171**: 808–812.

Kinde, H., Visvesvara, G.S., Barr, B.C., Nordhausen, R.W. and Chiu, P.H.W. (1998) Amebic meningoencephalitis caused by *Balamuthia mandrillaris* (leptomyxid ameba) in a horse. *J. Vet. Diagn. Invest.* **10**: 378–381.

Kodet, R., Nohynkova, E., Tichy, M., Soukup, J. and Visvesvara, G.S. (1998) Amebic encephalitis caused by *Balamuthia mandrillaris* in a Czech child: description of the first case from Europe. *Pathol. Res. Pract.* **194**: 423–430.

Louie, L.L., Ng, S., Hajjeh, R., Johnson, R., Vugia, D., Werner, S.B., Talbot, R. and Klitz, W. (1999) Influence of host genetics on the severity of coccidioidmycosis. *Emerg. Infect. Dis.* **5**: 672–680.

Martinez, A.J. and Visvesvara, G.S. (1997) Free-living, amphizoic and opportunistic amebas. *Brain Pathol.* **7**: 583–598.

Martinez, A.J. and Visvesvara, G.S. (2001) *Balamuthia mandrillaris* infection. *J. Med. Microbiol.* **50**: 205–207. (Editorial).

Matin, A., Jeong, S.R., Faull, J., Rivas, A.O. and Khan, N.A. (2006a) Evaluation of prokaryotic and eukaryotic cells as food source for *Balamuthia mandrillaris*. *Arch. Microbiol.* **186**: 261–271.

Matin, A., Jeong, S.R., Stins, M. and Khan, N.A. (2007) Effects of human serum on *Balamuthia mandrillaris* interactions with human brain microvascular endothelial cells. *J. Med. Microbiol.* **56**: 30–35.

Matin, A., Stins, M., Kim, K.S. and Khan, M.A. (2006b) *Balamuthia mandrillaris* exhibits metalloprotease activities. *FEMS Immunol. Med. Microbiol.* **47**: 83–91.

McClellan, K., Howard, K., Niederkorn, J.Y. and Alizadeh, H. (2001) Effect of steroids on *Acanthamoeba* cysts and trophozoites. *Invest. Ophthalmol. Vis. Sci.* **42**: 2885–2893.

Michel, R. and Janitschke, K. (1996) Axenic and monoxenic cultivation of *Balamuthia mandrillaris* (Visvesvara *et al.* 1993) Leptomyxidae. In: *Christian Gottfried Ehrenberg Festschrift*. Leipziger Universitätsverlag, Leipzig, Germany, pp. 100–102.

Mogi, G. and Maeda, S. (1982) Recurrent otitis media in association with immunodeficiency. *Arch. Otolaryngol.* **106**: 204–207.

Oddó D., Ciani, S. and Vial, P. (2006) Encefalitis amebiana granulomatosa por *Balamuthia mandrillaris*. Primer caso diagnosticado en Chile. *Rev. Chil. Infect.* **23**: 232–236. (In Spanish with English abstract).

Pritzker, A.S., Kim, B.K., Agrawal, D., Southern Jr, P.M. and Pandya, A.G. (2004) Fatal granulomatous amebic encephalitis caused by *Balamuthia mandrillaris* presenting as a skin lesion. *J. Am. Acad. Dermatol.* **50**: S38–S41.

Qvarnstrom, Y., Visvesvara, G.S., Sriram, R. and da Silva, A.J. (2006) Multiplex real-time PCR assay for simultaneous detection of *Acanthamoeba* spp., *Balamuthia mandrillaris*, and *Naegleria fowleri. J. Clin. Microbiol.* **44**: 3589–3595.

Rahimian, J., Kleinman, G. and Parta, M. (2005) Amebic encephalitis caused by *Balamuthia. Infect. Med.* **22**: 382–385.

Reed, R.P., Cooke-Yarborough, C.M., Jaquiery, A.L., Grimwood, K., Kemp, A.S., Su, J.C. and Forsyth, J.R.L. (1997) Fatal granulomatous amoebic encephalitis caused by *Balamuthia mandrillaris. Med. J. Aust.* **167**: 82–84.

Rideout, B.A., Gardiner, C.H., Stalis, I.H., Zuba, J.R., Hadfield, T. and Visvesvara, G.S. (1997) Fatal infections with *Balamuthia mandrillaris* (a free-living amoeba) in gorillas and other old world primates. *Vet. Pathol.* **34**: 15–22.

Rocha-Azevado, B., Jamerson, M., Cabral, G.A., Silva-Filho, F.C. and Marciano-Cabral, F. (2007) The interaction between the amoeba *Balamuthia mandrillaris* and extracellular glycoproteins *in vitro. Parasitology* **134**: 51–58.

Rowen, J.L., Doerr, C.A., Vogel, H. and Baker, C.J. (1995) *Balamuthia mandrillaris*: a newly recognized agent for amebic meningoencephalitis. *Pediatr. J. Infect. Dis.* **14**: 705–710.

Schumacher, D.J., Tien, R.D. and Lane, K. (1995) Neuroimaging findings in rare amebic infections of the central nervous system. *Am. J. Neuroradiol.* **16**: 930–935.

Schuster, F.L. (2002) Cultivation of pathogenic and opportunistic free-living amebas. *Clin. Microbiol. Rev.* **15**: 342–354.

Schuster, F.L. and Mandel, N. (1984) Phenothiazine compounds inhibit growth of pathogenic free-living amoebae. *Antimicrob. Agents Chemother.* **25**: 109–112.

Schuster, F.L. and Visvesvara, G.S. (1996) Axenic growth and drug sensitivity studies of *Balamuthia mandrillaris*, an agent of amebic meningoencephalitis in humans and other animals. *J. Clin. Microbiol.* **34**: 385–388.

Schuster, F.L. and Visvesvara, G.S. (2003) Amebic encephalitidies and amebic keratitis caused by pathogenic and opportunistic free-living amebas. *Curr. Treat. Opt. Infect. Dis.* **5**: 273–282.

Schuster, F.L. and Visvesvara, G.S. (2004a) Free-living amoebae as opportunistic and non-opportunistic pathogens of humans and animals. *Int. J. Parasitol.* **34**: 1001–1027.

Schuster, F.L. and Visvesvara, G.S. (2004b) Opportunistic amoebae: challenges in prophylaxis and treatment. *Drug Resist. Update* **7**: 41–51.

Schuster, F.L., Dunnebacke, T.H., Booton, G.C., *et al.* (2003) Environmental isolation of *Balamuthia mandrillaris* associated with a case of amebic encephalitis. *J. Clin. Microbiol.* **41**: 3175–3180.

Schuster, F.L., Glaser, C., Honarmand, S., Maguire, J.H. and Visvesvara, G.S. (2004) *Balamuthia* amebic encephalitis risk: Hispanic Americans. *Emerg. Infect. Dis.* **10**: 1510–1512.

Schuster, F.L., Guglielmo, B.J. and Visvesvara, G.S. (2006a) In-vitro activity of miltefosine and voriconazole on clinical isolates of free-living amoebas: *Balamuthia mandrillaris, Acanthamoeba* spp., and *Naegleria fowleri. J. Eukaryot. Microbiol.* **53**: 121–126.

Schuster, F.L., Honarmand, S., Visvesvara, G.S. and Glaser, C.A. (2006b) Detection of antibodies against free-living amoebae *Balamuthia mandrillaris* and *Acanthamoeba* species in a population of patients with encephalitis. *Clin. Infect. Dis.* **42**: 1260–1265.

Seas, C. and Bravo, F. (2006) Encefalitis amebiana granulomatosa por *Balamuthia mandrillaris*: una enfermedad fatal reconocida cada vez más frequentemente en América Latina. *Rev. Chil. Infect.* **23**: 197–199. (Editorial in Spanish).

Seidel, J.S., Harmatz, P., Visvesvara, G.S., Cohen, A., Edwards, J. and Turner, J. (1982) Successful treatment of primary amebic meningoencephalitis. *N. Engl. J. Med.* **306**: 346–348.

Shadrach, W.S., Radam, E., Flieger, A. and Kiderlen A.F. (2004) The pathogenic amoeba *Balamuthia mandrillaris* possesses cell-associated phospholipase A, lysophospholipase A, and lipase activities. *Int. J. Med. Microbiol.* **293**: 100. (Abstract).

Shadrach, W.S., Rydzewski, S., Laube, U., Holland, G., Özel, M., Kiderlen, A.F. and Flieger, A. (2005) *Balamuthia mandrillaris*, a free-living ameba and opportunistic agent of encephalitis, is a potential host for *Legionella pneumophila* bacteria. *Appl. Environ. Microbiol.* **71**: 2244–2249.

Shirabe, T., Monobe, Y. and Visvesvara, G.S. (2002) An autopsy case of amebic meningoencephalitis. The first Japanese case caused by *Balamuthia mandrillaris. Neuropathology* **22**: 213–217.

Stothard, D.R., Schroeder-Diedrich, J.M., Awwad, M.H., Gast, R.J., Ledee, D.R., Rodriguez-Zaragoza, S., Dean, C.L., Fuerst, P.A. and Byers, T.J. (1998) The evolutionary history of the genus *Acanthamoeba* and the identification of eight new 18S rRNA gene sequence types. *J. Eukaryot. Microbiol.* **45**: 45–54.

Taratuto, A.L., Monges, J., Acefe, J.C., Meli, F., Paredes, A. and Martinez, A.J. (1991) Leptomyxid amoeba encephalitis: report of the first case in Argentina. *Trans R. Soc. Trop. Med. Hyg.* **85**: 77.

Tata, P.S., Radam, E., Kayser, O. and Kiderlen, A.F. (2003) An *in vitro* assay for testing chemotherapeutic agents for inhibition of *Balamuthia mandrillaris* cytopathogencity. In: *Xth International Meeting on the Biology and Pathogenicity of Free-Living Amoebae Proceedings* (eds F. Lares-Villa, G.C. Booton and F. Marciano-Cabral). ITSON-DIEP, Ciudad Obregón, Mexico.

Tavares, M., da Costa, J.M.C., Carpenter, S.S., *et al.* (2006) Diagnosis of first case of *Balamuthia* amoebic encephalitis in Portugal by immunofluorescence and PCR. *J. Clin. Microbiol.* **44**: 2660–2663.

Uribe-Uribe, N.O., Bacerra-Lomelí, M., Alvarado-Cabrero, I., *et al.* (2001) Granulomatous amebic encephalitis by *Balamuthia mandrillaris. Patología* **39**: 141–148.

Valverde, J., Arrese, J.E. and Pierard, G.E. (2006) Granulomatoous cutaneous centrofacial and meningocerebral amebiasis. *Am. J. Dermatol.* **7**: 267–269.

Visvesvara, G.S., Martinez, A.J., Schuster, F.L., Leitch, G.J., Wallace, S.V., Sawyer, T.K. and Anderson, M. (1990) Leptomyxid ameba, a new agent of amebic meningoencephalitis in humans and animals. *J. Clin. Microbiol.* **28**: 2750–2756.

Visvesvara, G.S., Schuster, F.L. and Martinez, A.J. (1993) *Balamuthia mandrillaris*, N. G., N. sp., agent of amebic meningoencephalitis in humans and other animals. *J. Eukaryot. Microbiol.* **40**: 504–514.

Visvesvara, G.S., Moura, H. and Schuster, F.L. (2007) Pathogenic and opportunistic free-living amoebae: *Acanthamoeba* spp., *Balamuthia mandrillaris*, *Naegleria fowleri*, and *Sappinia diploidea*. *FEMS Immunol. Med. Microbiol.* **50**(1): 1–26.

Walochnik J., Duchene M., Seifert K., Obwaller A., Hottkowitz T., Wiedermann G., Eibl H. & Aspock H. (2002) Cytotoxic activities of alkylphosphocholines against clinical isolates of *Acanthamoeba* spp. *Antimicrob Agents Chemother* **46**(3): 695–701.

Yagi, S., Booton, G.C., Visvesvara, G.S. and Schuster, F.L. (2005) Detection of *Balamuthia* 16S rRNA gene DNA in clinical specimens by PCR. *J. Clin. Microbiol.* **43**: 3192–3197.

Zagardo, M.T., Castellani, R.J., Zoarski, G.H. and Bauserman, S.C. (1997) Granulomatous amebic encephalitis caused by leptomyxid amebae in an HIV-infected patient. *Am. J. Neuroradiol.* **18**: 903–908.

A3 *Naegleria fowleri*

Francine Marciano-Cabral
and Guy A. Cabral

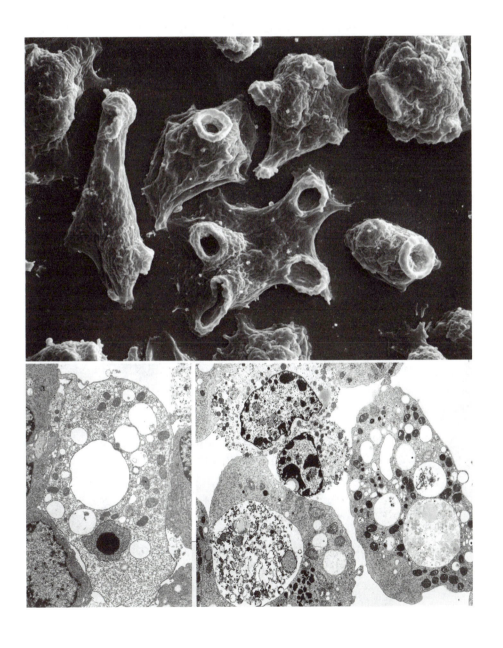

1 Introduction

The genus *Naegleria* consists of a group of free-living amoeboflagellates found in soil and freshwater habitats throughout the world. *Naegleria* have been termed 'amphizoic' to indicate that they can exist as free-living amoebae or as parasites (Page, 1974). Although pathogenic and nonpathogenic species of *Naegleria* have been isolated from environmental sources, of the more than 35 species that have been identified, only *N. fowleri* has been associated with human disease. Two other species, *N. australiensis* and *N. italica*, have been shown to be pathogenic for experimental animals but neither has been recovered from humans with disease (De Jonckheere, 2004). *N. lovaniensis*, a thermophilic species (i.e., species that grows at temperatures of 40°C or higher) that is antigenically similar to *N. fowleri*, is nonpathogenic in experimental animals (Stevens *et al.*, 1980).

N. fowleri can cause primary amoebic meningoencephalitis (PAM), a rapidly fatal disease of the central nervous system (CNS) that occurs generally in previously healthy children and young adults with a history of exposure to contaminated recreational, domestic, or environmental water sources (Butt *et al.*, 1968; Martinez, 1985). Human disease caused by this free-living amoeba was first reported in 1965 in four patients from South Australia who died from a CNS infection that was thought initially to be caused by a species of *Acanthamoeba* (Fowler and Carter, 1965). Subsequently, in the United States, Butt (1966) described three fatal CNS infections in Florida. In addition, Patras and Andujar (1966) reported a case from Texas and Callicott *et al.* (1968) described two cases from Virginia. Similar to the case reports from Australia, the victims had a history of swimming and diving prior to onset of the disease and it was proposed that infection was acquired by intranasal instillation of amoebae during swimming. A cluster of 16 fatal cases of PAM that occurred in Czechoslovakia (Czech Republic) from 1962 to 1965 also was associated with the affected individuals swimming in the same indoor pool (Cerva and Novak, 1968; Cerva *et al.*, 1968). A retrospective study conducted by dos Santos (1970) indicated that episodes of PAM also had occurred in Richmond, Virginia from 1951 to 1952.

In 1968, the first successful isolation and culture of *N. fowleri* from cerebrospinal fluid (CSF) and infected brain tissue were reported (Butt *et al.*, 1968; Callicott *et al.*, 1968; Carter, 1968). Shortly thereafter, Carter (1970) demonstrated that the agent was a previously undescribed species of *Naegleria*. The amoeba was named *N. fowleri* by Carter (1970) in honor of Malcom Fowler who had identified amoebic trophozoites in the olfactory bulbs and frontal lobes of patients who died from amoebic meningoencephalitis in Australia. Butt (1966) applied the term 'primary amebic meningoencephalitis' (PAM) to distinguish infection of the CNS in humans by free-living amoebae such as *N. fowleri* from infection of the brain by other amoebae. Following the description of this 'new disease', South

Australia became the focus of studies to determine the source of human infection. Mican and Cooter (Cooter, 2002) implicated the domestic water supply as the source of infection since all of the human cases recorded in Australia were from a specific locality to which water was piped a long distance in an aboveground pipeline. Anderson and Jamieson (1972) also isolated *N. fowleri* from pipeline water. Thereafter, sporadic cases of PAM were reported from Great Britain, Belgium, New Zealand, Panama, Brazil, Puerto Rico, and Nigeria (Fiordalisi *et al.*, 1992; Martinez, 1985). The majority of cases were those of individuals who had been swimming and diving in freshwater lakes and ponds. Thus, it became evident that *N. fowleri* could be acquired by swimming in pools that were not chlorinated adequately and from other water sources such as stream water heated by industrial discharge, freshwater lakes, ponds, stagnant pools, canals, and domestic water supplies (Schuster and Visvesvara, 2004a).

Aggressive control measures adopted by the Health Department and the Water Authority in Australia resulted in limiting the incidence of human infection and no new cases of PAM were reported in South Australia from 1985 to 1996 (Lugg, 1996). Control measures included a program of public education, closing of swimming pools containing amoeba or not containing appropriate levels of disinfectant, and the establishment of an 'amoeba alert system' based on monitoring water temperatures for exceeding a baseline set at 28°C. However, since the 1990s there has been an increase in cases worldwide (Taylor *et al.*, 1996) with reports of PAM in Mexico (Lares-Villa *et al.*, 1993), Italy (Cogo *et al.*, 2004), India (Hebbar *et al.*, 2005), New Zealand (Cursons *et al.*, 2003), Thailand (Siripanth, 2005; Wiwanitkit, 2004), Japan (De Jonckheere *et al.*, 1992; Sugita *et al.*, 1999), Venezuela (Cermeno *et al.*, 2006), Cuba (Cubero-Menendez and Cubero-Rego, 2004), Madagascar (Jaffar-Bandjee *et al.*, 2005), Hong Kong (Wang *et al.*, 1993), Africa (Schoeman *et al.*, 1993) and the United States (DeNapoli *et al.*, 1996; Gyori, 2003; Okuda *et al.*, 2004; Taylor *et al.*, 1996). In this context, an increased number of untreated water outbreaks caused by *N. fowleri* have been reported in recent years (Craun *et al.*, 2005; Dingley, 1996; Marciano-Cabral *et al.*, 2003).

2 Morphology of *Naegleria*

There are three morphological forms of *Naegleria*: a feeding trophozoite, a dormant cyst, and a transient swimming flagellate (Figure 1). In the environment, *Naegleria* spp. feed on bacteria (Figure 2) and yeast. Under conditions of nutrient deprivation, the trophozoite undergoes a transitory transformation to a flagellate stage, 'swims' to the water surface to seek a food source, and then reverts to the amoeboid form whereupon it feeds on bacteria (Preston and King, 2003). The trophozoite exhibits food-cups or amoebastomes, cytoplasmic extensions of the surface used to ingest bacteria or yeast in the environment or erythrocytes and brain tissue in the infected host (Figure 3A). Additionally *N. fowleri* produce pore-forming

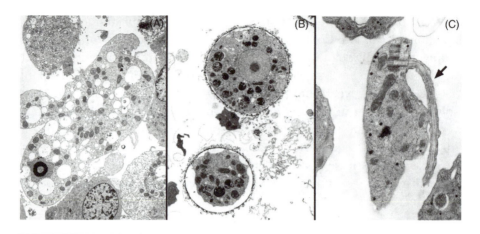

Figure 1 Transmission electron micrographs depicting *Naegleria fowleri* in its three states of transformation: (A) trophozoite, (B) cyst, and (C) flagellate. The arrow designates the flagellum.

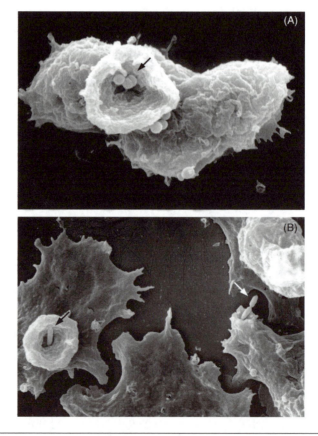

Figure 2 Scanning electron micrographs depicting *N. fowleri* trophozoites ingesting bacteria through 'food cups'. (A) *Staphylococcus aureus* (*arrow*). (B) *Escherichia coli* (*arrows*).

polypeptides, termed naegleriapores, that destroy ingested bacteria by permeabilizing the bacterial cell membrane (Leippe and Herbst, 2004) and lyse nerve cells and other mammalian cells on contact (Figure 3C; Herbst *et al.*, 2002, 2004). At the ultrastructural level, *Naegleria* trophozoites exhibit typical features of a eukaryotic cell with the exception that centrioles have yet to be observed (Fulton and Dingle, 1971). The amoeba is surrounded by a unit membrane and the cytoplasm contains large numbers of ribosomes both free and attached to the endoplasmic reticulum. A smooth endoplasmic reticulum is also present. A typical Golgi apparatus has not been observed but a primitive Golgi-like complex, made up of membranous components and coated vesicles, has been identified (Stevens *et al.*, 1980).

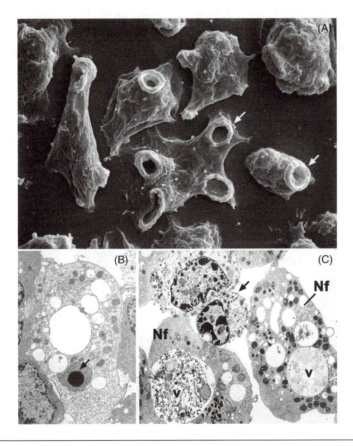

Figure 3 Electron micrographs illustrating typical features of the *N. fowleri* trophozoite. (A) Scanning electron micrograph depicting the presence of 'food cups' (*arrows*) extending from the amoebae. (B) Transmission electron micrograph of trophozoite depicting the prominent nucleus with a centrally located large nucleolus (*arrow*) that is typical of *N. fowleri*. (C) *N. fowleri* trophozoites (Nf) in contact with a target cell that has been lysed (*arrow*). Note the presence of cellular debris within vacuoles (v) of *N. fowleri*.

The cytoplasm is filled with membrane-bound food vacuoles, dumbbell-shaped or club-shaped mitochondria, and primary and secondary lysosomes (Feldman, 1977). A contractile vacuole also has been identified in the tail region of *Naegleria* (Vickerman, 1962). The nucleus of *Naegleria* is prominent and contains a large nucleolus (Figure 3B). The flagellate stage is transient and differentiation into this stage involves a change in cell shape and a change in synthesis of all organelles of the flagellum apparatus. During this process, cytoplasmic vacuoles decrease and basal bodies, a rootlet, and flagella are formed. Flagella exhibit the typical (9 + 2) arrangement of filaments surrounded by a sheath that is continuous with the cytoplasmic membrane. The rootlet is connected to the basal bodies by an intricate series of parallel and transverse microtubules. The mature flagellar apparatus usually consists of two flagella, two basal bodies, microtubules and a single striated rootlet (Fulton, 1983; Patterson *et al.*, 1981). The cyst of *N. fowleri* affords protection from adverse environmental conditions and consists of a single-walled structure with pores sealed with a mucoid plug (Figure 1B). Factors that induce cyst formation include food deprivation, crowding, desiccation, accumulation of waste products, exposure to toxic products of bacteria, pH changes, and salts (Marciano-Cabral, 1988).

3 Pathogenicity and Virulence Factors

Factors that account for free-living amoeba expression of pathogenic characteristics are unknown. It remains to be determined whether amoebae that are found in water that is enriched in certain bacterial species, pollutants, or specific elements such as iron are more pathogenic. Compromised specific and nonspecific host resistance resulting from deficiencies in secretory IgA or complement components may account, at least in part, for disease manifestations. Temperature tolerance is not a factor in pathogenicity since *N. lovaniensis* is thermotolerant but nonpathogenic. Virulence of *N. fowleri* has been shown to decline when the amoebae are grown axenically in media and to increase when the amoebae are passaged through experimental animals (Whiteman and Marciano-Cabral, 1989; Wong *et al.*, 1977). *In vitro*, *N. fowleri* destroy nerve cells and other mammalian cells by piecemeal ingestion, a process termed trogocytosis (Brown, 1979; Marciano-Cabral and John, 1983). *N. fowleri* also destroy mammalian cells by producing lytic substances upon contact with a target cell (Figure 3C; Marciano-Cabral *et al.*, 1982). Several investigators have suggested that *N. fowleri* releases cytolytic substances that account for invasiveness and tissue damage *in vivo* and cytopathogenicity *in vitro*. It has been proposed that the release of phospholipolytic enzymes by *N. fowleri* results in the rapid destruction of brain tissue (Chang, 1979). Cytotoxic enzymes such as phospholipase A, phospholipase C and proteases have been detected in the aqueous soluble cytolytic fraction derived from *N. fowleri* (Barbour and Marciano-Cabral, 2001; Chang, 1979; Cursons *et al.*, 1978; Fulford *et al.*, 1985; Mat Amin, 2004).

In addition, sphingomyelinase and lysophospholipase activities have been detected in the amoeba. The ability of these phospholipases to degrade phospholipids of human myelin has been reported (Hysmith and Franson, 1982). Hemolytic activity associated with the amoeba surface membrane also has been reported (Lowrey and McLaughlin, 1984). However, designation of these phospholipases and other enzymes as unique virulence factors *in vivo* has yet to be established.

Two major *in vitro* differences between pathogenic and nonpathogenic *Naegleria* species are the ability of pathogenic species to escape complement lysis and their ability to respond chemotactically to nerve cells and nerve cell factors. Pathogenic *N. fowleri* move in a directional manner toward nerve cell components whereas nonpathogenic *Naegleria* exhibit random migration (Brinkley and Marciano-Cabral, 1992). Antibodies to human CD59, a complement regulatory protein that protects mammalian cells from lysis by complement, reportedly react with a membrane protein of *N. fowleri* (Fritzinger *et al.*, 2006). It has been suggested that a 'CD59-like protein' on the surface of *N. fowleri* protects the amoeba from lysis by complement. Whether the amoeba membrane protein is homologous to mammalian CD59 remains to be determined.

4 Ecology and Methods of Detection from Environmental Sources

N. fowleri have been isolated from a variety of habitats including soil, freshwater lakes, ponds, thermal springs, air, and humidifier systems (Marciano-Cabral, 1988). However, moist soil is the preferred habitat of *Naegleria* with heavy rainfall washing the amoebae into nearby water sources (Kyle and Noblet, 1985; Singh and Dutta, 1984). Pathogenic *Naegleria*, as thermophilic amoebae, can be isolated from water at temperatures of 40°C and higher. Apparently, localized 'hot spots' of water serve as a propagation source. The propagation of *Naegleria* may be enhanced in these hot spots since high water temperatures eliminate nonthermophilic competitors (Brown *et al.*, 1983; Huizinga and McLaughlin, 1990; Tyndall *et al.*, 1989). In addition, it has been reported that iron in the environment has a positive effect on the growth of *N. fowleri* while copper inhibits its growth (Newsome and Wilhelm, 1983). Consistent with these observations, concentrations of *N. fowleri* have been shown to be almost 60 times higher in cooling systems that are equipped with stainless steel condensers as compared to those equipped with brass condensers due to the release of copper from brass into the water (Pernin and Palendakis, 2001).

Since *N. fowleri* is considered primarily a waterborne pathogen (Marshall *et al.*, 1997; Schuster and Visvesvara, 2004a) a variety of techniques and rapid methods have been developed for its detection and isolation from an aqueous environment. Samples (50 ml) in a sterile centrifuge tube can be obtained from surface water since amoebae are found in greater abundance at that location as they feed on bacteria (Preston and King, 2003). Sediment also can serve as a source of amoebae. Swab samples can be obtained by wiping the surface of rocks or floating objects with sterile cotton-tipped applicators. After collection, samples should be processed as quickly as possible. Water samples can be concentrated by centrifugation or filtration, with centrifugation yielding a higher recovery of trophozoites (Pernin *et al.*, 1998). The resultant pellet can be transferred onto 1.5% non-nutrient agar (NNA) that has been seeded with a lawn of *Escherichia coli* and maintained at 42°C for 48 h to obtain thermotolerant amoebae. Swab samples can be rolled directly over the agar using a cotton-tipped applicator. Amoebae that grow on the agar can be subcultured to new NNA-*E. coli* plates to avoid overgrowth by fungi. Amoebae then can be transferred from agar plates to liquid medium containing 200 U ml^{-1} penicillin/streptomycin (200 ug ml^{-1}; John and Howard, 1996). Transfer of amoebae from liquid growth medium into water can result in their transformation to flagellates.

Several tests are available for the identification of *Naegleria*. An enflagellation assay has been used to identify amoebae as *Naegleria*. However, enflagellation does not distinguish pathogenic species of *Naegleria* from nonpathogenic species. Other thermotolerant amoebae such as nonpathogenic *N. lovaniensis* also are generally present in the same environment as *N. fowleri* (Stevens *et al.*, 1980). In addition to an enflagellation assay, a mouse pathogenicity test can be performed by instilling 10–20 µl medium containing amoebae (10^4 to 10^5) into the nasal passages of 3- to 4-week-old mice. Following instillation, mice are observed for 28 to 30 days and brains from sick or dying mice can be cultured to recover amoebae (Martinez, 1985). However, a potential confound in the test is that it does not distinguish *N. australiensis* from *N. fowleri*. *N. australiensis* is not associated with human disease, but is pathogenic for mice.

For identification of *Naegleria* at the species level, enzyme-linked immunosorbent assay (ELISA) as well as polymerase chain reaction (PCR) can be performed. These two assay methods have been recommended for the identification of *N. fowleri* because the flagellation test, which is routinely used, may result in false negatives since under laboratory conditions not all *Naegleria* enflagellate (Behets *et al.*, 2003). The ELISA method that has been developed can identify *N. fowleri* in both clinical and environmental samples (Reveiller *et al.*, 2003). This immunoassay is based on the use of a monoclonal antibody (5D12) that recognizes a repeated glycosylated epitope present on proteins of *N. fowleri* (Reveiller *et al.*, 2000; Sparagano *et al.*, 1993, 1994). Various PCR assays also have been developed to identify *N. fowleri* isolated from environmental sources (Behets *et al.*, 2003; Kilvington and Beeching, 1995; McLaughlin *et al.*, 1991; Pelandakis *et al.*, 2000; Reveiller *et al.*, 2002; Figure 4). Kilvington and Beeching (1995) reported that as few as one trophozoite and one cyst could be detected

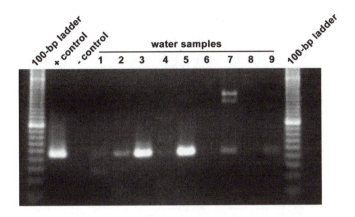

Figure 4 Ethidium-bromide-stained 1.5% agarose gel showing nested PCR amplification product from water samples using *N. fowleri*-specific primers. The assay was performed as described by Reveiller *et al.* (2002). DNA from clone Mp2cl5 served as a positive (+) control. The negative control consisted of PCR-grade water lacking DNA. Environmental water samples 2, 3, 5, 7, and 9 obtained from Virginia, Connecticut and Arizona are positive for *N. fowleri*.

from *N. fowleri* cultured from environmental samples after 45 cycles of amplification. Pelandakis and Pernin (2002), using sequences of the internal transcribed spacers (ITS) and the 5.8S ribosomal gene of *N. fowleri*, designed species-specific ITS region primers for *N. fowleri* genomic DNA. Using these primers in a PCR procedure applied to environmental isolates they were able to distinguish *N. fowleri* from other isolates and to explore the genetic diversity within the species. The ITS PCR protocol has been modified for use in a real-time PCR assay based on melting curve analysis (MCA; Behets *et al.*, 2006). The MCA of PCR products allowed for differentiation and quantification of several targeted free-living amoebae (FLA) in a single PCR assay. A triplex real-time PCR assay using specific primers for three FLA also has been developed based on the nuclear small subunit ribosomal (18S rRNA) gene (Qvarnstrom *et al.*, 2006). This assay was designed to identify *Naegleria*, *Acanthamoeba*, and *Balamuthia* within the same sample. However, the individual primer sets included in the assay can be used separately for selective identification of *Balamuthia*, *Acanthamoeba* spp., or *Naegleria*. A cDNA clone designated Mp2Cl5 derived from *N. fowleri* also has been used to design a nested PCR assay (Reveiller *et al.*, 2002). This assay was designed with two distinctive sets of primers (Table 1) to allow for discrimination of *N. fowleri* from the closely related *N. lovaniensis* and other species of amoebae that may be present in environmental samples. PCR amplification can be performed directly on a water sample without prior genomic DNA (gDNA) extraction. As few as five amoebae can be detected from a volume of 50 ml water (Figure 4). In addition, a quantitative single duplex PCR (qPCR) assay based on the

Mp2CL5 gene has been designed for amplification and quantitation of *N. fowleri* DNA from water samples. Using fluorescent Taqman technology, the qPCR was found to be 100% specific for *N. fowleri* and to have a detection limit of three cell equivalents (1.1 pg DNA; Behets *et al.*, 2007). A list of PCR primers that have been used to detect *N. fowleri* is shown in Table 1.

Table 1 Primers used for detection of *Naegleria fowleri* for diagnosis of PAM or for identification in environmental samples

Primers	Use	Reference
FP 5'-CGTATCTAGTAGATAGAACA-3' RP 5'-CGTAACGACACAAACCTACAGA-3' RP 5'-AACAAGTAGCCCACCATAC-3'	Diagnostic (internal reverse)	McLaughlin *et al.*, 1991
FP 5'-GCTATCGAATGGATTCAAGC-3' RP 5'-CACTACTCGTGGAAGGCTTA-3'	Environmental	Kilvington and Beeching, 1995
FP 5'-GAACCTGCGTAGGGATCATTT-3' RP 5'-TTTCTTTTCCTCCCCTTATTA-3' FP Fw1 5'–GTGAAAACCTTTTTTCCATTTACA-3' RP Fw2 5'-AAATAAAAGATTGACCATTTGAAA-3'	Environmental (ITS PCR) (all *Naegleria* spp. detected) Environmental/ diagnostic specific (*N. fowleri*)	Pelandakis *et al.*, 2000 Pelandakis and Pernin, 2002
ITS PCR – Real-time assay with melting curve analysis	Environmental	Behets *et al.*, 2006
F1 5'-TCTAGAGATCCAACCAATGG-3' R1 5'-ATTCTATTCACTCCACAATCC-3' F2 5'-GTACATTGTTTTTATTAATTTCC-3' R2 5'-GTCTTTGTGAAAACATCACC-3'	Environmental Nested PCR (MP2CL5 gene)	Reveiller *et al.*, 2002
FP 5'-AACCTGCGTAGGGATCATTT-3' RP 5'-TTTCCTCCCCTTATTAATAT-3'	Environmental consensus primers	Robinson *et al.*, 2006
F 3'-GTGCTGAAACCTAGCTATTGTAAC TCAGT-5' R 5'-CACTAGAAAAAGCAAACCTGAAAGG-3' HEX-AT AGCAATATA TTCAGGGGAGCTGGGC	Diagnostic Environmental (triplex real-time)	Qvarnstrom *et al.*, 2006
qPCR based on MP2CL5 gene (GenBank AY 049749) Taqman MGBprobe FP 5'-TGGAGAGAATCAGGAGGCAAA-3' RP 5'-TCTTGAGTCCAGGTGAAATGATGT-3' FAM probe 5'-TCTGGCACTGCACTC-3'	Environmental	Behets *et al.*, 2007

5 Clinical Presentation of Primary Amoebic Meningoencephalitis

The incubation period of PAM in humans is generally 3 to 10 days. In the early stages patients may present with headache, fever, and vomiting. The disease resembles purulent bacterial meningitis leading often to a misdiagnosis. Failure of antibiotic treatment and of identification of bacteria in CSF cultures suggests a fungal or viral etiology and, thus, amoebic meningoencephalitis generally is not considered. However, the appearance of motile cells in fresh wet mounts of CSF suggests amoebic encephalitis and the presence of amoebae can be confirmed upon brain biopsy (Figure 5). The characteristics of PAM caused by *N. fowleri* are sufficiently different from those of granulomatous amoebic encephalitis (GAE) that is caused by two other free-living amoebae, *Acanthamoeba* and *Balamuthia* (Figure 6). PAM is an acute, fulminant, and rapidly fatal infection of the brain whereas GAE may be chronic. Clinical manifestations of PAM include sudden onset of severe bilateral headache, followed by nausea, vomiting, fever, stiff neck, diplopia, loss of taste and smell, seizures and coma (Barnett *et al.*, 1996; Carter, 1968; Martinez, 1985). Laboratory findings indicate elevated white blood cells (WBC) and analysis of CSF reveals pleocytosis with elevated WBCs that are predominately polymorphonuclear leucocytes, and elevated protein. CSF is sometimes purulent (Carter, 1972; Taylor *et al.*, 1996). The portal of entry of *Naegleria* is through the nasal passages. The amoebae attach to the nasal mucosa, follow the olfactory nerves and migrate through the cribriform plate to the brain (Martinez, 1985). Once in the brain, amoebae disseminate throughout that compartment and death ensues from 7 to 10 days post infection. The factors that are responsible for the development of PAM are unknown. Strain differences in the amoebae,

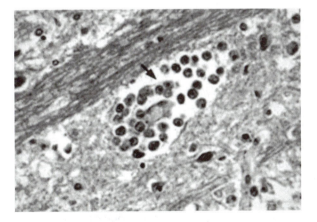

Figure 5 Light micrograph of human brain tissue section depicting the presence of *N. fowleri* trophozoites (*arrow*). The paraffinized section was stained with hematoxylin and eosin (section courtesy of Dr. A. J. Martinez).

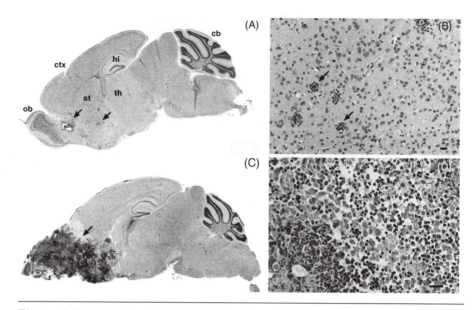

Figure 6 Light micrographs of $(B_6C_3)F_1$ mouse brain tissue illustrating comparison between granulomatous amoebic encephalitis (GAE) caused by *Acanthamoeba culbertsoni* and primary amoebic meningoencephalitis (PAM) caused by *Naegleria fowleri*. (A) Sagittal section illustrating focal areas of granulomas (*arrows*) proximal to the olfactory bulb (ob). (B) Higher magnification of tissue section denoting the focal areas of granulomas (*arrows*). (C) Sagittal section illustrating massive lesion in frontal portion of brain (*arrow*) as a result of infection with *N. fowleri*. (D) Higher magnification illustrating massive presence of amoeba trophozoites intermixed with erythrocytes, neutrophils, and mononuclear cells. All sections were stained with hematoxylin and eosin. cb, cerebellum; ctx, cerebral cortex; hi, hippocampus; ob, olfactory bulb; st, striatum; th, thalamus.

recent passage through an animal host in nature, presence of intracellular bacteria (Figure 7) or the immune status of the host are factors that may contribute to virulence (Marciano-Cabral, 1988). PAM is almost always fatal; however, there have been a few documented cases of survivors due to prompt diagnosis and early initiation of therapy (Table 3). Combined administration of amphotericin B intravenously, amphotericin B intrathecally and rifampin orally appears to improve survival rate (Brown, 1991; Seidel *et al.*, 1982).

6 Diagnostic Techniques

The majority of cases of PAM has been diagnosed post mortem by hematoxylin and eosin staining of brain tissue. Since disease progression is rapid, serological assays are not helpful as there is little possibility of measuring a rise in antibody titer. Also, computed tomography (CT) and magnetic resonance imaging (MRI) of the brain performed in response to symptoms of CNS

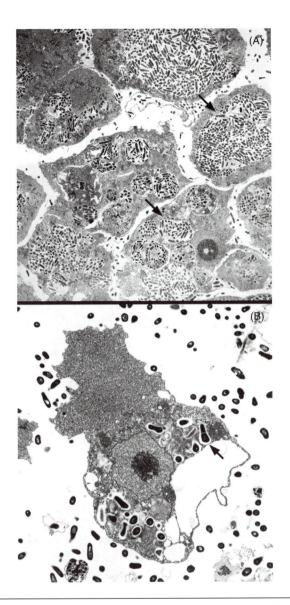

Figure 7 Transmission electron micrographs depicting uptake of bacteria by *N. fowleri*. (A) Trophozoites co-cultured with *Legionella pneumophila* harbor numerous large vesicular inclusions (*arrows*) replete with bacteria. (B) Trophozoites co-cultured with *Helicobacter pylori* contain intracytoplasmic bacteria that are surrounded by a 'halo' (*arrow*). Note that a cytoplasmic membrane does not circumscribe the bacteria.

infection and severe headache do not always indicate abnormalities. CT findings are typically nonspecific and may appear normal early in the infection process. However, CT may show evidence of brain edema and hydrocephalus (Kidney and Kim, 1998; Shumaker *et al.*, 1995) and of obliteration of cisterns with enhancing basilar exudates (Singh *et al.*, 2006). Martinez and Visvesvara

(1997) recommended that direct microscopic examination of CSF as a wet-mount preparation be performed. Currently, identification of motile trophozoites in CSF after lumbar puncture followed by immunofluorescent staining is the method of choice under conditions where CSF pressure is low and lumbar puncture can be performed. Trophozoites can be fixed with acetone, incubated with monoclonal or polyclonal anti-*N. fowleri* antibody and examined by indirect immunofluorescence (Hara and Fukuma, 2005). In addition, following identification of motile trophozoites in CSF, smears can be prepared, fixed with Schaudinn's fixative and stained with iron-hematoxylin or Giemsa. Alternatively, CSF cytospin preparations can be subjected to Wright's stain and examined by light microscopy (Benson *et al.*, 1985).

More recently, molecular techniques such as PCR and real-time PCR have been developed that allow for a more rapid, sensitive, and specific laboratory diagnosis (Hara and Fukuma, 2005; Table 1). PCR assays that have been developed for detection of *N. fowleri* in environmental samples have been adapted for laboratory diagnosis of clinical samples. In this context, amplification of repetitive DNA has been used for the identification of *N. fowleri* from purified nucleic acid extracted from infected mouse brain tissue (McLaughlin *et al.*, 1991). The specificity of this test, however, was dependent on the application of stringent hybridization conditions. Ribosomal ITS also have been reported as useful markers for specific detection of *N. fowleri*. Recently, a triplex real-time PCR assay based on the use of probes targeting regions of the nuclear small subunit ribosomal gene (18S rRNA gene) has been developed for detection of *Naegleria*, *Balamuthia*, and *Acanthamoeba* in the same sample (Qvarnstrom *et al.*, 2006). This assay has been evaluated using 22 well-characterized amoeba strains and nine clinical specimens that were characterized previously by *in vitro* culture and by immunofluorescence. Results were obtained within 5 h of receipt of the specimen and the detection limit was determined as one amoeba per sample. These results suggest that the triplex assay may be very useful in the diagnostic laboratory for rapid diagnosis of PAM or GAE as well as for identification of the three amoebae species in samples.

Recent reports indicate also that PCR can be applied to the detection of *Naegleria* in paraffinized tissues. This capability affords the opportunity to examine retrospectively neuropathology samples that could yield insight into the incidence of PAM in humans. Using this approach, Schild *et al.* (2007) were able to detect *N. fowleri* in fresh brain tissue as well as formalin-fixed paraffin embedded brain tissue.

7 Treatment and Therapeutic Agents

The mortality rate for PAM is greater than 95% (Barnett *et al.*, 1996). Major obstacles to effective therapy are the rapid progression of the disease and the paucity of drugs that have the ability to cross the blood-brain barrier (Schuster and Visvesvara, 2004b). Nevertheless, there have been documented recoveries from PAM (Seidel *et al.*, 1982; Wang *et al.*, 1993). Early recognition and treatment of the disease appear to be key elements in successful outcomes. To date, the drug of choice for treatment of human cases has been amphotericin B in combination with rifampin and other antifungal agents (Seidel *et al.*, 1982). Indeed, patients who have recovered from PAM have been treated with amphotericin B either alone or in combination (Anderson and Jamieson, 1972; Parija and Jayakeerthee, 1999; Singh *et al.*, 1998). Treatment with amphotericin B and fluconazole intravenously followed by oral administration of rifampicin led to successful treatment of a 10-year-old child who developed PAM (Vargas-Zepeda *et al.*, 2005). Optimal therapy for PAM, however, has yet to be developed since not all patients treated with amphotericin B survive (Cursons *et al.*, 2003; Shenoy *et al.*, 2002; Shrestha *et al.*, 2003; Stevens *et al.*, 1981). Poungvarin and Jariya (1991) suggested that a triple combination of low dose amphotericin B administered intravenously for 14 days with oral rifampacin and oral ketoconazole for 1 month would result in a more favorable outcome than when a high dose of amphotericin B was administered intrathecally. A summary of various approaches that have been applied to the treatment of PAM in humans is shown in Table 2.

In view of the limited armatorium of drugs available for treatment of human cases of PAM, a number of studies to assess the efficacy of therapeutic agents has been conducted *in vitro* and *in vivo* (Table 3). For *in vivo* studies the mouse model of PAM (Carter, 1969; Martinez *et al.*, 1971) has been used most extensively. However, use of this animal model has translational limitations to the human. For example, due to the faster rate of metabolism in the mouse, one may not obtain a true indication of whether the drugs that are effective in the mouse will also be effective in the human. Thong *et al.* (1979) treated PAM in Balb/c mice with a combination of amphotericin B and rifamycin. Rifamycin alone was found to be ineffective. However, a synergistic effect was observed when rifamycin was used in combination with amphotericin B resulting in increased survival in mice. Gupta *et al.* (1998) evaluated the activity of artemisinin, a naturally occurring sesquiterpene lactone, that is isolated from the Chinese herb *Artemisia annua*, and its derivative αβ-arteether, against *N. fowleri* in Swiss mice. Although these drugs cross the blood-brain barrier, neither was found to be effective against amoebae in experimental animals. The neuroleptics trifluoperazine and chlorpromazine, the antimycotics amphotericin B, ketoconazole and miconazole, and the antibiotics rifampicin, pentamidine, mepacrine and metronidazole also have been examined *in vitro* for their effect on trophozoites of *N. fowleri* (Ondarza *et al.*, 2006). It was determined that the most growth-inhibitory drugs were the antimycotics amphotericin B and ketoconazole and the neuroleptic antipsychotic drug trifluoperazine. In more recent studies, the capacity of these drugs to inhibit the NADPH-dependent disulfide reducing enzymes cysteine reductase and

Table 2 Successful treatment of primary amoebic meningoencephalitis in humans

Treatment	Dose[a]	Reference
Amp B	1 mg kg^{-1} day^{-1} IV for 5 days followed by 0.1 mg IT/IV on alternate days	Anderson and Jamieson, 1972 Carter, 1972
Amp B	1.5 mg kg $^{-1}$ 12-h^{-1} IV (3 days) 1.5 mg day^{-1} IT	Seidel et al., 1982
Miconazole	350 mg m^{-2} 8-h^{-1} IV (9 days) 10 mg day^{-1} IT (8 days)	
Rifampin	10 mg kg^{-1} 8-h^{-1} PO	
Sulfisoxazole	1 g 6-h^{-1} IV (3 days)	
Amp B/	75 mg day^{-1} initial IV	Brown, 1991
Rifampin	600 mg 12-h^{-1} (nasogastric tube)	
Amp B	0.1 mg alternate days IT (10 days)	
Dexamethasone	0.25 mg with 0.4 ml dextrose & water	
Amp B	0.5 mg kg^{-1} day^{-1} for 14 days IV	Poungvarin and
Rifampin	600 mg day^{-1} PO for 1 month	Jariya, 1991
Ketoconazole	800 mg day^{-1} PO for 1 month	
Amp B	60 mg day^{-1}	Wang et al., 1993
Rifampin	450 mg day^{-1}	
Chloramphenicol	1 g day^{-1}	
Amp B	10–50 mg IV 0.012–0.10 mg IT for 40 days	Loschiavo et al., 1993
Amp B	1 mg kg^{-1} day^{-1} IV for 40 days	Singh et al., 1998
Rifampin	10 mg kg^{-1} day^{-1} PO	
Amp B	1 mg kg^{-1} day^{-1}	Jain et al., 2002
Rifampin	450 mg day^{-1} PO	
Ornidazole	500 mg 8 h^{-1} for 3 weeks	
Amp B	0.25 mg kg^{-1} day^{-1} IV 1 mg kg^{-1} day^{-1} IV	Vargas-Zepeda et al., 2005
Fluconazole	10 mg kg^{-1} day^{-1} IV	
Rifampin	10 mg kg^{-1} day^{-1} IV	
Dexamethasone	0.6 mg kg^{-1} day^{-1} IV	
Ceftriaxone	100 mg kg^{-1} day^{-1} IV	

[a]IV, intravenous; IT, intrathecal; PO, per oral.

Table 3 Effect of treatment *in vivo* for PAM in experimental animals with various drugs

Drug	Dose	%Protection	Reference
Amp B[a]	7.5 mg kg^{-1} IP	60	Carter, 1969, 1972
Metronidazole	1.0 g kg^{-1} 10 days		
Rifamycin^{+}	150 mg kg^{-1} day^{-1}	40	Thong *et al.*, 1979
Amp B	2.5 mg day^{-1} for 10 days		
Artemisinin	120 mg kg^{-1}	0	Gupta *et al.*, 1998
Amp B	2.5 mg kg^{-1} for 5 days	100	
Amp B[b]	7.5 mg kg^{-1}	50	Goswick and Brenner, 2003a
Azithromycin	75 mg kg^{-1} day^{-1}[e]	100	
Liposomal AMB[c]	25 mg kg^{-1}	30	Goswick and Brenner, 2003b
Ketoconazole	25 mg kg^{-1}	30	
Quinupristin	150 mg kg^{-1}	50	
Minocycline	75 mg kg^{-1}	10	
Trifluoperazine	7.5 mg kg^{-1}	20	
Amp B in combination with azithromycin[d]	2.5 mg kg^{-1} 25 mg kg^{-1}	100	Soltow and Brenner, 2007

[a]Amp B: Amphotericin B aqueous solution (Sigma-Aldrich, Inc., St. Louis, MO).
[b]Amphotericin liposome for injection (AmBisome, Fujisawa Healthcare, Deerfield, IL).
[c]Per day: treatment for 5 days.
[d]Azithromycin (Zithromax, Pfizer, New York).
[e]Per day: treatment for 5 days.

glutathione reductase, and the trypanothione/trypanothione reductase system of *N. fowleri*, was investigated (Ondarza *et al.*, 2007). The rationale for these studies was that thiol compounds, such as glutathione and its reducing enzyme glutathione reductase, provide protection against oxygen toxicity and, in this capacity, have been suggested as therapeutic agents against trypanosomes. The antimycotics amphotericin B (34.6 µM) and miconazole (64 µM) were found to be the most potent inhibitors of the trypanothione/trypanothine reductase system of *N. fowleri*. These were followed in order by the neuroleptics trifluoperazine (65.5 µM) and chlorpromazine (89.6 µM) and by the antibiotic mepacrine (67.2 µM). It was proposed that combined use of drugs such as amphotericin B, miconazole, trifluoperazine, chlorpromazine and mepacrine that strongly inhibit disulfide-reducing activities may be an effective mode for treatment of *N. fowleri* infection (Ondarza *et al.*, 2007). The activity of azithromycin and amphotericin B against *N. fowleri* also has been studied *in vitro* and *in vivo* (Goswick and Brenner, 2003a, 2003b; Soltow and Brenner, 2007). Azithromycin alone was found to protect 40% of mice tested, while amphotericin B alone protected 27%. A combination of amphotericin B (2.5 mg kg^{-1}) and azithromycin (25 mg kg^{-1}) protected 100% of mice infected with *N. fowleri*. In summary, results of therapeutic approaches applied to animal models have differed. The use of various strains of *N. fowleri* or various strains of mice may account for differences obtained by various investigators.

8 Host-Parasite Interactions

A number of animal models including mice (Dempe *et al.*, 1982; Martinez *et al.*, 1971, 1973), monkeys (Wong *et al.*, 1975), guinea pigs (Diffley *et al.*, 1976), sheep (Simpson *et al.*, 1982), and rabbits (Smego and Durack, 1984), have been developed to study PAM. The mouse model has been used most extensively since it resembles the disease in humans and the immune system of the mouse is well-characterized. The nasal passages involving attachment to the olfactory epithelium can be used as the portal of entry by *N. fowleri* to mimic natural exposure in humans. Furthermore, migration via the olfactory nerves across the cribriform plate to the brain is similar in mouse and humans (Martinez *et al.*, 1973). Finally, mice infected intranasally develop a fatal disease resembling PAM in humans (Carter, 1970; Martinez, 1985). To produce PAM in the mouse, typically, a trophozoite suspension (10:1 containing 10^4 to 10^5 amoebae), in water or medium, can be instilled into the nasal passages using an Eppendorf pipette. Mice display symptoms which include ruffled fur, arched posture, loss of appetite, and a loss of equilibrium, within 4 to 5 days and up to 21 days post inoculation depending on the inoculum size and the virulence of the strain of amoebae. *N. fowleri* strains of low virulence can cause subacute or chronic encephalitis while highly virulent strains can cause death within 4 days post intranasal instillation (Dempe *et al.*, 1982; Whiteman and Marciano-Cabral, 1989). Following presentation of symptoms, amoebae can be isolated from the brain and occasionally from the lungs. The histologic picture of the brain indicates an acute inflammatory response associated with hemorrhage and edema (Figure 6), disintegration of neural structures and dissemination throughout the brain (Martinez *et al.*, 1971).

9 Immunity

The immune response to *N. fowleri* has been studied in experimental animals as well as in humans. Surveys for antibodies to *N. fowleri* in human sera have been conducted on subjects from Australia, New Zealand (Cursons *et al.*, 1980a), the United States (Marciano-Cabral *et al.*, 1987; Reilly *et al.*, 1983a) and the Czech Republic (Cerva, 1989). Antibodies have been identified using indirect immunofluorescence assays, agglutination assays, Western immunoblot analysis, ELISA and radial immunodiffusion. Antibodies have been found to be present in apparently healthy individuals (Marciano-Cabral *et al.*, 1987; Reilly *et al.*, 1983a), individuals with acute respiratory disease (Powell *et al.*, 1994), hospitalized patients (Dubray *et al.*, 1987), and patients with PAM (Cursons *et al.*, 1977; Seidel *et al.*, 1982). Although titers for these anti-*N. fowleri* antibodies have differed from study to study, they illustrate that almost all human sera tested are positive, indicating that exposure is common and widespread. Agglutinating activity in human serum from healthy individuals is due to production of IgM antibodies to surface antigens of *Naegleria* while IgG nonagglutinating antibodies reflect reactivity to internal antigens which are more abundant and show extensive cross-reactivity among species within the genus (Marciano-Cabral *et al.*, 1987; Reilly *et al.*, 1983a). Powell *et al.* (1994) examined serum samples from army recruits who suffered from acute respiratory disease for antibodies to free-living amoebae. Antibodies to six different species of free-living amoebae, including *N. fowleri*, were detected by immunoblot analysis (Powell *et al.*, 1994). Cursons *et al.* (1979) and Cain *et al.* (1979) obtained serum from two patients with fatal PAM prior to their death and noted, using an indirect immunofluorescence assay (IIF), that there was no elevation in specific antibody titer to *N. fowleri*. Notably, one of these patients had a history of recurring respiratory infections (Cursons *et al.*, 1980a). Although levels of serum IgA were measured rather than those for mucosal IgA, Cursons *et al.* (1977) suggested that low resistance to infections at the mucosal surface correlated with low levels of serum IgA which could, in turn, account for susceptibility to PAM. Rivera *et al.* (2001) and Rivera-Aguilar *et al.* (2000) used ELISA and Western immunoblot analysis to evaluate antibodies of the IgA and IgM classes to *N. fowleri* in serum and saliva. Samples were obtained from subjects with upper respiratory tract infections (URTI) living in endemic areas of Mexico as well as from healthy persons from endemic and nonendemic areas. The titers of IgA antibodies to *N. fowleri* were significantly higher in serum and saliva of patients with chronic bronchitis and rhinitis (URTIs), as compared to those of individuals who were healthy and living in endemic areas. The role of secretory IgA (sIgA) in adherence to collagen type 1, a component of connective tissue in the nasal passages that may be involved in amoeba passage into the brain, also has been studied. Shibiyama *et al.* (2003) using Western immunoblot, demonstrated that sIgA obtained from pooled human colostrum recognized several proteins in *N. fowleri* lysates and that it inhibited adhesion to collagen. However, although sIgA antibodies inhibited adherence, the antibodies were 'capped' by the amoeba and shed from the surface within a 2-h period. Seidel *et al.* (1982) identified by immunofluorescent assay (IFA) specific antibodies to *N. fowleri* in serum of a patient who survived PAM in the United States. Samples were shown to be positive for serum that was diluted up to 1:4096.

In summary, while various studies indicate that antibodies specific to *N. fowleri* decrease motility of amoebae, these antibodies may not be totally protective since amoebae have the ability to cap, internalize, and degrade the antibody (Ferrante and Thong, 1979). Thus, while infection with *Naegleria* results in elicitation of antibodies, to date, a correlation of susceptibility to PAM and the humoral immune status of the host has not been found. The absence of any correlation may be due to the rapid progression of PAM that may not allow for sufficient time for the host to mount an immune response of sufficient titer to exert a protective effect. Furthermore, the blood-brain barrier may preclude influx of antibodies elicited at peripheral sites that could exert a protective effect in the CNS.

The role of cell-mediated immunity (CMI) in resistance to PAM has been studied in experimental animals but not in humans. The delayed-type hypersensitivity (DTH) was assessed in guinea pigs sensitized by intradermal injection with *N. fowleri*, *N. gruberi* or *N. jadini* (Cursons *et al.*, 1977, 1980b; Diffley *et al.*, 1976). A reaction was detected in 24 h by measuring the diameter of hard nodular lesions elicited to the introduction of homologous or heterologous antigens of *Naegleria* species. Nevertheless, a protective role of CMI is uncertain. Diethylstilbesterol, which depresses DTH, was shown not to alter host resistance to *N. fowleri* in mice (Reilly *et al.*, 1983b). In addition, the incidence of mortality due to *N. fowleri* in congenitally athymic mice that are deficient in T-cell responses has been shown not to differ from that in euthymic mice (Newsome and Arnold, 1985).

Innate immunity may play a more important role than acquired immunity in resistance to *N. fowleri* infection (Marciano-Cabral, 1988). It has been reported that susceptibility to *N. fowleri* varies greatly among mouse strains but that the most sensitive mouse is the complement-deficient (C5) strain A/HeCr (Haggerty and John, 1978). Furthermore, mice depleted of complement using cobra venom factor are more susceptible to infection with *N. fowleri* (Reilly *et al.*, 1983b). *Naegleria* activate the alternative pathway of complement and it is thought that this process plays a role in inhibition of dissemination of amoebae from the CNS (Holbrook *et al.*, 1980). Whiteman and Marciano-Cabral (1989) and Toney and Marciano-Cabral (1994) have reported that highly pathogenic strains of *N. fowleri* are resistant to lysis by human complement while nonpathogenic species and weakly pathogenic species are sensitive to complement. It has been proposed that highly pathogenic *N. fowleri* evades complement-mediated lysis through the intervention of membrane proteins which protect it from being lysed (Toney and Marciano-Cabral, 1992). In addition, a vesiculation process may be operative that results in removal of the lytic membrane attack complex of complement (C5b-C9; Toney and Marciano-Cabral, 1994).

Cells of the immune system, such as neutrophils and macrophages, also, may play an important role in the innate immune response to *Naegleria*. In the mouse model of PAM, neutrophils represent the first responders to infection (Figure 8A). Selective depletion of neutrophils in infected mice

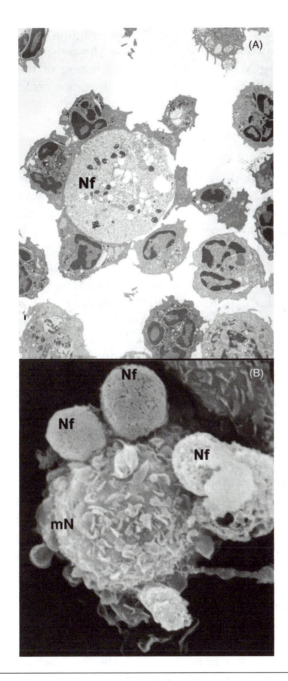

Figure 8 Transmission electron micrograph depicting immunocytes interacting with *Naegleria*. (A) Neutrophils circumscribing a trophozoite (Nf). (B) A murine peritoneal macrophage (mN) that has been activated *in vivo* with *C. parvum* exhibiting cell-contact-dependent destruction of *N. fowleri* (Nf). Note the presence of pores on the surface of *N. fowleri*.

with monoclonal antibodies results in increase mortality (Ferrante *et al.*, 1988). However, while neutrophils from *N. fowleri*-immunized mice are capable of killing *N. fowleri in vitro*, destruction of these amoebae by human neutrophils requires treatment with tumor necrosis factor-alpha (TNF-α; Ferrante and Mocatta, 1984). Michelson *et al.* (1990) reported that activation of neutrophils *in vitro* with this pro-inflammatory cytokine results in increased adherence of neutrophils to *N. fowleri* and elicits changes suggestive of cytotoxic damage to the amoebae. Fischer-Stenger and Marciano-Cabral (1992) demonstrated that TNF-α does not by itself lyse *Naegleria*, suggesting an indirect role for this cytokine in mediation of amoebic lysis. Macrophages, which inducibly express this cytokine and other pro-inflammatory cytokines, were found to destroy *N. fowleri in vitro* following their activation *in vivo* (Figure 8B; Cleary and Marciano-Cabral, 1986). However, the amoebicidal activity was found as attributed to nitric oxide. Thus, it appears that in contrast to macrophage-mediated killing of tumor cells, TNF-α and other pro-inflammatory cytokines such as inter-leukin-1α (IL-1α) and IL-1β, do not exert a direct lytic effect on *N. fowleri* (Fischer-Stenger and Marciano-Cabral, 1992; Fischer-Stenger *et al.*, 1992). Rather, the elicitation of pro-inflammatory cytokines by macrophages and macrophage-like cells in response to *N. fowleri* may actually contribute to the pathology of amoebic infection. In this context, it has been demonstrated that microglia, resident macrophages in the brain, produce robust levels of IL1-α, IL1-β, and IL-6 in response to *N. fowleri* (Marciano-Cabral *et al.*, 2001; Oh *et al.*, 2005). The production of high levels of pro-inflammatory cytokines in response to CNS infection has been reported to be a major contributive factor in a variety of neuropathies. Finally, the age and sex of the host may affect susceptibility to *N. fowleri* infection. Culbertson (1971) and Haggarty and John (1978) demonstrated, respectively, that older monkeys and mice are more resistant to infection than younger animals and that female mice are more resistant than male mice.

10 Pathogens Associated with *Naegleria*

Free-living amoebae interact with other organisms in the environment. Bacteria not only serve as a food source, but also may play a role as amoebic endosymbionts (Marciano-Cabral, 2003). Pathogenic bacteria also may use amoebae as a reservoir host. Early studies by Rowbotham (1980) suggested that *Naegleria* could act as a natural host in the environment for *Legionella pneumophila*, the agent of Legionnaire's disease. In addition, *L. pneumophila* and *Naegleria* sp. have been co-isolated from cooling tower water, implicating the amoebae as a reservoir for bacteria in nature and in the spread of legionellosis (Barbaree *et al.*, 1986). Newsome *et al.* (1985) demonstrated by electron microscopy that *N. fowleri* provides an intracellular environment that appears to be conducive to multiplication of bacteria. The intracellular events in amoebae associated with bacterial propagation were shown to be similar to those for bacteria growing in human monocytes. Collectively, these results suggest that infection of humans with *Legionella* and other bacteria could be acquired through inhalation of amoeba trophozoites or cysts. In this context, antibacterial therapy may be difficult to manage since the amoebae may sequester bacteria from the action of antibiotics. A host thus affected would be at increased risk of a poor disease outcome. The association of free-living amoebae and bacterial pathogens is becoming more widely recognized as amoebae containing pathogenic bacteria are being isolated from humans with disease and from the natural environment.

11 Summary and Conclusions

Naegleria fowleri is the causative agent of PAM, a disease for which over the past 15 years there has been increased reporting of cases from the United States and other countries (Craun *et al.*, 2005; Yoder *et al.*, 2004). The emergence of PAM caused by *N. fowleri* is likely influenced by a variety of factors, including greater recognition of amoebic infection and availability of novel, highly sensitive and specific methods for identification and laboratory diagnosis. However, a number of studies suggest that manmade factors also may play a role in the increased incidence of infection. *Naegleria* grow vigorously in the presence of Enterobacteriaceae and related organisms (Brown *et al.*, 1983; Kyle and Noblet, 1985; Singh and Dutta, 1984). Thermal pollution from industrial plants and cooling towers facilitate the growth of thermophilic *Naegleria* as well as that of bacteria that serve as its food source. Increased water pollution with 'raw' or inadequately treated sewage also presents a potential for propagation of *Naegleria*. Finally, human participation in water recreational or thermal therapeutic activities, such as swimming, diving, and use of 'hot tubs' and physiotherapy pools may contribute to exposure to *Naegleria*. Thus, with increased environmental temperatures and greater industrial development leading to thermal pollution, there is a potential for an increase in the incidence of PAM in humans.

12 Acknowledgments

This chapter is dedicated to the memory of Drs. A. Julio Martinez and Thomas Byers, two pioneers in the study of free-living amoebae, who shared their knowledge willingly with their colleagues.

13 References

Anderson, K. and Jamieson, A. (1972) Primary amoebic meningoencephalitis. *Lancet* **1**: 902.

Barbaree, J.M., Fields, B.S., Feeley, J.C., Gorman, G.W. and Martin, W.T. (1986) Isolation of protozoa from water associated with a legionellosis outbreak and demonstration of intracellular multiplication of *Legionella pneumophila. Appl. Environ. Microbiol.* **51**: 422–424.

Barbour, S.E. and Marciano-Cabral F. (2001) *Naegleria fowleri* amoebae express a membrane-associated calcium-independent phospholipase A(2). *Biochim. Biophys. Acta* **1530**: 123–133.

Barnett, N.D., Kaplan, A.M., Hopkin, R.J., Saubolle, M.A. and Rudinsky, M.F. (1996) Primary amoebic meningoencephalitis with *Naegleria fowleri*: clinical review. *Pediatr. Neurol.* **15**: 230–234.

Behets, J., Seghi, F., Declerck, D., Verelst, L., Duvivier, L., Van Damme, A. and Ollevier, F. (2003) Detection of *Naegleria* spp. and *Naegleria fowleri*: a comparison of flagellation tests, ELISA and PCR. *Water Sci. Technol.* **47**: 117–122.

Behets, J., DeClerck, P., Delaedt, Y., Verelst, L. and Ollevier, F. (2006) Quantitative detection and differentiation of free-living amoeba species using SYBR green-based real-time PCR melting curve analysis. *Curr. Microbiol.* **53**: 506–509.

Behets, J., DeClerck, P., Delaedt, Y., Verelst, L. and Ollevier, F. (2007) A duplex real-time PCR assay for the quantitative detection of *Naegleria fowleri* in water samples. *Water Res.* **41**: 118–126.

Benson, R.L., Ansbacher, L., Hutchison, R.E. and Rogers, W. (1985) Cerebrospinal fluid centrifuge analysis in primary amebic meningoencephalitis due to *Naegleria fowleri*. *Arch. Pathol. Lab. Med.* **109**: 668–671.

Brinkley, C. and Marciano-Cabral, F. (1992) A method for assessing the migratory response of *Naegleria fowleri* utilizing [³H]uridine-labeled amoebae. *J. Protozool.* **139**: 297–303.

Brown, T. (1979) Observations by immunofluorescence microscopy and electron microscopy on the cytopathogenicity of *Naegleria fowleri* in mouse embryo-cell cultures. *J. Med. Microbiol.* **12**: 363–371.

Brown, R.L. (1991) Successful treatment of primary amebic meningoencephalitis. *Arch. Intern. Med.* **151**: 1201–1202.

Brown, T.J., Cursons, R.T.M., Keys, E.A., Marks, M. and Miles, M. (1983) The occurrence and distribution of pathogenic free-living amoebae in thermal areas of the North Island of New Zealand. *NZ J. Mar. Freshw. Res.* **17**: 59–69.

Butt, C.G. (1966) Primary amebic meningoencephalitis. *New Engl. J. Med.* **274**: 1473–1476.

Butt, C.G., Baro, C. and Knorr, R.W. (1968) *Naegleria* sp. identified in amoebic encephalitis. *Am. J. Clin. Pathol.* **50**: 568–574.

Cain, A.R.R., Mann, P.G. and Warhurst, D.C. (1979) IgA and primary amoebic meningoencephalitis. *Lancet* **i**: 441.

Callicott, J.H., Nelson, E.C., Jones, M.M., dos Santos, J.G., Utz, J.P., Duma, R.J. and Morrison, J.V. (1968) Meningoencephalitis due to pathogenic free-living amoebae: report of two cases. *J. Am. Med. Assoc.* **206**: 579–582.

Carter, R.F. (1968) Primary amoebic meningo-encephalitis: clinical, pathological, and epidemiological features of six fatal cases. *J. Pathol. Bacteriol.* **96**: 1–25.

Carter, R.F. (1969) Sensitivity to amphotericin B of a *Naegleria* sp. isolated from a case of primary amoebic meningoencephalitis. *J. Clin. Pathol.* **22**: 470–474.

Carter, R.F. (1970) Description of a *Naegleria* species isolated from two cases of primary amoebic meningoencephalitis and of the experimental pathological changes induced by it. *J. Pathol.* **100**: 217–244.

Carter, R.F. (1972) Primary meningoencephalitis: an appraisal of present knowledge. *Trans R. Soc. Trop. Med. Hyg.* **66**: 193–213.

Cermeno, J.R., Hernandez, I., El Yasin, H., Tinedo, R., Sanchez, R., Perez, G., Gravano, R. and Ruiz, A. (2006) Meningoencephalitis by *Naegleria fowleri*: epidemiological study in Anzoategui State, Venezuela. *Rev. Soc. Bras. Med. Trop.* **39**: 264–268.

Cerva, L. (1989) *Acanthamoeba culbertsoni* and *Naegleria fowleri*: occurrence of antibodies in man. *J. Hyg. Epidemiol. Microbiol. Immunol.* **33**: 99–103.

Cerva, L. and Novak, K. (1968) Amebic meningoencephalitis: sixteen fatalities. *Science* **160**: 92.

Cerva, L., Novak, K. and Culbertson, C.G. (1968) An outbreak of acute fatal amebic meningoencephalitis. *Am. J. Epidemiol.* **88**: 436–444.

Chang, S.L. (1979) Pathogenesis of pathogenic *Naegleria* amoeba. *Folia Parasitol.* **26**: 195–200.

Cleary, S. and Marciano-Cabral, F. (1986) Activated macrophages demonstrate direct cytotoxicity, antibody-dependent cellular cytotoxicity, and enhanced binding of N*aegleria fowleri* amoebae. *Cell Immunol.* **98**: 125–136.

Cogo, P.E., Scagli, M., Gatti, S., Rossetti, F., Alaggio, R., Laverda, A.M., Zhou, L., Xiao, L. and Visvesvara, G.S. (2004) Fatal *Naegleria fowleri* meningoencephalitis, Italy. *Emerg. Infect. Dis.* **10**: 1835–1837.

Cooter, R. (2002) The history of the discovery of primary amoebic meningoencephalitis. *Aust. Fam. Physician* **31**: 399–400.

Craun, G.F., Calderon, R.L. and Craun, M.F. (2005) Outbreaks associated with recreational water in the United States. *Int. J. Environ. Health Res.* **15**: 243–262.

Cubero-Menendez, O. and Cubero-Rego, D. (2004) Meningoencefalitis amebiana primaria: comunicacion de un caso. *Nota Clinica* **38**: 336–338.

Culbertson C.G. (1971) The pathogenicity of soil amebas. *Annu. Rev. Microbiol.* **25**: 231–254.

Cursons, R., Brown, T. and Keys, E. (1977) Immunity to pathogenic free-living amoebae. *Lancet* **ii**: 875.

Cursons, R., Brown, T. and Keys, E. (1978) Virulence of pathogenic free-living amoebae. *J. Parasitol.* **64**: 744–745.

Cursons, R., Keys, E., Brown, T., Learmonth, J., Campbell, C. and Metcalf, P. (1979) IgA and primary amoebic meningoencephalitis. *Lancet* **i**: 223–224.

Cursons, R., Brown, T., Keys, E., Moriarty, K. and Till, D. (1980a) Immunity to pathogenic free-living amoebae: role of humoral antibody. *Infect. Immun.* **29**: 401–407.

Cursons, R., Brown, T., Keys, E., Moriarty, K. and Till, D. (1980b) Immunity to pathogenic free-living amoebae: role of cell mediated immunity. *Infect. Immun.* **29**: 408–410.

Cursons, R., Sleigh, J., Hood, D. and Pullon, D. (2003) A case of primary amoebic meningoencephalitis: North Island, New Zealand. *NZ Med. J.* **116**: U712.

De Jonckheere, J.F. (2004) Molecular definition and the ubiquity of species in the genus *Naegleria*. *Protist* **155**: 89–103.

De Jonckheere, J.F., Yagita, K. and Endo, T. (1992) Restriction-fragment-length polymorphism and variation in electrophoretic karyotype in *Naegleria fowleri* from Japan. *Parasitol. Res.* **78**: 475–478.

Dempe, S., Martinez, A.J. and Janitschke, K. (1982) Subacute and chronic meningoencephalitis in mice after experimental infection with a strain of *Naegleria fowleri* originally isolated from a patient. *Infection* **10**: 5–8.

DeNapoli, T.S., Rutman, J.Y., Robinson, J.R. and Rhodes, M.M. (1996) Primary amoebic meningoencephalitis after swimming in the Rio Grande. *Tex. Med.* **92**: 59–63.

Diffley, P., Skeels, M.R., Sogandares-Bernal, F. (1976) Delayed type hypersensitivity in guinea-pigs infected subcutaneously with *Naegleria fowleri* (Carter). *Z. Parasitenkd.* **49**: 133–137.

Dingley, D. (1996) Safe water practices can lower risk of contracting primary amoebic meningoencephalitis. *Tex. Med.* **92**: 28–29.

dos Santos, J.G. (1970) Fatal primary amebic meningoencephalitis: a retrospective study in Richmond, Virginia. *Am. J. Clin. Pathol.* **54**: 737–742.

Dubray, B.L., Wilhelm, W.E. and Jennings, B.R. (1987) Serology of *Naegleria fowleri* and *Naegleria lovaniensis* in a hospital survey. *J. Protozool.* **34**: 322–327.

Feldman, M.R. (1977) *Naegleria fowleri*: fine structural localization of acid phosphatase and heme proteins. *Exp. Parasitol.* **41**: 283–289.

Ferrante, A. and Mocatta, T.J. (1984) Human neutrophils require activation by mononuclear leucocyte conditioned medium to kill pathogenic free-living amoeba, *Naegleria fowleri*. *Clin. Exp. Immunol.* **56**: 559–566.

Ferrante, A. and Thong, Y.H. (1979) Antibody induced capping and endocytosis of surface antigens in *Naegleria fowleri*. *Int. J. Parasitol.* **9**: 599–601.

Ferrante, A., Carter, R.F., Lopez, A.F., Rowan-Kelly, B., Hill, N.L. and Vadas, M.A. (1988) Depression of immunity to *Naegleria fowleri* in mice by selective depletion of neutrophils with a monoclonal antibody. *Infect. Immun.* **56**: 2286–2291.

Fiordalisi, I., Christie, J. and Moffitt, C. (1992) Amebic meningoencephalitis-North Carolina 1991. *Morbid. Mortal. Weekly Rep.* **41**: 437–439.

Fischer-Stenger, K. and Marciano-Cabral, F. (1992) The arginine-dependent cytolytic mechanism plays a role in destruction of *Naegleria fowleri* amoebae by activated macrophages. *Infect. Immun.* **60**: 5126–5131.

Fischer-Stenger, K., Cabral, G.A. and Marciano-Cabral, F. (1992) Separation of soluble amoebicidal and tumoricidal activity of activated macrophages. *J. Protozool.* **39**: 235–241.

Fowler, M. and Carter, R.F. (1965) Acute pyogenic meningitis probably due to *Acanthamoeba* sp.: a preliminary report. *Brit. Med. J.* **2**: 740–742.

Fritzinger, A.E., Toney, D.M., MacLean, R.C. and Marciano-Cabral, F. (2006) Identification of a *Naegleria fowleri* membrane protein reactive with anti-human CD59 antibody. *Infect. Immun.* **74**: 1189–1195.

Fulford, D.E., Bradley, S.G. and Marciano-Cabral, F. (1985) Cytopathogenicity of *Naegleria fowleri* for cultured rat neuroblastoma cells. *J. Protozool.* **32**: 176–180.

Fulton, C. (1983) Macromolecular syntheses during the quick-change act of *Naegleria. J. Protozool.* **30**: 192–198.

Fulton, C. and Dingle, A.D. (1971) Basal bodies, but not centrioles, in *Naegleria. J. Cell Biol.* **51**: 826–836.

Goswick, S.M. and Brenner, G.M. (2003a) Activities of azithromycin and amphotericin B against *Naegleria fowleri* in vitro and in a mouse model of primary amebic meningoencephalitis. *Antimicrob. Agents Chemother.* **47**: 524–528.

Goswick, S.M. and Brenner, G.M. (2003b) Activities of therapeutic agents against *Naegleria fowleri in vitro* and in a mouse model of primary amebic meningoencephalitis. *J. Parasitol.* **89**: 837–842.

Gupta, S., Dutta, G.P. and Vishwakarma, R.A. (1998) Effect of alpha, beta-arteether against primary amoebic meningoencephalitis in Swiss mice. *Indian J. Exp. Biol.* **36**: 824–825.

Gyori, E. (2003) December 2002: 19-year-old male with febrile illness after jet ski accident. *Brain Pathol.* **13**: 237–239.

Haggerty, R.M. and John, D.T. (1978) Innate resistance of mice to experimental infection with *Naegleria fowleri. Infect. Immun.* **20**: 73–77.

Hara, T. and Fukuma, T. (2005) Diagnosis of the primary amoebic meningoencephalitis due to *Naegleria fowleri. Parasitol. Int.* **54**: 219–221.

Hebbar, S., Bairy, I., Bhaskaranand, N., Upadhyaya, S., Sarma, M.S. and Shetty, A.K. (2005) Fatal case of *Naegleria fowleri* meningo-encephalitis in an infant: case report. *Ann. Trop. Paediatr.* **25**: 223–226.

Herbst, R., Marciano-Cabral, F. and Leippe, M. (2004) Antimicrobial and pore-forming peptides of free-living and potentially highly pathogenic *Naegleria fowleri* are released from the same precursor molecule. *J. Biol. Chem.* **279**: 25955–25958.

Herbst, R., Ott, C., Jacobs, T., Marti, T., Marciano-Cabral, F. and Leippe, M. (2002) Pore-forming polypeptides of the pathogenic protozoon *Naegleria fowleri. J. Biol. Chem.* **277**: 22353–22360.

Holbrook, T.W., Boackle, R.J., Parker, B.W. and Vesely, J. (1980) Activation of the alternative complement pathway by *Naegleria fowleri. Infect. Immun.* **30**: 58–61.

Huizinga, H.W. and McLaughlin, G.L. (1990) Thermal ecology of *Naegleria fowleri* from a power plant cooling reservoir. *Appl. Environ. Microbiol.* **56**: 2200–2205.

Hysmith, R.M. and Franson, R.C. (1982) Degradation of human myelin phospholipids by phospholipase-enriched culture media of pathogenic *Naegleria fowleri. Biochim. Biophys. Acta* **712**: 698–701.

Jaffar-Bandjee, M.C., Alessandri, J.L., Molet, B., Clouzeau, J., Jacquemot, L., Samperiz, S. and Saly, J.C. (2005) Meningo-encephalite primitive a amibes libres: 1er cas observe a Madagascar. *Bull. Soc. Pathol. Exot.* **98**: 11–13.

Jain, R., Prabhakar, S., Modi, M., Bhatia, R. and Sehgal, R. (2002) *Naegleria* meningitis: a rare survival. *Neurol. India* **50**: 470–472.

John, D.T. and Howard, M.J. (1996) Techniques for isolating thermotolerant and pathogenic free-living amebae. *Folia Parasitol.* **43**: 261–271.

Kidney, D.D. and Kim, S.H. (1998) CNS infections with free-living amebas: neuroimaging findings. *Am. J. Roentgenol.* **171**: 809–812.

Kilvington, S. and Beeching, J. (1995) Development of a PCR for identification of *Naegleria fowleri* from the environment. *Appl. Environ. Microbiol.* **61**: 3764–3767.

Kyle, D.E. and Noblet, G.P. (1985) Vertical distribution of potentially pathogenic amoeba in freshwater lakes. *J. Protozool.* **32**: 99–105.

Lares-Villa, F., De Jonckheere, J.F., De Moura, H., Rechi-Iruretagoyena, A., Ferreira-Guerrero, E., Fernandez-Quintanilla, G., Ruiz-Matus, C. and Visvesvara, G.S. (1993) Five cases of primary amebic meningoencephalitis in Mexicali, Mexico: study of the isolates. *J. Clin. Microbiol.* **31**: 685–688.

Leippe, M. and Herbst, R. (2004) Ancient weapons for attack and defense: the pore-forming polypeptides of pathogenic enteric and free-living amoeboid protozoa. *J. Eukaryot. Microbiol.* **51**: 516–521.

Loschiavo, F., Ventura-Spangolo, T., Sessa, E. and Bramanti, P. (1993) Acute primary mengingoencephalitis from Entamoeba *Naegleria fowleri. Acta Neurol.* **15**: 333–340.

Lowrey, D.M. and McLaughlin, J. (1984) A multicomponent hemolytic system in the pathogenic amoeba *Naegleria fowleri. Infect. Immun.* **45**: 731–736.

Lugg, R. (1996) Public health programs for the prevention of amoebic meningitis in Western Australia, 1980–1995. In: *Proc. 7th Int. Conf. on Small Free-living Amoebae, North Adelaide, Australia.*

Marciano-Cabral, F. (1988) Biology of *Naegleria* spp. *Microbiol. Rev.* **52**: 114–133.

Marciano-Cabral, F. (2003) Introductory remarks: bacterial endosymbionts or pathogens of free-living amebae. *J. Eukaryot. Microbiol.* **51**: 497–501.

Marciano-Cabral, F. and John, D.T. (1983) Cytopathogenicity of *Naegleria fowleri* for rat neuroblastoma cell cultures: scanning electron microscopy study. *Infect. Immun.* **40**: 1214–1217.

Marciano-Cabral, F., Patterson, M., John, D.T. and Bradley, S.G. (1982) Cytopathogenicity of *Naegleria fowleri* and *Naegleria gruberi* for established cell cultures. *J. Parasitol.* **68**: 1110–1116.

Marciano-Cabral, F., Cline, M. and Bradley, S.G. (1987) Specific antibodies from human sera for *Naegleria* species. *J. Clin. Microbiol.* **25**: 692–697.

Marciano-Cabral, F., Ludwick, C. and Cabral, G.A. (2001) The interaction of *Naegleria fowleri* amoebae with brain microglial cells. In: *Proc. IXth Int. Meeting on the Biology and Pathogenicity of Free-living Amoebae, Paris, France* pp. 49–58.

Marciano-Cabral, F., MacLean, R., Mensah, A. and LaPat-Polasko, L. (2003) Identification of *Naegleria fowleri* in domestic water sources by nested PCR. *Appl. Environ. Microbiol.* **69**: 5864–5869.

Marshall, M.M., Naumovitz, D., Ortega, Y. and Sterling, C.R. (1997) Waterborne protozoan pathogens. *Clin. Microbiol. Rev.* **10**: 67–85.

Martinez, A.J. (1985) *Free-living amoebas: natural history, prevention, diagnosis, pathology, and treatment of disease.* CRC Press, Boca Raton, FL.

Martinez, A.J. and Visvesvara, G.S. (1997) Free-living, amphizoic and opportunistic amebas. *Brain Pathol.* **7**: 583–598.

Martinez, A.J., Nelson, E.C., Jones, M.M., Duma, R.J. and Rosenblum, W.I. (1971) Experimental *Naegleria* meningoencephalitis in mice: an electron microscopy study. *Lab. Invest.* **24**: 465–475.

Martinez, A.J., Duma, R.J., Nelson, E.C. and Moretta, F.L. (1973) Experimental *Naegleria* meningoencephalitis in mice. Penetration of the olfactory mucosal epithelium by *Naegleria* and pathogenic changes produced: a light and electron microscope study. *Lab. Invest.* **29**: 121–133.

Mat Amin, N. (2004) Proteinases in *Naegleria fowleri* (strain NF3), a pathogenic amoeba: a preliminary study. *Trop. Biomed.* **21**: 57–60.

McLaughlin, G.L., Vodkin, M.H. and Huizinga, H.W. (1991) Amplification of repetitive DNA for the specific detection of *Naegleria fowleri*. *J. Clin. Microbiol.* **29**: 227–230.

Michelson, M.K., Henderson Jr, W.R., Chi E., Fritsche, T.R. and Klebanoff, S.J. (1990) Ultrastructural studies on the effect of tumor necrosis factor on the interaction of neutrophils and *Naegleria fowleri*. *Am. J. Trop. Med. Hyg.* **42**: 225–233.

Newsome, A.L. and Arnold, R.R. (1985) Equivalent mortality in normal and athymic mice infected with *Naegleria fowleri*. *J. Parasit.* **71**: 678–679.

Newsome, A.L. and Wilhelm, W.E. (1983) Inhibition of *Naegleria fowleri* by microbial iron-chelating agents: ecological implications. *Appl. Environ. Microbiol.* **45**: 665–668.

Newsome, A.L., Baker, R.L., Miller, R.D. and Arnold, R.R. (1985) Interactions between *Naegleria fowleri* and *Legionella pneumophila*. *Infect. Immun.* **50**: 449–452.

Oh, Y.H., Jeong, S.R., Kim, J.H., Song, K.J., Kim, K., Park, S., Sohn, S. and Shin, H.J. (2005) Cytopathic changes and pro-inflammatory cytokines induced by *Naegleria fowleri* trophozoites in rat microglial cells and protective effects of an anti-Nfa1 antibody. *Parasite Immunol.* **27**: 453–459.

Okuda, D.T., Hanna, H.J., Coons, S.W. and Bodensteiner, J.B. (2004) *Naegleria fowleri* hemorrhagic meningoencephalitis: report of two fatalities in children. *J. Child Neurol.* **19**: 231–233.

Ondarza, R.N., Iturbe, A. and Hernandez, E. (2006) *In vitro* antiproliferative effects of neuroleptics, antimycotics and antibiotics on the human pathogens *Acanthamoeba polyphaga* and *Naegleria fowleri. Arch. Med. Res.* **37**: 723–729.

Ondarza, R.N., Iturbe, A. and Hernandez, E. (2007) The effects by neuroleptics, antimycotics and antibiotics on disulfide reducing enzymes from the human pathogens, *Acanthamoeba polyphaga* and *Naegleria fowleri. Exp. Parasitol.* **115**: 41–47.

Page, F.C. (1974) *Rosculus ithacus* Hawes, 1963, *Amoebida, Flabelluidea* and the amphizoic tendency in amoebae. *Acta Protozool.* **13**: 143.

Parija, S.C. and Jayakeerthee, S.R. (1999) *Naegleria fowleri*: a free living amoeba of emerging medical importance. *J. Commun. Dis.* **31**: 153–159.

Patras, D. and Andujar, J.J. (1966) Meningoencephalitis due to *Hartmannella* (*Acanthamoeba*). *Am. J. Clin. Pathol.* **46**: 226–233.

Patterson, M., Woodworth, T.W., Marciano-Cabral, F. and Bradley, S.G. (1981) Ultrastructure of *Naegleria fowleri* enflagellation. *J. Bacteriol.* **147**: 217–226.

Pelandakis, M. and Pernin, P. (2002) Use of multiplex PCR and PCR restriction enzyme analysis for detection and exploration of the variability in the free-living amoeba *Naegleria* in the environment. *Appl. Environ. Microbiol.* **68**: 2061–2065.

Pelandakis, M., Serre, S. and Pernin, P. (2000) Analysis of the 5.8S rRNA gene and the internal transcribed spacers in *Naegleria* spp. and in *N. fowleri. J. Eukaryot. Microbiol.* **47**: 116–121.

Pernin, P. and Pelandakis, M. (2001) About some aspects of the ecology and biodiversity of the *Naegleria* amoebae. In: *IXth Int. Meeting on the Biology and Pathogencity of Free-living Amoebae, Paris, France* pp. 81–85.

Pernin, P., Pelandakis, M., Rouby, Y., Faure, A. and Siclet, F. (1998) Comparative recoveries of *Naegleria fowleri* amoebae from seeded river water by filtration and centrifugation. *Appl. Environ. Microbiol.* **64**: 955–959.

Poungvarin, N. and Jariya, P. (1991) The fifth nonlethal case of primary amoebic meningoencephalitis. *J. Med. Assoc. Thai.* **74**: 112–115.

Powell, E., Newsome, A., Allen, S. and Knudson, G.B. (1994) Identification of antigens of pathogenic free-living amoebae by protein immunoblotting with rabbit immune and human sera. *Clin. Diagn. Lab. Immunol.* **1**: 493–499.

Preston, T.M. and King, C.A. (2003) Locomotion and phenotypic transformation of the amoeboflagellate *Naegleria gruberi* at the water-air interface. *J. Eukaryot. Microbiol.* **50**: 245–251.

Qvarnstrom, Y., Visvesvara, G.S., Sriram, R. and da Silva, A.J. (2006) Multiplex real-time PCR assay for simultaneous detection of *Acanthamoeba* spp. *Balamuthia mandrillaris*, and *Naegleria fowleri*. *J. Clin. Microbiol.* **44**: 3589–3595.

Reilly, M.F., Marciano-Cabral, F., Bradley D.W. and Bradley, S.G. (1983a) Agglutination of *Naegleria fowleri* and *Naegleria gruberi* by antibodies in human serum. *J. Clin. Microbiol.* **17**: 576–581.

Reilly, M.F., White Jr, K.L. and Bradley, S.G. (1983b) Host resistance of mice to *Naegleria fowleri* infections. *Infect. Immun.* **42**: 645–652.

Reveiller, F.L., Cabanes, P.A., Marciano-Cabral, F., Pernin, P., Cabanes, P.A. and Legastelois, S. (2000) Species specificity of a monoclonal antibody produced to *Naegleria fowleri* and partial characterization of its antigenic determinant. *Parasitol. Res.* **86**: 634–641.

Reveiller, F.L., Cabanes, P.A. and Marciano-Cabral, F. (2002) Development of a nested PCR assay to detect the pathogenic free-living amoeba *Naegleria fowleri*. *Parasitol. Res.* **88**: 443–450.

Reveiller, F.L., Varenne, M.P., Pougnard, C., Cabanes, P.A., Pringuez, E., Pourima, B., Legastelois, S. and Pernin, P. (2003) An enzyme-linked immunosorbent assay (ELISA) for the identification of *Naegleria fowleri* in environmental water samples. *J. Eukaryot. Microbiol.* **50**: 109–113.

Rivera-Aguilar, V., Hernandez-Martinez, D., Rojas-Hernandez, S., Oliver-Aguillon, G., Tsutsumi, V., Herrera-Gonzalez, N. and Campos-Rodriguez, R. (2000) Immunoblot analysis of IgA antibodies to *Naegleria fowleri* in human saliva and serum. *Parasitol. Res.* **86**: 775–780.

Rivera, V., Hernandez, D., Rojas, S., Oliver, G., Serrano, J., Shibayama, M., Tsutsumi, V. and Campos, R. (2001) IgA and IgM anti-*Naegleria fowleri* antibodies in human serum and saliva. *Can. J. Microbiol.* **47**: 464–466.

Robinson, B.S., Monis, P.T. and Dobson, P.J. (2006) Rapid, sensitive, and discriminating identification of *Naegleria* spp. by real-time PCR and melting-curve analysis. *Appl. Environ. Microbiol.* **72**: 5857–5863.

Rowbotham, T.J. (1980) Preliminary report on the pathogenicity of *Legionella pneumophila* for freshwater and soil amoebae. *J. Clin. Pathol.* **33**: 1179–1183.

Schild, M., Gianinazzi, C., Gottstein, B. and Muller, N. (2007) PCR-based diagnosis of *Naegleria* spp. infection in formalin-fixed and paraffin-embedded brain sections. *J. Clin. Microbiol.* **45**: 564-567.

Schoeman, C.J., van der Vyver, A.E. and Visvesvara, G.S. (1993) Primary amoebic meningo-encephalitis in southern Africa. *J. Infect.* **26**: 211–214.

Schuster, F.L. and Visvesvara, G.S. (2004a) Amebae and ciliated protozoa as causal agents of waterborne zoonotic disease. *Vet. Parasitol.* **126**: 91–120.

Schuster, F.L. and Visvesvara, G.S. (2004b) Opportunistic amoebae: challenges in prophylaxis and treatment. *Drug Resist. Update* **7**: 41–51.

Seidel, J.S., Harmatz, P., Visvesvara, G.S., Cohen, A., Edwards, J. and Turner, J. (1982) Successful treatment of primary amebic meningoencephalitis. *New Engl. J. Med.* **306**: 346–348.

Shenoy, S., Wilson, G., Prashanth, H.V., Vidyalakshmi, K., Dhanashree, B. and Bharath, R. (2002) Primary meningoencephalitis by *Naegleria fowleri*: first reported case from Mangalore, South India. *J. Clin. Microbiol.* **40**: 309–310.

Shibayama, M., Serrano-Luna, J., de J., Rojas-Hernandez, S., Campos-Rodriguez, R. and Tsutsumi, V. (2003) Interaction of secretory immunoglobulin A antibodies with *Naegleria fowleri* trophozoites and collagen type I. *Can. J. Microbiol.* **49**: 164–170.

Shrestha, N.K., Khanal, B., Sharma, S.K., Dhakal, S.S. and Kanungo, R. (2003) Primary amoebic meningoencephalitis in a patient with systemic lupus erythematosus. *Scand. J. Infect. Dis.* **35**: 514–516.

Shumaker, D.J., Tien, R.D. and Lane, K. (1995) Neuroimaging findings in rare amebic infections of the central nervous system. *Am. J. Neuroradiol.* **16**: 930–935.

Simpson, C.F., Willaert, E., Neal, F.C., Stevens, A.R. and Young, M.D. (1982). Experimental *Naegleria fowleri* meningoencephalitis in sheep: light and electron microscopic studies. *Am. J. Vet. Res.* **43**: 154–157.

Singh, B.N. and Dutta, G.D.P. (1984) Small free-living aerobic amoebae: soil as a suitable habitat, isolation, culture, classification, pathogenicity, epidemiology, and chemotherapy. *Indian J. Parasitol.* **8**: 1–23.

Singh, S.N., Patwari, A.K., Dutta, R., Taneja, N. and Anand, V.K. (1998) *Naegleria* meningitis. *Indian Pediatr.* **35**: 1012–1015.

Singh, P., Kochhar, R., Vashista, R.K., Khandelwal, N., Prabhakar, S., Mohindra, S. and Singhi, P. (2006) Amebic meningoencephalitis: spectrum of imaging findings. *Am. J. Neuroradiol.* **27**: 1217–1221.

Siripanth, C. (2005) Amphizoic amoebae: pathogenic free-living protozoa; review of the literature and review of cases in Thailand. *J. Med. Assoc. Thai.* **88**: 701–707.

Smego Jr, R.A. and Durack, D.T. (1984) An experimental model for *Naegleria fowleri*-induced primary amebic meningoencephalitis in rabbits. *J. Parasitol.* **70**: 78–81.

Soltow, S.M. and Brenner, G.M. (2007) Synergistic activity of azithromycin and amphotericin B against *Naegleria fowleri in vitro* and in a mouse model of primary amebic meningoencephalitis. *Antimicrob. Agents Chemother.* **51**: 23-27.

Sparagano, O., Drouet, E., Brebant, R., Manet, E., Denoyel, G.A. and Pernin, P. (1993) Use of monoclonal antibodies to distinguish pathogenic *Naegleria fowleri* (cysts, trophozoites, or flagellate forms) from other *Naegleria* species. *J. Clin. Microbiol.* **31**: 2758–2763.

Sparagano, O., Drouet, E., Denoyel, G., Pernin, P. and Ruchaud-Sparagano, M.H. (1994) Differentiation of *Naegleria fowleri* from other species of *Naegleria* using monoclonal antibodies and the polymerase chain reaction. *Trans R. Soc. Trop. Med. Hyg.* **88**: 119–120.

Stevens, A.R., DeJonckheere, J. and Willaert, E. (1980) *Naegleria lovaniensis* new species: isolation and identification of six thermophilic strains of a new species found in association with *Naegleria fowleri. Int. J. Parasitol.* **10**: 51–64.

Stevens, A.R., Shulman, S.T., Lansen, T.A., Cichon, M.J. and Willaert, E. (1981) Primary amoebic meningoencephalitis: a report of two cases and antibiotic and immunologic studies. *J. Infect. Dis.* **143**: 193–199.

Sugita, Y., Fujii, T., Hayashi, I., Aoki, T., Yokoyama, T., Morimatsu, M., Fukuma, T. and Takamiya, Y. (1999) Primary amebic meningoencephalitis due to *Naegleria fowleri*: an autopsy case in Japan. *Pathol. Int.* **49**: 468–470.

Taylor, J.P., Hendricks, K.A. and Dingley, D.D. (1996) Amoebic meningoencephalitis. *Infect. Med.* **13**: 1021–1024.

Thong, Y.H., Rowan-Kelly, B. and Ferrante, A. (1979) Treatment of experimental *Naegleria* meningoencephalitis with a combination of amphotericin B and rifamycin. *Scand. J. Infect. Dis.* **11**: 151–153.

Toney, D.M. and Marciano-Cabral, M. (1992) Alterations in protein expression and complement resistance of pathogenic *Naegleria* amoebae. *Infect Immun.* **60**: 2781–2790.

Toney, D.M. and Marciano-Cabral, M. (1994) Membrane vesiculation of *Naegleria fowleri* amoebae as a mechanism for resisting complement damage. *J. Immunol.* **152**: 2952.

Tyndall, R.L., Ironside, K.S., Metler, P.L., Tan, E.L., Hazen, T.C. and Fliermans, C.B. (1989) Effect of thermal additions on the density and distribution of thermophilic amoebae and pathogenic *Naegleria fowleri* in a newly created cooling lake. *Appl. Environ. Microbiol.* **55**: 722–732.

Vargas-Zepeda, J., Gomez-Alcala, A.V., Vasquez-Morales, J.A., Licea-Amaya, L., De Jonckheere, J.F. and Lares-Villa, F. (2005) Successful treatment of *Naegleria fowleri* meningoencephalitis by using intravenous amphotericin B, fluconazole and rifampicin. *Arch. Med. Res.* **36**: 83–86.

Vickerman, K. (1962) Patterns of cellular organization in Limax amoebae. An electron microscope study. *Exp. Cell Res.* **26**: 497–519.

Wang, A., Kay, R., Poon, W.S. and Ng, H.K. (1993) Successful treatment of amoebic meningoencephalitis in a Chinese living in Hong Kong. *Clin. Neurol. Neurosurg.* **95**: 249–252.

Whiteman, L.Y. and Marciano-Cabral, F. (1989) Resistance of highly pathogenic *Naegleria fowleri* amoebae to complement-mediated lysis. *Infect. Immun.* **57**: 3869–3875.

Wiwanitkit, V. (2004) Review of clinical presentations in Thai patients with primary amoebic meningoencephalitis. *Med. Gen. Med.* **6**: 2–8.

Wong, M.M., Karr Jr, S.L. and Balamuth, W.B. (1975) Experimental infections with pathogenic free-living amebae in laboratory primate hosts. I (A) A study on susceptibility to *Naegleria fowleri. J. Parasitol.* **61**: 199–208.

Wong, M.M., Karr Jr, S.L. and Chow, C.K. (1977) Changes in the virulence of *Naegleria fowleri* maintained *in vitro. J. Parasitol.* **63**: 872–878.

Yoder, J.S., Blackburn, B.J., Craun, G.F., Hill, V., Levy, D.A., Chen, N., Lee, S.H., Calderon, R.L. and Beach, M.J. (2004) Surveillance for waterborne-disease outbreaks associated with recreational water – United States, 2001–2002. *MMWR Surveill. Summ.* **53**: 1–22.

A4 Blastocystis spp.

Kevin S. W. Tan

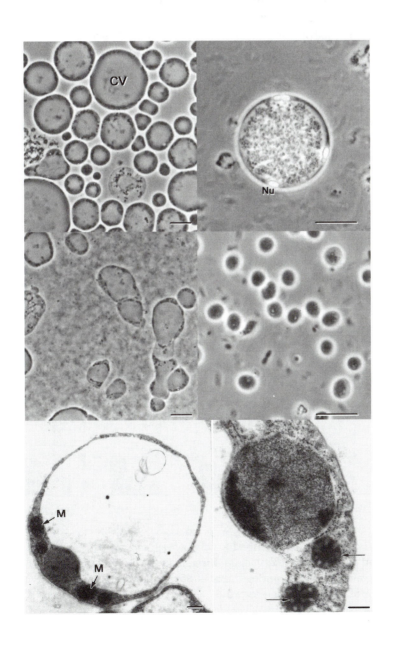

1 Introduction

The inclusion of *Blastocystis* as a chapter in a book on emerging protozoan pathogens is timely as, over the last decade, accumulating evidence suggests that, this parasite is a cause of, or is at least associated with, intestinal and other disorders. *Blastocystis* is a common enteric protozoan of humans and animals. The species found in humans, *B. hominis*, has been shown to comprise isolates that display extreme genetic variability (Clark, 1997; Noël *et al.*, 2005). Further analyses revealed that many of these genotypes were similar to those isolated from animals, which led researchers to conclude that humans are in fact inhabited by a number of zoonotic *Blastocystis* species (Abe, 2004; Abe *et al.*, 2003c; Noël *et al.*, 2005; Yoshikawa *et al.*, 2004a, 2004b). Previously, most reports on *Blastocystis* were descriptions of the organism's morphology or were epidemiological in nature, but little was known about *Blastocystis* molecular and cell biology, life cycle and pathogenesis. However, in recent years, there has been significant progress in our understanding of these aspects of *Blastocystis* biology, and this has been a result of increasing attention from the scientific community on this enigmatic organism (Figure 1). This chapter will focus on these recent advances, highlighting studies that strongly suggest *Blastocystis*' zoonotic and pathogenic potential. Excellent reviews on various topics in *Blastocystis*

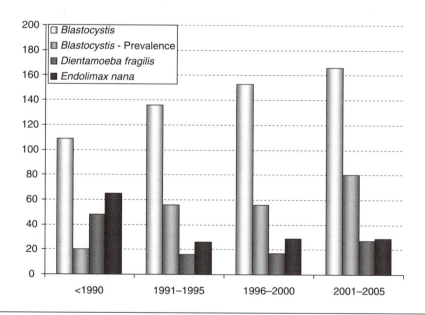

Figure 1 The number of published articles on enteric protozoa with little-understood pathogenesis (*Blastocystis, Dientamoeba fragilis* and *Endolimax nana*) but which are gaining interest among the scientific community. Note the increasing proportion of studies relating to prevalence of *Blastocystis*. Data obtained from PubMed (http://www.ncbi.nlm.nih.gov/entrez/query.fcgi).

biology can be found elsewhere (Boreham and Stenzel, 1993; Stenzel and Boreham, 1996; Tan, 2004; Tan *et al.*, 2002; Zierdt, 1991).

2 Taxonomy and Speciation

The taxonomic classification of *Blastocystis* has seen numerous twists and turns over the last two decades. Initial characterization of the organism was based on morphological and physiological criteria. Early workers erroneously described *Blastocystis* as the cyst of a flagellate, vegetable material, yeast and fungus (Zierdt, 1991). *Blastocystis* was subsequently described as a protist based on the presence of one or more nuclei, smooth and rough endoplasmic reticulum, Golgi complex, and mitochondrion-like organelles; it failed to grow on fungal media and was not killed by antifungal drugs, but was sensitive to some antiprotozoal drugs (Zierdt, 1991; Zierdt *et al.*, 1988). Zierdt (1991) classified the organism initially as a sporozoan based on morphology, cultural characteristics and schizogony-like cell division, and latter reclassified it as a sarcodine.

More recently, molecular analysis of *Blastocystis* small-subunit rRNA (ssrRNA) and elongation factor-1α (EF-1α) gene sequences have provided more information on the organism's taxonomic and phylogenetic affiliations. Partial ssrRNA analysis revealed that *B. hominis* is not monophyletic with the yeasts (*Saccharomyces*), fungi (*Neurospora*), sarcodines (*Naegleria*, *Acanthamoeba*, and *Dictyostelium*) or sporozoans (*Sarcocystis* and *Toxoplasma*; Johnson *et al.*, 1989). A later study, which involved the analysis of the complete ssrRNA gene, placed *Blastocystis* within the Stramenopiles (Silberman *et al.*, 1996). This was in contrast to phylogenetic studies involving EF-1α (Ho *et al.*, 2000; Nakamura *et al.*, 1996), which suggested that *Blastocystis* diverged before the Stramenopiles. A recent study involving multiple molecular sequence data sought to clarify this apparent contradiction (Arisue *et al.*, 2002). The phylogenetic analysis of *Blastocystis* ssrRNA, cytosolic-type 70-kDa heat-shock protein, translation elongation factor-2, and the noncatalytic 'B' subunit of vacuolar ATPase clearly demonstrated that *Blastocystis* is a Stramenopile. This study highlights the importance of analyzing multiple candidate genes in order to place organisms into a more precise taxonomic framework. The Stramenopiles, synonymous with Heterokonta and Chromista (Cavalier-Smith, 1997), are a complex assemblage of 'botanical' protists with both heterotrophic and photosynthetic representatives. This group, comprising slime nets, water molds and brown algae, are characterized by possessing flagella with mastigonemes (hair-like structures which extend laterally from the flagellum) that are hollow or straw-like. Molecular phylogenetic studies revealed that *Blastocystis* is closely related to *Proteromonas lacertae* (Arisue *et al.*, 2002; Silberman *et al.*, 1996), a flagellate of the hindgut of lizards and amphibia. Interestingly, *Blastocystis* does not possess flagella and mastigonemes, and is therefore placed in a newly created class

Blastocystea, subphylum Opalinata, infrakingdom Heterokonta, subkingdom Chromobiota, kingdom Chromista (Cavalier-Smith, 1998). This classification makes *Blastocystis* the first chromist known to colonize humans.

Blastocystis has been isolated from a wide range of hosts including primates, rodents, birds, pigs, amphibians and insects, as reviewed previously (Boreham and Stenzel, 1993). These organisms could be relatively easily established in laboratory culture and often display similar morphologies. Hence, morphological criteria is not a method of choice for speciation, and other approaches, such as optimal growth conditions, karyotyping and molecular phylogenetic analysis have been employed to investigate if *Blastocystis* from different hosts represented different species. However, early studies did base speciation on hosts of origin and organism ultra-structure, and was used to describe new species from domestic hens, geese and turkeys (Belova, 1992a, 1992b; Belova and Kostenko, 1990). Subsequent work focused on karyotyping by pulsed-field gel electrophoresis (PFGE), and was used to speciate *Blastocystis* isolated from rats (*B. ratti*; Chen *et al.*, 1997b), and reptiles including a sea snake (*B. lapemi*; Teow *et al.*, 1991), a reticulated python (*B. pythoni*), a red-footed tortoise (*B. geocheloni*) and a rhino iguana (*B. cycluri*; Singh *et al.*, 1996). There are a number of limitations to these approaches since *Blastocytis* is a highly polymorphic organism and a single isolate may exist in a variety of morphological types. Furthermore, various forms of *Blastocystis* may be found in isolates from diverse host types. Karyotypic profiles have been shown to vary significantly between *B. hominis* isolates (Carbajal *et al.*, 1997; Yoshikawa *et al.*, 2004b), and substantial intraspecies variations have also been observed for other protozoa (*Giardia intestinalis*, *Trypanosoma cruzi* and *T. brucei*; Bhattacharya *et al.*, 2000; El-Sayed *et al.*, 2000; Myler, 1993; Upcroft *et al.*, 1996). It was therefore suggested that more discriminatory approaches, such as polymerase chain reaction-restriction length polymorphism (PCR-RFLP), and phylogenetic analysis of gene sequences be used to distinguish between species and strains of *Blastocystis* (Yoshikawa *et al.*, 2004b). In fact, we now know that *Blastocystis* from avian hosts and rodents cluster together with human isolates when subjected to ssrRNA phylogenetic analysis (Abe, 2004; Noël *et al.*, 2005), and so caution is indeed warranted when deciding on methods for speciation.

Molecular phylogenetics, the study of evolutionary relationships among organisms or genes by a combination of molecular biology and statistical techniques, have been recently employed to gain a clearer understanding of *Blastocystis* isolated from humans and animals. An earlier study (Clark, 1997), revealed that *B. hominis* isolates exhibited extreme genetic diversity. Based on sequence variation in the ssrRNA gene, it was suggested that *Blastocystis* species in reality, consist of at least seven morphologically identical but genetically distinct organisms. Later studies revealed that these ribotypes (genotypes) were also present in *Blastocystis* isolated from numerous animal hosts (Abe *et al.*, 2003a, 2003b, 2003c; Noël *et al.*, 2005; Thathaisong *et al.*, 2003; Yoshikawa *et al.*, 2004a, 2004b), strongly suggesting that this parasite has zoonotic potential. In a recent, extensive study (Noël *et al.*, 2005), phylogenetic relationships were inferred for all *Blastocystis* ssrRNA genes available in GenBank at the time of study. The results showed that most of the 78 isolates could be unambiguously clustered into seven clades, with *P. lacertae* as outgroup. These clades were referred to as groups I to VII, which correlated to subtypes 1, 5, 3, 7, 6, 4 and 2 respectively of previous studies that employed a PCR-based genotyping approach (Abe, 2004; Arisue *et al.*, 2003; Yoshikawa *et al.*, 2003b, 2004b). Interestingly, *Blastocystis* from both humans and animals could be found in six of the seven groups. Because the evolutionary distances between each group was comparable to or greater than those observed for other Stramenopile species, it is likely that humans are in fact hosts for at least six different species of *Blastocystis* (Arisue *et al.*, 2003, Noël *et al.*, 2005). Based on a recent prevalence study on the geographical distribution of each *Blastocystis* genotype (Yoshikawa *et al.*, 2004b), Noël *et al.* (2005) commented on the possible origins of some of these groups. The authors suggest that group I comprised zoonotic isolates of mammalian origin, group III represented a genotype of human origin, group IV comprised zoonotic isolates of rodent origin, and groups VI and VII were likely zoonotic isolates of avian origins. These studies also suggested that transmission of *Blastocystis* can occur via animal-to-animal, human-to-animal and animal-to-human routes.

3 Cryptic, Extensive Genetic Diversity

Blastocystis isolated from humans and animals are morphologically similar. However, it is noteworthy that there have not been any systematic studies to verify if there are minute morphological variations between isolates. For example, *Blastocystis* isolated from reptiles are morphologically distinct from those isolated form humans, varying in cell size and thickness of cytoplasm, depending on the species (Singh *et al.*, 1996). However, extensive genetic variation has been observed in *Blastocystis* isolates when analyzed by random amplified polymorphic DNA (RAPD; Tan *et al.*, 2006; Yoshikawa *et al.*, 1996, 1998), PCR-RFLP (Abe *et al.*, 2003a, 2003b, 2003c; Böhm-Gloning *et al.*, 1997; Clark, 1997; Ho *et al.*, 2001; Hoevers *et al.*, 2000; Kaneda *et al.*, 2001; Snowden *et al.*, 2000; Yoshikawa *et al.*, 2003b) and PCR with genotype-specific primers (Yan *et al.*, 2006; Yoshikawa *et al.*, 2004a, 2004b). Not surprisingly, there have also been a number of studies indicating that *Blastocystis* exhibits antigenic heterogeneity. These observations were based on approaches including sodium dodecyl sulfate-polyacrylamide gel electrophoresis (SDS-PAGE) and immunoblotting (Boreham *et al.*, 1992; Kukoschke and Müller, 1991; Lanuza *et al.*, 1999; Puthia *et al.*, 2005; Tan *et al.*, 2001b), immunodiffusion (Lanuza *et al.*, 1999; Müller, 1994) and isoenzyme analysis (Gericke *et al.*, 1997; Mansour *et al.*, 1995). Clark (1997) investigated the ssrRNA sequence diversity of 30 randomly selected isolates by PCR-RFLP. The RFLP profiles (riboprints) showed that these isolates could be clustered into seven distinct genotypes (ribodemes). This study revealed the extensive genetic diversity of *Blastocystis* isolated from humans, with over 7% divergence between ribodemes 1 and 2, which is approximately four times the genetic distance between the homologous genes of *Entamoeba histolytica* and *E. dispar* (Clark, 1997). This has led to the suggestion that the conflicting reports on *Blastocystis* pathogenecity could have stemmed from different researchers studying different genotypes of the organism. A recent study employing RAPD on 16 *Blastocystis* isolates comprising eight asymptomatic and eight symptomatic isolates revealed that this technique of PCR fingerprinting was able to distinguish between isolates of varying pathogenic potential (Tan *et al.*, 2006). However, two other groups (Böhm-Gloning *et al.*, 1997; Yoshikawa *et al.*, 2004b), working on larger sample sizes could not show, by PCR-RFLP or PCR using genotype-specific primers, that a particular genotype correlated with pathogenesis of *Blastocystis*.

4 Morphology

Blastocystis is a highly polymorphic and pleomorphic organism (Figure 2A–D). The most commonly observed forms in *in vitro* culture are the vacuolar form, sometimes referred to as the central vacuole form, and the granular form (Stenzel and Boreham, 1996; Tan, 2004). Other less commonly seen forms are the amoeboid, cystic, multivacuolar and avacuolar forms (Stenzel and Boreham, 1996). In reality, cultures may present with a bewildering array of morphological forms, and one may find it difficult to assign a particular form to the cells in question. Most of the current literature on *Blastocystis* morphology contains descriptions of the major forms, by light or electron microscopy. Apart from a handful of studies describing the transition of the cyst to the vacuolar forms (Chen *et al.*, 1999; Moe *et al.*, 1999; Zaman *et al.*, 1999b), we have limited information on the roles of each form in the parasite's life cycle, and the transitions of one form to another.

Blastocystis contains organelles typically seen in eukaryotes. The most observable structures by TEM are the nuclei, Golgi apparatus and mitochondria-like organelles (Figure 2E, F). The nuclei of *Blastocystis* are spherical to ovoid, generally range from 1 to 2 μm in diameter, and are often characterized by a crescent-shaped chromatin mass at one end of the organelle (Tan *et al.*, 2001a). The presence of mitochondria-like organelles in an anaerobe such as *Blastocystis* is unusual and this warrants some attention. Some groups (Boreham and Stenzel, 1993; Tan *et al.*, 2002) have suggested that these may instead be hydrogenosomes, because biochemical analyses have shown that a number of typical mitochondrial enzymes were absent from *Blastocystis*. Hydrogenosomes were originally described in trichomonads (Lindmark and

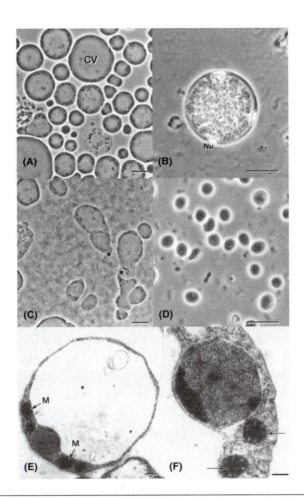

Figure 2 Morphological forms of *Blastocystis* by phase-contrast (A–D) and transmission electron (E, F) microscopy. (A) Vacuolar forms from *in vitro* axenic culture, displaying extensive size variation. (B) Granular form with distinct granular inclusions within central vacuole. (C) Amoeboid forms from colonies grown in soft agar with pseudopod-like cytoplasmic extensions (*). (D) Ovoid fecal cysts; note small size and refractile appearance. (E) Transmission electron micrograph of vacuolar form revealing large central vacuole resulting in a thin band of peripheral cytoplasm. The nucleus is flanked with electron dense mitochondria-like organelles (MLOs). (F) At higher magnification, the nucleus reveals features typical of *Blastocystis* with course nucleoplasm and peripheral chromatin mass. The MLOs possess distinct saccate cristae (arrows). CV, central vacuole; Nu, nucleus; M, MLO; bar, A–D, 10 μm; bar, E, F, 0.5 μm.

Müller, 1973) but were later shown to be present in a broad phylogenetic range of organisms. If *Blastocystis* does indeed harbor hydrogenosomes, this would make it the first description of such a structure in a Stramenophile.

The vacuolar form is spherical and may vary widely in size, ranging from 2 to 200 μm within a single culture, with diameters averaging 4–15 μm (Stenzel and Boreham, 1996; see Figure 2A). Such extreme variation in size is seldom encountered in other luminal protists such as *Entamoeba*, *Giardia* and *Trichomonas*, and the significance of such size variation in *Blastocystis* is currently unknown. There is a possibility that this variation is an artifact of *in vitro* culture, since this phenomenon has not been reported *in vivo*. The vacuolar form is characterized by a large central vacuole, which may occupy up to 90% of the cell's volume. This cellular organization results in a thin band of peripheral cytoplasm. Organelles, such as nuclei, mitochondria-like structures and Golgi apparatus, are localized to thickened regions within this cytoplasmic rim. The vacuolar form is covered by a fibrillar surface coat of varying thickness (Dunn *et al.*, 1989), and has been shown to contain a variety of carbohydrates (Lanuza *et al.*, 1996; Tan *et al.*, 1996a). The fibrillar nature of the surface coat is evident in both transmission (Zaman *et al.*, 1997a) and scanning (Tan *et al.*, 2000; Zaman *et al.*, 1999a) electron microscopy, and individual fibrils have been observed to extend up to 10 μm from the periphery of the parasite (Tan *et al.*, 2000). The exact function of the surface coat is unknown but has been suggested to play a role in trapping and degrading bacteria for nutrition (Zaman *et al.*, 1997a, 1999a), protecting against osmotic shock (Cassidy *et al.*, 1994) or to provide a mechanical barrier for functionally important plasma-membrane proteins from the immune system (Tan *et al.*, 1997). The surface coat is usually thicker in cells examined from fresh fecal matter, as compared with that of cells from *in vitro* culture (Cassidy *et al.*, 1994; Stenzel *et al.*, 1991), suggesting that it is more important for survival in the host than in culture. The central vacuole of the vacuolar form may appear empty or may contain fine to flocculant material. The vacuole may contain carbohydrates, evidenced by positive staining with periodate acid-Schiff and alcian blue (Yoshikawa *et al.*, 1995b), or lipids, by Sudan black B and Nile blue staining (Yoshikawa *et al.*, 1995c), suggesting a storage function for the organelle. There have been some descriptions of cytoplasmic contents invaginating into the central vacuole, giving rise to membrane-bound filament- or vesicle-like inclusions, sometimes containing organelles (Nasirudeen *et al.*, 2004; Pakandl, 1999; Stenzel *et al.*, 1991; Suresh *et al.*, 1995; Tan *et al.*, 2001a). The exact significance of this process is unclear although it has been postulated to be a mechanism of apoptotic-body deposition in *Blastocystis* undergoing programmed cell death (Nasirudeen *et al.*, 2001; Tan and Nasirudeen, 2005).

The granular form resembles the vacuolar form except that granules are present in the cytoplasm, or more commonly, within the central vacuole of the organism (Figure 2B). Granular forms are more commonly observed in nonaxenized, older and antibiotic-treated cultures (Boreham and Stenzel, 1993; Stenzel and Boreham, 1996; Zierdt *et al.*, 1967), and are often described as vacuolar forms containing granules, rather than a distinct entity (Boreham and Stenzel, 1993; Stenzel and Boreham, 1996). These intracellular granules are heterogeneous and have been described as myelin-like inclusions, small vesicles, crystalline granules and lipid droplets (Dunn *et al.*, 1989). It is likely that the central vacuole is multifunctional, and may be important for nutrition, storage or cell death. There are also reports that the central vacuole contains reproductive granules, which are in fact miniscule *Blastocystis* progeny (Singh *et al.*, 1995; Suresh *et al.*, 1994, 1997; Zierdt, 1991; Zierdt *et al.*, 1967), which led to the proposition that *Blastocystis* undergoes a schizogeny-like form of asexual reproduction (Govind *et al.*, 2002; Singh *et al.*, 1995; Suresh *et al.*, 1998). This has been refuted by others on the grounds that these are in fact electron-dense inclusions commonly observed in transmission electron micrographs of granular forms (Stenzel and Boreham, 1996; Tan and Stenzel, 2003). The suggestion that vacuolar and granular forms are degenerated cells or fixation artifacts (Vdovenko, 2000) has not been substantiated by further studies.

The amoeboid form is only rarely reported, and there are conflicting definitions on what constitutes this morphological type (Dunn *et al.*, 1989; McClure *et al.*, 1980; Tan *et al.*, 1996b, 2001a; Zierdt, 1991). An early description of amoeboid forms from culture reported small cells, ranging from 2.6 to 7.8 μm, with extended pseudopodia. These forms contained ingested bacteria within lysosomal-like compartments, but lacked a central vacuole, Golgi complex, surface coat and mitochondria (Dunn *et al.*, 1989). Colony growth of *Blastocystis* has been reported to induce the formation of numerous amoeba-like forms (Tan *et al.*, 1996b, 1996c, 2001a), exhibiting irregular outlines and distinct pseudopodia (Figure 2C). In contrast to Dunn *et al.*'s study, a transmission electron microscopy (TEM) study (Tan *et al.*, 2001a) described these amoeboid forms as possessing a central vacuole, surface coat with numerous Golgi bodies and mitochondria within the cytoplasmic extensions of the pseudopods. This suggests that the amoeboid forms are involved in highly active, energy-requiring processes. Taking into account the genotypic diversity of *Blastocystis*, it is not unreasonable that these differing observations may be due to isolate variations. The presence of bacteria and bacterial remnants within the amoeboid form more likely indicates a nutritional role for this form. The amoeboid form has also been postulated to play a role in pathogenesis (Tan and Suresh, 2006a, 2006b; Zierdt and Tan, 1976). However, the micrographs in two of these reports (Tan and Suresh, 2006a, 2006b) are unconvincing for amoeboid forms and, in some cases, appear more like irregular-shaped vacuolar or granular forms, a common artifact of TEM processing. The identification of stage-specific molecular markers would be useful for studies on various developmental forms of the parasite, and would obviate the problem of identifying the various forms of this pleomorphic organism.

The cyst form (Figure 2D) of *Blastocystis* is the most recently described form (Mehlhorn, 1988) and this is probably due to its infrequent occurrence in laboratory culture, small size and unique morphology when compared to the other forms (Chen *et al.*, 1999; Moe *et al.*, 1996, 1999; Stenzel and Boreham, 1991; Yoshikawa *et al.*, 2003a; Zaman *et al.*, 1995, 1997b). It is generally much smaller (2–5 μm) than the other forms, although cysts as large as 15 μm in diameter have been described in fecal material from *Macaca* monkeys (Stenzel *et al.*, 1997). *Blastocystis* cysts are variable in shape but are mostly spherical or ovoid. It is protected by a multilayered cyst wall which may or may not be covered by a loose surface coat (Moe *et al.*, 1996; Zaman *et al.*, 1995, 1997b). The cytoplasmic region of the cyst usually contains one to four nuclei, mitochondria, glycogen deposits and a variable number of small vacuoles (Abou El Naga and Negm, 2001; Moe *et al.*, 1996; Zaman *et al.*, 1995). The cyst form has been observed to survive in water for up to 19 days at normal temperatures but is fragile at extreme temperatures and in common disinfectants (Moe *et al.*, 1996). This is in contrast to the vacuolar and granular forms, which are sensitive to temperature changes, osmotic shock and exposure to air (Matsumoto *et al.*, 1987; Zierdt, 1991). Experimental infection studies on BALB/c mice (Moe *et al.*, 1997), Wistar rats (Yoshikawa *et al.*, 2004c) and a variety of bird species (Tanizaki *et al.*, 2005) with the cyst indicates that the cyst form is undoubtedly the transmissible stage of the parasite. It has been reported that the vacuolar forms can be induced to encyst *in vitro* using a variety of encystation media (Suresh *et al.*, 1993; Villar *et al.*, 1998). These encysted forms appear very similar in size and morphology to the granular form, except for the presence of a thick osmiophilic cyst-wall-like structure. Curiously, these 'cysts' were resistant to osmotic shock (Villar *et al.*, 1998) and could establish experimental infections in Wistar rats (Suresh *et al.*, 1993). Due to the significant morphological differences of these *in vitro*-derived cysts when compared to the numerous reports on fecal cysts, it is likely that the reported encystation medium induced the formation of an artifactual cell type not seen *in vivo*. A later study (Abou El Naga and Negm, 2001) described the encystation of trophic stages by storing samples in potassium dichromate solution for 2 weeks. These *in vitro*-derived cysts were apparently morphologically similar to fecal cysts, and were infective to laboratory mice. However, this method of encystation for *Blastocystis* needs to be validated, since it is possible that potassium dichromate could have enriched fecal cysts already present in culture by cytotoxic activity of the trophic stages.

Other morphological forms have been described, albeit less frequently. Avacuolar forms have been observed from a patient discharging copious volumes of diarrheal fluid (Zierdt and Tan, 1976) and from a sample taken at colonoscopy (Stenzel *et al.*, 1991). These forms lacked a central vacuole and surface coat, which probably led to the observed small size of the organism (5 μm). Unfortunately the colonoscopy sample could not be established in laboratory culture, precluding confirmatory studies on its significance (Stenzel *et al.*, 1991). Multivacuolar forms are occasionally observed in laboratory culture (Dunn *et al.*, 1989) and in stool samples (Stenzel *et al.*, 1991). It is presently unclear whether these arose from central vacuole forms, whether these vacuoles eventually coalesce to form the single central vacuole, or whether they represent a distinct stage of parasite development.

5 Programmed Cell Death

Programmed cell death (PCD) is an important feature of multicellular organisms and functions in development, regulation of cell numbers, response to external stimuli, and eradication of intracellular pathogens. The propensity for unicellular eukaryotes to undergo PCD has been well documented in recent years. This fascinating yet somewhat counterintuitive phenomenon has been reported to occur for a number of unicellular microorganisms including the parasitic protozoa. These include bacteria, yeast, trypanosomes, *Leishmania*, *Tetrahymena*, and *Dictyostelium*, as reviewed previously (Ameisen, 2002). It has been proposed that unicellular organisms do not live in isolation but interact as complex communities very much like their multicellular counterparts (Dosreis and Barcinski, 2001). Hence, roles of PCD in survival, development and regulation of numbers can conceptually be applied to unicellular organisms at the population level. The most well studied phenotype of PCD is apoptosis, which is characterized by cell shrinkage, maintenance of plasma membrane integrity, caspase activation, DNA fragmentation, and formation of apoptotic bodies, followed by ordered removal via phagocytosis. *Blastocystis* has been shown to undergo PCD, with apoptotic and nonapoptotic features (Nasirudeen and Tan, 2004, 2005; Nasirudeen *et al.*, 2001, 2004; Tan and Nasirudeen, 2005). We have shown that classical apoptotic pathways involving caspases and mitochondria exist in *Blastocystis* cells undergoing PCD. More interestingly, caspase- and mitochondria-independent PCD also occurs in *Blastocystis*, suggesting that PCD pathways in protozoa may be as complex as those of mammalian cells (Nasirudeen and Tan, 2005). The identification of novel mediators of PCD in protozoa should provide exciting new targets for antiparasite strategies.

6 Life Cycle

There have been numerous life cycles proposed for *Blastocystis* (Alexeieff, 1911; Boreham and Stenzel, 1993; Burghelea and Radulescu, 1991; Silard, 1979; Singh *et al.*, 1995; Stenzel and Boreham, 1996; Tan, 2004; Zierdt, 1973). The differences in these proposals can be rather stark and this is due mainly to the confusion about the reproductive stages seen (Govind *et al.*, 2002). Descriptions of plasmotomy and a schizogony-like mode of reproduction resulting in numerous daughter cells within the central vacuole have been proposed (Silard, 1979; Singh *et al.*, 1995; Zierdt, 1973). This cannot be accepted for lack of evidence that these stages are indeed reproductive, and these views have been cautioned by others (Tan, 2004; Tan and

Stenzel, 2003; Windsor *et al.*, 2003). These 'schizonts' are most likely variants of the classical granular forms, containing numerous spherical granule inclusions, which were misconstrued as *Blastocystis* progeny. The plethora of morphological forms seen in *in vitro* cultures and in stool samples has thus led to this confusing array of life cycles. The recent report that *Blastocystis* exhibits multiple modes of reproduction, namely, binary fission, budding, schizogony and sac-like pouches illustrates how microscopic data can be misleading (Govind *et al.*, 2002). Currently, the only accepted mode of division is binary fission. There is still a lack of information on the transition of one form to the other. The only exceptions are TEM studies on the development of cysts to vacuolar forms, which were elegantly demonstrated with a human and rat isolate (Chen *et al.*, 1999; Moe *et al.*, 1999). In these reports, fecal cysts from both humans and rat develop similarly and dramatically into vacuolar forms within 24 h of inoculation into growth media. In the most recent description of the *Blastocystis* life cycle (Tan, 2004), infection in humans and animals is initiated when the fecal cysts are ingested. These develop into vacuolar forms in the large intestines, which subsequently reproduce via binary fission. Some vacuolar forms encyst and loose their surface coat during maturation. The environmentally resistant cyst is then transmitted to humans and animals via the fecal-oral route and the cycle is repeated. Recent molecular typing data have shown that it is likely that *Blastocystis* isolated from humans actually comprise human and zoonotic genotypes (or species) of varying host specificities (Abe, 2004; Noël *et al.*, 2003, 2005). Here, we revise the life cycle, taking into account these recent observations (Figure 3).

7 Laboratory Culture

Xenic or monoxenic cultures of *Blastocystis* can be maintained in Jones's (Jones, 1946) or Boeck and Drbohlav's inspissated egg (Boeck and Drbohlav, 1925) medium. Once axenized, *Blastocystis* shows luxuriant growth in a variety of defined media such as Iscove's Modified Dulbecco's Medium (IMDM) and minimal essential medium (MEM) that have been supplemented with 10% horse serum and pre-reduced for 24–48 h prior to culture (Ho *et al.*, 1993). The optimal temperature for growth appears to be 37°C for human isolates (Zierdt and Williams, 1974) while reptilian isolates grow well at room temperature (Singh *et al.*, 1996; Yoshikawa *et al.*, 1988). The optimal temperatures for growth of other isolates have not been extensively investigated. Colony growth of *Blastocystis* has also been described (Tan *et al.*, 1996b, 1996c, 2000). *Blastocystis* colonies can be grown within soft agar (0.36%) with IMDM using the pour plate method (Tan *et al.*, 1996b, 1996c) or can be grown on the surface of solid agar (1%) as discrete colonies (Tan *et al.*, 2000). Interestingly, surface colonies were viable for up to 14 days while liquid cultures enter the death phase around day 7. The authors suggested that colony growth could be a convenient and

Figure 3 Revised life cycle for *Blastocystis* taking into account recent molecular data (Noël *et al.*, 2005) suggesting existence of zoonotic genotypes (Groups I–IV, VI and VII) with varying host specificities. Humans and animals are infected by fecal cysts which develop into vacuolar forms in the large intestines. Group I is cross-infective among mammalian and avian isolates, Groups II, III and IV comprises primate, human and rodent isolates respectively, and Groups VI and VII include avian isolates. The proposed scheme suggests that humans are potentially infected by six or more species of *Blastocystis* and that certain animals represent reservoirs for transmission to humans.

time-saving method for maintaining cultures (Tan *et al.*, 2000). Clonal growth as colonies was achieved when the agar was supplemented with sodium thioglycollate, a common reducing agent for the culture of anaerobes (Tan *et al.*, 1996b). Clonal growth using micromanipulation has also been described for *Blastocystis* isolated from the cecal contents of turkeys (Hess *et al.*, 2006). Unlike *T. vaginalis* and *G. intestinalis*, *Blastocystis* cannot be established directly into axenic cultures and usually lengthy (over a month) culturing in the presence of antibiotic cocktails is necessary before axenization is achieved (Ho *et al.*, 1993; Lanuza *et al.*, 1997; Ng and Tan, 1999). Colony growth, in the presence of antibiotics, to physically separate *Blastocystis* from accompanying bacterial flora has been useful as a step towards axenization of a human, reptilian and rat isolate (Chen *et al.*, 1997b; Ng and Tan, 1999). This was achieved by using a sterile Pasteur pipette to isolate individual colonies for expansion in liquid IMDM, with little or no bacterial contamination. So far, the most systematic axenization protocol for *Blastocystis* was described by Lanuza *et al.* (1997). Stool samples were cultured in pre-reduced modified Boeck-Drbohlav medium supplemented with antibiotics (0.4% ampicillin, 0.1% streptomycin, 0.0006% amphotericin B). Axenization was performed by the combination of partial purification of *Blastocystis* by Ficoll-metrizoic acid gradient and inoculation in fresh medium containing active antibiotics against the remaining bacteria. A total of 25 strains were obtained by this procedure with the time required for axenization ranging between 3 and 5 weeks. It is noteworthy that not all *Blastocystis* isolates can be axenized (Boreham and Stenzel, 1993; Lanuza *et al.*, 1997); this may be due to the greater dependence of some strains on the presence of bacteria, or the lack of an efficient, standardized axenization protocol for the parasite. The latter problem may never be solved considering the genetic and therefore biological heterogeneity of the organism.

8 Clinical Aspects

8.1 Epidemiology and prevalence

In an earlier review on *Blastocystis*, Boreham and Stenzel (1993) stated 'Nowhere is our knowledge of *Blastocystis* more lacking than in the study of its epidemiology. There are very few studies where the topic is even mentioned and we can only speculate about such important practical issues as the mode of transmission, importance of animal reservoirs and prevalence in symptomatic and asymptomatic individuals.' This sentiment was echoed in a later review by the same authors (Stenzel and Boreham, 1996). As if in response to this observation, the number of prevalence studies has increased by about 40% in 2001–2005 when compared to 1996–2000 and 1991–1995 (Figure 1). Questions on the mode of transmission, animal reservoirs and prevalence among various groups have been addressed and will be covered in this and other sections.

It is now clear that *Blastocystis* is an extremely common parasite with a worldwide distribution (Tan, 2004). In many surveys, it is not uncommon for *Blastocystis* to be the most frequently isolated parasite (Baldo *et al.*, 2004; Cirioni *et al.*, 1999; Florez *et al.*, 2003; Herwaldt *et al.*, 2001; Pegelow *et al.*, 1997; Saksirisampant *et al.*, 2003; Taamasri *et al.*, 2000; Wang, 2004; Windsor *et al.*, 2002). In general, prevalence of infection is higher in developing than in developed countries (Stenzel and Boreham, 1996), although carriage values vary widely from country to country, and within various communities of the same country. Carriage can be low with values of 0.5% and 3.3% among healthy individuals in Japan (Horiki *et al.*, 1997) and Singapore (Tan, unpublished observations) respectively, moderate with prevalence of 14–21% and 23% in Thailand (Yaicharoen *et al.*, 2005) and United States (Amin, 2002) respectively, or can be as high as 40–60% in developing countries such as Venezuala (46.9%; Velasquez *et al.*, 2005), Philippines (40.7%; Baldo *et al.*, 2004), and Indonesia (60%; Pegelow *et al.*, 1997). In Thailand, asymptomatic carriage ranges from 14% to 21%

(Yaicharoen *et al.*, 2005); however, Thai military personnel have been reported to harbor the parasite in the range of 36.9–44% (Leelayoova *et al.*, 2004; Taamasri *et al.*, 2002). In Thai children from different provinces, prevalence can be extremely variable, with 0.8% of children from Nan province (Waikagul *et al.*, 2002) and 45.2% from Pathum Thani province (Saksirisampant *et al.*, 2003) harboring the parasite. However, there is no standardized method of isolation and identification of the parasite, and this problem is compounded with the inherent difficulty in identifying such a polymorphic organism. Taking these into account, these comparisons of *Blastocystis* prevalence should be viewed with some caution.

Blastocystis is common in both healthy individuals and in patients suffering from bowel disease. A previous report (Clark, 1997) observed that there were approximately equal numbers of papers that either implicate or exonerate *Blastocystis* as a cause of intestinal disease. In recent years (2003–2006), this balance has tipped dramatically, with a majority of epidemiological and other studies suggesting that the parasite is pathogenic or associated with a variety of disorders (Andiran *et al.*, 2006; Barahona Rondon *et al.*, 2003; Cassano *et al.*, 2005; Chen *et al.*, 2003; Cimerman *et al.*, 2003; El-Shazly *et al.*, 2005; Florez *et al.*, 2003; Graczyk *et al.*, 2005; Gupta and Parsi, 2006; Hailemariam *et al.*, 2004; Leelayoova *et al.*, 2004; Mahmoud and Saleh, 2003; Miller *et al.*, 2003; Minvielle *et al.*, 2004; Nassir *et al.*, 2004; Nimri and Meqdam, 2004; Pasqui *et al.*, 2004; Puthia *et al.*, 2005, 2006; Rao *et al.*, 2003; Rossignol *et al.*, 2005; Tan and Suresh, 2006b; Valsecchi *et al.*, 2004; Yakoob *et al.*, 2004b; Zali *et al.*, 2004), as compared to a handful of reports that suggested otherwise (Chen *et al.*, 2003; Leder *et al.*, 2005; Tungtrongchitr *et al.*, 2004). Certain populations appear to be more susceptible to infection with *Blastocystis*, and risk factors include compromised health, poor hygiene practices, and exposure to or consumption of contaminated food or water. There are accumulating reports that *Blastocystis* is associated with intestinal disorders in individuals immunocompromised by HIV or immunosuppressive therapy (Cimerman *et al.*, 2003; Cirioni *et al.*, 1999; Florez *et al.*, 2003; Gassama *et al.*, 2001; Hailemariam *et al.*, 2004; Lebbad *et al.*, 2001; Noureldin *et al.*, 1999; Ok *et al.*, 1997; Rao *et al.*, 2003; Tasova *et al.*, 2000), suggesting that *Blastocystis* is an opportunistic pathogen. The parasite is also common among food and animal handlers (Amin, 1997; Danchaivijitr *et al.*, 2005; Khan and Alkhalife, 2005; Rajah Salim *et al.*, 1999; Requena *et al.*, 2003, Sadek *et al.*, 1997), although only one (Rajah Salim *et al.*, 1999) of these surveys included a comparison with a control population.

Besides humans, *Blastocystis* is also very common among members of the animal kingdom. Numerous animal isolates have been phylogenetically characterized which had led to the conclusion that humans are in fact hosts for a variety of *Blastocystis* species of animal origin (Noël *et al.*, 2005). The parasite has been reported in annelids, arthropods, amphibians, reptiles, birds and mammals (Boreham and Stenzel, 1993). Although isolated from a wide variety of animal hosts, the parasite appears to be frequently associated with certain animal types. The parasite is common among avian hosts, notably with domestic chickens exhibiting very high prevalence (80–100%; Belova and Kostenko, 1990; Lee and Stenzel, 1999; Yamada *et al.*, 1987). Laboratory rats are also frequently infected with *Blastocystis* (60%; Chen *et al.*, 1997a), and it was recently shown that rats can be experimentally infected with the cyst stage of the parasite (Yoshikawa *et al.*, 2004c). Others have shown that pigs exhibit extremely high prevalence for *Blastocystis* (60–90%; Burden, 1976; Pakandl, 1991). In light of recent phylogenetic data (Abe, 2004; Noël *et al.*, 2005) showing that *Blastocystis* from these hosts are similar to human isolates, it is likely that pigs, rodents and domestic birds represent huge reservoirs of the parasite for zoonotic transmission.

8.2 Infection and disease
Perhaps one of the greatest controversies surrounding *Blastocystis* is its equivocal association with bowel disease. It was, until recently, unclear if *Blastocystis* was pathogenic, caused disease under special circumstances or was merely a harmless commensal. Such confusion arose mainly from conflicting epidemiological reports which either implicated (Carrascosa *et al.*, 1996; Levy *et al.*, 1996; Logar *et al.*, 1994; Nimri and Batchoun, 1994) or absolved (Albrecht *et al.*, 1995; Junod, 1995; Shlim *et al.*, 1995) the parasite as a cause of intestinal disease. In recent years, however, numerous epidemiological, immunological and molecular studies have helped clarify the pathogenic potential of *Blastocystis*. It can be concluded that *Blastocystis* is a potential pathogen, and recent data suggests that the clinical outcome of an infection is dependent on parasite genotype, parasite load, and host immune status. The recent reports on experimental infections of laboratory rats (Yoshikawa *et al.*, 2004c) and domestic chickens (Tanizaki *et al.*, 2005) will better help ascertain the parasite's role in disease, as these bring us closer to identifying a suitable animal model.

The clinical consequences of *Blastocystis* infection are mainly diarrhea and abdominal pain as well as nonspecific gastrointestinal symptoms such as nausea, anorexia, bloating, vomiting, weight loss, lassitude, dizziness, and flatulence. A number of recent epidemiological studies have shown that *Blastocystis* is associated with disease. Most of these surveys showed the parasite is associated with intestinal disorders, most commonly diarrhea, in immunocompetent individuals (Andiran *et al.*, 2006; Barahona Rondon *et al.*, 2003; El-Shazly *et al.*, 2005; Graczyk *et al.*, 2005; Leelayoova *et al.*, 2004; Miller *et al.*, 2003; Minvielle *et al.*, 2004; Nassir *et al.*, 2004; Nimri and Meqdam, 2004; Rossignol *et al.*, 2005; Yakoob *et al.*, 2004b). Individuals compromised by HIV infection and immunosuppressive therapy also appear to be susceptible to *Blastocystis*-associated diarrhea (Florez *et al.*,

2003; Hailemariam *et al.*, 2004; Rao *et al.*, 2003). There are also accumulating reports of the association of *Blastocystis* with irritable bowel syndrome (IBS; Giacometti *et al.*, 1999; Hussain *et al.*, 1997; Markell and Udkow, 1986; Yakoob *et al.*, 2004b), a functional gastrointestinal disorder in which abdominal pain is associated with a defect or a change in bowel habits, although one study suggested that *Blastocystis* was not associated with IBS (Tungtrongchitr *et al.*, 2004). Interestingly, there are also increasing case reports implicating *Blastocystis* as a cause of cutaneous lesions, notably urticaria (Armentia *et al.*, 1993; Barahona Rondon *et al.*, 2003; Biedermann *et al.*, 2002; Cassano *et al.*, 2005; Giacometti *et al.*, 2003; Gupta and Parsi, 2006; Pasqui *et al.*, 2004; Valsecchi *et al.*, 2004). The concept of luminal protozoa as causative agents of allergy-like cutaneous lesions is interesting and has been suggested to be linked to the activation, by parasite molecules, of certain specific immune-cell subsets, such as interleukin 3-, 4-, 5- or 13-secreting Th2 cells, which mediate IgE allergic responses (Pasqui *et al.*, 2004). It was also suggested that *Blastocystis* molecules may activate the complement pathway with the generation of anaphylotoxins C3a and C5a. The interactions of these molecules with mast cells and basophils induce histamine release and subsequent related skin disorders (Valsecchi *et al.*, 2004). Taken together these studies suggest that *Blastocystis* can cause a variety of disorders, not necessarily confined to the intestinal tract.

The uncertainty about *Blastocystis* pathogenesis will be better addressed once a suitable animal model is identified. There are only a handful of studies of experimental infections in animal models, including laboratory rats, mice, guinea pigs and chicks (Moe *et al.*, 1997; Phillips and Zierdt, 1976; Silard *et al.*, 1977; Suresh *et al.*, 1995; Tanizaki *et al.*, 2005; Yoshikawa *et al.*, 2004c). BALB/c mice were successfully infected with *Blastocystis* although infections were self-limiting and produced mild symptoms (Moe *et al.*, 1997), despite intracecal inoculation with high numbers 4–8 $\times 10^6$ of the parasite. This may be due to the observation that mice do not naturally harbor *Blastocystis* (Chen *et al.*, 1997a) or that infectivity may be genotype-related. It has been observed that rats are especially susceptible to *Blastocystis* infections (Chen *et al.*, 1997a; Yoshikawa *et al.*, 2004c). This has led to the idea that rats may be ideal animal models to study various aspects of *Blastocysis* biology, including pathogenesis. In a recent study (Yoshikawa *et al.*, 2004c), specific pathogen free (SPF) Wistar rats were orally infected with *Blastocystis* cells obtained from another infected rat. With doses varying from $1 \times 10^6 - 4 \times 10^7$, all five SPF rats showed positive infection 3–4 days post inoculation. It was also observed from this study that *Blastocystis* exhibits a very low infectious dose; only 10 cysts were required to establish infection. It is not possible to generalize about infectivity at this stage, as these values may be specific to the *Blastocystis* isolate used in that particular study.

In the absence of an animal model, *in vitro* studies are useful in understanding the pathobiology of *Blastocystis*. *Blastocystis* isolates from symptomatic and asymptomatic individuals were exposed to monolayers of
Chinese hamster ovary (CHO) and adeno carcinoma HT-29 cells
(Walderich *et al.*, 1998). It was observed that parasite cells and lysates
caused significant cytopathic effects on CHO but not HT-29 cells. However,
the mode of CHO cell death was not determined in this study. In a later
study (Long *et al.*, 2001), *Blastocystis* was shown to induce inflammatory
cytokine responses in colonic epithelial cell lines HT-29 and T-84, despite
the lack of cytopathic effects. After a 24-h incubation with the cell lines,
parasite cells induced significant increases in the release of the cytokines
IL-8 and GM-CSF. Interestingly, after 6 h of incubation, it was observed that
the production of IL-8 was not increased in HT-29 cells, and even reduced
when *Escherichia coli* (bacteria or lipopolysaccharide) was co-incubated
with *Blastocystis*. The authors suggest that *Blastocystis* induces as well as
modulates the immune response in intestinal epithelial cells. The down-
regulation of inflammatory response in the presence of pro-inflammatory
factors (e.g. LPS), may serve as a survival mechanism for the parasite, as has
been shown for other enteric pathogens such as *Cryptosporidium parvum*
(Laurent *et al.*, 1997) and *Toxoplasma gondii* (Denney *et al.*, 1999). In a
recent study (Puthia *et al.*, 2006), *Blastocystis* was shown to induce contact-
independent apoptosis, F-actin rearrangement, and barrier function disruption in intestinal epithelial cells (IEC-6). These studies were facilitated
by exposing parasites, parasite secretory products or lysates to IEC-6
monolayers grown on 0.6-cm² Millicell-HA filter membrane inserts placed
within wells of tissue culture plates. This study showed that *Blastocystis* was
able to induce moderate but significant levels of apoptosis in IEC-6 cells.
The parasite was also shown to compromise IEC-6 barrier function,
evidenced by increased permeability to Lucifer yellow and decreased
transepithelial resistance. Such increase in epithelial permeability may be
linked to *Blastocystis*-associated diarrhea, and this association has been
shown for other diseases (Madara *et al.*, 1986). Interestingly, prevention of
apoptosis with caspase inhibitors failed to rescue the IEC from barrier
function disruption. The authors concluded that programmed cell death of
host cells plays a minor role in *Blastocystis* pathogenesis (Puthia *et al.*,
2006), and that other, as yet unidentified parasite factors had contributed
to loss of barrier function.

Parasite proteases, predominantly of the cysteine type, have been shown
to be important for virulence (Sajid and McKerrow, 2002). Protease activity
was characterized in an isolate (B) of *B. hominis* by azocasein and gelatin
SDS-PAGE analysis (Sio *et al.*, 2006). The parasites showed high levels of
protease activity and, by inhibition studies, were shown to be predominantly cysteine proteases. Gelatin gel SDS-PAGE revealed parasitic lysates
comprised nine protease bands of low (20–33 kDa) and high (44–75 kDa)

molecular weights. Proteases were pH-dependent and highest proteolytic activity was observed at neutral pH. From the protease profile of *B. hominis* isolate B, the majority of proteases were of the cysteine type with molecular weight ranging between 20 and 33 kDa. This study did not assay for secretory proteases though it would be worthwhile to investigate, if different genotypes or isolates of *Blastocystis* produce different levels of cysteine proteases. Such results may shed light on the pathogenesis of *Blastocystis*, as pathogenic *E. histolytica* has been shown to secrete 10- to 1000-fold more proteases when compared to the avirulent *E. dispar* (Reed *et al.*, 1989). Evidence of human IgA degradation by protozoan proteinases has previously been reported for *E. histolytica* (Kelsall and Ravdin, 1993) and *T. vaginalis* (Provenzano and Alderete, 1995). It was recently reported that proteinases from *Blastocystis* isolated from a human (isolate B) and rat (isolate WR1) contributed to the breakdown of secretory IgA, suggesting that this is a mechanism by which the parasite can persist in gut (Puthia *et al.*, 2005). Conditioned (spent) medium also showed the ability to cleave IgA suggesting active secretion of parasitic IgA proteinases. IgA proteases from isolates B and WR1 were inhibited by cysteine and aspartic-specific inhibitors respectively, suggesting that distinct variations in protease types exist among *Blastocystis* isolates.

In summary, recent reports have provided us with a better understanding of *Blastocystis* pathogenesis (Figure 4). There are now compelling case studies, and epidemiological and *in vitro* biological data, to indicate that *Blastocystis* is pathogenic, and virulence factors may include proteases, a subset of which cleaves secretory immunoglobulin A, while other parasite factors may be involved in disruption of barrier function. Exposure of parasite to intestinal epithelia induces pro- or anti-inflammatory immune responses, depending on experimental set-up. Future studies should focus on specific virulence factors and on the effect of different genotypes on host cells. Once an animal model is identified, *in vitro* findings can be correlated with *in vivo* studies.

8.3 Diagnosis and treatment

Blastocystis poses considerable challenges to the diagnostic laboratory, for a number of reasons. The uncertain pathogenesis of the protozoan does not encourage microbiologists to look for the organism in specimens. The organism can be rather nondescript even in stained preparations, and can be confused as yeast, *Cyclospora* or fat globules. The pleomorphic nature of the parasite complicates identification. Lastly, the fecal cyst may predominate in cultures but are extremely difficult to identify without concentration methods due to their small size (3–5 μm). *Blastocystis* is traditionally identified by looking for vacuolar forms in direct stool specimens (Andiran *et al.*, 2006; Katz and Taylor, 2001; Windsor *et al.*, 2002), although this approach has been shown to be rather insensitive (Leelayoova *et al.*, 2002;

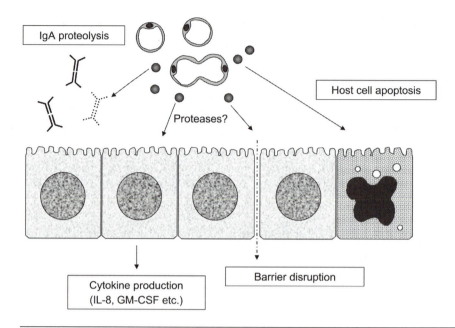

Figure 4 Model for pathogenesis of *Blastocystis*. *Blastocystis* infection may result in a variety of pathological outcomes such as secretory IgA degradation, barrier function compromise, host cell apoptosis and induction of pro-inflammatory cytokines. Ig A degradation and barrier disruption may promote the growth and invasion of neighboring pathogens.

Termmathurapoj *et al.*, 2004). Present-day diagnostic laboratories should also include fecal cysts as an indicator of infection. These can be selectively concentrated by density-gradient approaches to increase sensitivity (Zaman, 1996). A number of reports have shown that *in vitro* cultures of fecal specimens in Jones' medium can increase the pick-up rate of *Blastocystis* infections, by looking for vacuolar forms after 24–48 h incubation (Leelayoova *et al.*, 2002; Suresh and Smith, 2004; Termmathurapoj *et al.*, 2004). In a recent report (Termmathurapoj *et al.*, 2004), it was observed that *in vitro* cultures of stool specimens were six and two times more sensitive when compared to simple smears and trichrome staining respectively. However, the same study revealed that the culture method did fail to detect some parasites that simple smears and trichrome staining did, indicating that not all *Blastocystis* isolates can be easily cultured *in vitro*. It was also shown that culture expanded parasites were more amenable to PCR amplification of ssurDNA genes. This suggests that the use of subtype specific PCR primers described previously (Yoshikawa *et al.*, 2004a) in conjunction with DNA extracted from positive stool cultures may be useful for identifying and genotyping the parasite from patient samples. Genotype information may become increasingly relevant if certain genotypes are found to be associated with disease, as suggested previously (Clark, 1997).

Taken together, the accurate identification of *Blastocystis* should be carried out with a combination of direct microscopy and *in vitro* culture. Other methods of laboratory identification have not been extensively explored. ELISA- and immunofluorescence (IF)-based detection methods have been established for other enteric protozoans such as *Giardia* and *Cryptosporidia* species. There have been a handful of reports on the use of ELISA to study the host immune response to *Blastocystis* (Mahmoud and Saleh, 2003; Zierdt *et al.*, 1995), and monoclonal antibodies against *Blastocystis* have been described (Tan *et al.*, 1996a; Yoshikawa *et al.*, 1995a), which can potentially be developed for IF. However, the antigenic diversity of *Blastocystis* may pose a major obstacle to these endeavors.

The need to treat individuals infected with *Blastocystis* has been equivocal, due to the uncertain pathogenesis of the organism and the observation that the disease is often mild and self-limiting. In cases where treatment is warranted, metronidazole (Flagyl), is the most commonly prescribed antibiotic (Cassano *et al.*, 2005; Moghaddam *et al.*, 2005; Nassir *et al.*, 2004; Nigro *et al.*, 2003; Tasova *et al.*, 2000; Valsecchi *et al.*, 2004). Various drug regimens for metronidazole have been prescribed ranging from 250–750 mg three times a day for 10 days (Moghaddam *et al.*, 2005; Nassir *et al.*, 2004; Valsecchi *et al.*, 2004), 1.5 mg a day for 10 days (Cassano *et al.*, 2005; Tasova *et al.*, 2000), or metronidazole may be used in combination with other drugs such as paromomycin (Pasqui *et al.*, 2004) or co-trimoxazole (trimethoprim/sulfamethoxazole; Andiran *et al.*, 2006). The mode of action for metronidazole on *Blastocystis* is unknown but may be similar to that described for other anaerobic luminal protozoa, such as *Entamoeba*, *Giardia* and *Trichomonas*. In these protozoa, the low redox potential environment is required for normal functioning of the key metabolic enzyme pyruvate ferredoxin oxidoreductase (PFOR; Upcroft and Upcroft, 2001). This requirement also reduces the metronidazole nitro group which results in the production of cytotoxic radicals (R-NO$_2^-$; Kulda, 1999). Whether a similar mode of action is operating in *Blastocystis* is currently unknown although we have observed that the drug induces programmed cell death with a number of features resembling apoptosis in higher eukaryotes (Nasirudeen *et al.*, 2004). Apart from metronidazole, other drugs have also been used to treat blastocystosis, with variable success. These include iodoquinol (Grossman *et al.*, 1992), co-trimoxazole (Moghaddam *et al.*, 2005; Ok *et al.*, 1999) and paromomycin (Armentia *et al.*, 1993; Valsecchi *et al.*, 2004). More recently, some studies have indicated that the broad spectrum antiparasitic drug nitazoxanide is effective against *Blastocystis* (Cimerman *et al.*, 2003; Diaz *et al.*, 2003; Rossignol *et al.*, 2005). More studies should be carried out before any conclusions on the effectiveness of these drugs can be made. Until then, metronidazole should be the first line of drug defense against blastocystosis. However, treatment failures have been reported for metronidazole (Garavelli, 1991; Moghaddam *et al.*, 2005; Wilson *et al.*, 1990). In such circumstances, co-trimoxazole and nitozoxanide, may be considered as second-choice drugs. *In vitro* drug activity studies have shown that different *Blastocystis* isolates exhibit large variations in sensitivity to metronidazole (Haresh *et al.*, 1999; Yakoob *et al.*, 2004a), and the cyst form has been shown to be resistant (up to 5 mg ml^{-1}) to the cytotoxic effect of the drug (Zaman and Zaki, 1996). These observations, together with the extensive genetic heterogeneity of the organism, may offer explanations for the variability in drug susceptibilities and treatment failures.

There have been a handful of *in vitro* studies investigating the effects of drugs on *Blastocystis*. An early study (Zierdt *et al.*, 1983) tested 10 antiprotozoal drugs against four axenic strains of *Blastocystis*. Inhibitory drugs in order of effectiveness were emetine, metronidazole, furazolidone, cotrimoxazole, 5-chloro-8-hydroxy-7-iodo-quinoline (Entero-Vioform), and pentamidine. Moderately inhibitory were chloroquine and 5,7-diiodo-8-hydroxy-quinoline (Floraquin). Interestingly, paromomycin was not inhibitory. In a later study (Dunn and Boreham, 1991), the activity of 42 compounds was tested against a single isolate of *Blastocystis*, by ^{3}H-hypoxanthine incorporation. The authors showed that the 5-nitroimidazoles were generally effective against *Blastocystis*. There have been two reports on the *in vitro* activities of traditional medicine against *Blastocystis* (Sawangjaroen and Sawangjaroen, 2005; Yang *et al.*, 1996). In the earlier study (Yang *et al.*, 1996), 20 crude extracts of various traditional Chinese medicines (TCM) were tested on three axenized isolates of *Blastocystis*, while a recent study (Sawangjaroen and Sawangjaroen, 2005) tested extracts from five antidiarrheic Thai medicinal plants against fresh isolates of *Blastocystis*. Yang *et al.* (1996) observed that two of the extracts (from *Coptis chinensis* and *Brucea javanica*) were inhibitory to parasite growth at concentrations of 100 and 500 μg ml^{-1}, respectively, while Sawangjaroen and Sawangjaroen (2005) observed that extracts from the *B. javanica* seed and from *Quercus infectoria* nut gall showed the highest activity at concentration of 2000 μg ml^{-1}. These are relatively high inhibitory concentrations considering that the ID$_{50}$ of a metronidazole-sensitive isolate of *Blastocystis* can be as low as 0.06 μg ml^{-1} (Dunn and Boreham, 1991).

In summary, metronidazole appears to be the most effective drug for *Blastocystis* chemotherapy, despite some evidence for treatment failures. Treatment should be considered if diarrhea is persistent and no other pathogen apart from *Blastocystis* is identified in fecal specimens. Future studies should investigate if there is an association between genotype and variations in drug sensitivity, and should also focus on the mechanism/s of action of and resistance to metronidazole.

9 Concluding Remarks

In the last decade or so, there has been significant progress in our understanding of *Blastocystis* biology in relation to phylogenetic affiliations, genotypic diversity, clinical relevance and pathogenic potential. The parasite is genetically heterogeneous and, among human infective isolates, can be potentially divided into at least six different species. It is tempting to speculate that genotypic/species differences account for the conflicting reports on *Blastocystis* pathogenesis, and we are now in the position to test this hypothesis using *in vitro* tissue culture systems. We now have more information on the pathogenic potential of the parasite from case reports, epidemiological studies, and animal infection and *in vitro* studies. Laboratory rats appear to be good candidates for development of an animal model and major efforts should be directed at this endeavor. Studies should also focus on the identification and characterization of parasite virulence factors. The organism displays unusual morphological types and organelles (e.g. mitochondrial-like organelles) and can undergo programmed cell death, providing a treasure trove of potential discoveries for the discerning cell biologist. Research on parasitic protozoa has been aided by partial or complete sequencing of their genomes (http://www.sanger.ac.uk/Projects/Protozoa/), and these databases provide us with avenues for the identification of functional homologues. However, the current lack of a fully sequenced genome for *Blastocystis* is undesirable for research progress, but this status will inevitably change in due course.

10 Acknowledgments

I am grateful to Ng Geok Choo, N.P. Ramachandran, Manoj Kumar and A.M.A. Nasirudeen for research input and laboratory support pertaining to our work mentioned in this chapter. Research from K.S.W. Tan's laboratory has been supported by generous grants from the Academic Research Fund, National Medical Research Council and the Biomedical Research Council.

11 References

Abe, N. (2004) Molecular and phylogenetic analysis of *Blastocystis* isolates from various hosts. *Vet. Parasitol.* **120**: 235–242.

Abe, N., Wu, Z. and Yoshikawa, H. (2003a) Molecular characterization of *Blastocystis* isolates from birds by PCR with diagnostic primers and restriction fragment length polymorphism analysis of the small subunit ribosomal RNA gene. *Parasitol. Res.* **89**: 393–396.

Abe, N., Wu, Z. and Yoshikawa, H. (2003b) Molecular characterization of *Blastocystis* isolates from primates. *Vet. Parasitol.* **113**: 321–325.

Abe, N., Wu, Z. and Yoshikawa, H. (2003c) Zoonotic genotypes of *Blastocystis hominis* detected in cattle and pigs by PCR with diagnostic primers and restriction fragment length polymorphism analysis of the small subunit ribosomal RNA gene. *Parasitol. Res.* **90**: 124–128.

Abou El Naga, I.F. and Negm, A.Y. (2001) Morphology, histochemistry and infectivity of *Blastocystis hominis* cyst. *J. Egypt. Soc. Parasitol.* **31**: 627–635.

Albrecht, H., Stellbrink, H.J., Koperski, K. and Greten, H. (1995) *Blastocystis hominis* in human immunodeficiency virus-related diarrhea. *Scand. J. Gastroenterol.* **30**: 909–914.

Alexeieff, A. (1911) Sur la nature des formations dites kystes de *Trichomonas intestinalis. C. R. Soc. Biol.* **71**: 296–298.

Ameisen, J.C. (2002) On the origin, evolution, and nature of programmed cell death: a timeline of four billion years. *Cell Death Differ.* **9**: 367–393.

Amin, A.M. (1997) *Blastocystis hominis* among apparently healthy food handlers in Jeddah, Saudi Arabia. *J. Egypt. Soc. Parasitol.* **27**: 817–823.

Amin, O.M. (2002) Seasonal prevalence of intestinal parasites in the United States during 2000. *Am. J. Trop. Med. Hyg.* **66**: 799–803.

Andiran, N., Acikgoz, Z.C., Turkay, S. and Andiran, F. (2006) *Blastocystis hominis* – an emerging and imitating cause of acute abdomen in children. *J. Pediatr. Surg.* **41**: 1489–1491.

Arisue, N., Hashimoto, T., Yoshikawa, H., Nakamura, Y., Nakamura, G., Nakamura, F., Yano, T.A. and Hasegawa, M. (2002) Phylogenetic position of *Blastocystis hominis* and of stramenopiles inferred from multiple molecular sequence data. *J. Eukaryot. Microbiol.* **49**: 42–53.

Arisue, N., Hashimoto, T. and Yoshikawa, H. (2003) Sequence heterogeneity of the small subunit ribosomal RNA genes among *Blastocystis* isolates. *Parasitology* **126**: 1–9.

Armentia, A., Mendez, J., Gomez, A., Sanchis, E., Fernandez, A., De La Fuente, R. and Sanchez, P. (1993) Urticaria by *Blastocystis hominis*. Successful treatment with paromomycin. *Allergol. Immunopathol. (Madr.)* **21**: 149–151.

Baldo, E.T., Belizario, V.Y., De Leon, W.U., Kong, H.H. and Chung, D.I. (2004) Infection status of intestinal parasites in children living in residential institutions in Metro Manila, the Philippines. *Korean J. Parasitol.* **42**: 67–70.

Barahona Rondon, L., Maguina Vargas, C., Naquira Velarde, C., Terashima, I.A. and Tello, R. (2003) [Human blastocystosis: prospective study symptomatology and associated epidemiological factors]. *Rev. Gastroenterol. Peru* **23**: 29–35.

Belova, L.M. (1992a) [A finding of *Blastocystis galli* (Rhizopoda, Lobosea) in domestic turkeys]. *Parazitologiia* **26**: 166–168.

Belova, L.M. (1992b) [A new species of *Blastocystis anseri* (Protista: Rhizopoda) from domestic geese]. *Parazitologiia* **26**: 80–82.

Belova, L.M. and Kostenko, L.A. (1990) [*Blastocystis galli* sp. n. (Protista: Rhizopoda) from the intestines of domestic hens]. *Parazitologiia* **24**: 164–168.

Bhattacharya, A., Satish, S., Bagchi, A. and Bhattacharya, S. (2000) The genome of *Entamoeba histolytica. Int. J. Parasitol.* **30**: 401–410.

Biedermann, T., Hartmann, K., Sing, A. and Przybilla, B. (2002) Hypersensitivity to non-steroidal anti-inflammatory drugs and chronic urticaria cured by treatment of *Blastocystis hominis* infection. *Br. J. Dermatol.* **146**: 1113–1114.

Boeck, W.C. and Drbohlav, J. (1925) The cultivation of *Entamoeba histolytica. Am. J. Hyg.* **5**: 371–407.

Böhm-Gloning, B., Knobloch, J. and Walderich, B. (1997) Five subgroups of *Blastocystis hominis* from symptomatic and asymptomatic patients revealed by restriction site analysis of PCR-amplified 16S-like rDNA. *Trop. Med. Int. Health* **2**: 771–778.

Boreham, P.F. and Stenzel, D.J. (1993) *Blastocystis* in humans and animals: morphology, biology, and epizootiology. *Adv. Parasitol.* **32**: 1–70.

Boreham, P.F., Upcroft, J.A. and Dunn, L.A. (1992) Protein and DNA evidence for two demes of *Blastocystis hominis* from humans. *Int. J. Parasitol.* **22**: 49–53.

Burden, D.J. (1976) *Blastocystis* sp.: a parasite of pigs. *Parasitology* **73**: 4–5.

Burghelea, B. and Radulescu, S. (1991) Ultrastructural evidence for a possible differentiation way in the life-cycle of *Blastocystis hominis. Rom. Arch. Microbiol. Immunol.* **50**: 231–244.

Carbajal, J.A., del Castillo, L., Lanuza, M.D., Villar, J. and Borras, R. (1997) Karyotypic diversity among *Blastocystis hominis* isolates. *Int J. Parasitol.* **27**: 941–945.

Carrascosa, M., Martinez, J. and Perez-Castrillon, J.L. (1996) Hemorrhagic proctosigmoiditis and *Blastocystis hominis* infection. *Ann. Intern. Med.* **124**: 278–279.

Cassano, N., Scoppio, B.M., Loviglio, M.C. and Vena, G.A. (2005) Remission of delayed pressure urticaria after eradication of *Blastocystis hominis. Acta Derm. Venereol.* **85**: 357–358.

Cassidy, M.F., Stenzel, D.J. and Boreham, P.F. (1994) Electron microscopy of surface structures of *Blastocystis* sp. from different hosts. *Parasitol. Res.* **80**: 505–511.

Cavalier-Smith, T. (1997) Sagenista and Bigyra, two phyla of heterotrophic heterokont Chromists. *Arch. Protistenkd.* **148**: 253–267.

Cavalier-Smith, T. (1998) A revised six-kingdom system of life. *Biol. Rev. Camb. Philos. Soc.* **73**: 203–266.

Chen, X.Q., Singh, M., Ho, L.C., Moe, K.T., Tan, S.W. and Yap, E.H. (1997a) A survey of *Blastocystis* sp. in rodents. *Lab. Anim. Sci.* **47**: 91–94.

Chen, X.Q., Singh, M., Ho, L.C., Tan, S.W., Ng, G.C., Moe, K.T. and Yap, E.H. (1997b) Description of a *Blastocystis* species from *Rattus norvegicus. Parasitol. Res.* **83**: 313–318.

Chen, X.Q., Singh, M., Howe, J., Ho, L.C., Tan, S.W. and Yap, E.H. (1999) In vitro encystation and excystation of *Blastocystis ratti. Parasitology* **118** (Pt 2): 151–160.

Chen, T.L., Chan, C.C., Chen, H.P., Fung, C.P., Lin, C.P., Chan, W.L. and Liu, C.Y. (2003) Clinical characteristics and endoscopic findings associated with *Blastocystis hominis* in healthy adults. *Am. J. Trop. Med. Hyg.* **69**: 213–216.

Cimerman, S., Ladeira, M.C. and Iuliano, W.A. (2003) [Blastocystosis: nitazoxanide as a new therapeutic option]. *Rev. Soc. Bras. Med. Trop.* **36**: 415–417.

Cirioni, O., Giacometti, A., Drenaggi, D., Ancarani, F. and Scalise, G. (1999) Prevalence and clinical relevance of *Blastocystis hominis* in diverse patient cohorts. *Eur. J. Epidemiol.* **15**: 389–393.

Clark, C.G. (1997) Extensive genetic diversity in *Blastocystis hominis. Mol. Biochem. Parasitol.* **87**: 79–83.

Danchaivijitr, S., Rongrungruang, Y., Kachintorn, U., Techasathit, V., Pakaworavuthi, S. and Kachintorn, K. (2005) Prevalence and effectiveness of an education program on intestinal pathogens in food handlers. *J. Med. Assoc. Thai.* **88** (Suppl 10): S31–S35.

Denney, C.F., Eckmann, L. and Reed, S.L. (1999) Chemokine secretion of human cells in response to *Toxoplasma gondii* infection. *Infect. Immun.* **67**: 1547–1552.

Diaz, E., Mondragon, J., Ramirez, E. and Bernal, R. (2003) Epidemiology and control of intestinal parasites with nitazoxanide in children in Mexico. *Am. J. Trop. Med. Hyg.* **68**: 384–385.

Dosreis, G.A. and Barcinski, M.A. (2001) Apoptosis and parasitism: from the parasite to the host immune response. *Adv. Parasitol.* **49**: 133–161.

Dunn, L.A. and Boreham, P.F. (1991) The in-vitro activity of drugs against *Blastocystis hominis. J. Antimicrob. Chemother.* **27**: 507–516.

Dunn, L.A., Boreham, P.F. and Stenzel, D.J. (1989) Ultrastructural variation of *Blastocystis hominis* stocks in culture. *Int. J. Parasitol.* **19**: 43–56.

El-Sayed, N.M., Hegde, P., Quackenbush, J., Melville, S.E. and Donelson, J.E. (2000) The African trypanosome genome. *Int. J. Parasitol.* **30**: 329–345.

El-Shazly, A.M., Abdel-Magied, A.A., El-Beshbishi, S.N., El-Nahas, H.A., Fouad, M.A. and Monib, M.S. (2005) *Blastocystis hominis* among symptomatic and asymptomatic individuals in Talkha Center, Dakahlia Governorate, Egypt. *J. Egypt. Soc. Parasitol.* **35**: 653–666.

Florez, A.C., Garcia, D.A., Moncada, L. and Beltran, M. (2003) [Prevalence of microsporidia and other intestinal parasites in patients with HIV infection, Bogota, 2001]. *Biomedica* **23**: 274–282.

Garavelli, P.L. (1991) The therapy of blastocystosis. *J. Chemother.* **3** (Suppl 1): 245–246.

Gassama, A., Sow, P.S., Fall, F., Camara, P., Gueye-N'diaye, A., Seng, R., Samb, B., M'boup, S. and Aidara-Kane, A. (2001) Ordinary and opportunistic enteropathogens associated with diarrhea in Senegalese adults in relation to human immunodeficiency virus serostatus. *Int. J. Infect. Dis.* **5**: 192–198.

Gericke, A.S., Burchard, G.D., Knobloch, J. and Walderich, B. (1997) Isoenzyme patterns of *Blastocystis hominis* patient isolates derived from symptomatic and healthy carriers. *Trop. Med. Int. Health* **2**: 245–253.

Giacometti, A., Cirioni, O., Fiorentini, A., Fortuna, M. and Scalise, G. (1999) Irritable bowel syndrome in patients with *Blastocystis hominis* infection. *Eur. J. Clin. Microbiol. Infect. Dis.* **18**: 436–439.

Giacometti, A., Cirioni, O., Antonicelli, L., D'amato, G., Silvestri, C., Del Prete, M.S. and Scalise, G. (2003) Prevalence of intestinal parasites among individuals with allergic skin diseases. *J. Parasitol.* **89**: 490–492.

Govind, S.K., Khairul, A.A. and Smith, H.V. (2002) Multiple reproductive processes in *Blastocystis. Trends Parasitol.* **18**: 528.

Graczyk, T.K., Shiff, C.K., Tamang, L., Munsaka, F., Beitin, A.M., Moss, W.J. (2005) The association of *Blastocystis hominis* and *Endolimax nana* with diarrheal stools in Zambian school-age children. *Parasitol. Res.* **98**: 38–43.

Grossman, I., Weiss, L.M., Simon, D., Tanowitz, H.B. and Wittner, M. (1992) *Blastocystis hominis* in hospital employees. *Am. J. Gastroenterol.* **87**: 729–732.

Gupta, R. and Parsi, K. (2006) Chronic urticaria due to *Blastocystis hominis. Aust. J. Dermatol.* **47**: 117–119.

Hailemariam, G., Kassu, A., Abebe, G., Abate, E., Damte, D., Mekonnen, E. and Ota, F. (2004) Intestinal parasitic infections in HIV/AIDS and HIV seronegative individuals in a teaching hospital, Ethiopia. *Jpn J. Infect. Dis.* **57**: 41–43.

Haresh, K., Suresh, K., Khairul Anus, A. and Saminathan, S. (1999) Isolate resistance of *Blastocystis hominis* to metronidazole. *Trop. Med. Int. Health* **4**: 274–277.

Herwaldt, B.L., De Arroyave, K.R., Wahlquist, S.P., De Merida, A.M., Lopez, A. S. and Juranek, D.D. (2001) Multiyear prospective study of intestinal parasitism in a cohort of Peace Corps volunteers in Guatemala. *J. Clin. Microbiol.* **39**: 34–42.

Hess, M., Kolbe, T., Grabensteiner, E. and Prosl, H. (2006) Clonal cultures of *Histomonas meleagridis, Tetratrichomonas gallinarum* and a *Blastocystis* sp. established through micromanipulation. *Parasitology* **133**: 547–554.

Ho, L.C., Singh, M., Suresh, G., Ng, G.C. and Yap, E.H. (1993) Axenic culture of *Blastocystis hominis* in Iscove's modified Dulbecco's medium. *Parasitol. Res.* **79**: 614–616.

Ho, L.C., Armiugam, A., Jeyaseelan, K., Yap, E.H. and Singh, M. (2000) *Blastocystis* elongation factor-1 alpha: genomic organization, taxonomy and phylogenetic relationships. *Parasitology* **121** (Pt 2): 135–144.

Ho, L.C., Jeyaseelan, K. and Singh, M. (2001) Use of the elongation factor-1 alpha gene in a polymerase chain reaction-based restriction-fragment-length polymorphism analysis of genetic heterogeneity among *Blastocystis* species. *Mol. Biochem. Parasitol.* **112**: 287–291.

Hoevers, J., Holman, P., Logan, K., Hommel, M., Ashford, R. and Snowden, K. (2000) Restriction-fragment-length polymorphism analysis of small-subunit rRNA genes of *Blastocystis hominis* isolates from geographically diverse human hosts. *Parasitol. Res.* **86**: 57–61.

Horiki, N., Maruyama, M., Fujita, Y., Yonekura, T., Minato, S. and Kaneda, Y. (1997) Epidemiologic survey of *Blastocystis hominis* infection in Japan. *Am. J. Trop. Med. Hyg.* **56**: 370–374.

Hussain, R., Jaferi, W., Zuberi, S., Baqai, R., Abrar, N., Ahmed, A. and Zaman, V. (1997) Significantly increased IgG2 subclass antibody levels to *Blastocystis hominis* in patients with irritable bowel syndrome. *Am. J. Trop. Med. Hyg.* **56**: 301–306.

Johnson, A.M., Thanou, A., Boreham, P.F. and Baverstock, P.R. (1989) *Blastocystis hominis*: phylogenetic affinities determined by rRNA sequence comparison. *Exp. Parasitol.* **68**: 283–288.

Jones, W.R. (1946) The experimental infection of rats with *Entamoeba histolytica*. *Ann. Trop. Med. Parasitol.* **40**: 130.

Junod, C. (1995) [*Blastocystis hominis*: a common commensal in the colon. Study of prevalence in different populations of Paris]. *Presse Med.* **24**: 1684–1688.

Kaneda, Y., Horiki, N., Cheng, X.J., Fujita, Y., Maruyama, M. and Tachibana, H. (2001) Ribodemes of *Blastocystis hominis* isolated in Japan. *Am. J. Trop. Med. Hyg.* **65**: 393–396.

Katz, D.E. and Taylor, D.N. (2001) Parasitic infections of the gastrointestinal tract. *Gastroenterol. Clin. N. Am.* **30**: 797–815, x.

Kelsall, B.L. and Ravdin, J.I. (1993) Degradation of human IgA by *Entamoeba histolytica*. *J. Infect. Dis.* **168**: 1319–1322.

Khan, Z.A. and Alkhalife, I.S. (2005) Prevalence of *Blastocystis hominis* among 'healthy' food handlers in Dammam, Saudi Arabia. *J. Egypt. Soc. Parasitol.* **35**: 395–401.

Kukoschke, K.G. and Müller, H.E. (1991) SDS-PAGE and immunological analysis of different axenic *Blastocystis hominis* strains. *J. Med. Microbiol.* **35**: 35–39.

Kulda, J. (1999) Trichomonads, hydrogenosomes and drug resistance. *Int. J. Parasitol.* **29**: 199–212.

Lanuza, M.D., Carbajal, J.A. and Borras, R. (1996) Identification of surface coat carbohydrates in *Blastocystis hominis* by lectin probes. *Int. J. Parasitol.* **26**: 527–532.

Lanuza, M.D., Carbajal, J.A., Villar, J. and Borras, R. (1997) Description of an improved method for *Blastocystis hominis* culture and axenization. *Parasitol. Res.* **83**: 60–63.

Lanuza, M.D., Carbajal, J.A., Villar, J., Mir, A. and Borras, R. (1999) Soluble-protein and antigenic heterogeneity in axenic *Blastocystis hominis* isolates: pathogenic implications. *Parasitol. Res.* **85**: 93–97.

Laurent, F., Eckmann, L., Savidge, T.C., Morgan, G., Theodos, C., Naciri, M. and Kagnoff, M.F. (1997) *Cryptosporidium parvum* infection of human intestinal epithelial cells induces the polarized secretion of C-X-C chemokines. *Infect. Immun.* **65**: 5067–5073.

Lebbad, M., Norrgren, H., Naucler, A., Dias, F., Andersson, S. and Linder, E. (2001) Intestinal parasites in HIV-2 associated AIDS cases with chronic diarrhoea in Guinea-Bissau. *Acta Trop.* **80**: 45–49.

Leder, K., Hellard, M.E., Sinclair, M.I., Fairley, C.K. and Wolfe, R. (2005) No correlation between clinical symptoms and *Blastocystis hominis* in immunocompetent individuals. *J. Gastroenterol. Hepatol.* **20**: 1390–1394.

Lee, M.G. and Stenzel, D.J. (1999) A survey of *Blastocystis* in domestic chickens. *Parasitol. Res.* **85**: 109–117.

Leelayoova, S., Taamasri, P., Rangsin, R., Naaglor, T., Thathaisong, U. and Mungthin, M. (2002) In-vitro cultivation: a sensitive method for detecting *Blastocystis hominis. Ann. Trop. Med. Parasitol.* **96**: 803–807.

Leelayoova, S., Rangsin, R., Taamasri, P., Naaglor, T., Thathaisong, U. and Mungthin, M. (2004) Evidence of waterborne transmission of *Blastocystis hominis. Am. J. Trop. Med. Hyg.* **70**: 658–662.

Levy, Y., George, J. and Shoenfeld, Y. (1996) Severe *Blastocystis hominis* in an elderly man. *J. Infect.* **33**: 57–59.

Lindmark, D.G. and Müller, M. (1973) Hydrogenosome, a cytoplasmic organelle of the anaerobic flagellate *Tritrichomonas* foetus, and its role in pyruvate metabolism. *J. Biol. Chem.* **248**: 7724–7728.

Logar, J., Andlovic, A. and Poljsak-Prijatelj, M. (1994) Incidence of *Blastocystis hominis* in patients with diarrhoea. *J. Infect.* **28**: 151–154.

Long, H.Y., Handschack, A., Konig, W. and Ambrosch, A. (2001) *Blastocystis hominis* modulates immune responses and cytokine release in colonic epithelial cells. *Parasitol. Res.* **87**: 1029–1030.

Madara, J.L., Barenberg, D. and Carlson, S. (1986) Effects of cytochalasin D on occluding junctions of intestinal absorptive cells: further evidence that the cytoskeleton may influence paracellular permeability and junctional charge selectivity. *J. Cell Biol.* **102**: 2125–2136.

Mahmoud, M.S. and Saleh, W.A. (2003) Secretory and humoral antibody responses to *Blastocystis hominis* in symptomatic and asymptomatic human infections. *J. Egypt. Soc. Parasitol.* **33**: 13–30.

Mansour, N.S., Mikhail, E.M., El Masry, N.A., Sabry, A.G. and Mohareb, E.W. (1995) Biochemical characterisation of human isolates of *Blastocystis hominis. J. Med. Microbiol.* **42**: 304–307.

Markell, E.K. and Udkow, M.P. (1986) *Blastocystis hominis*: pathogen or fellow traveler? *Am. J. Trop. Med. Hyg.* **35**: 1023–1026.

Matsumoto, Y., Yamada, M. and Yoshida, Y. (1987) Light-microscopical appearance and ultrastructure of *Blastocystis hominis*, an intestinal parasite of man. *Zentralbl. Bakteriol. Mikrobiol. Hyg. A* **264**: 379–385.

Mcclure, H.M., Strobert, E.A. and Healy, G.R. (1980) *Blastocystis hominis* in a pig-tailed macaque: a potential enteric pathogen for nonhuman primates. *Lab. Anim. Sci.* **30**: 890–894.

Mehlhorn, H. (1988) *Blastocystis hominis*, Brumpt 1912: are there different stages or species? *Parasitol. Res.* **74**: 393–395.

Miller, S.A., Rosario, C.L., Rojas, E. and Scorza, J.V. (2003) Intestinal parasitic infection and associated symptoms in children attending day care centres in Trujillo, Venezuela. *Trop. Med. Int. Health* **8**: 342–347.

Minvielle, M.C., Pezzani, B.C., Cordoba, M.A., De Luca, M.M., Apezteguia, M.C. and Basualdo, J.A. Epidemiological survey of *Giardia* spp. and *Blastocystis hominis* in an Argentinian rural community. *Korean J. Parasitol.* **42**: 121–127.

Moe, K.T., Singh, M., Howe, J., Ho, L.C., Tan, S.W., Ng, G.C., Chen, X.Q. and Yap, E.H. (1996) Observations on the ultrastructure and viability of the cystic stage of *Blastocystis hominis* from human feces. *Parasitol. Res.* **82**: 439–444.

Moe, K.T., Singh, M., Howe, J., Ho, L. C., Tan, S.W., Chen, X.Q., Ng, G.C. and Yap, E.H. (1997) Experimental *Blastocystis hominis* infection in laboratory mice. *Parasitol. Res.* **83**: 319–325.

Moe, K.T., Singh, M., Howe, J., Ho, L.C., Tan, S.W., Chen, X.Q. and Yap, E.H. (1999) Development of *Blastocystis hominis* cysts into vacuolar forms in vitro. *Parasitol. Res.* **85**: 103–108.

Moghaddam, D.D., Ghadirian, E. and Azami, M. (2005) *Blastocystis hominis* and the evaluation of efficacy of metronidazole and trimethoprim/sulfamethoxazole. *Parasitol. Res.* **96**: 273–275.

Müller, H.E. (1994) Four serologically different groups within the species *Blastocystis hominis*. *Zentralbl. Bakteriol.* **280**: 403–408.

Myler, P.J. (1993) Molecular variation in trypanosomes. *Acta Trop.* **53**: 205–225.

Nakamura, Y., Hashimoto, T., Yoshikawa, H., Kamaishi, T., Nakamura, F., Okamoto, K. and Hasegawa, M. (1996) Phylogenetic position of *Blastocystis hominis* that contains cytochrome-free mitochondria, inferred from the protein phylogeny of elongation factor 1 alpha. *Mol. Biochem. Parasitol.* **77**: 241–245.

Nasirudeen, A.M. and Tan, K.S. (2004) Caspase-3-like protease influences but is not essential for DNA fragmentation in *Blastocystis* undergoing apoptosis. *Eur. J. Cell Biol.* **83**: 477–482.

Nasirudeen, A.M. and Tan, K.S. (2005) Programmed cell death in *Blastocystis hominis* occurs independently of caspase and mitochondrial pathways. *Biochimie* **87**: 489–497.

Nasirudeen, A.M., Tan, K.S., Singh, M. and Yap, E.H. (2001) Programmed cell death in a human intestinal parasite, *Blastocystis hominis*. *Parasitology* **123**: 235–246.

Nasirudeen, A.M., Hian, Y.E., Singh, M. and Tan, K.S. (2004) Metronidazole induces programmed cell death in the protozoan parasite *Blastocystis hominis*. *Microbiology* **150**: 33–43.

Nassir, E., Awad, J., Abel, A.B., Khoury, J., Shay, M. and Lejbkowicz, F. (2004) *Blastocystis hominis* as a cause of hypoalbuminemia and anasarca. *Eur. J. Clin. Microbiol. Infect. Dis.* **23**: 399–402.

Ng, G.C. and Tan, K.S. (1999) Colony growth as a step towards axenization of *Blastocystis* isolates. *Parasitol. Res.* **85**: 678–679.

Nigro, L., Larocca, L., Massarelli, L., Patamia, I., Minniti, S., Palermo, F. and Cacopardo, B. (2003) A placebo-controlled treatment trial of *Blastocystis hominis* infection with metronidazole. *J. Travel Med.* **10**: 128–130.

Nimri, L. and Batchoun, R. (1994) Intestinal colonization of symptomatic and asymptomatic schoolchildren with *Blastocystis hominis. J. Clin. Microbiol.* **32**: 2865–2866.

Nimri, L.F. and Meqdam, M. (2004) Enteropathogens associated with cases of gastroenteritis in a rural population in Jordan. *Clin. Microbiol. Infect.* **10**: 634–639.

Noel, C., Peyronnet, C., Gerbod, D., Edgcomb, V.P., Delgado-Viscogliosi, P., Sogin, M.L., Capron, M., Viscogliosi, E. and Zenner, L. (2003) Phylogenetic analysis of *Blastocystis* isolates from different hosts based on the comparison of small-subunit rRNA gene sequences. *Mol. Biochem. Parasitol.* **126**: 119–123.

Noël, C., Dufernez, F., Gerbod, D., *et al.* (2005) Molecular phylogenies of *Blastocystis* isolates from different hosts: implications for genetic diversity, identification of species, and zoonosis. *J. Clin. Microbiol.* **43**: 348–355.

Noureldin, M.S., Shaltout, A.A., El Hamshary, E.M. and Ali, M.E. (1999) Opportunistic intestinal protozoal infections in immunocompromised children. *J. Egypt. Soc. Parasitol.* **29**: 951–961.

Ok, U.Z., Cirit, M., Uner, A., Ok, E., Akcicek, F., Basci, A. and Ozcel, M.A. (1997) Cryptosporidiosis and blastocystosis in renal transplant recipients. *Nephron* **75**: 171–174.

Ok, U.Z., Girginkardesler, N., Balcioglu, C., Ertan, P., Pirildar, T. and Kilimcioglu, A.A. (1999) Effect of trimethoprim-sulfamethaxazole in *Blastocystis hominis* infection. *Am. J. Gastroenterol.* **94**: 3245–3247.

Pakandl, M. (1991) Occurrence of *Blastocystis* sp. in pigs. *Folia Parasitol. (Praha)* **38**: 297–301.

Pakandl, M. (1999) *Blastocystis* sp. from pigs: ultrastructural changes occurring during polyxenic cultivation in Iscove's modified Dulbecco's medium. *Parasitol. Res.* **85**: 743–748.

Pasqui, A.L., Savini, E., Saletti, M., Guzzo, C., Puccetti, L. and Auteri, A. (2004) Chronic urticaria and *Blastocystis hominis* infection: a case report. *Eur. Rev. Med. Pharmacol. Sci.* **8**: 117–120.

Pegelow, K., Gross, R., Pietrzik, K., Lukito, W., Richards, A.L. and Fryauff, D.J. (1997) Parasitological and nutritional situation of school children in the Sukaraja district, West Java, Indonesia. *Southeast Asian J. Trop. Med. Public Health* **28**: 173–190.

Phillips, B.P. and Zierdt, C.H. (1976) *Blastocystis hominis*: pathogenic potential in human patients and in gnotobiotes. *Exp. Parasitol.* **39**: 358–364.

Provenzano, D. and Alderete, J.F. (1995) Analysis of human immunoglobulin-degrading cysteine proteinases of *Trichomonas vaginalis. Infect. Immun.* **63**: 3388–3395.

Puthia, M.K., Vaithilingam, A., Lu, J. and Tan, K.S. (2005) Degradation of human secretory immunoglobulin A by *Blastocystis. Parasitol. Res.* **97**: 386–389.

Puthia, M.K., Sio, S.W., Lu, J. and Tan, K.S. (2006) *Blastocystis ratti* induces contact-independent apoptosis, F-actin rearrangement, and barrier function disruption in IEC-6 cells. *Infect. Immun.* **74**: 4114–4123.

Rajah Salim, H., Suresh Kumar, G., Vellayan, S., Mak, J.W., Khairul Anuar, A., Init, I., Vennila, G.D., Saminathan, R. and Ramakrishnan, K. (1999) *Blastocystis* in animal handlers. *Parasitol. Res.* **85**: 1032–1033.

Rao, K., Sekar, U., Iraivan, K.T., Abraham, G. and Soundararajan, P. (2003) *Blastocystis hominis* – an emerging cause of diarrhoea in renal transplant recipients. *J. Assoc. Physicians India* **51**: 719–721.

Reed, S.L., Keene, W.E. and Mckerrow, J.H. (1989) Thiol proteinase expression and pathogenicity of *Entamoeba histolytica. J. Clin. Microbiol.* **27**: 2772–2777.

Requena, I., Hernandez, Y., Ramsay, M., Salazar, C. and Devera, R. (2003) [Prevalence of *Blastocystis hominis* among food handlers from Caroni municipality, Bolivar State, Venezuela]. *Cad. Saude Publica* **19**: 1721–1727.

Rossignol, J.F., Kabil, S.M., Said, M., Samir, H. and Younis, A.M. (2005) Effect of nitazoxanide in persistent diarrhea and enteritis associated with *Blastocystis hominis. Clin. Gastroenterol. Hepatol.* **3**: 987–991.

Sadek, Y., El-Fakahany, A.F., Lashin, A.H. and El-Salam, F.A. (1997) Intestinal parasites among food-handlers in Qualyobia Governorate, with reference to the pathogenic parasite *Blastocystis hominis. J. Egypt. Soc. Parasitol.* **27**: 471–478.

Sajid, M. and Mckerrow, J.H. (2002) Cysteine proteases of parasitic organisms. *Mol. Biochem. Parasitol.* **120**: 1–21.

Saksirisampant, W., Nuchprayoon, S., Wiwanitkit, V., Yenthakam, S. and Ampavasiri, A. (2003) Intestinal parasitic infestations among children in an orphanage in Pathum Thani province. *J. Med. Assoc. Thai.* **86** (Suppl 2): S263–S270.

Sawangjaroen, N. and Sawangjaroen, K. (2005) The effects of extracts from anti-diarrheic Thai medicinal plants on the in vitro growth of the intestinal protozoa parasite: *Blastocystis hominis. J. Ethnopharmacol.* **98**: 67–72.

Shlim, D.R., Hoge, C.W., Rajah, R., Rabold, J.G. and Echeverria, P. (1995) Is *Blastocystis hominis* a cause of diarrhea in travelers? A prospective controlled study in Nepal. *Clin. Infect. Dis.* **21**: 97–101.

Silard, R. (1979) Contributions to *Blastocystis hominis* studies. Aspects of degenerescence. *Arch. Roum. Pathol. Exp. Microbiol.* **38**: 105–114.

Silard, R., Petrovici, M., Panaitescu, D. and Stoicescu, V. (1977) *Blastocystis hominis* in the liver of *Cricetus auratus. Arch. Roum. Pathol. Exp. Microbiol.* **36**: 55–60.

Silberman, J.D., Sogin, M.L., Leipe, D.D. and Clark, C.G. (1996) Human parasite finds taxonomic home. *Nature* **380**: 398.

Singh, M., Suresh, K., Ho, L.C., Ng, G.C. and Yap, E.H. (1995) Elucidation of the life cycle of the intestinal protozoan *Blastocystis hominis. Parasitol. Res.* **81**: 446–450.

Singh, M., Ho, L.C., Yap, A.L., Ng, G.C., Tan, S.W., Moe, K.T. and Yap, E.H. (1996) Axenic culture of reptilian *Blastocystis* isolates in monophasic medium and speciation by karyotypic typing. *Parasitol. Res.* **82**: 165–169.

Sio, S.W., Puthia, M.K., Lee, A.S., Lu, J. and Tan, K.S. (2006) Protease activity of *Blastocystis hominis. Parasitol. Res.* **99**: 126–130.

Snowden, K., Logan, K., Blozinski, C., Hoevers, J. and Holman, P. (2000) Restriction-fragment-length polymorphism analysis of small-subunit rRNA genes of *Blastocystis* isolates from animal hosts. *Parasitol. Res.* **86**: 62–66.

Stenzel, D.J. and Boreham, P.F. (1991) A cyst-like stage of *Blastocystis hominis. Int. J. Parasitol.* **21**: 613–615.

Stenzel, D.J. and Boreham, P.F. (1996) *Blastocystis hominis* revisited. *Clin. Microbiol. Rev.* **9**: 563–584.

Stenzel, D.J., Boreham, P.F. and Mcdougall, R. (1991) Ultrastructure of *Blastocystis hominis* in human stool samples. *Int. J. Parasitol.* **21**: 807–812.

Stenzel, D.J., Lee, M.G. and Boreham, P.F. (1997) Morphological differences in *Blastocystis* cysts – an indication of different species? *Parasitol. Res.* **83**: 452–457.

Suresh, K. and Smith, H. (2004) Comparison of methods for detecting *Blastocystis hominis. Eur. J. Clin. Microbiol. Infect. Dis.* **23**: 509–511.

Suresh, K., Ng, G.C., Ramachandran, N.P., Ho, L.C., Yap, E.H. and Singh, M. (1993) In vitro encystment and experimental infections of *Blastocystis hominis. Parasitol. Res.* **79**: 456–460.

Suresh, K., Chong, S.Y., Howe, J., Ho, L.C., Ng, G.C., Yap, E.H. and Singh, M. (1995) Tubulovesicular elements in *Blastocystis hominis* from the caecum of experimentally-infected rats. *Int. J. Parasitol.* **25**: 123–126.

Suresh, K., Howe, J., Ng, G.C., Ho, L.C., Ramachandran, N.P., Loh, A.K., Yap, E.H. and Singh, M. (1994) A multiple fission-like mode of asexual reproduction in *Blastocystis hominis. Parasitol. Res.* **80**: 523–527.

Suresh, K., Mak, J.W., Chuong, L.S., Ragunathan, T. and Init, I. (1997) Sac-like pouches in *Blastocystis* from the house lizard *Cosymbotus platyurus. Parasitol. Res.* **83**: 523–525.

Suresh, K., Init, I., Reuel, P.A., Rajah, S., Lokman, H. and Khairul Anuar, A. (1998) Glycerol with fetal calf serum – a better cryoprotectant for *Blastocystis hominis. Parasitol. Res.* **84**: 321–322.

Taamasri, P., Mungthin, M., Rangsin, R., Tongupprakarn, B., Areekul, W. and Leelayoova, S. (2000) Transmission of intestinal blastocystosis related to the quality of drinking water. *Southeast Asian J. Trop. Med. Public Health* **31**: 112–117.

Taamasri, P., Leelayoova, S., Rangsin, R., Naaglor, T., Ketupanya, A. and Mungthin, M. (2002) Prevalence of *Blastocystis hominis* carriage in Thai army personnel based in Chonburi, Thailand. *Mil. Med.* **167**: 643–646.

Tan, K.S. (2004) *Blastocystis* in humans and animals: new insights using modern methodologies. *Vet. Parasitol.* **126**: 121–144.

Tan, K.S. and Nasirudeen, A.M. (2005) Protozoan programmed cell death – insights from *Blastocystis* deathstyles. *Trends Parasitol.* **21**: 547–550.

Tan, K.S. and Stenzel, D.J. (2003) Multiple reproductive processes in *Blastocystis*: proceed with caution. *Trends Parasitol.* **19**: 290–291.

Tan, T.C. and Suresh, K.G. (2006a) Amoeboid form of *Blastocystis hominis* – a detailed ultrastructural insight. *Parasitol. Res.* **99**: 737–742.

Tan, T.C. and Suresh, K.G. (2006b) Predominance of amoeboid forms of *Blastocystis hominis* in isolates from symptomatic patients. *Parasitol. Res.* **98**: 189–193.

Tan, S.W., Ho, L.C., Moe, K.T., Chen, X.Q., Ng, G.C., Yap, E.H. and Singh, M. (1996a) Production and characterization of murine monoclonal antibodies to *Blastocystis hominis*. *Int. J. Parasitol.* **26**: 375–381.

Tan, S.W., Singh, M., Thong, K.T., Ho, L.C., Moe, K.T., Chen, X.Q., Ng, G.C. and Yap, E.H. (1996b) Clonal growth of *Blastocystis hominis* in soft agar with sodium thioglycollate. *Parasitol. Res.* **82**: 737–739.

Tan, S.W., Singh, M., Yap, E.H., Ho, L.C., Moe, K.T., Howe, J. and Ng, G.C. (1996c) Colony formation of *Blastocystis hominis* in soft agar. *Parasitol. Res.* **82**: 375–377.

Tan, S.W., Singh, M., Ho, L.C., Howe, J., Moe, K.T., Chen, X.Q., Ng, G.C. and Yap, E.H. (1997) Survival of *Blastocystis hominis* clones after exposure to a cytotoxic monoclonal antibody. *Int. J. Parasitol.* **27**: 947–954.

Tan, K.S., Ng, G.C., Quek, E., Howe, J., Ramachandran, N.P., Yap, E.H. and Singh, M. (2000) *Blastocystis hominis*: a simplified, high-efficiency method for clonal growth on solid agar. *Exp. Parasitol.* **96**: 9–15.

Tan, K.S., Howe, J., Yap, E.H. and Singh, M. (2001a) Do *Blastocystis hominis* colony forms undergo programmed cell death? *Parasitol. Res.* **87**: 362–367.

Tan, K.S., Ibrahim, M., Ng, G.C., Nasirudeen, A.M., Ho, L.C., Yap, E.H. and Singh, M. (2001b) Exposure of *Blastocystis* species to a cytotoxic monoclonal antibody. *Parasitol. Res.* **87**: 534–538.

Tan, K.S., Singh, M. and Yap, E.H. (2002) Recent advances in *Blastocystis hominis* research: hot spots in terra incognita. *Int. J. Parasitol.* **32**: 789–804.

Tan, T.C., Suresh, K.G., Thong, K.L. and Smith, H.V. (2006) PCR fingerprinting of *Blastocystis* isolated from symptomatic and asymptomatic human hosts. *Parasitol. Res.* **99**: 459–465.

Tanizaki, A., Yoshikawa, H., Iwatani, S. and Kimata, I. (2005) Infectivity of *Blastocystis* isolates from chickens, quails and geese in chickens. *Parasitol. Res.* **96**: 57–61.

Tasova, Y., Sahin, B., Koltas, S. and Paydas, S. (2000) Clinical significance and frequency of *Blastocystis hominis* in Turkish patients with hematological malignancy. *Acta Med. Okayama* **54**: 133–136.

Teow, W.L., Zaman, V., Ng, G.C., Chan, Y.C., Yap, E.H., Howe, J., Gopalakrishnakone, P. and Singh, M. (1991) A *Blastocystis* species from the sea-snake, *Lapemis hardwickii* (Serpentes: Hydrophiidae). *Int. J. Parasitol.* **21**: 723–726.

Termmathurapoj, S., Leelayoova, S., Aimpun, P., Thathaisong, U., Nimmanon, T., Taamasri, P. and Mungthin, M. (2004) The usefulness of short-term in vitro cultivation for the detection and molecular study of *Blastocystis hominis* in stool specimens. *Parasitol. Res.* **93**: 445–447.

Thathaisong, U., Worapong, J., Mungthin, M., Tan-Ariya, P., Viputtigul, K., Sudatis, A., Noonai, A. and Leelayoova, S. (2003) *Blastocystis* isolates from a pig and a horse are closely related to *Blastocystis hominis. J. Clin. Microbiol.* **41**: 967–975.

Tungtrongchitr, A., Manatsathit, S., Kositchaiwat, C., Ongrotchanakun, J., Munkong, N., Chinabutr, P., Leelakusolvong, S. and Chaicumpa, W. (2004) *Blastocystis hominis* infection in irritable bowel syndrome patients. *Southeast Asian J. Trop. Med. Public Health* **35**: 705–710.

Upcroft, P. and Upcroft, J.A. (2001) Drug targets and mechanisms of resistance in the anaerobic protozoa. *Clin. Microbiol. Rev.* **14**: 150–164.

Upcroft, J.A., Chen, N. and Upcroft, P. (1996) Mapping variation in chromosome homologues of different *Giardia* strains. *Mol. Biochem. Parasitol.* **76**: 135–143.

Valsecchi, R., Leghissa, P. and Greco, V. (2004) Cutaneous lesions in *Blastocystis hominis* infection. *Acta Derm. Venereol.* **84**: 322–323.

Vdovenko, A.A. (2000) *Blastocystis hominis*: origin and significance of vacuolar and granular forms. *Parasitol. Res.* **86**: 8–10.

Velasquez, V., Caldera, R., Wong, W., Cermeno, G., Fuentes, M., Blanco, Y., Aponte, M. and Devera, R. (2005) [Blastocystosis: a high prevalence of cases found in patients from Health Center of Soledad, Anzoategui State, Venezuela]. *Rev. Soc. Bras. Med. Trop.* **38**: 356–357.

Villar, J., Carbajal, J.A., Lanuza, M.D., Munoz, C. and Borras, R. (1998) In vitro encystation of *Blastocystis hominis*: a kinetics and cytochemistry study. *Parasitol. Res.* **84**: 54–58.

Waikagul, J., Krudsood, S., Radomyos, P., Radomyos, B., Chalemrut, K., Jonsuksuntigul, P., Kojima, S., Looareesuwan, S. and Thaineau, W. (2002) A cross-sectional study of intestinal parasitic infections among schoolchildren in Nan Province, Northern Thailand. *Southeast Asian J. Trop. Med. Public Health* **33**: 218–223.

Walderich, B., Bernauer, S., Renner, M., Knobloch, J. and Burchard, G.D. (1998) Cytopathic effects of *Blastocystis hominis* on Chinese hamster ovary (CHO) and adeno carcinoma HT29 cell cultures. *Trop. Med. Int. Health* **3**: 385–390.

Wang, L.C. (2004) Changing patterns in intestinal parasitic infections among Southeast Asian laborers in Taiwan. *Parasitol. Res.* **92**: 18–21.

Wilson, K.W., Winget, D. and Wilks, S. (1990) *Blastocystis hominis* infection: signs and symptoms in patients at Wilford Hall Medical Center. *Mil. Med.* **155**: 394–396.

Windsor, J.J., Macfarlane, L., Hughes-Thapa, G., Jones, S.K. and Whiteside, T.M. (2002) Incidence of *Blastocystis hominis* in faecal samples submitted for routine microbiological analysis. *Br. J. Biomed. Sci.* **59**: 154–157.

Windsor, J.J., Stenzel, D.J. and Macfarlane, L. (2003) Multiple reproductive processes in *Blastocystis hominis. Trends Parasitol.* **19**: 289–290.

Yaicharoen, R., Sripochang, S., Sermsart, B. and Pidetcha, P. (2005) Prevalence of *Blastocystis hominis* infection in asymptomatic individuals from Bangkok, Thailand. *Southeast Asian J. Trop. Med. Public Health* **36** (Suppl 4): 17–20.

Yakoob, J., Jafri, W., Jafri, N., Islam, M. and Asim Beg, M. (2004a) In vitro susceptibility of *Blastocystis hominis* isolated from patients with irritable bowel syndrome. *Br. J. Biomed. Sci.* **61**: 75–77.

Yakoob, J., Jafri, W., Jafri, N., Khan, R., Islam, M., Beg, M.A. and Zaman, V. (2004b) Irritable bowel syndrome: in search of an etiology: role of *Blastocystis hominis*. *Am. J. Trop. Med. Hyg.* **70**: 383–385.

Yamada, M., Yoshikawa, H., Tegoshi, T., Matsumoto, Y., Yoshikawa, T., Shiota, T. and Yoshida, Y. (1987) Light microscopical study of *Blastocystis* spp. in monkeys and fowls. *Parasitol. Res.* **73**: 527–531.

Yan, Y., Su, S., Lai, R., Liao, H., Ye, J., Li, X., Luo, X. and Chen, G. (2006) Genetic variability of *Blastocystis hominis* isolates in China. *Parasitol. Res.* **99**: 597–601.

Yang, L.Q., Singh, M., Yap, E.H., Ng, G.C., Xu, H.X. and Sim, K.Y. (1996) In vitro response of *Blastocystis hominis* against traditional Chinese medicine. *J. Ethnopharmacol.* **55**: 35–42.

Yoshikawa, H., Yamada, M. and Yoshida, Y. (1988) Freeze-fracture study of *Blastocystis hominis*. *J. Protozool.* **35**: 522–528.

Yoshikawa, H., Katagiri, I., Li, X.H., Beckerhapak, M. and Graves, D.C. (1995a) Antigenic differences between *Blastocystis hominis* and *Blastocystis* sp. revealed by polyclonal and monoclonal antibodies. *J. Protozool. Res.* **5**: 118–128.

Yoshikawa, H., Kuwayama, N. and Enose, Y. (1995b) Histochemical detection of carbohydrates of *Blastocystis hominis*. *J. Eukaryot. Microbiol.* **42**: 70–74.

Yoshikawa, H., Satoh, J. and Enose, Y. (1995c) Light and electron microscopic localization of lipids in *Blastocystis hominis*. *J. Electron Microsc. (Tokyo)* **44**: 100–103.

Yoshikawa, H., Nagono, I., Yap, E.H., Singh, M. and Takahashi, Y. (1996) DNA polymorphism revealed by arbitrary primers polymerase chain reaction among *Blastocystis* strains isolated from humans, a chicken, and a reptile. *J. Eukaryot. Microbiol.* **43**: 127–130.

Yoshikawa, H., Nagano, I., Wu, Z., Yap, E.H., Singh, M. and Takahashi, Y. (1998) Genomic polymorphism among *Blastocystis hominis* strains and development of subtype-specific diagnostic primers. *Mol. Cell Probes* **12**: 153–159.

Yoshikawa, H., Nagashima, M., Morimoto, K., Yamanouti, Y., Yap, E.H. and Singh, M. (2003a) Freeze-fracture and cytochemical studies on the in vitro cyst form of reptilian *Blastocystis pythoni*. *J. Eukaryot. Microbiol.* **50**: 70–75.

Yoshikawa, H., Wu, Z., Nagano, I. and Takahashi, Y. (2003b) Molecular comparative studies among *Blastocystis* isolates obtained from humans and animals. *J. Parasitol.* **89**: 585–594.

Yoshikawa, H., Abe, N. and Wu, Z. (2004a) PCR-based identification of zoonotic isolates of *Blastocystis* from mammals and birds. *Microbiology* **150**: 1147–1151.

Yoshikawa, H., Wu, Z., Kimata, I., Iseki, M., Ali, I.K., Hossain, M.B., Zaman, V., Haque, R. and Takahashi, Y. (2004b) Polymerase chain reaction-based genotype classification among human *Blastocystis hominis* populations isolated from different countries. *Parasitol. Res.* **92**: 22–29.

Yoshikawa, H., Yoshida, K., Nakajima, A., Yamanari, K., Iwatani, S. and Kimata, I. (2004c) Fecal-oral transmission of the cyst form of *Blastocystis hominis* in rats. *Parasitol. Res.* **94**: 391–396.

Zali, M.R., Mehr, A.J., Rezaian, M., Meamar, A.R., Vaziri, S. and Mohraz, M. (2004) Prevalence of intestinal parasitic pathogens among HIV-positive individuals in Iran. *Jpn. J. Infect. Dis.* **57**: 268–270.

Zaman, V. (1996) The diagnosis of *Blastocystis hominis* cysts in human faeces. *J. Infect.* **33**: 15–16.

Zaman, V. and Zaki, M. (1996) Resistance of *Blastocystis hominis* cysts to metronidazole. *Trop. Med. Int. Health* **1**: 677–678.

Zaman, V., Howe, J. and Ng, M. (1995) Ultrastructure of *Blastocystis hominis* cysts. *Parasitol. Res.* **81**: 465–469.

Zaman, V., Howe, J. and Ng, M. (1997a) Observations on the surface coat of *Blastocystis hominis*. *Parasitol. Res.* **83**: 731–733.

Zaman, V., Howe, J. and Ng, M. (1997b) Variation in the cyst morphology of *Blastocystis hominis*. *Parasitol. Res.* **83**: 306–308.

Zaman, V., Howe, J., Ng, M. and Goh, T.K. (1999a) Scanning electron microscopy of the surface coat of *Blastocystis hominis*. *Parasitol. Res.* **85**: 974–976.

Zaman, V., Zaki, M., Manzoor, M., Howe, J. and Ng, M. (1999b) Postcystic development of *Blastocystis hominis*. *Parasitol. Res.* **85**: 437–440.

Zierdt, C.H. (1973) Studies of *Blastocystis hominis*. *J. Protozool.* **20**: 114–121.

Zierdt, C.H. (1991) *Blastocystis hominis* – past and future. *Clin. Microbiol. Rev.* **4**: 61–79.

Zierdt, C.H. and Tan, H.K. (1976) Ultrastructure and light microscope appearance of *Blastocystis hominis* in a patient with enteric disease. *Z. Parasitenkd.* **50**: 277–283.

Zierdt, C.H. and Williams, R.L. (1974) *Blastocystis hominis*: axenic cultivation. *Exp. Parasitol.* **36**: 233–243.

Zierdt, C.H., Rude, W.S. and Bull, B.S. (1967) Protozoan characteristics of *Blastocystis hominis*. *Am. J. Clin. Pathol.* **48**: 495–501.

Zierdt, C.H., Swan, J.C. and Hosseini, J. (1983) In vitro response of *Blastocystis hominis* to antiprotozoal drugs. *J. Protozool.* **30**: 332–334.

Zierdt, C.H., Donnolley, C.T., Müller, J. and Constantopoulos, G. (1988) Biochemical and ultrastructural study of *Blastocystis hominis*. *J. Clin. Microbiol.* **26**: 965–970.

Zierdt, C.H., Zierdt, W.S. and Nagy, B. (1995) Enzyme-linked immunosorbent assay for detection of serum antibody to *Blastocystis hominis* in symptomatic infections. *J. Parasitol.* **81**: 127–129.

B Apicomplexans

B1 *Cryptosporidium spp.*

Stanley Dean Rider Jr.
and Guan Zhu

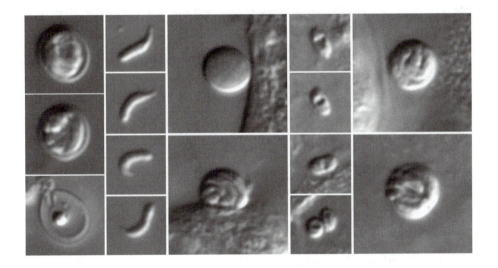

1 Introduction

Cryptosporidium species have become increasingly recognized as important pathogens of humans and domestic and wild animals. Major emphasis has been placed on these species in the last two or three decades with *Cryptosporidium parvum* and *C. hominis* being recognized as major causes of severe and sometimes fatal diarrheal diseases in humans. This review focuses on cryptosporidiosis in humans and animals, and serves as a guide that covers our basic understanding of these protozoan pathogens – beginning with a historical perspective on the group and culminating with our current knowledge on the disease and future directions for developing much-needed methods to control these diseases. A number of outstanding reviews on *Cryptosporidium* have been published in recent years, and are recommended throughout the text for additional reading. As it is not possible to include every relevant reference, readers are also encouraged to seek information present in the primary literature to gain greater insight into the details of *Cryptosporidium* biology that pique their own interests.

2 An Emerging Disease

What makes *Cryptosporidium* species an emerging threat is a complex issue, but a major driving force appears to be awareness. Our knowledge is still limited, but our understanding of *Cryptosporidium* continues to grow. As discussed later, *Cryptosporidium* gained recognition as a prevalent cause of diarrhea in humans, sometimes being present in stool samples from as much as 37% of the individuals in a population. Although cryptosporidiosis is of worldwide concern, the data presented here is biased toward information available in the US. However, similar trends in awareness and research focus can be seen around the world.

A review of the literature represented in PubMed (http://www.ncbi.nlm.nih.gov) indicated over 3,500 articles containing '*Cryptosporidium*' or 'Cryptosporidiosis' in the title. Prior to the mid 1980s *Cryptosporidium* was poorly represented in the primary scientific and medical literature (Figure 1). This contrasts what is seen for giardiasis, a common protozoan diarrhea, which has been more extensively studied for many years (Figure 1). About one third of the publications on *Cryptosporidium* were generated within the last 6 years. Word frequency counts of titles from *Cryptosporidium* articles indicated (as one might expect) content associated with disease, symptoms of the disease, or host species. Assuming that the words present in the title are a reflection of the textual content of the manuscript, 'water' and 'detection' as well as '*Giardia*', 'AIDS' and 'HIV' are also among the major areas of emphasis.

The emphasis areas were partly spurred by the discovery of *C. parvum* as a cause of chronic or fatal disease in immunocompromised patients, particularly those also infected with HIV and presenting AIDS. Reports covering cryptosporidiosis in HIV/AIDS patients rose throughout the 1980s until

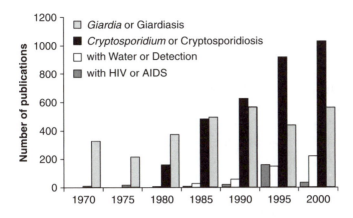

Figure 1 Publications with various key words in their titles. Binning is done every 5 years, except for the first bin, which represents all publications prior to and including 1970. The PubMed database search was performed on 18 September 2006. Publications listing 'water' or 'detection' and publications listing 'HIV' or 'AIDS' in their titles are subsets of those containing '*Cryptosporidium*' or 'cryptosporidiosis' in their titles.

the mid 1990s before declining somewhat (Figure 1). The awareness of *Cryptosporidium* as an opportunistic pathogen associated with AIDS, however, also coincided with other epidemiological studies that revealed *Cryptosporidium* was a major cause of human and animal disease in cases where immunocompetent individuals displayed symptoms of diarrhea.

In the 1990s, three significant events occurred that might underlie the subsequent changing trends in *Cryptosporidium* awareness and research reports: (i) The number of new cases (annually) of HIV/AIDS began to decline in the United States, (ii) antiretroviral therapies for HIV were emerging, and (iii) an infamous waterborne outbreak of cryptosporidosis occurred due to contamination in the water supply of the US city of Milwaukee, Wisconsin. Public and governmental awareness of *Cryptosporidium* as a general human health concern beyond that of an AIDS opportunistic agent began to take hold. Numerous smaller outbreaks have since been reported for both foodborne and waterborne cryptosporidosis and the number of publications regarding water quality and monitoring have steadily increased (Figure 1). Although *Giardia lamblia* is the most frequently reported cause of nonbacterial diarrhea in the United States (Hlavsa *et al.*, 2005), the number of manuscripts covering *Cryptosporidium* and cryptosporidiosis now appears to rival the number covering *Giardia* and giardiasis.

Today, *Cryptosporidium* is recognized as a major public concern and is currently listed as a 'Nationally notifiable infectious disease' by the US Centers for Disease Control and Prevention (CDC). Both the US National Institutes of Health (NIH) and the CDC currently list *C. parvum* as a priority pathogen. The US Environmental Protection Agency (EPA) and the

Drinking Water Inspectorate (DWI) of the UK are among the groups that have established standardized testing protocols and guidelines to protect public water supplies against *Cryptosporidium*. Additionally, focused reviews of *Cryptosporidium* incidence in different geographical regions and over long periods of time are beginning to appear. Recently, the World Health Organization (WHO) added both *Cryptosporidium* and *Giardia* to their list of 'neglected diseases' and have also drafted safe drinking water and risk assessment guides for *Cryptosporidium*. From these reports, it is clear that *Cryptosporidium* species are ubiquitous and important parasites around the world.

3 The Apicomplexa

Cryptosporidium species belong to the phylum Apicomplexa. Members within the phylum (formerly Sporozoa) share the morphological distinction of having a group of specialized organelles (the apical complex) that is present during the invasive stage of the life cycle (Figure 2). During infection this apical complex provides many of the necessary components for both attachment to and invasion of a host cell as well as locomotion. Although all members within the phylum share this essential feature, considerable diversity exists among the Apicomplexa. Some species are monoxenous (single host, e.g., *Eimeria* and

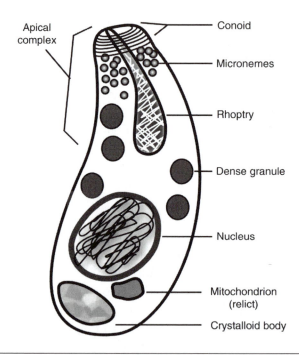

Figure 2 Artistic rendering of a *Cryptosporidium* sporozoite with various organelles depicted. The apical complex is comprised of the conoid, micronemes, rhoptry and dense granules.

Isospora) while others are heteroxenous (multiple hosts, e.g., *Plasmodium* and *Toxoplasma*). Some apicomplexans are blood parasites, while others infect the digestive system or various organs. The most diverse group is represented by the gregarines, which are restricted to invertebrate hosts.

To provide a context for understanding the apicomplexan clade, several important genera are worth mentioning. Of the Apicomplexa, malaria parasites (*Plasmodium* spp. within the class Haemosporida) are world-renowned and malaria (as a disease) has been known since ancient times. Other well-known blood parasite species lie within the genera *Theileria* and *Babesia* (class Piroplasmida). These bloodborne species require arthropod vectors for dissemination and to complete their life cycles. In addition to cryptosporidial species, intestinal parasites of humans and animals include members of the *Cyclospora* and *Isospora* genera and many different *Eimeria* species (Emeriidae) may infect and cause economic losses in farmed chickens and other animals. Members of the genera *Neospora* and *Sarcocystis* (Sarcosystidae) are known to induce birth defects or abortion in domesticated animals. While *Toxoplasma gondii* (Sarcosystidae) can cause similar teratogenicity in humans and domesticated animals, it is also capable of causing encephalitis in immunocompromised individuals. In short, these examples embody the point that all members of the Apicomplexa are parasitic, and many are of great medical or veterinary importance.

The classification of apicomplexans has been the subject of a number of debates. A classification is presented in Figure 3, but this representation

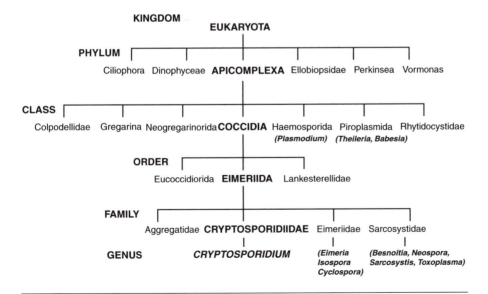

Figure 3 Classification of *Cryptosporidium*. Several other genera are listed in parentheses below the classes or families, including many other parasites of medical or veterinary importance. Please refer to the text for more information.

may not reflect the opinions of various experts within the scientific community. Recent molecular evidence indicates that the evolutionary relationships among the Apicomplexa may be significantly different from their taxonomic classification. *Cryptosporidium*, for example, may have a closer relationship to the gregarines, rather than to other eimeriid coccidia (Barta and Thompson, 2006). It is generally not debated that *Cryptosporidium* species are eukaryotes (protozoa, alveolates) within the phylum Apicomplexa. However, classification schemes within the Apicomplexa continue to invoke substantial scrutiny. Despite intensive efforts, a recent authoritative text on the classification of the protozoa indicates that subdivisions that include *Cryptosporidium* and related species are artificial, not monophyletic, and presently unresolved (Adl *et al.*, 2005).

4 *Cryptosporidium* spp.

The family Cryptosporidiidae contains a single genus – *Cryptosporidium*, which contains less than 20 different well-described species. Dr. Ernest Tyzzer established this family approximately 100 years ago (Tyzzer, 1907). The type species (*C. muris*) was discovered in the gastric glands in the stomachs of mice, where it was observed to complete its life cycle. This species was described in this seminal work because it did not match any previously reported parasite of domesticated mice. The basic morphology, life cycle (discussed later), and sessile nature led to it being described as coccidialike. However, in its original description, other aspects of the parasite, such as the presence of an 'attachment organ' and 'idiophilic granules', were referred to as resembling the gregarines. Oocysts of this new species, present in the mouse feces, were different from other eimeriid coccidia in that they were fully sporulated, but lacked any apparent sporocyst within the oocyst. However, the reasoning behind the choice to use '*Cryptosporidium*' (hidden spore) was not revealed until a later publication. *Cryptosporidium parvum* was then described a few years later.

Morphological descriptions of oocysts from different cryptosporidial species are all very similar, but can be important for species identification when combined with host information (Xiao *et al.*, 2004). Oocysts are roughly spherical and range from 4 to 8 μm in length or width. Length/width ratios range from 1.0 to 1.2 depending upon species. Within the oocyst, sporozoites may be visible, along with a structure that is referred to as the residual body (Figure 4). Although oocysts are considered to be the infective stage of the parasite, it is the individual zoite that bears the apical complex and invades a host cell.

An excellent review of cryptosporidial species has been generated, along with a proposal that molecular, as well as morphological and host range data be generated before proposing a new species (Xiao *et al.*, 2004). At the time of that publication, the authors accepted 13 described species as meeting criteria sufficient enough to warrant a novel species designation.

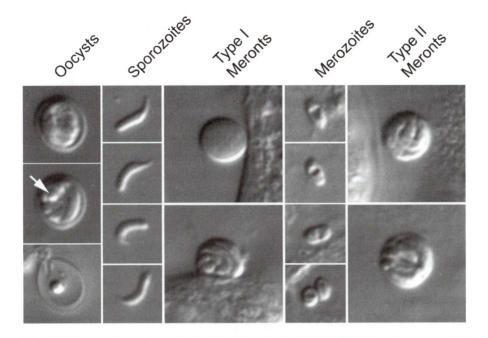

Oocysts   Sporozoites   Type I Meronts   Merozoites   Type II Meronts

Figure 4 The *C. parvum* asexual life cycle. The top oocyst contains four sporozoites surrounded by the oocyst wall. The residual body can be seen in the middle oocyst (*arrow*). The bottom oocyst has excysted – only an empty shell with an open suture and the residual body remains. Excysted sporozoites then develop into type-I meronts. The top meront has not yet undergone cellularization, while portions of the eight merozoites can be seen within the mature meront in the bottom image. Upon emergence, merozoites invade nearby cells and frequently form clusters of two or more merozoites (as seen in the bottom image). Type-II meronts may form from type-I merozoites and contain only four developing merozoites. The sexual stages (not pictured) are then thought to develop from type-II merozoites. The meronts and merozoites were from *in vitro* cultures at 24 h post infection with HCT-8 cells as hosts.

At least 15 more species had been proposed but were deprecated for various reasons. Since then, at least two additional species have been proposed, and each proposal includes the criteria that were recommended to justify a new species definition. The most straightforward method for discriminating these species is through the use of molecular (DNA-based) markers. Of the 15 or so cryptosporidial species (Table 1), many have been found to infect human beings as minor hosts, or as opportunistic parasites (Caccio, 2005). Humans are commonly infected by *C. parvum* and *C. hominis* that are morphologically indistinguishable. However, the former is zoonotic, while the latter is more specific to humans. The two species also have overlapping host ranges and both appear to infect humans with similar frequency (Table 2, and references included therein). *C. meleagridis* was originally isolated from turkeys, but is also found

Table I Characteristics of *Cryptosporidium* species affecting humans and animals[a]

Species	Human Occurrence	Major Hosts	Oocyst Dimensions
C. hominis	Primary	Primates, dugongs, sheep	5.0 × 4.2
C. parvum	Primary	Mammals	5.0 × 4.5
C. meleagridis	Common	Birds (intestinal)	5.2 × 4.6
C. felis	Infrequent	Cats, mammals	5.0 × 4.5
C. canis	Infrequent	Dogs	4.9 × 4.7
C. muris	Infrequent	Rodents, mammals	8.4 × 6.3
C. suis	Infrequent	Pigs	4.6 × 4.2
C. andersoni	Infrequent	Ruminant mammals	7.4 × 5.5
C. baileyi	Opportunistic	Birds (respiratory)	6.2 × 4.6
C. wrairi	NR	Guinea pigs	5.4 × 4.6
C. galli	NR	Birds (intestinal)	8.2 × 6.3
C. serpentis	NR	Reptiles	6.2 × 5.3
C. saurophilum	NR	Reptiles	5.0 × 4.7
C. molnari	NR	Fish	4.7 × 4.4
C. bovis	NR	Cattle	4.8 × 4.6

[a]Data is adapted from Xiao *et al.*, 2004, Caccio *et. al.*, 2005 and other references listed in the text. *C. parvum* and *C. hominis* are the primary (most frequent) pathogens associated with human infection while other parasites may also be common, infrequent, or opportunistic pathogens (present in a small fraction of infections). Species not reported to specifically infect humans are marked NR: however genotypes resembling some of these species may have been reported as *C. parvum* subtypes. Oocyst dimensions are length × width in microns.

infecting humans. Unless specifically mentioned, the majority of this text is focused on *C. parvum*, but is probably applicable to *C. hominis*.

5 Ecological Distribution

Cryptosporidium has a worldwide distribution. The organism is not free-living, and the only stage known to be capable of surviving outside a host is the oocyst. The oocyst can survive in the environment for several months and remain infective (Fayer *et al.*, 1998). However, the oocyst is vulnerable to desiccation, freezing, high temperature and ultraviolet light. Thus, cool, dark, moist conditions favor oocyst survival. Oocysts are shed in a host's feces. Therefore any environment contaminated by a host's feces may harbor cryptosporidial oocysts. Transmission is through the fecal-oral route,

Table 2 Epidemiological studies for which *Cryptosporidium* genotypes were determined[a]

Population studied	Reference	Country of Origin	C. hominis % of Total	C. parvum % of Total
Children less than 5 years old	Gatei *et al.*, 2006	Kenya	87	9
Day Care Center (outbreak)	Goncalves *et al.*, 2006	Brazil	100	0
Children with diarrhea	Glaeser *et al.*, 2004	Switzerland	73.3	20
Children & ambulatory patients with diarrhea	Neira-Otero *et al.*, 2005	Chile	50	50
Children or AIDS patients with diarrhea	Raccurt *et al.*, 2006	Hati	59	38
Patients with gastroenteritis	Dalle *et al.*, 2003	Switzerland	0	100
Patients with cryptosporidiosis	El-Osta *et al.*, 2003	UK	49	46
AIDS patients with diarrhea	Cama *et al.*, 2003	Peru	67.5	11.3*
HIV patients with cryptosporidiosis	Alves *et al.*, 2003	Portugal	55.2	24.1
HIV patients with cryptosporidiosis	Gatei *et al.*, 2002	Thailand	50	14.7*
Patients with suspected cryptosporidiosis	Mathieu *et al.*, 2004	US	31.2	39.5

[a]The percentage of *C. hominis* and *C. parvum* genotypes present in the population of genotyped *Cryptosporidium* samples is presented. An asterisk indicates that *C. meleagridis* was more abundant than *C. parvum*. This list is presented as an example and is not intended to be exhaustive.

and contaminated food or water is a typical source of infection. *Cryptosporidium* has recently been reported being found in apple cider, wildlife centers, sewage, farms, day care centers, water parks, and pet stores. Several of these sources were discovered as a result of a *Cryptosporidium* outbreak. Although not exhaustive, these reports highlight the surprising diversity of places that may be contaminated with infective oocysts.

6 Life Cycle

The *C. parvum* life cycle has been examined in detail (Current and Reese, 1986). It begins with the ingestion of an oocyst by the host. Sporozoites (which are haploid) emerge from the ingested oocyst via excystation in the gut and invade host epithelial cells. Replication of *C. parvum* in humans typically takes place in the small intestine, or in the colon. An individual sporozoite develops into a trophozoite and occupies territory near the surface of the host cell – being enveloped in a parasitophorous vacuolar membrane (PVM), but not surrounded by the host cell's cytoplasm. At the attachment site, the trophozoite develops a series of invaginated membranes and a vacuole that are collectively referred to as the feeder organelle. It is suspected that nutrients are taken in by the feeder organelle, and that various nutrients or compounds may also enter the parasite through the lumen side of the PVM. The trophozoite grows, and through a process called merogony, typically produces eight haploid merozoites within a single meront (maturing trophozoite). The distinction between a trophozoite and a young meront is unclear (the two terms are sometimes interchanged), but the presence of multiple nuclei implies that merogony is taking place. This initial round of replication is called type-I merogony, which is typically followed by a second generation of merogony to produce type-II meronts containing four merozoites (Figure 4). It is possible that some repetitive type-I merogony may occur, contributing to persistent infections (Current and Reese, 1986). Type-II merogony produces merozoites that subsequently develop into the sexual stages of the parasite: microgamonts or macrogamonts. Microgamonts grow and produce up to 16 (perhaps more) microgametes – the male counterpart of the parasite. Macrogamonts are considered to be the female equivalent, and grow, but apparently do not divide. Fertilization (fusion) between a single microgamete and a macrogamete produces a zygote – the only known diploid portion of the *C. parvum* life cycle. Recombination presumably could occur during the sexual phase, but this has not been thoroughly investigated. Zygotes may then develop into oocysts of the thick-walled or thin-walled type. Approximately 80% of total oocysts become thick-walled (bearing two membranes) and are the environmentally resistant form released in the feces (Current and Reese, 1986). Thin-walled oocysts (bearing a single membrane) are believed to be capable of autoinfection through sporozoite release and may also contribute to persistent infections (Current and Reese, 1986). Additionally, extracellular forms of *Cryptosporidium* have been reported that may be analogous to the extracellular gamonts produced by gregarine species (Hijjawi *et al.*, 2004). However, whether they are truly extracellular stages (or just deformed sporozoites or merozoites) is still debatable and needs to be carefully confirmed and investigated further.

7 Cell Biology

Little is known about the construction of the oocyst beyond the presence of dual membranes, specific oocyst wall proteins, and certain sugar modifications. Sugar modifications include *N*-acetyl galactosamine and can be identified by lectin binding assays (Luft *et al.*, 1987). Within the oocyst, sporozoites and a residual body can be observed. The residual body is comprised primarily of amylopectin-like granules, lipid, and crystalline protein (Harris *et al.*, 2004). The oocyst appears to have a suture through which sporozoites will emerge but this suture may be derived from the sporocyst wall.

Cryptosporidial sporozoites and merozoites contain a nucleus (apparently surrounded by ribosomes), crystalloid bodies, an apical complex, and other membrane-bound structures (Tetley *et al.*, 1998). The apical complex contains a conoid structure, and three types of organelles: two rhoptries (a singular rhoptry is present in a sporozoite), more than 100 micronemes, and around 12 or so dense granules. While the rhoptries and micronemes are specifically present in the apical complex, dense granules may be distributed at various locations within the parasite. The three apical complex organelles are secretory in nature, and are used for gliding motility and for host cell invasion.

Sporozoites invade host cells and undergo merogony that takes approximately 12–14 h or more for completion when parasites are cultivated *in vitro*. Mature meronts and merozoites can be observed after 12–24 h in culture. Immature oocysts can be observed beginning 36–48 h post infection. *In vitro* cultures are asynchronous, meaning individual parasites will mature at different times.

Using light and fluorescence microscopy, nuclei of *C. parvum* parasites can be observed to undergo dramatic changes during development (Figure 5). Some parasites have compact nuclei that can stain brightly with DAPI (4′,6-diamidino-2-phenylindole), while other parasites may show

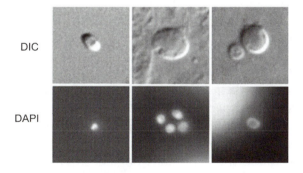

Figure 5 Parasite nuclei. Top panels are differential interference contrast images and bottom panels are fluorescence microscopy with DAPI stain. A compact bright nucleus is seen in the merozoite (left column). Larger diffuse nuclei are seen in the meront pictured in the middle column, and in the last column, both a low-staining nucleus (smaller meront) and a toroidal-shaped nucleus can be seen (larger meront). The DAPI signal seen at the edges of the meront images is from host cell nuclei.

larger, more diffuse staining, nuclei. Some nuclei have a distinct toroidal shape with the central nucleolar region lacking any apparent DAPI staining. In addition, some parasite's nuclear DNA stains only weakly with DAPI. For developing meronts that contain multiple nuclei, the individual nuclei appear to be synchronized. The changes in DAPI staining and nuclear morphology probably coincide with changes in the cell cycle, although thorough studies on cell cycle changes have yet to be made for these species.

The chromosomes are distinguishable by pulsed field gel electrophoresis. Certain chromosomes comigrate, but when digested with rare cutting restriction enzymes, eight chromosomes can be revealed (Caccio *et al.*, 1998). The total genome size is roughly 9.1 Mb and the eight chromosomes range from 0.9 to 1.4 Mb in size. The *C. parvum* and *C. hominis* genomes have been sequenced (Abrahamsen *et al.*, 2004; Xu *et al.*, 2004). The gene complement of *C. hominis* appears to be identical to *C. parvum*, and the two species are about 90–99% identical at the nucleotide sequence level. The DNA is about 70% A + T with about 75% of it apparently coding for proteins. The number of predicted genes is approximately 3,900 and the vast majority does not appear to contain introns. Cryptic introns, however, have been experimentally identified. Unlike the tandem arrays of numerous ribosomal RNA genes found in other eukaryotes, only five copies exist that are dispersed throughout the genome. *C. parvum* also lacks certain repetitive gene families encoding variant surface antigens that are observed in *Plasmodium* species. Overall, the genome size and gene number are rather small compared to *P. falciparum* (~22.9 Mb, 5,300 genes) and *T. gondii* (~65 Mb, 8,000 genes), but of comparable size to the available *Thelieria* and *Babeisa* genomes.

Many apicomplexans have mitochondria, and/or nonphotosynthetic plastids. Both appear to be lacking in *C. parvum*, and extrachromosomal DNA has not been observed. Although a relict mitochondrion that lacks a genome (sometimes referred to as a mitosome) may be present, its function has yet to be resolved. It has been proposed that this organelle functions in iron and sulfide (Fe-S) sequestration or metabolism (LaGier *et al.*, 2003; Putignani, 2005). The lack of a need to maintain a plastid or mitochondrion is believed to be partly responsible for the reduced nuclear gene content in *C. parvum* compared to *P. falciparum* and *T. gondii*. In addition, *C. parvum*, however, does possess a pair of partiti-virus-like double-stranded RNAs that encode an RNA-dependent RNA polymerase and a protein that probably functions as the capsid (Khramtsov and Upton, 2000).

8 Methods of Isolation

For public drinking water, the primary goal is protecting the water supply by removal of contaminants like infective oocysts. A recent review of the water-treatment processes used in the US has been prepared and should be consulted for further reading (Betancourt and Rose, 2004). These processes remove contaminants in general, and are not specifically designed for cryptosporidia. Flocculation or coagulation using inorganic aluminum or iron salts (e.g., Alum, ferric chloride) or organic polymers typically represents the first step in oocyst removal. The efficiency of this step is critical to downstream portions of the process that include removal of the flocculated materials by sedimentation or removal of fragile particles by dissolved air flotation. The final steps of water purification are filtration through porous beds of material typically comprised of sand, diatomaceous earth, or anthracite coal plus sand. The purified water is then subjected to disinfection that typically involves chlorination. Chlorination however, is not effective on *Cryptosporidium* oocysts and other methods are being sought. Chlorine dioxide, ultraviolet light, and ozone treatments are among the methods being evaluated as disinfectants.

To test for a healthy water supply, methods for the isolation of *Cryptosporidium* oocysts from water for the purposes of identification have been developed. The US EPA has standardized protocols (e.g., Method 1623). The standardized methods are based upon membrane filtration of approximately a 10-l sample, followed by immunomagnetic collection of oocysts. Oocysts are identified by immunofluorescence and microscopic techniques. Method 1623 and related methods have also been used to monitor the water supplies in other countries.

For medical or veterinary practices, biopsies of the intestine or colon have been used to isolate samples for diagnostic purposes. However, less invasive procedures are more routinely used (e.g., the collection of fecal samples from patients). Fecal flotation and/or smears are common, with specimens being subjected to one of several staining procedures or to molecular-based diagnostics (Garcia and Current, 1989). Although the priority is normally for diagnoses, cryptosporidial oocysts have also been collected for research use from chronically infected human AIDS patients.

For scientific research on the biology of *Cryptosporidium*, purified viable oocysts need to be prepared. Excellent reviews of *Cryptosporidium* cultivation for laboratory purposes, and using various animal models have been created (Arrowood, 2002). Fecal material is usually collected over a period of several days from experimentally infected newborn calves (piglets for *C. hominis*) that are reared in isolation and are free of other infections. The fecal material is treated with potassium dichromate solutions to sterilize it and remove odors. The oocyst fecal mixture is filtered and subjected to gradient purifications – usually discontinuous sucrose gradients. Oocysts harvested in this manner can then be surface sterilized (e.g., using weak bleach solutions), washed thoroughly with sterile water, and separated from remaining microscopic debris using Percoll gradient centrifugation.

9 Oocyst Storage

Purified oocysts (usually derived from animal sources) are useful for both *in vitro* and *in vivo* studies. When stored around 4°C, viability can be reproducibly maintained for up to 6 months, although viability up to 18 months has been recorded (Fayer *et al.*, 1998). Viability, however is rapidly lost upon freezing of the parasites, and the effective shelf life is dramatically reduced when the temperature is greater than 15–20°C. Oocysts may be stored in purified water. The addition of antibiotics is not necessary, but may be beneficial for preventing microbial growth if the oocysts have not been surface sterilized and rigorously cleaned.

10 Methods for Excystation

An excellent review of what is known (or not known) about natural excystation has been compiled (Smith *et al.*, 2005). For laboratory investigations, *in vitro* excystation of oocysts is useful for generating sporozoites for further study. The *in vitro* excystation methods attempt to mimic conditions that might be responsible for excystation *in vivo*. Acidic pH, warmer temperature, and proteases comprise the basis for these methods. One excystation medium that works well comprises 0.25% trypsin (containing some chymotrypsin) and 0.75% taurodeoxycholic acid in phosphate buffered saline (Arrowood 2002; Current and Haynes, 1984). However, a great number of *C. parvum* oocysts may also undergo excystation in culture medium or PBS, a feature that is very useful in infecting cultured cells (i.e., by directly adding bleached oocysts into the culture). Incubation from 30 to 90 min at 37°C is sufficient for essentially complete excystation of sporozoites from viable oocysts. Once excystation is complete, it may be useful to pellet the sporozoites and wash them to remove the excystation medium so that the proteases do not destroy the sporozoites. If purified sporozoites are needed, Percoll gradient centrifugation may be useful for separating sporozoites from unexcysted oocysts and empty shells.

11 Laboratory Cultivation

Currently, the best *in vitro* procedures for cultivating *Cryptosporidium* species result in the production of very few viable oocysts and continuous cultivation in the laboratory has not been reproducibly achieved. The entire life cycle can be observed using *in vitro* cultivation systems, although few groups have reported observing gamont-like extracellular forms (Hijjawi *et al.*, 2004). Axenic cultivation has not been widely achieved (Girouard *et al.*, 2006). The more robust protocols that are currently used involve the cultivation of host cells, inoculation of the host cells with oocysts or sporozoites, and maintenance of the inoculated culture for up to several days under conditions that favor host cell growth. If oocysts are used, the removal of spent oocyst shells a few hours following inoculation, and the addition of fresh medium improves subsequent growth of the parasites (Eggleston *et al.*, 1994). It has been proposed that oocysts may contain a toxin that inhibits growth of the parasites. Additionally, parasites have been reported to preferentially invade dividing cells (Widmer *et al.*, 2006) and the addition of fresh culture media can facilitate host cell proliferation.

Improvements in the cultivation of *Cryptosporidium in vitro* have occurred in incremental steps, largely as a result of tests with different additives in the culture medium, or attempting cultivation with different host cell types. Side-by-side comparisons indicate that the best growth identified to date probably occurs when using human ileocecal adenocarcinoma (HCT-8) cells as a host (Upton *et al.*, 1994). HEPES-buffered RPMI 1640 medium, when supplemented with antibiotics and heat-inactivated fetal calf serum, forms the basis of the culture medium as described in the original *in vitro* culture system (Current and Haynes, 1984). Selected vitamins, sugars, insulin, and nucleoside analogs have all been reported to improve *C. parvum* development *in vitro*. In spite of these advances, large-scale production of *C. parvum* oocysts still requires the use of animal models of infection. Experimentally infected mice can be used for smaller scale maintenance of *C. parvum*, but newborn ruminant animals (calves) can provide up to 10×10^9 oocysts from a single infection (Arrowood, 2002).

The *in vitro* culture systems, despite their limitations, provide ample opportunity to study aspects of *C. parvum* growth and development, to study the effects of candidate drugs, and for small-scale biochemical or genetic analyses. From a practical standpoint, using host cultures that are far from confluent, as well as using a low multiplicity of infection, facilitates studies on morphology and development using light microscopy. Cultures prepared in this manner allow parasites to become situated at the edges of host cells and if there are spaces between the host cells, allows for unobstructed examination of the parasites during light and immunofluorescence microscopy (e.g., see Figure 4). Additionally, the level of host cell confluence and the multiplicity of infection may be adjusted according to the experiments to be performed. This is of special concern if the experiment is continued for an extended period of time since both the host cells and the parasites will multiply at different rates during the course of the experiment.

12 Classification and Genotyping

Cryptosporidium species can be classified from samples using oocyst measurements, but the utility of this alone is generally insufficient. Host species and host range are also useful, but of limited value for preserved (e.g., dead) specimens. Determining cross-species infectivity is generally beyond the capabilities of most laboratories. Once a species has been designated and characterized at the molecular level, this facilitates more routine identification of a particular species or subtype. Molecular methods are more useful for a greater range of sample types. Immunological and isozyme methods have been used to discriminate species based on oocyst proteins or antigens. Unfortunately, antibodies against different apicomplexan parasites often display significant cross-species reactivity. PCR-based detection has been commonly used to distinguish *Cryptosporidium* species and subspecies. The variety of PCR methods employed have included: randomly amplified polymorphic DNA (RAPD), PCR combined with restriction fragment length polymorphism (PCR/RFLP), single strand conformation polymorphisms (SSCP), direct sequencing of PCR amplicons, and various real-time PCR methods. The loci used for discrimination are equivalently varied, and have included micro-/minisatellites, coding and intergenic sequences for ribosomal RNAs, genes for oocyst wall protein, glycoprotein 40/15 (GP40/15), dihydrofolate reductase (DHFR), actin, the heat-shock protein 70 gene (HSP70) and several other loci. Although significant overlap exists among the loci tested by different research groups, no standardized methods for genotypic discrimination have been developed. Although more costly than other methods, genotyping multiple loci at the DNA sequence level seems to be a reliable method for comparing unidentified samples to known species for putative identification (Caccio, 2005). It is likely that with more loci and more populations being examined, classification at or below the species level will become more complex as new genotypes are identified.

13 Infection and Disease

13.1 Epidemiology

The first human cryptosporidiosis was reported more than 30 years ago (Nime *et al.*, 1976). This discovery was made after examining a rectal biopsy specimen from a three-year-old child with atypical entorocolitis. By 1985, an early review indicated that this 'new' pathogen was becoming more widely recognized not only as a veterinary pathogen, but also as a human pathogen associated with AIDS (Casemore *et al.*, 1985). Additionally, less invasive testing procedures were available for diagnosis of the condition and person-to-person transmission was identified as a likely mode of transmission (Casemore *et al.*, 1985). Although *Cryptosporidium* is primarily transmitted through the fecal-oral route, experimental inoculations into the bloodstream or body cavity of laboratory animals have resulted in intestinal cryptosporidiosis. As mentioned previously, human infections can originate from a variety of sources, but often from contaminated food or water supplies.

Cryptosporidia have been reported worldwide, in numerous countries, and on all continents except Antarctica. Within the US, *Cryptosporidium* has been reported in every state in immunocompromised as well as immunocompetent individuals. The rate of incidence for infection ranges from not being detected in a given place during a given time period, to more than 37% of individuals that are presenting gastrointestinal symptoms (Hlavsa *et al.*, 2005). In general, *Cryptosporidium* incidence is usually on the level of a few percent but this often varies geographically and seasonally. Children and AIDS patients from the same geographic regions may have a higher incidence of *Cryptosporidium* infections (Hlavsa *et al.*, 2005). Although outbreaks of *Cryptosporidium* are frequently cited, unexplained, sporadic, individual infections may actually represent the majority of cryptosporidiosis cases (McLauchlin *et al.*, 2000).

While species-specific cryptosporidia may predominate, *C. parvum* can be a zoonotic pathogen, and dozens of mammalian species are known or suspected of being hosts (i.e., carriers) to *C. parvum* or *C. parvum*-like parasites. In addition to humans and other mammals, nonhost species may offer additional routes for dissemination of infective oocysts. For example, *C. parvum* oocysts have been demonstrated to pass through various bird species intact and remain infective to mammalian hosts (Graczyk *et al.*, 1997). Additional studies have suggested that filth flies, cockroaches, and dung beetles may mechanically vector cryptosporidial oocysts (Graczyk *et al.*, 2004; Mathison and Ditrich, 1999; Zerpa and Huicho, 1994). For dung beetles, the majority of ingested oocysts are destroyed, but a few may remain intact (Mathison and Ditrich, 1999). The nonhost mechanical vectors represent a potentially difficult problem since they may travel great distances from their point of origin and can be easily overlooked as a source of inoculum.

Several recent studies have indicated that shellfish, which are filter feeders, may accumulate substantial numbers of cryptosporidial oocysts from contaminated seawater (Graczyk and Schwab, 2000). These oocysts can retain their infectivity, and may contribute to human disease if the shellfish are not properly cooked or pasteurized. Fresh water rotifers, ciliated protozoa, and acanthomoeba may also ingest infective oocysts, but it was suggested that these organisms may facilitate destruction of oocysts present in fresh water (Fayer *et al.*, 2000; Stott *et al.*, 2003).

Although it is clear that nonhost organisms may play a role as mechanical vectors, serve as biomagnifiers, or help to destroy infective oocysts, more research is needed to determine their significance in the epidemiology of *Cryptosporidium* infections.

13.2 Risk factors

Cryptosporidium infections often display clear seasonal and demographic trends that are dependent upon geographical location. One of the clearest examples is the increase in cases or outbreaks reported during the summer months in several countries. These cases are frequently associated with recreational activities involving water (Craun *et al.*, 2005; Hlavsa *et al.*, 2005). Thus, exposure to *Cryptosporidium* is more likely to occur for individuals that visit water parks, spas, swimming pools, lakes and reservoirs used for outdoor summer activities. A greater risk comes from swallowing recreational water or having ones head submerged. In other countries, where water is scarce and *C. parvum* zoonotic transmission appears to be most prevalent, cryptosporidiosis has been shown to peak during the cooler season (Iqbal *et al.*, 2001). The zoonotic feature of many cryptosporidial species makes individuals who are frequently exposed to animal feces (farm workers, veterinarians, zookeepers) at an increased risk of infections (Keusch *et al.*, 1995).

Age-related occurrences of cryptosporidial infections are also apparent, with children being the most likely to become infected (Hlavsa *et al.*, 2005). Thus, children attending childcare facilities, caregivers at those facilities, and family members of young children may be at increased risk of being exposed to *Cryptosporidium*. Additionally, children fed formula rather than being breast-fed are more likely to become exposed to *Cryptosporidium* and become infected (Hlavsa *et al.*, 2005).

Eating uncooked food or unpasteurized food that is likely to become contaminated with oocysts also presents a risk. For example, shellfish collected from oocyst-contaminated areas may represent a risk (Graczyk and Schwab, 2000). Vegetables that are irrigated using surface waters, washed with oocyst-contaminated water, and vegetables exposed to fresh manure or farm runoff may be more likely to become contaminated (Chaidez *et al.*, 2005).

Lastly, anyone with a compromised immune system may be at a greater risk of infection and may experience more severe cryptosporidiosis. Immunocompromised individuals may also acquire opportunistic infections in organs other than the intestinal tract, and also may be more likely to become infected with cryptosporidium species other than *C. parvum* and *C. hominis*.

13.3 Clinical diagnosis

If cryptosporidial infection is suspected due to unexplained persistent watery diarrhea, diagnosis is most frequently done by detecting antigens in stool samples by ELISA or oocysts by a modified acid-fast or immunofluorescence (IF) labeling assay (Johnston *et al.*, 2003). Oocysts appear pink or bright red when stained with the acid-fast assay. However, some oocysts may not be stained and the background is generally high. In addition, other microbes (isospora, cyclospora) may be detected with the acid-fast technique and must be ruled out using overall shape and measurements of oocyst size. Direct fluorescent assays provide a lower background, and are available to simultaneously detect cryptosporidial oocysts and giardial cysts from unconcentrated stool samples. PCR and other molecular-based methods remain largely within the realm of research. Typically, two techniques such as ELISA and an IF assay may be performed to confirm results.

For any test, individual stools may display false negatives (or false positives), and multiple stools may be required for diagnosis.

13.4 Host susceptibility

All people should be considered as potentially susceptible to infection by cryptosporidia. Once an infection takes place, the immune system is the primary defense against cryptosporidiosis. In both immunocompetent and immunocompromised patients, the gastrointestinal tract is the major site of infection. Reports of other infected organs have been made, but usually in immunocompromised patients. Organ specificity is also dependent upon the animal species infected.

Cross-species infection experiments have suggested that the number of *Cryptosporidium* species was initially overestimated and that *C. parvum* was capable of infecting a variety of mammalian species (Xiao *et al.*, 2004). However, it is also clear that host-adapted species and races of *Cryptosporidium* do exist. In particular, *C. hominis* appears to have a limited host range compared to *C. parvum*. At least four additional cryptosporidial species have been discovered infecting humans, often as opportunistic infections in immunocompromised individuals (Xiao *et al.*, 2004). Thus, host and tissue specificity may be partially dependent upon the immune systems of the host.

13.5 Pathophysiology

Severe, life-threatening cryptosporidiosis is usually only associated with infants and children or individuals with poor immune systems (e.g. elderly individuals, AIDS and leukemia patients). The pathophysiology of cryptosporidial infection has been assessed in healthy human volunteers that were immunologically naive to *Cryptosporidium* (Chappell *et al.*, 1996). Asymptomatic infections can occur. Different isolates may display different levels of virulence. Infection and diarrhea can be caused by as few as 10 oocysts (Okhuysen *et al.*, 1999). The prepatent period can be as little as a few days, but 7–10 days seems to be around the average. Oocyst shedding usually occurs within 1 week and may last for a few days to over 1 week. Diarrhea has been reported to last from 6 to 223 h, with averages in the 60- to 90-h range (3–4 days). *Cryptosporidium* typically causes a watery diarrhea and may be accompanied by other symptoms such as fever, nausea, vomiting, abdominal pain or cramps, and flatulence. The number of unformed stools has been reported to be as high as 11 in a single 24-h period with seven being near the average. Recurrence of symptoms may happen in more than 50% of infected individuals. Symptoms usually resolve without intervention, although in severe cases, dehydration may require medical intervention.

Similar experiments have been performed using volunteers that displayed anti-*Cryptosporidium* antibodies (Chappell *et al.*, 1999). Larger doses of oocysts were required to produce symptoms similar to those of immunologically naive individuals. Individuals that had anticryptosporidia antibodies prior to experimental infection also shed fewer oocysts. Thus, previous exposure to cryptosporidia appears to provide some protection during subsequent exposure to the parasite.

The diarrheal illness caused by cryptosporidia has been the subject of the most intensive studies. However, evidence is accumulating that suggests exposure to or infection by cryptosporidia can have longer-term consequences or side effects. Headaches, eye pain, joint pain, dizziness and fatigue are among the symptoms individuals may report experiencing after apparently recovering from cryptosporidiosis (Hunter *et al.*, 2004). This is more frequently reported with *C. hominis* infections. It is also becoming clear that nutritional deficits occur in individuals infected with cryptosporidia and animal tests support this hypothesis (Topouchian *et al.*, 2005). Malnutrition, even for short duration, can be associated with lasting negative consequences on cognitive and motor skills, especially for young children – the group most frequently affected by *Cryptosporidium* infections (Grantham-McGregor and Baker-Henningham, 2005).

13.6 Treatment

Immunocompetent individuals typically recover without intervention. Dehydration is the most frequent symptom requiring attention in severely infected individuals (Keusch *et al.*, 1995). Treatment of cryptosporidiosis may involve oral or intravenous liquids to replace water and electrolytes if patients are suffering from dehydration. Re-establishing a healthy immune system is the most effective method for combating chronic cryptosporidiosis in immunocompromised patients (Schmidt *et al.*, 2001). However, this may be difficult or impossible for severely immunocompromised individuals. Antiretroviral therapies for AIDS patients, if successful in restoring the immune system (increased CD4 counts), can lead to recovery from cryptosporidiosis. However, options for treating the cryptosporidium infection itself are limited.

Nitazoxanide (NTZ) is currently the only drug available in the US market listed specifically for the treatment of cryptosporidiosis (FDA Electronic Orange Book available at http://www.fda.gov/). The efficacy of NTZ has been examined. Using a double-blind study, efficacy was found to range from 63 to 67% compared to about 25% in a control (placebo) group, where effective treatment was defined as a lack of parasites on day 15 (Rossignol *et al.*, 1998). In that study, drug or placebo treatments lasted for 14 days followed by placebo for an additional 14 days. Of those that were free of parasites, 80–92% did not display recurrence of diarrhea between days 15–29, while a similar proportion (80–86%) of the patients that harbored parasites also did not display diarrhea. Although encouraging, this treatment regime was not completely effective in eliminating parasites or diarrhea.

NTZ is approved (by the FDA) under the trade name of Alinia (Romark Laboratories). Outside of the US, nitazoxanide may be available as an antiparasitic agent in Mexico and other countries under the trade name of Daxon. Additional synonyms include: Colufase, Heliton, Taenitaz, and Cryptaz. Alinia is available in the US by prescription in tablet and suspension form. Suspension form is listed for use in children 1 year of age and older. Tablet form is available for persons aged 12 years and older. Although Alinia is only approved for persons that do not have AIDS, it has been used experimentally to treat cryptosporidiosis in AIDS patients during a compassionate use clinical trial (Rossignol, 2006). In this intent-to-treat study, 59% of AIDS patients treated with NTZ displayed clinical improvements. Improvement was defined as parasite-negative stools while on treatment and corresponded with reduced symptoms associated with cryptosporidiosis. However, loose stools still represented a substantial proportion of the total number of stools. Treatment times varied and lasted up to 4 years (median 62 days).

NTZ belongs to the thiazole family and is reported to display broad-spectrum activity against helminths, several anaerobic microorganisms, and some viruses (Hemphill *et al.*, 2006). NTZ rarely causes any severe side effects in patients. However, it could have negative effects on commensal gut microbes. NTZ is believed to target pyruvate-ferredoxin oxidoreductase (PFO) in some organisms, but may have additional mechanisms of action in others. The *C. parvum* and *C. hominis* genomes do possess a gene that encodes a bifunctional enzyme containing the N-terminal PFO domain and a C-terminal cytochrome P450 reductase (i.e., pyruvate:NADP$^+$ oxidoreductase [CpPNO]; Ctrnacta *et al.*, 2006). This protein can be detected by immunolabeling as a cytosolic protein in sporozoites. Thus, CpPNO could be the target of NTZ in cryptosporidial species. On the other hand, NTZ has antiparasitic activity against *Neospora canium* (a related apicomplexan) that is apparently independent of its capacity to interact with PFO (Esposito *et al.*, 2005). Therefore, an unidentified target may also exist in *Cryptosporidium*.

Paromomycin sulfate is an older antibiotic that was approved (by the FDA) under the trade name of Humatin (FDA Electronic Orange Book available at http://www.fda.gov/). It is widely available with additional synonyms that include: Aminosidine, Catenulin, Estomycin, Hydroxymycin, Neomycin E, and Gabbromycin. Paromomycin is available by prescription in tablet form for adults and pediatric patients. Paromomycin has been used experimentally to treat cryptosporidiosis in AIDS patients with several trials giving differing results. In one controlled clinical study, 47.1% of AIDS patients treated with paromomycin displayed clinical improvements while 35.7% of patients using a placebo displayed improvements (Hewitt *et al.*, 2000). Thus, minimal benefit was seen through the use of paromomycin in that study. Improvement was defined as a lower number of stools and reduced need for antidiarrheal drugs compared to what occurred before receiving treatment. Treatment or placebo times were for 21 days, followed by 21 additional days with all patients receiving paromomycin. Many other studies have indicated good improvement or recovery from different types of cryptosporidiosis using other definitions of improvement. Some of these studies were conducted in combination with highly active antiretroviral therapy (HAART). These confounding results are being followed up. In a more recent study, *in vitro* tests using paromomycin alone or in combination with certain protease inhibitors (those used for HAART) indicate that paromomycin combined with antiretroviral drugs is more effective against *C. parvum* than either treatment alone (Hommer *et al.*, 2003). Paromomycin, which is an amino glycoside, is reported to display activity against a wide array of organisms. Paromomycin interferes with protein synthesis by binding to the aminoacyl tRNA site of ribosomes, but recent evidence indicates a possible role in inhibiting tRNA maturation. *C. parvum* and *C. hominis*, of course, synthesize proteins and do possess tRNAs and ribosomes that could be the targets of this drug.

Other drugs, including some antiretroviral agents, have been tested against cryptosporidiosis with some promising results. However, nitazoxanide and paromomycin remain as the two most well-recognized treatment options. Additionally, anticoccidia drugs have been tested or used in animals, but their utility remains unclear.

14 Opportunistic Infections

Cryptosporidium infections in humans are typically associated with the small intestine and colon, but in immunocompromised individuals, a number of other tissues may become infected. The tissues that may become infected have included: the biliary system, pulmonary system/respiratory tract, the appendix and the middle ear. The biliary system is comprised of the gall bladder and associated ducts that lead to the liver pancreas and stomach. Cholangitis is an inflammation with infection of the bile duct system, which is typically caused by bacterial infections in the biliary system and often accompanied by an obstruction (e.g., gallstones) or constriction that may be caused by scarring. In biliary cryptosporidiosis, cholangitis may occur and may be complicated by co-infection or other pathogens. Reports of this condition are rare and have been difficult to diagnose. Tracheal or pulmonary cryptosporidiosis may occur where parasites infect and develop in the epithelium of the lungs and trachea (Dupont *et al.*, 1996). This often occurs in connection with an existing intestinal infection and may be confused with other respiratory ailments.

Additionally, species other than *C. parvum* and *C. hominis* are more likely to be encountered in immunocompromised patients. *Cryptosporidium felis*, *C. canis*, *C. meleagridis*, *C. muris*, *C. baileyii*, *C. suis*, and *C. andersoni* have all been found to infect humans (Xiao *et al.*, 2004). Because these organisms display host species specificity that does not typically include humans, they often represent zoonotic and/or opportunistic infectious agents of man.

15 Virulence Factors

Cryptosporidial virulence is poorly understood. Much of the work done in this area is focused on the sporozoite, with particular emphasis on its interaction with the host cell – gliding motility, attachment to, and invasion of the host cell. The approaches used to understand these interactions rely on the identification of antibodies or chemicals that disrupt or augment these processes. Antibodies that are capable of inhibiting adherence of parasites to a host cell or that inhibit invasion have been used to identify the corresponding sporozoite or merozoite antigens. Analyses of these antigens revealed that they are composed of mucin-like and other glycoproteins. Gp900, gp1300, gp40/15 and a manosyl glycolipid (CPS-500) are among the antigens identified that are required for host invasion (Boulter-Bitzer *et al.*, 2007). In addition, an 18- to 20-kDa *C. parvum* glycoprotein that alters host cell ion flux has been identified. Ion flux is abrogated using antibodies against this antigen during *in vitro* cultivation of the antigen with host cells. Thus, this antigen encodes a putative enterotoxin. Immunolocalizations demonstrated that many of these glycoproteins are localized to the apical complex and it has been demonstrated that some of them are secreted during parasite gliding. Genetic variation in the glycoprotein genes exists among different cryptosporidium genotypes; however a relationship to host specificity or virulence remains to be determined.

Gliding motility, host cell attachment, and host cell invasion all require the discharge of the contents of the apical organelles. The discharge of these contents is influenced by temperature (efficient at 37°C) and calcium is essential for apical organelle discharge (Chen *et al.*, 2004a). In addition, disrupting the polymerization of actin and tubulin or myosin function prevents host cell invasion (Chen *et al.*, 2004a; Wetzel *et al.*, 2005). Thus, the cytoskeleton of the parasite appears to play an important role in host cell invasion.

Phospholipases are recognized virulence factors in other pathogenic microbes. The addition of exogenous phospholipases to *C. parvum* sporozoites or host cells increases the number of intracellular parasites present during *in vitro* culture (Pollok *et al.*, 2003). Phospholipase inhibitors and antibodies raised against phospholipases also reduce host cell invasion when added to cultures *in vitro* (Pollok *et al.*, 2003). In addition, *C. parvum* sporozoites possess phospholipase activity (Pollok *et al.*, 2003). Thus, there appears to be increasing evidence that parasite lipases are involved in host cell invasion.

16 Host Factors

The immune system is the primary defense against cryptosporisiosis and normally allows immunocompetent individuals to recover from acute infections. A number of host cell factors involved in the infection process have also been identified. Attachment of cryptosporidial parasites to host cells presumably requires recognition of something on the host cell. Evidence exists that implicates lipids in this process. Bile salts, which emulsify and carry lipids in the intestine, enhance host cell invasion by *C. parvum in vitro* (Feng *et al.*, 2006). Lipid-like molecules from bovine MDBK cells are able to inhibit *C. parvum* binding to cells, implicating a potential for lipids in the invasion process (Johnson *et al.*, 2004). Sphingolipid microdomains co-localize with *C. parvum* infection sites, and acid sphingomyelinase (which causes aggregation of sphingolipid microdomains) is activated in response to *C. parvum* infection (Nelson *et al.*, 2006). In addition, knockdown by RNAi of acid sphingomyelinase reduced host cell infection by *C. parvum* (Nelson *et al.*, 2006). Thus, lipid domains present on host cells are clearly implicated in *C. parvum* infection.

In addition to lipid microdomains, certain receptors on the host cell appear to be important to the invasion process. SGLT1, a sodium ion and glucose transporter, appears to be recruited to the attachment site during *C. parvum* invasion (Chen *et al.*, 2005). The transporter, along with aquaporin1, facilitates invasion by a localized increase in cell volume that leads to protrusions of the cell surface that encompass the invading sporozoite (Chen *et al.*, 2005). Knockdown of aquaporin1, or chemical inhibition of SGLT1, reduce *C. parvum* invasion (Chen *et al.*, 2005).

Host cell-signaling pathways also mediate pathogenesis. Phosphatidylinositol 3,4,5-triphosphate [PI(3,4,5)-P3] is a cellular messenger involved in a number of pathways including membrane ruffling, actin polymerization, glucose transport and cell survival. Phosphorylation of this molecule is part of the signaling pathways it is involved in and the phosphorylation of PI(3,4,5)-P3 is performed by PI(3,4,5)-P3 kinase (PI3K). PI3K activity then may act through guanine exchange factor pathways (Rac or Cdc42) or protein kinase C (PKC) pathways to facilitate, among other things, cytoskeletal changes. Disruption of the PI3K pathway reduces cryptosporidium invasion, and activation of the pathway facilitates cryptosporidial invasion (Chen *et al.*, 2004b). In the context of cryptosporidium invasion, the frabin-cdc42 pathway has been shown to be involved in the process (Chen *et al.*, 2004b). PKC has also been shown to be involved in cryptosporidial invasion of host cells (Hashim *et al.*, 2006).

Cytokines and their associated pathways have been shown to be involved in host cell responses to *Cryptosporidium* infection. Tumor necrosis factor alpha (TNF-α) and interferon gamma (IFN-γ) are inflammatory cytokines that activate the JAK–STAT pathway resulting in gene regulation. Both TNF-α and IFN-γ have been shown to hinder invasion by

Cryptosporidium (Lean *et al.*, 2002; Pollok *et al.*, 2001). In certain cells, IFN-γ signaling pathways can lead to divalent iron starvation, tryptophan catabolism and starvation, or nitric oxide production as a means of preventing infection. In general, tryptophan catabolism can be ameliorated by exogenous tryptophan and iron starvation can be ameliorated by the addition of exogenous iron. In experiments on *C. parvum* infection *in vitro*, only exogenous iron promoted the infection process of *C. parvum* (Pollok *et al.*, 2001). Recent evidence suggests that the parasite may be able to synthesize tryptophan from indole-3-glycerol phosphate (Abrahamsen *et al.*, 2004). Inhibition of NO production did not promote *C. parvum* infection of enterocytes (Pollok *et al.*, 2001). Transforming growth factor beta (TGF-β) abrogates the negative effects of interferon, while interleukin-4 enhances the inhibitory effects of interferon on *C. parvum* invasion of enterocytes (Lean *et al.*, 2003).

The host cell cytoskeleton may also play a substantial role in the pathogenesis of *C. parvum*. Treatment of host cells with cytoskeleton disrupting agents reduces cryptosporidial invasion or infection and demonstrates that the cytoskeleton plays a role in those processes (Wiest *et al.*, 1993). A number of studies have implicated actin in cryptosporidial invasion (Chen *et al.*, 2004c; Elliott and Clark, 2000). Additionally, actin and tubulin transcripts are up-regulated in host cells during cryptosporidial infection (Deng *et al.*, 2004). Reorganization of the cytoskeletal proteins tropomyosin, villin, ezrin and F-actin have been observed in *Cryptosporidium*-infected cells (Bonnin *et al.*, 1999; O'Hara and Lin, 2006). Thus, it appears that there is ample evidence to conclude that the host cell cytoskeleton plays an important role during cryptosporidium development.

The role of programmed cell death (apoptosis) in *C. parvum* infection has been investigated. In this situation, the goal for the parasite is to prevent apoptosis so it can use the cell as a host, while the host attempts to commit suicide in an effort to kill existing immature parasites. *C. parvum* infection appears to modulate host cell gene expression in a manner that is designed to prevent apoptosis by activating heat-shock protein genes (Deng *et al.*, 2004). Additionally NF-κB appears to be activated during cryptosporidial infection (Chen *et al.*, 2001; Gookin *et al.*, 2006; McCole *et al.*, 2000). More recent evidence, however, suggests that when the parasite matures, it may actually promote host cell apoptosis (Mele *et al.*, 2004). Other authors suggest that following emergence from the host cell, host cells are killed in a necrotic fashion (Elliott and Clark, 2003). Although apoptosis may be an important host factor, its role may change depending on the developmental stage of the parasite.

Several miscellaneous host factors may promote pathogenesis and spread of cryptosporidiosis. Host-derived mucus secreted by intestinal cells normally acts as a protective barrier. This barrier apparently promotes re-invasion of nearby host cells by *Cryptosporidium* merozoites because the parasites travel beneath this mucous layer, rather than being released directly into the intestinal lumen. Substance P is a host-derived tachykinin neuropeptide that regulates chloride ion release from intestinal cells. Substance P levels are increased during cryptosporidiosis and may be partly responsible for the presence of diarrheal symptoms that promote oocyst shedding (Robinson *et al.*, 2003).

17 Drug Targets and Development

Drug discovery efforts for cryptosporidiosis have been more challenging perhaps than for other apicomplexans. The discovery of the apicoplast in several apicomplexan parasites led to the apicoplast being a notable target for drug discovery efforts (Fichera and Roos, 1997). However, translating this to *C. parvum* was difficult. With the genome sequence in hand and other seminal work, it became clear that *C. parvum* does not possess a plastid (Abrahamsen *et al.*, 2004; Zhu *et al.*, 2000). Thus, focusing on plastid metabolism for identifying potential drugs to use against cryptosporidiosis was ultimately found to be unwarranted. A number of putative targets are now being identified where *C. parvum* homologs of known drug targets have been predicted based upon the complete genome sequence (Abrahamsen *et al.*, 2004; Zhu *et al.*, 2000).

Drug delivery is still a potential pitfall for all rational drug discovery and design efforts. *Cryptosporidium* prefers the small intestine, an area that is notably difficult to target due to the short time in which drug delivery can take place (Streubel *et al.*, 2006). Drugs present in the host intestinal lumen may have a difficult time entering the parasite before being metabolized by commensal microbes or before being washed away by gastric emptying. The unique intracellular, but extracytoplasmic, niche filled by *C. parvum* also makes drug delivery an especially interesting problem. It has been shown that some compounds enter the parasite without first entering the host cell (Griffiths *et al.*, 1998). This means that drugs delivered through the host cell or prodrugs that require host cell machinery for activation may not reach the parasite. Therefore, *C. parvum* may have found an effective hiding place and has the potential to resist chemotherapy simply by means of escape.

18 Conclusions

Cryptosporidiosis is an emerging problem around the world. The major economic and social challenges toward combating cryptosporidiosis include protection of the environment (e.g., water supplies) from contamination with oocysts that arise from human and animal waste and investing in better measures to prevent parasitic infections. In the absence of preventive measures, diligently monitoring water supplies and the incidence of cryptosporidiosis can help to identify sources of contamination during or following outbreaks. Unfortunately, many infections are sporadic in nature. These sporadic infections will still probably represent a major challenge to the medical and veterinary communities. A 'cure' or vaccine remains elusive and the current therapies range from essentially supportive care to moderately successful treatment. New (and inexpensive) therapies for both human and animal cryptosporidiosis are needed, especially for immunocompromised individuals. As these challenges are met, it will be important to consider that resistance to control practices has always been a recurring theme in the pursuit to reduce the burden of human and animal disease.

19 References

Abrahamsen, M.S., Templeton, T.J., *et al.* (2004) Complete genome sequence of the apicomplexan, *Cryptosporidium parvum. Science* **304**(5669): 441–445.

Adl, S.M., Simpson, A.G., *et al.* (2005) The new higher level classification of eukaryotes with emphasis on the taxonomy of protists. *J. Eukaryot. Microbiol.* **52**(5): 399–451.

Alves, M., Matos, O., *et al.* (2003) Microsatellite analysis of *Cryptosporidium hominis* and C. *parvum* in Portugal: a preliminary study. *J. Eukaryot. Microbiol.* **50**(Suppl): 529–530.

Arrowood, M.J. (2002) In vitro cultivation of *Cryptosporidium* species. *Clin. Microbiol. Rev.* **15**(3): 390–400.

Barta, J.R. and Thompson, R.C. (2006) What is *Cryptosporidium*? Reappraising its biology and phylogenetic affinities. *Trends Parasitol.* **22**(10): 463–468.

Betancourt, W.Q. and Rose, J.B. (2004) Drinking water treatment processes for removal of *Cryptosporidium* and *Giardia. Vet. Parasitol.* **126**(1–2): 219–234.

Bonnin, A., Lapillonne, A., *et al.* (1999) Immunodetection of the microvillous cytoskeleton molecules villin and ezrin in the parasitophorous vacuole wall of *Cryptosporidium parvum* (Protozoa: Apicomplexa). *Eur. J. Cell Biol.* **78**(11): 794–801.

Boulter-Bitzer, J.I., Lee, H., *et al.* (2007) Molecular targets for detection and immunotherapy in *Cryptosporidium parvum. Biotechnol. Adv.* **25**: 13–44.

Caccio, S.M. (2005) Molecular epidemiology of human cryptosporidiosis. *Parassitologia* **47**(2): 185–192.

Caccio, S., Camilli, R., *et al.* (1998) Establishing the *Cryptosporidium parvum* karyotype by NotI and SfiI restriction analysis and Southern hybridization. *Gene* **219**(1–2): 73–79.

Cama, V.A., Bern, C., *et al.* (2003) *Cryptosporidium* species and genotypes in HIV-positive patients in Lima, Peru. *J. Eukaryot. Microbiol.* **50**(Suppl): 531–533.

Casemore, D.P., Sands, R.L., *et al.* (1985) *Cryptosporidium* species a 'new' human pathogen. *J. Clin. Pathol.* **38**(12): 1321–1336.

Chaidez, C., Soto, M., *et al.* (2005) Occurrence of *Cryptosporidium* and *Giardia* in irrigation water and its impact on the fresh produce industry. *Int. J. Environ. Health Res.* **15**(5): 339–345.

Chappell, C.L., Okhuysen, P.C., et al. (1996) *Cryptosporidium parvum*: intensity of infection and oocyst excretion patterns in healthy volunteers. *J. Infect. Dis.* **173**(1): 232–236.

Chappell, C.L., Okhuysen, P.C., et al. (1999) Infectivity of *Cryptosporidium parvum* in healthy adults with pre-existing anti-*C. parvum* serum immunoglobulin G. *Am. J. Trop. Med. Hyg.* **60**(1): 157–164.

Chen, X.M., Levine, S.A., et al. (2001) *Cryptosporidium parvum* activates nuclear factor kappaB in biliary epithelia preventing epithelial cell apoptosis. *Gastroenterology* **120**(7): 1774–1783.

Chen, X.M., O'Hara, S.P., et al. (2004a) Apical organelle discharge by *Cryptosporidium parvum* is temperature, cytoskeleton, and intracellular calcium dependent and required for host cell invasion. *Infect. Immun.* **72**(12): 6806–6816.

Chen, X.M., Splinter, P.L., et al. (2004b) Phosphatidylinositol 3-kinase and frabin mediate *Cryptosporidium parvum* cellular invasion via activation of Cdc42. *J. Biol. Chem.* **279**(30): 31671–31678.

Chen, X.M., Huang, B.Q., et al. (2004c) Cdc42 and the actin-related protein/neural Wiskott-Aldrich syndrome protein network mediate cellular invasion by *Cryptosporidium parvum. Infect. Immun.* **72**(5): 3011–3021.

Chen, X.M., O'Hara, S.P., et al. (2005) Localized glucose and water influx facilitates *Cryptosporidium parvum* cellular invasion by means of modulation of host-cell membrane protrusion. *Proc. Natl Acad. Sci. USA* **102**(18): 6338–6343.

Craun, G.F., Calderon, R.L., et al. (2005) Outbreaks associated with recreational water in the United States. *Int. J. Environ. Health Res.* **15**(4): 243–262.

Ctrnacta, V., Ault, J.G., et al. (2006) Localization of pyruvate:NADP+ oxidoreductase in sporozoites of *Cryptosporidium parvum. J. Eukaryot. Microbiol.* **53**(4): 225–231.

Current, W.L. and Haynes, T.B. (1984) Complete development of *Cryptosporidium* in cell culture. *Science* **224**(4649): 603–605.

Current, W.L. and Reese, N.C. (1986) A comparison of endogenous development of three isolates of *Cryptosporidium* in suckling mice. *J. Protozool.* **33**(1): 98–108.

Dalle, F., Roz, P., et al. (2003) Molecular characterization of isolates of waterborne *Cryptosporidium* spp. collected during an outbreak of gastroenteritis in South Burgundy, France. *J. Clin. Microbiol.* **41**(6): 2690–2693.

Deng, M., Lancto, C.A., et al. (2004) *Cryptosporidium parvum* regulation of human epithelial cell gene expression. *Int. J. Parasitol.* **34**(1): 73–82.

Dupont, C., Bougnoux, M.E., et al. (1996) Microbiological findings about pulmonary cryptosporidiosis in two AIDS patients. *J. Clin. Microbiol.* **34**(1): 227–229.

Eggleston, M.T., Tilley, M., et al. (1994) Enhanced development of *Cryptosporidium parvum* in vitro by removal of oocyst toxins from infected cell monolayers. *J. Helminth. Soc. Washington* **61**(1): 118–121.

El-Osta, Y.G., Chalmers, R.M., *et al.* (2003) Survey of *Cryptosporidium parvum* genotypes in humans from the UK by mutation scanning analysis of a heat shock protein gene region. *Mol. Cell. Probes* **17**(4): 127–134.

Elliott, D.A. and Clark, D.P. (2000) *Cryptosporidium parvum* induces host cell actin accumulation at the host-parasite interface. *Infect. Immun.* **68**(4): 2315–2322.

Elliott, D.A. and Clark, D.P. (2003) Host cell fate on *Cryptosporidium parvum* egress from MDCK cells. *Infect. Immun.* **71**(9): 5422–5426.

Esposito, M., Stettler, R., *et al.* (2005) In vitro efficacies of nitazoxanide and other thiazolides against *Neospora caninum* tachyzoites reveal antiparasitic activity independent of the nitro group. *Antimicrob. Agents Chemother.* **49**(9): 3715–3723.

Fayer, R., Trout, J.M., *et al.* (1998) Infectivity of *Cryptosporidium parvum* oocysts stored in water at environmental temperatures. *J. Parasitol.* **84**(6): 1165–1169.

Fayer, R., Trout, J.M., *et al.* (2000) Rotifers ingest oocysts of *Cryptosporidium parvum. J. Eukaryot. Microbiol.* **47**(2): 161–163.

Feng, H., Nie, W., *et al.* (2006) Bile acids enhance invasiveness of *Cryptosporidium* spp. into cultured cells. *Infect. Immun.* **74**(6): 3342–3346.

Fichera, M.E. and Roos D.S. (1997) A plastid organelle as a drug target in apicomplexan parasites. *Nature* **390**(6658): 407–409.

Garcia, L.S. and Current W.L. (1989) Cryptosporidiosis: clinical features and diagnosis. *Crit. Rev. Clin. Lab. Sci.* **27**(6): 439–460.

Gatei, W., Suputtamongkol, Y., *et al.* (2002) Zoonotic species of *Cryptosporidium* are as prevalent as the anthroponotic in HIV-infected patients in Thailand. *Ann. Trop. Med. Parasitol.* **96**(8): 797–802.

Gatei, W., Wamae, C.N., *et al.* (2006) Cryptosporidiosis: prevalence, genotype analysis, and symptoms associated with infections in children in Kenya. *Am. J. Trop. Med. Hyg.* **75**(1): 78–82.

Girouard, D., Gallant, J., *et al.* (2006) Failure to propagate *Cryptosporidium* spp. in cell-free culture. *J. Parasitol.* **92**(2): 399–400.

Glaeser, C., Grimm, F., *et al.* (2004) Detection and molecular characterization of *Cryptosporidium* spp. isolated from diarrheic children in Switzerland. *Pediatr. Infect. Dis. J.* **23**(4): 359–361.

Goncalves, E.M., da Silva, A.J., *et al.* (2006) Multilocus genotyping of *Cryptosporidium hominis* associated with diarrhea outbreak in a day care unit in Sao Paulo. *Clinics* **61**(2): 119–126.

Gookin, J.L., Chiang, S., *et al.* (2006) NF-kappaB-mediated expression of iNOS promotes epithelial defense against infection by *Cryptosporidium parvum* in neonatal piglets. *Am. J. Physiol. Gastrointest. Liver Physiol.* **290**(1): G164–G174.

Graczyk, T.K. and Schwab, K.J. (2000) Foodborne infections vectored by molluscan shellfish. *Curr. Gastroenterol. Rep.* **2**(4): 305–309.

Graczyk, T.K., Cranfield, M.R., et al. (1997) Infectivity of *Cryptosporidium parvum* oocysts is retained upon intestinal passage through a migratory water-fowl species (Canada goose, *Branta canadensis*). *Trop. Med. Int. Health* **2**(4): 341–347.

Graczyk, T.K., Grimes, B.H., et al. (2004) Mechanical transmission of *Cryptosporidium parvum* oocysts by flies. *Wiad. Parazytol.* **50**(2): 243–247.

Grantham-McGregor, S. and Baker-Henningham, H. (2005) Review of the evidence linking protein and energy to mental development. *Public Health Nutr.* **8**(7A): 1191–1201.

Griffiths, J.K., Balakrishnan, R., et al. (1998) Paromomycin and geneticin inhibit intracellular *Cryptosporidium parvum* without trafficking through the host cell cytoplasm: implications for drug delivery. *Infect. Immun.* **66**(8): 3874–3883.

Harris, J.R., Adrian, M., et al. (2004) Amylopectin: a major component of the residual body in *Cryptosporidium parvum* oocysts. *Parasitology* **128**(Pt 3): 269–282.

Hashim, A., Mulcahy, G., et al. (2006) Interaction of *Cryptosporidium hominis* and *Cryptosporidium parvum* with primary human and bovine intestinal cells. *Infect. Immun.* **74**(1): 99–107.

Hemphill, A., Mueller, J., et al. (2006) Nitazoxanide, a broad-spectrum thiazolide anti-infective agent for the treatment of gastrointestinal infections. *Expert Opin. Pharmacother.* **7**(7): 953–964.

Hewitt, R.G., Yiannoutsos, C.T., et al. (2000) Paromomycin: no more effective than placebo for treatment of cryptosporidiosis in patients with advanced human immunodeficiency virus infection. AIDS Clinical Trial Group. *Clin. Infect. Dis.* **31**(4): 1084–1092.

Hijjawi, N.S., Meloni, B.P., et al. (2004) Complete development of *Cryptosporidium parvum* in host cell-free culture. *Int. J. Parasitol.* **34**(7): 769–777.

Hlavsa, M.C., Watson, J.C., et al. (2005) Cryptosporidiosis surveillance – United States, 1999–2002 and giardiasis surveillance – United States, 1998–2002. *MMWR Surveill. Summ.* **54**(1): 1–16.

Hommer, V., Eichholz, J., et al. (2003) Effect of antiretroviral protease inhibitors alone, and in combination with paromomycin, on the excystation, invasion and in vitro development of *Cryptosporidium parvum*. *J. Antimicrob. Chemother.* **52**(3): 359–364.

Hunter, P.R., Hughes, S., et al. (2004) Health sequelae of human cryptosporidiosis in immunocompetent patients. *Clin. Infect. Dis.* **39**(4): 504–510.

Iqbal, J., Hira, P.R., et al. (2001) Cryptosporidiosis in Kuwaiti children: seasonality and endemicity. *Clin. Microbiol. Infect.* **7**(5): 261–266.

Johnson, J.K., Schmidt, J., et al. (2004) Microbial adhesion of *Cryptosporidium parvum* sporozoites: purification of an inhibitory lipid from bovine mucosa. *J. Parasitol.* **90**(5): 980–990.

Johnston, S.P., Ballard, M.M., et al. (2003) Evaluation of three commercial assays for detection of *Giardia* and *Cryptosporidium* organisms in fecal specimens. *J. Clin. Microbiol.* **41**(2): 623–626.

Keusch, G.T., Hamer, D., *et al.* (1995) Cryptosporidia – who is at risk? *Schweiz Med. Wochenschr.* **125**(18): 899–908.

Khramtsov, N.V. and Upton, S.J. (2000) Association of RNA polymerase complexes of the parasitic protozoan *Cryptosporidium parvum* with virus-like particles: heterogeneous system. *J. Virol.* **74**(13): 5788–5795.

LaGier, M.J., Tachezy, J., *et al.* (2003) Mitochondrial-type iron-sulfur cluster biosynthesis genes (IscS and IscU) in the apicomplexan *Cryptosporidium parvum*. *Microbiology* **149**(Pt 12): 3519–3530.

Lean, I.S., McDonald, V., *et al.* (2002) The role of cytokines in the pathogenesis of *Cryptosporidium* infection. *Curr. Opin. Infect. Dis.* **15**(3): 229–234.

Lean, I.S., McDonald, S.A., *et al.* (2003) Interleukin-4 and transforming growth factor beta have opposing regulatory effects on gamma interferon-mediated inhibition of *Cryptosporidium parvum* reproduction. *Infect. Immun.* **71**(8): 4580–4585.

Luft, B.J., Payne, D., *et al.* (1987) Characterization of the *Cryptosporidium* antigens from sporulated oocysts of *Cryptosporidium parvum*. *Infect. Immun.* **55**(10): 2436–2441.

Mathieu, E., Levy, D.A., *et al.* (2004) Epidemiologic and environmental investigation of a recreational water outbreak caused by two genotypes of *Cryptosporidium parvum* in Ohio in 2000. *Am. J. Trop. Med. Hyg.* **71**(5): 582–589.

Mathison, B.A. and Ditrich, O. (1999) The fate of *Cryptosporidium parvum* oocysts ingested by dung beetles and their possible role in the dissemination of cryptosporidiosis. *J. Parasitol.* **85**(4): 678–681.

McCole, D.F., Eckmann, L., *et al.* (2000) Intestinal epithelial cell apoptosis following *Cryptosporidium parvum* infection. *Infect. Immun.* **68**(3): 1710–1713.

McLauchlin, J., Amar, C., *et al.* (2000) Molecular epidemiological analysis of *Cryptosporidium* spp. in the United Kingdom: results of genotyping *Cryptosporidium* spp. in 1705 fecal samples from humans and 105 fecal samples from livestock animals. *J. Clin. Microbiol.* **38**(11): 3984–3990.

Mele, R., Gomez Morales, M.A., *et al.* (2004) *Cryptosporidium parvum* at different developmental stages modulates host cell apoptosis in vitro. *Infect Immun.* **72**(10): 6061–6067.

Neira-Otero, P., Munoz-Saldias, N., *et al.* (2005) Molecular characterization of *Cryptosporidium* species and genotypes in Chile. *Parasitol. Res.* **97**(1): 63–67.

Nelson, J.B., O'Hara, S.P., *et al.* (2006) *Cryptosporidium parvum* infects human cholangiocytes via sphingolipid-enriched membrane microdomains. *Cell Microbiol.* **8**(12): 1932–1945.

Nime, F.A., Burek, J.D., *et al.* (1976) Acute enterocolitis in a human being infected with the protozoan *Cryptosporidium*. *Gastroenterology* **70**(4): 592–598.

O'Hara S.P. and Lin, J.J. (2006) Accumulation of tropomyosin isoform 5 at the infection sites of host cells during *Cryptosporidium* invasion. *Parasitol. Res.* **99**(1): 45–54.

Okhuysen, P.C., Chappell, C.L., et al. (1999) Virulence of three distinct *Cryptosporidium parvum* isolates for healthy adults. *J. Infect. Dis.* **180**(4): 1275–1281.

Pollok, R.C., Farthing, M.J., et al. (2001) Interferon gamma induces enterocyte resistance against infection by the intracellular pathogen *Cryptosporidium parvum. Gastroenterology* **120**(1): 99–107.

Pollok, R.C., McDonald, V., et al. (2003) The role of *Cryptosporidium parvum*-derived phospholipase in intestinal epithelial cell invasion. *Parasitol. Res.* **90**(3): 181–186.

Putignani, L. (2005) The unusual architecture and predicted function of the mitochondrion organelle in *Cryptosporidium parvum* and *hominis* species: the strong paradigm of the structure-function relationship. *Parasitologia* **47**(2): 217–225.

Raccurt, C.P., Brasseur, P., et al. (2006) [Human cryptosporidiosis and *Cryptosporidium* spp. in Haiti]. *Trop. Med. Int. Health* **11**(6): 929–934.

Robinson, P., Okhuysen, P.C., et al. (2003) Substance P expression correlates with severity of diarrhea in cryptosporidiosis. *J. Infect. Dis.* **188**(2): 290–296.

Rossignol, J.F. (2006) Nitazoxanide in the treatment of acquired immune deficiency syndrome-related cryptosporidiosis: results of the United States compassionate use program in 365 patients. *Aliment Pharmacol. Ther.* **24**(5): 887–894.

Rossignol, J.F., Hidalgo, H., et al. (1998) A double-'blind' placebo-controlled study of nitazoxanide in the treatment of cryptosporidial diarrhoea in AIDS patients in Mexico. *Trans R. Soc. Trop. Med. Hyg.* **92**(6): 663–666.

Schmidt, W., Wahnschaffe, U., et al. (2001) Rapid increase of mucosal CD4 T cells followed by clearance of intestinal cryptosporidiosis in an AIDS patient receiving highly active antiretroviral therapy. *Gastroenterology* **120**(4): 984–987.

Smith, H.V., Nichols, R.A., et al. (2005) *Cryptosporidium* excystation and invasion: getting to the guts of the matter. *Trends Parasitol.* **21**(3): 133–142.

Stott, R., May, E., et al. (2003) Predation of *Cryptosporidium* oocysts by protozoa and rotifers: implications for water quality and public health. *Water Sci. Technol.* **47**(3): 77–83.

Streubel, A., Siepmann, J., et al. (2006) Drug delivery to the upper small intestine window using gastroretentive technologies. *Curr. Opin. Pharmacol.* **6**(5): 501–508.

Tetley, L., Brown, S.M., et al. (1998) Ultrastructural analysis of the sporozoite of *Cryptosporidium parvum. Microbiology* **144**(Pt 12): 3249–3255.

Topouchian, A., Kapel, N., et al. (2005) *Cryptosporidium* infection impairs growth and muscular protein synthesis in suckling rats. *Parasitol. Res.* **96**(5): 326–330.

Tyzzer, E.E. (1907) A sporozoan found in the peptic glands of the common mouse. *Proc. Soc. Exp. Biol. Med.* **5**: 12–13.

Upton, S.J., Tilley, M., et al. (1994) Comparative development of *Cryptosporidium parvum* (Apicomplexa) in 11 continuous host cell lines. *FEMS Microbiol. Lett.* **118**(3): 233–236.

Wetzel, D.M., Schmidt, J., *et al.* (2005) Gliding motility leads to active cellular invasion by *Cryptosporidium parvum* sporozoites. *Infect. Immun.* **73**(9): 5379–5387.

Widmer, G., Yang, Y.L., *et al.* (2006) Preferential infection of dividing cells by *Cryptosporidium parvum*. *Parasitology* **133**(Pt 2): 131–138.

Wiest, P.M., Johnson, J.H., *et al.* (1993) Microtubule inhibitors block *Cryptosporidium parvum* infection of a human enterocyte cell line. *Infect. Immun.* **61**(11): 4888–4890.

Xiao, L., Fayer, R., *et al.* (2004) *Cryptosporidium* taxonomy: recent advances and implications for public health. *Clin. Microbiol. Rev.* **17**(1): 72–97.

Xu, P., Widmer, G., *et al.* (2004) The genome of *Cryptosporidium hominis*. *Nature* **431**(7012): 1107–1112.

Zerpa, R. and Huicho, L. (1994) Childhood cryptosporidial diarrhea associated with identification of *Cryptosporidium* sp. in the cockroach *Periplaneta americana*. *Pediatr. Infect. Dis. J.* **13**(6): 546–548.

Zhu, G., Marchewka, M.J., *et al.* (2000) *Cryptosporidium parvum* appears to lack a plastid genome. *Microbiology* **146**(Pt 2): 315–321.

B2 *Toxoplasma gondii*

Marie-Laure Dardé, Dominique
Aubert, Francis Derouin,
Aurélien Dumétre, Hervé
Pelloux and Isabella Villena

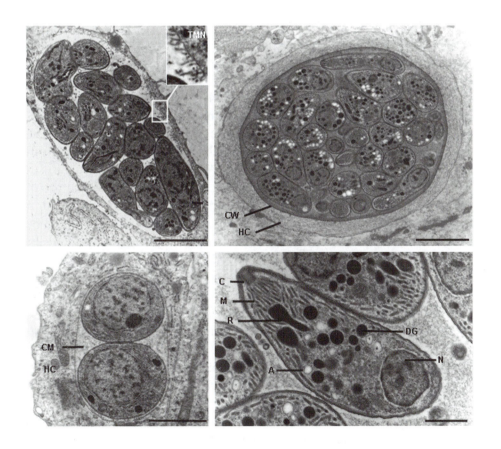

1 Introduction

Toxoplasma gondii is a protozoan intracellular parasite, the most wide-spread among parasites of the phylum Apicomplexa, with a broad host range including many birds and mammals and a geographic range that is nearly worldwide. Infection is mainly acquired by ingestion of food or water that is contaminated with oocysts shed by cats or by eating under-cooked or raw meat containing tissue cysts. It is commonly estimated that up to one third of the world's population is chronically infected. While infection of healthy adults is usually relatively mild, serious disease can result from congenital transmission or when the host is immunocompromised. Due to its medical and veterinary importance, a great deal of research has been done in the last 30 years about host-parasite relationship of this sophisticated intracellular parasite, about genomic organization and genetic diversity.

2 Discovery and Taxonomy

Nicolle and Manceaux in the Pasteur Institute of Tunis described *Toxoplasma gondii* in 1908 in a North African rodent (*Ctenodactylus gundi*). In the same year, Splendore noted identical forms in a laboratory rabbit in Brazil. It was not before the 1960s that transmission through ingestion of infected meat was clearly established. Finally, the full life cycle was completed only in 1970 with the description of oocysts and the discovery of the role of felids in transmission, simultaneously by Frenkel and by Hutchison (Dubey and Beattie, 1988).

The description of the complete life cycle and of the ultrastructure of the different stages allows the classification of *Toxoplasma* in the phylum Apicomplexa, characterized by an obligate intracellular parasitism and by apical organelles and structures involved in invasion and survival in cells. Only one species, *T. gondii*, is recognized in the genera *Toxoplasma*. It resides among the group of tissue-cyst-forming coccidia parasites together with animal pathogens *Sarcocystis* spp., *Neospora* spp. and the largely non-pathogenic genera *Hammondia* and *Besnoitia*.

The first case of human congenital toxoplasmosis with eye involvement was described in 1923 by Janku. But the responsibility of *Toxoplasma* in a case of congenital infection in a human child was really demonstrated by Wolf *et al.* (1939). The classical signs of acquired symptomatic toxoplasmosis (lymphadenopathies) were described in 1956 by Siims. After the development of the serologic dye test by Sabin and Feldman in 1948, it became clear that asymptomatic *T. gondii* infections are widely prevalent in humans in many countries. The latest step in this clinical history of human toxoplasmosis was the recognition of reactivation of latent infections in immunosuppressed adults, notably since the onset of the acquired immunodeficiency syndrome (AIDS) epidemic.

3 Life Cycle

Tissue-cyst-forming coccidia have complex two-host life cycles functioning in a prey-predator system that alternate between definitive (sexual reproduction) and intermediate hosts (asexual replication; Figure 1). *T. gondii* is unique among this group because the parasite can transmit not only between intermediate and definitive hosts (sexual cycle), but also between intermediate hosts via carnivorism (asexual cycle) or even between definitive hosts (although less efficiently in this last case). It is also remarkable for the extremely wide range of birds and mammals that serve as intermediate hosts. There are three infective stages: two asexual stages, a rapidly dividing, invasive tachyzoite, and a slowly dividing bradyzoite in tissue cysts; and one stage resulting from a sexual reproduction, the sporozoite, protected inside an oocyst, highly resistant in the environment.

Sexual reproduction occurs only in felids (domestic and wild cats). After ingestion of cysts present in tissues of an intermediate host, the cyst wall is destroyed by gastric enzymes (Figure 2A–C) and pepsin-resistant bradyzoites reach the small intestine. They settle in a parasitophorous vacuole within enterocytes where they undergo a self-limiting number of asexual multiplications. This first step of asexual reproduction is characterized by the development of merozoites within schizonts of five ultrastructural

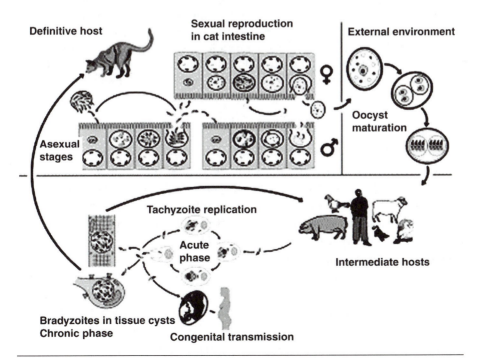

Figure 1 Life cycle of *Toxoplasma gondii*. Copyright Ferguson (2002).

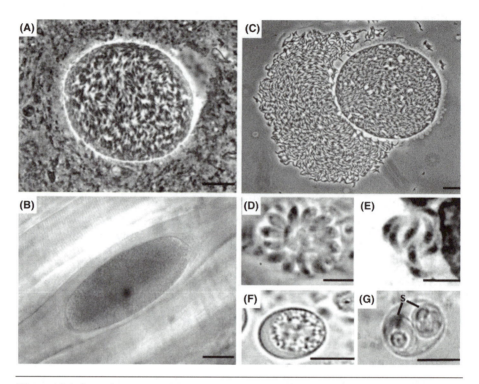

Figure 2 Life-cycle stages of *T. gondii*. Scale bars, 10 μm. (A) Tissue cyst in the brain of an infected mouse containing hundreds of bradyzoites. (B) Tissue cyst in a muscular cell of an infected mouse. Note the elongated aspect of the cyst. (C) Tissue cyst liberating hundreds of bradyzoites after pepsin digestion; phase contrast. (D) Colonies of tachyzoites inside a fibroblast; phase contrast. (E) Extracellular tachyzoites; Giemsa stain. (F) Unsporulated oocyst in cat feces. (G) Sporulated oocyst with two sporocysts (S), each containing four sporozoites and a residual body.

types (A to E; Dubey *et al.*, 1998). It is followed by a conversion from asexual to sexual development with the formation of male and female gametocytes (gametogony). A female gametocyte matures in only one macrogamete while a male gametocyte gives rise to several flagellated microgametes. After fertilization, oocysts are formed within enterocytes and excreted as an unsporulated form in cat feces (Figure 2F). These unsporulated oocysts are the only diploid stage of *Toxoplasma*. The process of sporogony occurs after a few days in the external environment. It implies a meiotic reduction and morphological changes leading to the formation of a sporulated oocyst with two sporocysts, each containing four haploid sporozoites (Figure 2G). Shedding of oocysts begins 3–7 days after ingestion of tissue cysts and may continue for up to 20 days. Infected cats can shed more than 100 million oocysts in their feces. Oocysts are extremely resistant to environmental conditions and can infect a wide range of intermediate hosts, virtually all warm-blooded animals, from mammals to birds, when ingested with food or water.

Within the intermediate host, the parasite undergoes only asexual development. After oocyst excystation, sporozoites penetrate the intestinal epithelium and differentiate into tachyzoites. Tachyzoites rapidly replicate by endodyogeny inside any kind of cells, and disseminate throughout the organism (acute infection; Figure 2D, E). As a result of differentiation to the bradyzoite stage, tissue cysts arise 7–10 days post infection and may persist for the life of the host, predominantly in the brain or musculature. The infection is usually latent and chronic. Reactivation may occur in immunodeficient hosts.

Upon ingestion of these tissue cysts in raw or undercooked meat from a chronically infected host, cysts are ruptured as they pass through the digestive tract, causing bradyzoite release. The bradyzoites will infect intestinal epithelium of the new host. If the ingesting animal is a cat, the bradyzoites can differentiate into the sexual stages, as already described. If the ingesting animal is an intermediate host, a new asexual cycle will take place: bradyzoites differentiate back to the rapidly dividing tachyzoite stage for dissemination throughout the body, thereby completing the asexual cycle (Black and Boothroyd, 2000). The asexual phase can theoretically cycle between intermediate hosts *ad infinitum*. In addition, if the acute phase occurs during pregnancy, the parasite can cross the placenta and infect the fetus (congenital transmission). A role for this vertical transmission in maintaining high levels of infection in certain species has been suggested.

Several aspects of this cycle necessary to understand the genetic population structure of *T. gondii* or its epidemiology can be underlined: (i) self-fertilization is possible meaning that all macrogametes and microgametes of a strain will be genetically identical, and (ii) many macrogametes remain unfertilized but seem capable of forming oocysts and undergoing sporulation, leading to the hypothesis of a form of parthenogenesis (Ferguson, 2002). The part of the sexual and asexual cycle in nature is still unknown and may vary according to environmental conditions and to the population of both intermediate and definitive hosts.

4 Cell Organization

The zoites (tachyzoites, bradyzoites or sporozoites), infective stages of *T. gondii*, are crescent-shaped cells, approximately 5 μm long and 2 μm wide with a pointed apical end and a rounded posterior end. They are limited by a complex membrane named the pellicle and possess numerous secretory organelles (rhoptries, dense granules and micronemes), a multiple-membrane-bound plastid-like organelle called the apicoplast, a mitochondrion, a Golgi complex located anterior to the nucleus, ribosomes, and an endoplasmic reticulum continuous to the nuclear envelope (Dubey *et al.*, 1998; Figure 3).

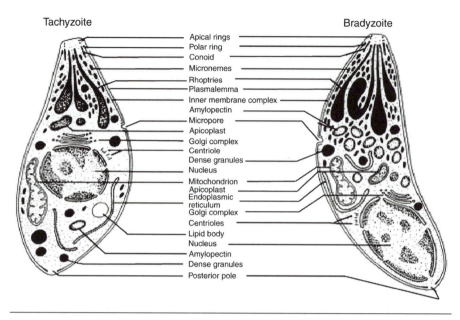

Tachyzoite   Bradyzoite

- Apical rings
- Polar ring
- Conoid
- Micronemes
- Rhoptries
- Plasmalemma
- Inner membrane complex
- Amylopectin
- Micropore
- Apicoplast
- Golgi complex
- Centriole
- Dense granules
- Nucleus
- Mitochondrion
- Apicoplast
- Endoplasmic reticulum
- Golgi complex
- Centrioles
- Lipid body
- Nucleus
- Amylopectin
- Dense granules
- Posterior pole

Figure 3 Schematic drawing of the cell organization of a tachyzoite (*on the left*) and a bradyzoite (*on the right*). Copyright Dubey *et al.* (1998).

4.1 Trimembrane pellicle

The pellicle of *T. gondii* consists of three membranes: the plasma membrane and an inner membrane complex (IMC) of two membranes formed by a patchwork of flattened cisternae, sutured together in a spiral, that lie just beneath the plasmalemma. The inner membrane is interrupted at the anterior end above the polar rings, at a micropore situated in the middle of the parasite body, and at the posterior pore at the extreme posterior tip. The micropore is implicated in endocytosis and vesicles have been observed in this region (Nichols and Chiappino, 1987; Figure 4).

4.2 Cytoskeletal network

Two preconoidal rings at the anterior end surround a truncated cone or conoid. The conoid consists of 14 microtubular elements arranged like a compressed spring with a counterclockwise spiral, active during cell invasion. Two microtubules extend through the center of the conoid, closely associated to rhoptries and micronemes, from the preconoidal rings to the body of the cell. They may serve as a scaffold directing rhoptries through the conoid to secrete their contents from the apical tip (Black and Boothroyd, 2000). The cytoskeleton is beneath the IMC and closely associated to it. It consists of 22 subpellicular microtubules, originating from a polar ring, just posterior to the conoid and forming a spiral down two-thirds of the body. They are involved in structural integrity and motility of the cell. Microtubule-associated proteins (MAPs) and intramembranous

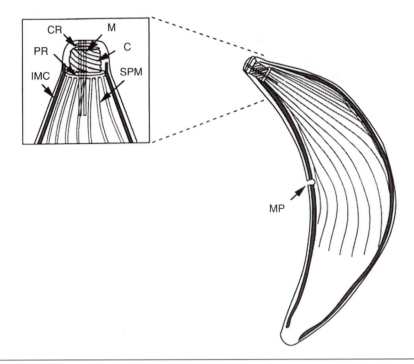

Figure 4 The diverse structures of the *Toxoplasma* cytoskeleton: preconoidal rings (CR), the conoid (C), the two apical microtubules (M), polar ring (PR), subpellicular microtubules (SPM), internal membrane complex (IMC) and the micropore (MP). Copyright Black and Boothroyd (2000).

particles (IMPs) are found within the IMC, both exhibiting a 32-nm periodicity longitudinally. Actin has been detected in the conoid, preconoidal rings, and subpellicular microtubules by immunoelectron microscopy. *Toxoplasma* also has an actin-binding protein, named toxofilin, which plays a role in the creation and function of actin filaments (Poupel *et al.*, 2000). Myosin, a mechanoprotein that interacts with actin, colocalizes with actin in the anterior portion of the parasite as well as along the inner membrane complex (Schwartzman and Pfefferkorn, 1983.). Three unconventional myosins (*TgM-A*, *TgM-B*, *TgM-C*) have been cloned from *Toxoplasma*. The use of this actin-myosin cytoskeleton, associated with the calcium-dependent secretion of adhesins (microneme proteins) which recognize host cell receptors, and the shedding of GPI-anchored surface proteins on the substratum contributes to the gliding mobility of *Toxoplasma* and to attachment to host cell and penetration (Black and Boothroyd, 2000).

4.3 Surface antigens

The surface of *T. gondii* is coated with developmentally expressed, glyco-sylphosphatidylinositol (GPI)-linked proteins structurally related to the highly immunogenic surface antigen SAG1. Collectively, these surface antigens are known as the SRS (SAG1-related sequences) superfamily of proteins. SAG1 and SAG2A are prototypic members of this superfamily that includes at least 20 homologous proteins. All SRS proteins have an N-terminal signal peptide and are tethered to the outer surface membrane of the parasite by the GPI anchor. They typically share between 24 and 99% amino-acid sequence identity. Tachyzoites and bradyzoites express distinct, largely nonoverlapping sets of SRS antigens, with SAG1 and SAG2A expressed only on tachyzoites, whereas SAG2C/D, BSR4, and SRS9 are found only on bradyzoites (Jung *et al.*, 2004). Bioinformatically, 161 unique SRS DNA sequences were identified in the *T. gondii* type II Me49 genome based on TBLASTN analyses of the *Toxoplasma* genome sequence database (www.toxodb.org).

4.4 Organelles and secretory proteins

As for other apicomplexans, the apical part concentrates different secretory organelles: rhoptries, micronemes, and dense granules.

Rhoptries

Rhoptries are eight to 10 club-shaped organelles with a neck extending through the conoid. The bulbous posterior end of the rhoptries presents a more or less labyrinthine structure depending on its secretory activity. At the moment of cellular entry, they look like small ovoid saccules, with amorphous residues of their former dense content (Nichols *et al.*, 1983). This aspect corresponds to the release of the contents of the rhoptries into the nascent parasitophorous vacuole during invasion into the host cell. Rhoptries contain unusual lipids (cholesterol, phosphatidylcholine, phosphatic acid, and lysophospholipids) and numerous proteins specialized for intracellular parasitism. Several ROP (ROP1 to ROP7) antigens have been described. A detailed proteomic analysis using mass spectrometry has identified 38 novel proteins (Bradley *et al.*, 2005). Among them, were detected toxofilin (the actin-binding protein), Rab11, and also kinases, phosphatases, and proteases that are likely to play a role in the ability of the parasite to invade the host cell. Immunofluorescence staining with monoclonal antibodies reveals their localization either in the rhoptry necks (TgRON) and/or the bulbous base (TgROP). Analysis of the rhoptry proteome identified proteins unique to *Toxoplasma*, but also homologues common to *Toxoplasma* and to *Plasmodium*.

Micronemes

The rod-like structures named micronemes are mostly located between rhoptries. Micronemes are involved in the trafficking and storage of ligands (MICs) for host-cell receptors. There is a large repertoire of microneme proteins that contribute to attachment to the host cell surface and to invasion. They belong to a family of transmembrane proteins, known as TRAP (thrombospondin-related anonymous proteins), conserved among the Apicomplexan. They can contain adhesive modules within their ectodomains, composed of combinations of an integrin-like I/A domain and various type 1 repeats of thrombospondin (TSR). The repertoire includes adhesins such as TgMIC2 and escorters such as TgMIC6, which forms a complex with the soluble adhesins, TgMIC1 and TgMIC4. Escorters bridge host-cell receptors to the parasite membrane during invasion. Most TgMICs are secreted apically. They are proteolytically cleaved either during their transport along the secretory pathway and/or after exocytosis. Gliding motility and host-cell penetration involve the translocation of the micronemes toward the posterior pole, via interaction between the MICs-receptor complexes and the actomyosin system of the parasite cytoskeleton (Soldati *et al.*, 2001).

Dense granules

Dense granules are microspheres of approximately 200 nm in diameter, surrounded by a unique membrane, and dispersed throughout the cell. Dense granule proteins (GRA) are involved in the maturation of the parasitophorous vacuole and of the cyst wall. Nine dense granule proteins were detected in *Toxoplasma*, GRA1–8, for which no clear function is yet defined, and GRA9, for which an enzymatic activity (NTPase) was shown (Mercier *et al.*, 2005).

4.5 Apicoplast and fatty acid synthesis

The apicoplast is a nonphotosynthetic plastid organelle, enclosed by four membranes as a result of a possible acquisition by the parasite via secondary endosymbiosis of a free-living red alga (Roos *et al.*, 1999). It contains many plant-specific metabolic pathways, of cyanobacterial origin, not found in mammalian cells, and so is a known and potential drug target. Genomic analysis revealed apicoplast pathways for the synthesis of fatty acids, isoprenoids, heme, and iron–sulfur cluster biogenesis. Apicoplast prokaryotic type-II fatty acid synthesis (FAS II) differs in structure, kinetics, and inhibitor susceptibility from the eukaryotic FAS I pathway found in the mammalian host. In addition to FAS II, *Toxoplasma* also harbors fatty acylelongases and a FAS I pathway. FAS II is required for the activation of pyruvate dehydrogenase, an important source of the metabolic precursor acetyl-CoA (Mazumdar *et al.*, 2006). Many of the functions of the apicoplast have yet to be fully elucidated. The apicoplast seems essential for parasite viability, and tachyzoites lacking an apicoplast are able to invade, but not multiply, in new host cells.

5 Differences Between the Infective Stages

The various infective stages (tachyzoite, bradyzoite and sporozoite) can be identified by their structure, distinctive developmental processes, host-parasite relationship, and molecular markers.

5.1 Ultrastructural differences

Sporozoites, tachyzoites, and bradyzoites of *T. gondii* differ in certain organelles and inclusion bodies (Figure 3). Bradyzoites are more slender than tachyzoites. Their nucleus is situated toward the posterior end, whereas the nucleus in tachyzoites is more centrally located (Figure 5D). They have similar numbers of rhoptries but the rhoptries of tachyzoites are uniformly labyrinthine; sporozoites usually contain both labyrinthine and uniformly electron-dense rhoptries. The contents of rhoptries in bradyzoites vary with the age of the tissue cyst: bradyzoites in younger tissue cysts

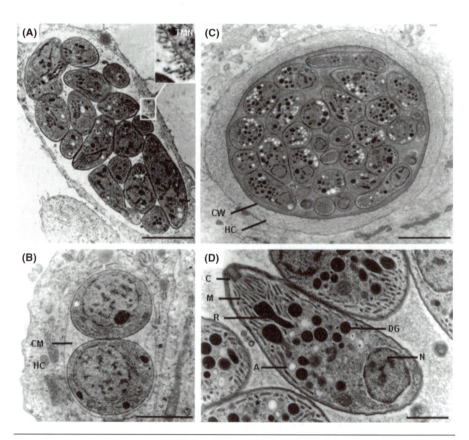

Figure 5 Transmission electron micrographs. (A) Colony of tachyzoites within a parasitophorous vacuole (fibroblast in cell culture). Note the empty space between zoites and the tubulovesicular membranous network (TMN; inset). (B) Early cyst observed in cell culture, containing only two *Toxoplasma* cells. The space between the two *Toxoplasma* cells is filled with a granular material corresponding to the cyst matrix (CM; HC, host cell). (C) Young cyst in the brain of an AIDS patient. Note the bradyzoites with numerous amylopectin granules (electrolucent spots). CW, cyst wall. (D) Bradyzoites in a tissue cyst. Bradyzoites are surrounded by the cyst matrix. Note the posteriorly located nucleus (N), the numerous micronemes (M), the dense rhoptries (R), the dense granules (DG), amylopectin granules (A) and the conoid (C).

may have labyrinthine rhoptries, whereas those in older tissue cysts are electron dense. Also, most bradyzoites have one to three rhoptries, which are looped back on themselves. Tachyzoites have few micronemes, sporozoites have an intermediate number, and bradyzoites have many. Dense granules are more numerous in sporozoites and tachyzoites (5–17) than in bradyzoites (1–5). Dense granules of bradyzoites have a smaller mean diameter than those observed at the tachyzoite stage (Speer *et al.*, 1998). The slowly dividing encysted bradyzoites and the sporozoites inside the oocyst contain a high number of large amylopectin granules, while the rapidly replicating tachyzoites are nearly devoid of amylopectin granules. Lipid bodies are not seen in bradyzoites, but are numerous in sporozoites and occasionally seen in tachyzoites (Dubey *et al.*, 1998).

5.2 Metabolic differences

Given the different rate of growth and location of bradyzoites, tachyzoites and sporozoites, it is likely that their energy metabolism is different. The bradyzoites and the sporozoites synthesize a genuine amylopectin storage polysaccharide of glucose. The biosynthetic pathway of amylopectin in *T. gondii* is characterized by the presence of genes that are of plant origin (Guerardel *et al.*, 2005). Tachyzoites utilize the glycolytic pathway with the production of lactate as their major source of energy; however, mitochondria with a functional TCA cycle exist and contribute to energy production. Bradyzoites lack a functional TCA cycle and respiratory chain. Stage-specific differences have been demonstrated in the activity of various glycolytic enzymes. Lactate dehydrogenase (LDH) and pyruvate kinase activity was higher in bradyzoites than tachyzoites while PP_i-phosphofructokinase activity was higher in tachyzoites than bradyzoites. Different isoforms of LDH were identified, which are stage-specifically expressed during tachyzoite-bradyzoite interconversion: LDH1 is the only isoenzyme produced by the tachyzoites, while the transcription of LDH2 is suppressed during transition from the bradyzoite to the tachyzoite stage. Two other enzymes (glucose 6-phosphate isomerase and enolases) possess isoforms which are stage-specific. The bradyzoite enzymes were resistant to acidic pH, maybe as the result of an accumulation of the glycolytic products derived from the catabolism of amylopectin to lactate.

6 Developmental Processes and Host-Parasite Relationship

6.1 Parasitophorous vacuole

After host cell invasion, tachyzoites are contained in a parasitophorous vacuole (PV) limited by a membrane (Figure 5A). They typically divide into this PV during a 6- to 9-h cycle by a process called endodyogeny, leading to the formation of two daughter cells within each mother cell and to the exponential production of approximately 128 parasites from a single vacuole. The PV membrane is initially formed by the secretion of rhoptry products (lipids and proteins) (Nichols *et al.*, 1983). These products lead to the formation of small vesicles, which then collapse with the invaginated host-cell membrane. So, lipids of PV membranes are derived from both the parasite and the host-cell neutral lipids. Active secretion of membrane proteins from both rhoptries (ROP2/4/8) (Bradley *et al.*, 2005) and dense granules contribute to the acquisition of new intramembrane particles during the first hour following invasion. Vacuolar membrane is therefore a hybrid membrane, clearly different from the membrane of a conventional phagosome. ROP2, one of the rhoptry proteins associated with the PV membrane, participates in the recruitment of host-cell mitochondria, which are closely associated to the PV and contribute to the acquisition of nutriments by the dividing parasites. Within the host cell, the PV is wrapped up by vimentin-like intermediate filaments of the host cell. Soon after penetration, a tubulovesicular membranous network (TMN) develops within the vacuolar space. This network is constituted of two standard membrane bilayers forming tubules from 40 to 60 nm in diameter. Some of the TMN membranes are connected to the PV membrane. It will persist during the entire development of the parasite within the vacuole, progressively filling up the entire space available between parasites. This TMN is hypothesized to have a role in developing exchanges between the parasite and the host cell, bringing small nutrients from the host-cell cytosol to the parasite or exporting proteins and/or lipids from the parasite to the PV membrane and/or the host cell (Mercier *et al.*, 2005).

A different type of PV has been described after oral infection by sporozoites. Entry of the sporozoite into the enterocyte is characterized by formation of a large, impermeable PV (PV1) without TMN and surrounded by host-cell mitochondria. The parasite does not divide within this PV1 but moves freely inside it. The dense granule proteins detected in the sporozoite are not incorporated into PV1 membrane (Mercier *et al.*, 2005). Approximately 24 h later, the parasite escapes PV1 and enters a second PV (PV2), tightly adjusted to the parasite and in which soon appears a TMN, correlated with dense-granule-protein secretion.

6.2 Tissue cyst

Tissue cysts are intracellular structures. In the brain, they are often spheroidal and rarely reach a diameter of 70 μm, whereas intramuscular cysts are elongated and may be up to 100 μm long (Figure 2A, B). They vary in size: young tissue cysts may be as small as 5 μm in diameter and contain only two bradyzoites (Figure 5B), while older ones may contain hundreds of densely packed organisms (Figure 5C). Cysts remain intracellular throughout their life span, wrapped in host-cell intermediate filaments. No accumulation of host-cell mitochondria or endoplasmic reticulum was noted around the cyst wall. The limiting membrane, derived from the

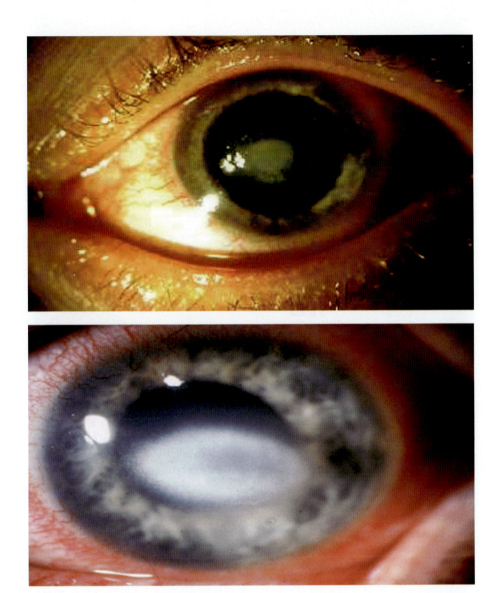

Section A1, Figure 6 *Acanthamoeba*-infected eye. Note the ulcerated epithelium and stromal infiltration exhibiting corneal opacity in acute *Acanthamoeba* keratitis (published with permission from Elsevier).

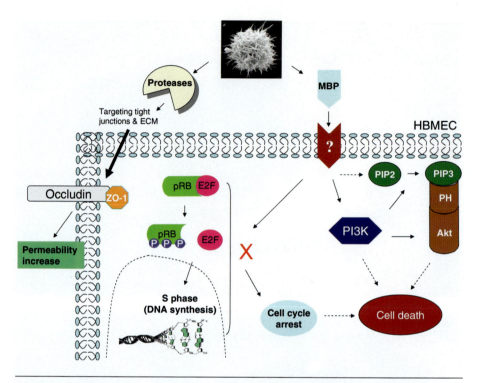

Section A1, Figure 12 Host intracellular signaling in response to *Acanthamoeba*. Note that *Acanthamoeba* induces cell-cycle arrest in the host cells by altering expression of genes as well as by modulating protein retinoblastoma (pRb) phosphorylations. In addition, *Acanthamoeba* have also been shown to induce host cell death via phosphatidylinositol 3-kinase (PI3K). By secreting proteases, amoebae disrupt tight junctions by targeting zonula-1 and occludin proteins. MBP, mannose-binding protein; E2F, a transcription factor that controls cell proliferation through regulating the expression of essential genes required for cell-cycle progression; PIP2, phosphatidylinositol-4,5-bisphosphate; PIP3, phosphatidylinositol-3,4,5-trisphosphate; Akt (protein kinase B)-PH domain, a serine/threonine kinase – a critical enzyme in signal transduction pathways involved in cell proliferation, apoptosis, angiogenesis, and diabetes.

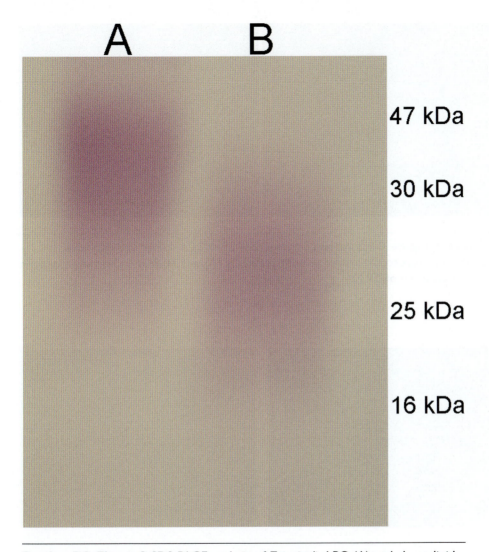

Section D2, Figure 3 SDS-PAGE analysis of *T. vaginalis* LPG (A) and glycan-lipid inositol core of LPG (B) released by mild acid treatment of LPG. Gel was stained with periodic acid-Schiff (PAS) reagent.

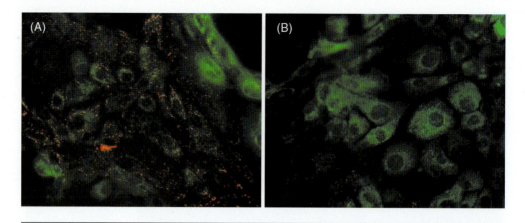

Section D2, Figure 8 Fluorescence microscopy of HVECs stained with JC-1 dye. Fluorescence microscopic images of normal HVECs (A), and HVECs undergoing apoptosis (B) in the presence of CP30.

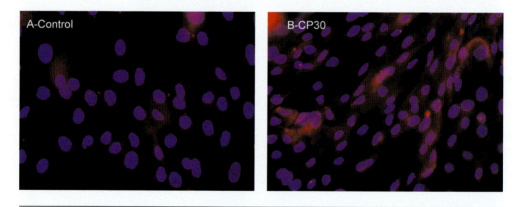

Section D2, Figure 9 Activation of HVEC caspase-3 by CP30. HVECs treated with CP30, washed, formalin-fixed and labeled with anti-ACTIVE Caspase-3 pAb (Promega Corp., WI, USA), followed by fluorescence-labeled anti-rabbit secondary antibody. The cell nuclei were counterstained with DAPI (40× magnification).

parasitophorous vacuole membrane, has a ruffled appearance, presenting numerous invaginations (Ferguson and Hutchison, 1987). This limiting membrane and an underlying layer of electron-dense granular material constitute the cyst wall. The thickness of the wall and the amount of branching of invaginations of the limiting membrane increase in older cysts. Polysaccharides and chitin are major components of this cyst wall as suggested by the presence of lectin-binding sugars (sugar haptens *N*-acetylgalactosamine and *N*-acetylglucosamine) and by disruption of the cyst wall with chitinase (Weiss and Kim, 2000). A granular material is also dispersed between the bradyzoites and constitutes the cyst matrix. However, in older cysts, the space between the bradyzoites may appear electron-lucent.

Whereas the endodyogeny process of the tachyzoites is synchronous, bradyzoites divide into the cysts by asynchronous endodyogeny or schizogony coupled to a missegregation of the apicoplast. It is one of the earliest hallmarks of bradyzoite differentiation. The proportion of dividing bradyzoites decreased with increasing size and age of the cysts. Fully differentiated bradyzoites are arrested in G_0 of the cell cycle.

6.3 Oocyst

The oocyst wall of *T. gondii* consists of an electron-dense outer layer and an electron-lucent inner layer. It is surrounded by an outer loose veil which is probably destroyed after excretion. The veil-forming bodies (VFB) and the wall-forming bodies, type 1 (WFB1) and type 2 (WFB2), are three cytoplasmic structures of the mature macrogamete involved in the development of the veil, the outer and the inner layer, respectively. A micropyle is randomly distributed at the surface of the oocyst wall of sporulated oocysts. Its function is unknown, but it might be sensible to CO_2 and enzymes during sporozoite excystation. The sporocyst wall consists of an electron-dense outer layer and an electron-lucent inner layer composed of four curved plates tightened with thick sutures (Speer *et al.*, 1998). The precise biochemical composition of oocyst and sporocyst walls is unclear. Their natural UV autofluorescence (exciter filter, 330–385 nm; dichroic beam splitting, 400 nm; barrier filter, 420 nm) suggests the presence of tyrosine-protein cross-links in one or both layers as described in the oocyst of *Eimeria maxima* (Belli *et al.*, 2006). Monoclonal antibodies have been raised against oocyst or sporocyst walls (Dumètre and Dardé, 2005) but antigenic structures have been not precisely identified. The veil may be stained with a polyclonal antibody to the apple domains of the MIC4 protein (Belli *et al.*, 2006).

7 Markers for Identification of Stage

Classical markers for differentiation of bradyzoites from tachyzoites were: (i) staining with periodic acid-Schiff (PAS) which is positive only for bradyzoites, (ii) higher resistance of bradyzoites to acid pepsin (1- to 2-h survival in pepsin-HCl) than that of tachyzoites (10 min survival), and (iii) a shorter prepatent period (time to oocyst shedding) in cats following feeding of bradyzoites (3 to 7 days) than that following feeding of tachyzoites (over 14 days). More recently, a set of tools (monoclonal antibodies, cytochemistry, and proteome analysis) was described to detect markers useful to identify tachyzoite, bradyzoite and coccidian development in tissue sections or in cell culture. They are reviewed by Ferguson (2004; Tables 1 and 2).

Tachyzoites are positive for surface antigen 1 (SAG1), enolase isoenzyme 2 (ENO2), lactic dehydrogenase isoenzyme 1 (LDH1) and negative for bradyzoite antigen 1 (BAG1). Bradyzoites are negative for SAG1 but positive for BAG1, ENO1 and LDH2. At the early stage of cyst formation, 10 to 15 days

Table 1 The expression of various markers in the different infectious forms of *T. gondii* as identified by immunocytochemistry; adapted from Ferguson (2004)

	Tachyzoites	Bradyzoites	Merozoites	Sporozoites
SAG1	+	−	−	−
SRS2	+	−	−	NE[a]
BSR4	−	+	+	NE
SAG2C	−	+	−	NE
SAG2D	−	+	−	NE
BAG1	−	+	−	−
Enolase 1	−	+	−	NE
Enolase 2	+	−	+	NE
LDH1	+	−	+	NE
LDH2	−	+	−	NE
GRA1	+	+	−	+
GRA2	+	+	−	+
GRA3	+	+	−	−
GRA4	+	+	−	+
GRA5	+	+	−	+
GRA6	+	+	−	+
GRA7	+	+	−	+
GRA8	+	+	−	NE
NTPase	+	+/−	+	−

[a] NE, not examined.

Table 2 Details of the cytochemical labelling of the parasitophorous vacuole (PV) with various lectins and antibodies to secreted molecules; adapted from Ferguson (2004)

	Tachyzoite PV	Bradyzoites PV	Coccidian PV
Lectins			
Dolichos biflorus	−	+	−
Glycin max	−	+	−
Wheat germ Agg.[a]	−	+	−
Antigens			
CST1	−	+	NE[b]
MAG1	+	+	+
GRA1	+++	++	−
GRA2	+++	+	−
GRA3	+++	++	−
GRA4	+++	−	−
GRA5	+++	+++	−
GRA6	+++	+	−
GRA7	+++	++	+++
GRA8	+++	−	−
NTPase	+++	−	+++

[a]Agg., agglutinin.
[b]NE, not examined.

post infection, a number of intermediate organisms displaying tachyzoite and certain bradyzoite markers were observed. In older cysts, small groups of organisms displaying only bradyzoite markers were also present. MAG1, a 65-kDa protein expressed in the matrix of the cyst between bradyzoites, is also seen in the cyst wall.

The nine GRA proteins have also been identified in both tachyzoites and bradyzoites but there were differences in their location during parasite development. All the dense granule proteins label the parasitophorous vacuole during tachyzoite development. During conversion from tachyzoite to bradyzoite, GRA4, GRA8 and NTPases disappear from the cyst wall, while the same proteins were still detected in bradyzoite dense granules. All the other GRA proteins were shown to be present within the cyst wall (GRA3, GRA5 and GRA7 associated to the exterior membrane of the cyst wall; GRA1, GRA2 and GRA6 associated with the inner layer). The tissue cyst wall also displays an additional unique cyst-wall protein. For the merozoites observed during the development in the enterocytes of the cat small intestine, no specific markers were available, but this stage was characterized by the absence of SAG1 and BAG1 markers, the presence of ENO2 and LHD1, and the expression of only two (GRA7 and NTPase) of the nine dense-granule proteins. Sporozoite markers include at least GRA1, 2, 4, 5, 6, 7, ROP2, 3, 4, and MIC4 proteins and specific surface antigens SporoSAG, Sp67 and Sp190, which are absent in tachyzoites.

Besides those molecular markers, cytochemical tools are available to differentiate cyst wall and matrix from the parasitophorous vacuole containing tachyzoites. The cyst wall is PAS-positive, stains with some silver stains, and can bind both *Dolichos biflorus* (DBA) and succinylated wheat-germ agglutinin (S-WGA).

8 Genomic Organization

T. gondii has a complex nuclear genome and two organellar genomes, the mitochondrial genome and the apicoplast. The nuclear genome is haploid at all the infective stages. The only diploid stage is the fertilized macrogamete. It consists of about 65 mb of DNA, organized into 14 distinct chromosomes or linkage groups, as determined by analyzing the segregation of RFLP markers among the progeny of several experimental genetic crosses between the three common lineages (http://toxomap.wustl.edu/genomes.htm).

The annotated genome of *Toxoplasma* (sequencing of a clone of Me49, a type-II strain) was recently released. Furthermore, due to a large-scale EST sequencing project for the *Toxoplasma* genome, a cDNA library has been constructed for the three main lineages of *Toxoplasma* and for the different stages (tachyzoite, bradyzoite, and oocyst; Kissinger *et al.*, 2003; http://www.toxodb.org/toxo/). Between the three main lineages, polymorphisms occur at a frequency of around 1 in 100 bp on all chromosomes with the exception of chromosome Ia (genome polymorphism of only 0.03%; Khan *et al.*, 2006a). Large chromosomal regions are dominated by one of the three single nucleotide polymorphism (SNP) types characterizing the three main lineages (Boyle *et al.*, 2006). Over all, about 16% of genes are expressed in a stage-specific manner.

The mitochondrial genome consists of a repeated element of approximately 6–7 kb in length and encodes subunits I and III of cytochrome *c* oxidase, cytochrome *b*, and a number of short fragments representing the small and the large subunit rRNAs (Esseiva *et al.*, 2004). The apicoplast (Maréchal and Cesbron-Delauw, 2001) contains a streamlined 35-kb circular genome. It expresses a small number of genes, as many genes formerly encoded on the apicoplast have been transferred into the nucleus. The products of these nuclear genes are post-translationally targeted to the organelle via the secretory pathway courtesy of a bipartite N-terminal leader sequence. The mitochondrion and the apicoplast are inherited uniparentally during sexual crosses, as evidenced by the absence of apicoplast in the microgamete (Ferguson *et al.*, 2005).

9 Genetic Diversity, Population Structure and Phylogeny

There are marked biological differences between *Toxoplasma* isolates in terms of their pathogenicity to mice. Most are avirulent producing chronic asymptomatic infections; however, a few are highly virulent and result in fatal acute toxoplasmosis. Given these biological differences, the world-wide distribution of *T. gondii* and its capacity to infect virtually all mammals and birds, one would also predict high levels of genetic variation, particularly given the potential for meiotic recombination in this protozoan with a well-described sexual cycle.

9.1 Markers and tools for analyzing genetic diversity

After the first studies based on isoenzyme markers, multiple genetic markers have been developed to differentiate *Toxoplasma* strains and analyze the parasite population structure (Dardé, 2004). Single nucleotide polymorphisms constitute the most abundant group of genetic markers. They have been found in genes coding for major antigens located on the surface of the parasite, in dense granules or rhoptries or within the matrix of the cyst, in genes encoding structural proteins, in those encoding enzymes, or in genes of unknown function. Recently, Khan *et al.* (2005a and 2005b) have designed 200 new SNP-RFLP markers which, together with 50 existing markers, map to approximate 300-kb intervals across the 14 chromosomes of *T. gondii* genome. Multilocus sequencing is one of the best approaches to detect polymorphism and to analyze the structure of a population. But the less informative PCR-RFLP of single-copy genes remains the most commonly used method for typing *T. gondii* isolates. *SAG2* gene polymorphism analyzed by PCR-RFLP using two restriction sites (Howe and Sibley, 1995) was the first method described to differentiate, with a single PCR, the three main genotypes of *T. gondii*. Strain-typing strategies relying entirely on the *SAG2* locus were adopted in many studies, leading to the misidentification of atypical or recombinant strains. A multiplex strategy, based on 11 loci analyzed by PCR-RFLP, was recently described (Su *et al.*, 2006).

Microsatellite sequences are considered to be rapidly evolving sequences. However, in *T. gondii*, some microsatellite markers appear to evolve rather slowly, as demonstrated by the relatively low number of alleles detected and their stability in a large number of isolates of the three main *Toxoplasma* lineages. These low-polymorphic microsatellites can be used for genotyping isolates with a good correlation with the patterns observed via multilocus PCR-RFLP on single-copy genes, although they are more resolutive than these last markers (Ajzenberg *et al.*, 2002a, 2004). A multiplex PCR for five microsatellites allowing multilocus analysis of isolates following a single PCR amplification was recently described (Ajzenberg *et al.*, 2005). An association of more polymorphic microsatellites is also useful for analyzing the population structure (Ajzenberg *et al.*,

2002a; Lehmann *et al.*, 2006) or for individual identification of isolates (epidemiological tracking, ratification of genotypes of laboratory strains, detection of mixed infection even if they are caused by isolates belonging to the same lineage; Ajzenberg *et al.*, 2004).

Synthetic peptides derived from polymorphic sites of the genes coding for *Toxoplasma* antigens (mainly GRA6 and GRA7; Kong *et al.*, 2003) were designed to develop an identification of strains via serotyping, with no need for strain isolation or DNA extraction. This method is well suited for large epidemiological studies, but not for individual strain identification due to the variability of individual immunological response.

9.2 Toxoplasma *population structure*
Clonal population structure and the three main lineages

After the first studies performed with several independent genetic markers on collections comprising well-established laboratory strains and more recent isolates originating mainly from human toxoplasmosis cases in France and in USA (Dardé, 1992; Howe and Sibley, 1995), it was established that *T. gondii* has a highly clonal population structure with three main lineages designated types I, II and III. The main criteria for a clonal population structure in *T. gondii* are the isolation of identical multilocus genotypes over large geographic areas and at intervals of several years, and the high linkage disequilibrium. Several of the biological properties of *Toxoplasma* may account for this perplexing predominant clonal structure in a parasite with a sexual cycle and an experimentally proven potential for recombination:

(i)   The majority of infections in cats likely involve only a single *Toxoplasma* isolate derived from a single prey source. This means that autofertilisation would be common and would limit the flow of genes between strains (Howe and Sibley, 1995).

(ii) Many macrogametes of the parasite remain unfertilized but are capable of forming oocysts in the small intestine of cats by parthenogenesis (Ferguson, 2002).

(iii) The parasite does not have an obligatory sexual cycle and can be transmitted asexually through carnivory (Su *et al.*, 2003).

(iv) *Toxoplasma* is known to induce a strong immune response both in intermediate and in definitive hosts, meaning that superinfection leading to mixed infection is rare in nature.

This simple clonal structure is accompanied by a low genetic divergence between the three main lineages (only 1% divergence at the DNA sequence level between lineages). Khan *et al.* (2005), mapping the genome with 250 SNP markers to approximate 300-kb intervals across the 14 chromosomes of the *T. gondii* genome, found that in the three predominant strains, each locus contained only two allelic types, referred to as A (Adam) and E (Eve). In fact, more recently, it was shown that a large majority (84%) of the 4,324 SNPs identified among the three types are type I and II SNPs (Boyle *et al.*, 2006). Large chromosomal regions are dominated by one of the three SNP types, type-III SNPs (only 16% of SNPs) being localized mainly on chromosome IV. The assymetric distribution of type-I, -II or -III SNPs on the chromosomes revealed by this genome-wide analysis of SNPs indicates that types I and III are second and first generation offspring, respectively, of a cross between a type-II strain and one of two ancestral strains (Boyle *et al.*, 2006). The relative lack of intratypic variation associated with a low divergence between lineages strongly suggests that these three clonal lineages have emerged as the dominant strains relatively recently. It was proposed that these three clonal lineages shared a common ancestor 10 000 years ago (Su *et al.*, 2003).

Recombination events and genetic diversity

The clonal population structure with the three main genotypes was observed with isolates predominantly obtained from humans and domesticated species in Europe and North America, leading to an epidemiological bias. Extension of epidemiological screening across a wider geographical and host range and multilocus analysis of isolates revealed a more complex population structure than initially described, with higher levels of variation and recombination among some parasite populations (Ajzenberg *et al.*, 2004; Lehmann *et al.*, 2004).

Multilocus genotyping of isolates from South America with PCR-RFLP markers or with microsatellites revealed that the majority of them possess type-I, -II or -III alleles (mainly I and III), identical to those found in the three major lineages, but these have segregated differently among the loci analyzed (Ajzenberg *et al.*, 2004). These isolates, presenting different mixtures of classical alleles, can be considered as recombinant genotypes. They are related to the three main lineages but the inclusion of these mixed genotypes in phylogenetic analysis decreases the robustness of association between type-I, -II and -III strains. More rarely, 'atypical' or 'exotic' strains are discovered (Ajzenberg *et al.*, 2004; Su *et al.*, 2003). At some loci, these atypical strains show evidence of the dimorphic allele patterns that typify the clonal lineages, but they also have many unique polymorphisms and 'novel' alleles. Phylogenetic analysis shows that the atypical genotypes cannot be related to the three main lineages. The classification of an isolate either as a recombinant or an atypical isolate may be difficult to establish and may depend on the number and on the resolutive power of the genetic markers used. The increasing number of genetic markers used for genotyping may reveal unique polymorphisms or mixtures of classical alleles in isolates previously classified as belonging to one of the three classical types. Despite the existence of unique polymorphisms and a higher allelic diversity, global observation of the genetic diversity indices for these atypical strains showed that the level of genetic polymorphism of sequences remained very low.

The best example of atypical genotypes described in the literature are the nine French Guianan strains obtained from severe cases of human toxoplasmosis acquired in the wild Amazonian forest in French Guiana. Multiple unique polymorphisms were reported via sequencing of genes such as *SAG1*, *SAG2A*, *SAG3*, *SAG4*, *BSR4* and *GRA6*, and a multilocus microsatellite sequencing revealed that each French Guianan strain had a unique multilocus genotype (Ajzenberg *et al.*, 2004). Another example of atypical strains is represented by type X, isolated from marine mammals (Conrad *et al.*, 2005).

A more complex pattern of the *Toxoplasma* population is emerging from these ongoing studies:

(i) *There is a worldwide geographical structuration of the* Toxoplasma *population.*

a. *Toxoplasma* strains observed in Europe and USA belong to three main lineages. Among these three lineages, the type II is largely predominant. In France, it is observed in more than 90% of human congenital toxoplasmosis, but also in all isolates originating from a large variety of animals.

b. Strains circulating in South America are different from the archetype strains and more diverse. They exhibit different mixtures of type-I and -III alleles, and unique polymorphisms. Type-II alleles are less frequently detected. Unique polymorphisms are present especially in strains isolated in the wild part of the Amazonian forest. Some alleles found only in South American strains were not detected in Europe or North America. Genetically distinct isolates are found in different regions of South America (Dubey *et al.*, 2007), suggesting the absence of genetic flow between them. The few strains isolated from Africa or from Caribbean islands also possess a mixture of type-I and -III alleles (Ajzenberg *et al.*, 2004).

c. A study using hypervariable microsatellites for determining the geographical structure of *Toxoplasma* suggests that most genotypes are locale-specific, but some, closely related to each other, are found across continents, indicating a recent radiation of a pandemic group. One population is confined to South America and another is found worldwide. The North American population is similar to those of Africa and Eurasia. South American and Eurasian populations may have evolved separately until migration opportunities between continents for *Toxoplasma* due to ships populated with rats, mice and cats (Lehmann *et al.*, 2006). This has to be confirmed by a larger number of isolates sampled in Asia or in Africa.

(ii) *Clonal or sexual propagation may be selected under different environmental conditions.*

    a. Strains with clonal propagation and peri-domestic transmission dominate in North America and Europe. Only a limited number of host species such as cats, a few meat-producing animals and peri-domestic mammals and birds are involved in the *T. gondii* domestic cycle, which may limit the complexity of the parasite genetic pool in this cycle. Among many genotypes, the combination of alleles characterizing the three clonal lineages (and particularly type II) seemed to be most successfully adapted to these hosts. They would have diverged about 10 000 years ago, which coincides with the domestication of companion and agricultural animals (Su *et al.*, 2003). In Europe or North America, intensive breeding of a narrow range of domestic meat-producing animals, together with cat domestication, offered a major niche to these three lineages. Furthermore, farms are reservoirs of infection (small peri-domestic rodents and birds, and cats), from which transmission of clonal types can radiate to the surrounding wild environment (Lehmann *et al.*, 2003), leading to an impoverishment of genetic diversity. Human activities in North America and Europe, which diminished recombinations and thus gene flow in *Toxoplasma*, may favor genetic drift for *Toxoplasma* evolution. This would reduce the parasite biodiversity.

    b. Sexual recombination may be more common between *Toxoplasma* isolates in the wild life cycle (e.g., French Guiana), and in areas where breeding is recent or not intensive and cat domestication recently introduced (e.g., Brazil, West or Central Africa, the Caribbean). A few recombinant isolates have also been described from wildlife (bears and deer), but very uncommonly from domestic species, in North America (Ajzenberg *et al.*, 2004; Howe and Sibley, 1995). The lower linkage disequilibrium found in different studies comparing samples from Brazil or French Guiana to European or North American samples favored this hypothesis of a higher rate of outcrossing in tropical areas. The most parsimonious hypothesis is that *T. gondii* presents a complex global population structure with a mix of clonal and sexual propagation. The greater diversity of parasite genotypes would allow it to colonize the maximum of ecological niches in a species-rich environment like the Amazonian rain forest. Mixed infections or even re-infection due to different genotypes might not be a rare event in this environment leading to new crosses in definitive hosts. Although atypical genotypes are phylogenetically unrelated to the three main lineages, they may represent 'unknown' clones which are adapted to wild hosts. For example, there are good arguments that type X isolated from various sea mammals in California (Conrad *et al.*, 2005) should be considered a new emerging type.

c. The above hypothesis of a different use of sexual crosses and of a different allelic diversity in a wild and in a domestic environment suggests an influence of the host on the parasite genotype. If *Toxoplasma* strains have co-evolved with particular host species, then it may also be true that density and diversity of host species is linked to parasite diversity. However, to date, there are no clear arguments for a host specificity of genotypes, although the same genotype could express different virulence according to the hosts.

Several questions remain to be resolved about population structure of *Toxoplasma* and the answers lie in more intensive epidemiological sampling of this environment and analyzing isolates with multiple markers. The population structure has consequences in terms of possible links between phenotypes and genotypes. In case of a clonal structure, biological information obtained from one isolate may successfully predict the others because of their identical genetic background. If strains have different genetic make up, the association between genotype and phenotype cannot be predicted and continued recombination may lead to strains that acquire new pathogenic mechanisms, as suggested by the severity of cases of toxoplasmosis acquired in French Guiana. The new genotypes resulting from these crosses can expand in the population leading to emergent diseases. The priority for the future is to bring together population genetic and phenotypic analysis in order to predict the distribution, transmission cycle in the environment and pathogenesis of strains in humans.

10 Epidemiology

Toxoplasma gondii is a widely prevalent protozoan in warm-blooded animals. Many host species (birds, rodents, carnivorous or herbivorous animals) from polar to tropical areas have been identified by serology or bioassay. As in humans, toxoplasmosis is usually asymptomatic but can be associated with abortions or clinical diseases. In meat-producing animals, toxoplasmosis is of veterinary and economic importance.

Cats are essential for the maintenance of *T. gondii* in the environment since infections are virtually absent from areas lacking cats. A single oocyst is able to infect pigs or mice but pathogenicity depends on the strain, inoculum and infection route (Dubey and Beattie, 1988). Oocysts are less infectious and pathogenic for cats than for intermediate hosts (Tenter *et al.*, 2000). They are responsible for most of the *T. gondii* infections in noncarnivorous mammals and birds whereas other species are mainly contaminated by eating animals harboring tissue cysts.

In humans, the parts of each route (tissue cysts or oocysts) in parasite transmission remains undetermined even by serologic investigations.

The consumption of raw or undercooked meat is associated with toxoplasmosis in industrialized countries whereas environmentally resistant oocysts are responsible for contamination in people with little meat in their diet. Due to their resistance to disinfecting agents such as chlorine, oocysts have been involved in waterborne outbreaks of toxoplasmosis in several countries (Dubey, 2004). Current efforts to control and prevent infection in humans are made at two levels. First, intensive farm management with appropriate measures of hygiene and cat control aims to reduce the infection of meat-producing animals. This is especially true for fattening pigs in which seroprevalence has decreased significantly in Europe and USA in the past 20 years (Tenter *et al.*, 2000). Second, microorganism inactivation methods used in the food industry (meat enhancement with salt solutions, high pressure processing, gamma irradiation) and in water treatment plants (sand filtration, ultrafiltration, ozonation, UV irradiation) are under investigation to evaluate the safety of food and water supplied (Dubey, 2004; Dumètre and Dardé, 2003). The identification of parasite reservoirs is essential to design specific methods for controlling animal contamination and preventing human toxoplasmosis.

10.1 Environmental reservoir
Oocyst excretion by domestic cat (*Felis catus*) and wild felids

Based on serologic surveys, up to 74% of the adult cat (*Felis catus*) population may be infected by *T. gondii* (Tenter *et al.*, 2000). Antibodies are usually detected soon after weaning. Seroprevalence increases with age and is higher in male free-roaming cats which hunt for food, than in indoor cats fed with preserved food. In nature, cats are infected by eating small prey harboring tissue cysts, or by ingesting oocysts from soil. Fatal toxoplasmosis is rare and occurs more often in immunocompromised cats and in kittens. Vertical transmission is uncommon and is probably not important for parasite propagation. Cats excrete millions of unsporulated oocysts in their feces after ingesting any of the three infectious stages. The prepatent period is 3–10 days after ingesting bradyzoites and more than or equal to 18 days after ingesting oocysts or tachyzoites. The patent period is only 1–3 weeks but a re-excretion is possible, at least experimentally, after a second challenge with *T. gondii*, after corticoid treatment or superinfection by *Isospora felis*. The proportion of domestic cats excreting oocysts at any one time is less than 2% in most countries. Therefore, seroprevalence data are more useful than feces examination for studying *T. gondii* infection in cats.

Seventeen species of wild felids have been recognized as definitive hosts for *T. gondii* (Tenter *et al.*, 2000). Jaguarundi (*Herpailurus yaguarondi*), ocelot (*Leopardus pardalis*), bobcat (*Lynx rufus*), cougar (*Puma concolor*), and Asian leopard cat (*Prionailurus bengalensis*) experimentally infected with a type-III domestic strain excrete oocysts more inconstantly than domestic cats (Dubey and Beattie, 1988). Natural infection in wild felids is relatively unknown. Based on limited studies on species such as cougars and bobcats, seroprevalence varies from 15% to 75% in arctic and tropical areas respectively. Oocysts shed by wild felids have been responsible for waterborne outbreaks of acute toxoplasmosis in humans in Panama and Canada (Dubey, 2004; Dumètre and Dardé, 2003). The role of wild felids in parasite transmission to humans may be important especially in areas where the domestic cat, *Felis catus*, is absent (e.g. tropical forest).

Oocyst survival in the environment

Oocysts sporulate under definite temperatures and favorable aerobic and hygrometric conditions (Dubey, 2004; Dubey *et al.*, 1970; Dumètre and Dardé, 2003). They sporulate in 2 days at 25°C and in 21 days at 11°C but not after freezing (1 day at −21°C or 7 days at −6°C) or heating (50°C, 10 min). Cool temperatures (4°C for 6–11 weeks) do not prevent sporulation. Experimentally, only 36% of oocysts sporulate in water, whereas 74–80% sporulate in aqueous sulphuric acid or potassium bichromate solutions (Dubey *et al.*, 1970). Therefore, successful sporulation in nature probably depends on multiple and complex parameters which are still unknown. Oocysts sporulate in saline water at 15 and 32 ppt (part per thousand) in 3 days at 24°C (Dubey, 2004).

Sporulated oocysts remain infective for long periods under most ordinary environmental conditions: 548 days in soil between −20°C and 35°C and 28, 1620, 548, 107, and 28 days in water at −21, 4, 20–22, 30, and 40°C respectively (Dubey, 2004). Oocysts survive 56 days on berries stored at 4°C. In contrast, they are very sensitive to desiccation and heating (>60°C). Cats burying feces and moist climates promote oocyst survival in soil. Soil macrofauna including earthworms and dung beetles, and flies and cockroaches can spread oocysts mechanically and carry them on to food (Dubey and Beattie, 1988). Surface and underground waters are likely contaminated by soil washing after rainfall. Coastal freshwater runoff and sewage outfall allow oocysts to enter the marine environment where they can survive up to 180 days depending on water salinity (Dubey, 2004). Bivalve mollusks can act as paratenic hosts for *T. gondii* by filtering contaminated water. Infective oocysts are retained up to 85 days in tissues of oysters and mussels, which are preyed on by some marine mammal species (Fayer *et al.*, 2004).

Detection of oocysts in water and soil

Interest for detecting oocysts in the environment is emerging due to recent cases of waterborne toxoplasmosis in humans and animals (Dubey, 2004; Dumètre and Dardé, 2003). Detecting oocysts in water and soil is difficult because their number in random samples is probably very low (Dumètre and Dardé, 2003). For detection in water, available methods include filtration or flocculation, flotation, followed by bioassay and polymerase chain reaction (PCR; Dubey, 2004; Dumètre and Dardé, 2003). Their sensitivity depends mainly on water quality (i.e. turbidity). In the study of Villena *et al.* (2004), public drinking water and raw surface waters experimentally contaminated with oocysts were filtered, purified by sucrose flotation and oocysts were recovered by PCR and mouse bioassay. Sensitivity was better by PCR than by bioassay. With this method, 1 oocyst l^{-1} was detected from public drinking water samples by PCR and bioassay, whereas 100 and 1000 oocysts l^{-1} were detected from raw surface water samples by PCR and bioassay respectively. So far, oocysts have also been recovered in water by bioassay and PCR after an outbreak in Brazil (de Moura *et al.*, 2006) and by PCR in waters of various qualities in Poland (Sroka *et al.*, 2006). For detection in solid matrices (soil, vegetables, bivalve mollusks), flotation is currently the only method to separate oocysts from particles. However, flotation is not specific to *T. gondii* and various substances (humic acids, polysaccharides, phenols) are co-extracted with oocysts and may inhibit downstream applications such as PCR. In the past 20 years, oocysts have been isolated in soil or on pig feed but never on fruits, vegetables or seafood.

The development of specific and sensitive methods for oocyst detection is still in progress. Investigations focus on oocyst purification from sample particles by immunomagnetic separation using monoclonal antibodies directed against the oocyst or sporocyst walls, or by flotation on sucrose or Percoll-sucrose density gradients. Mouse bioassay is the current method to assess oocyst viability, but this technique is influenced by the amount of material to inoculate into mice. The development of more reliable methods, such as RT-PCR and sporozoite staining with fluorogenic vital markers, could help to refine data on oocyst survival under various conditions.

Oocyst resistance to water- and food-treatment processes

Oocysts can survive various inactivation procedures especially those using chemical reagents (Dubey *et al.*, 2004). Oocysts remain viable in an aqueous 2% sulfuric acid for at least 18 months at 4°C. Oocysts are also resistant to detergents or disinfectant solutions such as sodium hypochlorite solutions. A 1-log diminution of infectivity (based on the oral dose of oocysts able to infect 50% of mice) is only reported after 1–2 h of contact between oocysts and a 6% sodium hypochlorite solution, and a 2-log diminution only after 4 h. Drinking-water treatment plants using chlorination as single disinfecting method could potentially supply drinking water containing infective oocysts. This mode of dissemination has been involved in two waterborne outbreaks in Victoria (Canada) in 1995 and in Santa Isabel do Ivai (Brazil) in 2001–02 (Dubey, 2004). Resistance to other chemical (ozone), physical (UV irradiation) or physical/chemical (flocculation) disinfection processes found in drinking-water plants remains to evaluate, but is highly probable when comparing with data on the oocyst of *Cryptosporidum parvum*. In the food industry, gamma radiation and high pressure processing (Lindsay *et al.*, 2005) could be used routinely for oocyst inactivation on fruits and vegetables with no undesirable changes in food features.

10.2 Animal reservoir
Diagnosis of *Toxoplasma* infection in animals

Diagnosis of *Toxoplasma* infection in animals is based on serology with detection of specific antibodies (IgG). Many serologic tests have been used: Sabin-Feldman dye test, indirect hemagglutination, indirect fluorescent antibody, complement fixation, modified agglutination test (MAT) and enzyme-linked immunosorbent assay (ELISA). Sensitivities of techniques depend on species but MAT seems to be the most adapted to a large number of species. Serological surveys alone do not provide information about the prevalence of viable *T. gondii*. *T. gondii* was recovered from approximately 3% of serological negative pigs (Dubey *et al.*, 2002). Toxoplasmosis can rarely be diagnosed by direct observation of the parasites in tissues. PCR techniques can be helpful (classical, nested or real-time PCR) with results depending on the quantity of meat analyzed. *T. gondii* can also be isolated from muscle, brain or blood using mouse inoculation or, more sensitive but expensive, ingestion by cats. The number of tissue cysts that may develop inside a certain host and their locations vary with the intermediate host species.

Toxoplasma in meat-producing animals

Prevalence of toxoplasmosis is highest in sheep and toxoplasmosis is implicated in 10–20% of sheep flocks with an abortion problem (Buxton, 1990). The prevalence increases with age, reaching more than 90% in some studies (Tenter *et al.*, 2000). The prevalence in ewes is more than twice that in lamb. Viable *T. gondii* has been recovered from as many as 67% of sheep samples. Seroprevalence levels were lower in cattle and very variable in pig (1–60%) and poultry (0–30%), depending on their lifestyle (indoor or outdoor). *T. gondii* has rarely been isolated from bovine tissue. It is unclear whether this is associated with fast elimination of cysts from cattle tissues or with inconsistent cyst formation following infection. In a study analyzing more than 2,000 samples of beef, *T. gondii* was not isolated by bioassay in cat (Dubey *et al.*, 2005). The prevalence of *T. gondii* infection has decreased significantly with changes in pig production. It was suggested that infected pork products cause 50–75% of all cases of human toxoplasmosis in the United States, but the recent study of Dubey *et al.* (2005) showed a seroprevalence of 0.57% and parasite isolation in only eight cases (0.38%). The prevalence of viable *T. gondii* in chickens produced in intensive farming is usually very low, but may be high in free-range chickens.

It is not known how many tissue cysts result in the infection of humans, but ingestion of one cyst is sufficient for a cat to become infected. During the production of various meat products, meat of many animals is mixed, which also amplifies the risk in cases where only few animals would be infected (Aspinall *et al.*, 2003).

Cyst resistance to food treatment processes

Although tissue cysts are less resistant to environmental conditions than oocysts, they are relatively resistant to changes in temperature but remain infectious in refrigerated carcasses (1–4°C) or minced meat for up to 3 weeks, and probably as long as the meat remains suitable for human consumption (Dubey, 2000). They survive at 4–6°C for up to 2 months. Tissue cysts also survive between −1 and −8°C for longer than a week (Kotula *et al.*, 1991). Tissue cysts are usually killed at temperatures of −12°C or less and by heating to 67°C (Dubey, 2000). Isolated tissue cysts were destroyed by heating at 55°C for 30 min. Some tissue cysts will remain infectious after cooking in a microwave oven, possibly due to uneven heating of the meat (Lunden and Uggla, 1992). Some studies suggested that tissue cysts are killed by commercial procedures of curing with salt, sucrose or low temperature smoking, but Warnekulasuriya *et al.* (1998) was able to detect viable *T. gondii* in one of 67 ready-to-eat cured meat products indicating a failure of the commercial curing process. Under laboratory conditions, tissue cysts in pork loins were killed in a 2% sodium chloride or 1.4% potassium or sodium lactate solution (Hill *et al.*, 2006). Salting alone is probably not sufficient to prevent transmission to humans via tissue cysts. Tissue cysts are killed by gamma irradiation at a dose of 0.7 kGy (Kuticic and Wikerhauser, 1996). Recent study indicates that high-pressure processing treatment of ground pork with 300 mPa pressure will render tissue cysts nonviable and make pork safe for human consumption (Lindsay *et al.*, 2006).

11 Pathogenesis and Immunology of *Toxoplasma gondii* Infection

The pathogenesis of *Toxoplasma gondii* infection is a very wide topic, which is almost impossible to cover in a few book pages. The complexity of the phenomena involved is to a large extent due to the fact that the pathogenesis of the disease is the result of the interaction between two eukaryotic cells, the parasite and the host cell. The conflict can become a 'gentleman's agreement', if possible, in the case of the existence of a host-parasite equilibrium. This is, in human medicine, the most frequent situation. However, this equilibrium can be inexistent or can be ruptured by different mechanisms arising from the host, the parasite, or both. This is the key to the understanding of *T. gondii* pathogenesis: how can the host control parasite virulence, and how can the parasite avoid this control?

Some important mechanisms have been evidenced: the parasite stage conversion (tachyzoite/bradyzoite) and the ability of *T. gondii* to form dormant cysts (that can be ruptured and reactivate) are the most important of them. However, the respective role of the parasite or the host in the control of such phenomena is largely unknown to date. On the parasite side, the genotype and the strain differences seem to be of paramount importance to drive the virulence in the host and could be a strong determinant of the clinical lesions. On the other side, the host response to parasitic aggression is obviously the basis of control of the infection. The host response is very complex and, to date, none of the researchers working on this topic can affirm that he (or she) has discovered the unique key to control by the host. In fact, 'the host' means nothing since the potential hosts for *T. gondii* are numerous and diverse. Furthermore, in human medicine, the hosts can be very different depending on their immune status: immunocompetent adult, immunocompromised patients, fetuses, newborns, and so on. Finally, in one host, the different host cells can exhibit very different behaviors in reaction to *T. gondii*, depending on their stimulation and on control by the immune system.

From a clinical point of view, in human medicine, toxoplasmosis is the cause of serious lesions in primo infection in fetuses (congenital infection), and rarely in immunocompetent adults, or due to reactivation of a latent infection previously acquired in immunocompromised patient such as AIDS or graft patients (Ambroise-Thomas and Pelloux, 1993). The pathogenesis of the two clinical situations presents some similarities and some discrepancies. In both cases, the central nervous system is the main, but not the only, target of the toxoplasmic infection, and the clinical lesions are mainly located in the brain. In primarily acquired infection, the lesions are due to tachyzoites that do not switch to bradyzoites, while in the case of reactivation, dormant bradyzoites in cysts switch to virulent tachyzoites.

11.1 Parasitic factors
Role of *Toxoplasma* strain

One major point concerning the 'parasitic side' of the *Toxoplasma* pathogenesis is the importance of the strain characteristics. An understanding of how particular genotypes can exhibit different behaviors in the host, and thus how they can orientate the pathogenicity of the parasite, is needed. In fact, the three main genotypes (I, II, III) differ by their virulence and their ability to form cysts in mice: type I is highly virulent, leading to widespread parasite dissemination and death of mice less than 10 days after inoculation with less than 10 tachyzoites; in contrast, mice survived infection with a type-II or -III strain (50% lethal dose $>10^3$). Some other strains lead to encephalitis a few months after mouse inoculation. Atypical and naturally recombinant strains are usually more virulent in mice than type II or III.

But due to their genomic diversity, they cannot be directly compared. However, this approach is limited since mice are not the only host for *T. gondii*. Strain virulence is not the same across host species: for example, type-I strains, which are highly virulent in mice, are not pathogenic in rats. Thus, if it is clear that the repartition of the different genotypes helps to understand the population biology of *T. gondii*, it is less obvious whether this classification is directly linked to the virulence mechanisms or not. In fact, it has been evidenced that isolates of the same genotype can show different behavior in terms of virulence, ability to cause encephalitis or cytokine induction (Saeij *et al.*, 2005). Thus, it is necessary to improve the genotyping of the parasite before drawing definitive conclusions concerning the role of the genotype in the pathogenesis of *T. gondii* in human medicine. The fact that isolates from the same genotype can express different phenotypes could be one key phenomenon. This could be due to the fact that the distinct lineage of the parasite that infects humans, at least in Europe and North America, differs genetically by 1% or less. Furthermore, recent works have evidenced the role of epigenetic regulation in *Toxoplasma* growth and differentiation. These epigenetic mechanisms can overlap the genotypic ones and explain the fact that one genotype can be 'inhomogeneous' for some phenotypic aspects. However, the relationship between genotype and some virulence traits remains, such as growth rate (parasite burden), translocation of the NF-κB factor, induction of IL-12, IL-10, IL-1β or IL-6, and attraction and recruitment of different cell types.

Stage conversion

The tachyzoite/bradyzoite stage conversion and cystogenesis are essential in the understanding of *Toxoplasma* pathogenesis (Lyons *et al.*, 2002). This is a very complex phenomenon, largely misunderstood to date. The data obtained rely mainly on *in vitro* techniques and on the genetic manipulation of *T. gondii*. Cyst or cyst-like structures had been reported in tissue cultures for a long time, but the precise analysis of the development of bradyzoites and stage conversion is quite recent. The more frequently used technique is to grow parasites in alkaline medium, but other methods stressing the parasite can be used: heat shock, acid medium, inhibition of mitochondrial functions, induction of oxidative stress.

It is very difficult to analyze physiological stage conversion *in vivo*. However, some mechanisms have been described to be involved, such as nitric oxide or heat shock (Figure 6). Immunological factors such as IFN-γ, TNF-α, IL-12 and T-cells, which play an important role in controlling tachyzoite growth, could indirectly control stage differentiation. The stage conversion is associated with molecular, biological and morphological modifications such as stage-specific antigen expression and alterations to metabolism (see section 2.2.4). Heat-shock proteins (HSP 60 mainly) are involved in stress-induced interconversion since the inhibition of their

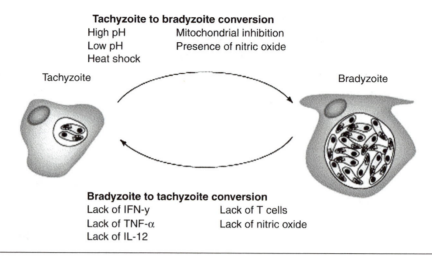

Tachyzoite to bradyzoite conversion
High pH   Mitochondrial inhibition
Low pH   Presence of nitric oxide
Heat shock

Tachyzoite   Bradyzoite

Bradyzoite to tachyzoite conversion
Lack of IFN-y   Lack of T cells
Lack of TNF-α   Lack of nitric oxide
Lack of IL-12

Figure 6 Factors associated with tachyzoite and bradyzoite conversion. The process of stage conversion is reversible, and is responsible for establishing chronic disease associated with bradyzoites or for reactivation in case of immune deficiency. Copyright Lyons *et al.* (2002).

synthesis inhibits *in vitro* bradyzoite development, and one specific bradyzoite antigen, BAG1, is homologous to the small HSP of plants but is not absolutely necessary for cyst formation *in vitro*.

The stage conversion is characterized by the appearance/disappearance of stage-specific surface antigens (see sections 5, 6 and 7). Surface antigens of the tachyzoite and bradyzoite are GPI-anchored, and most are members of the SRS (SAG1-related sequences) superfamily of proteins. These antigens play a role in host cell invasion, immune modulation and virulence attenuation. It has been hypothesized that the SRS superfamily exists to function as a set of quasi-redundant receptors that facilitate parasite attachment and entry into the broad spectrum of cell types and hosts that *Toxoplasma* is able to infect (Jung *et al.*, 2004). SAG1 and SAG2A are highly immunogenic and their expression at the tachyzoite stage is thought to activate host immunity to regulate the virulence of infection. The quasi-redundant antigenic variation may attract immunity against SAG1 and SAG2A and distract immunity away from other sets of SRS proteins simultaneously expressed. Furthermore, as sets of SRS proteins are regulated in a development-specific manner, it is possible that SRS-encoded variant T-cell epitopes modulate B and T-cell response during infection and constitute a mechanism of immune evasion by *Toxoplasma*, promoting bradyzoite survival and the establishment of chronic infections.

Host cell invasion

The kinetics of secretion of antigens from the three organelles is the following: first, binding to the host cell triggers the release of microneme proteins at the tight attachment zone between the parasite and the host cell; second, discharge of rhoptry proteins initiates invagination of the host-cell plasma membrane to form a nascent parasitophorous vacuole; finally, GRA proteins are released when the parasite is inside the parasitophorous vacuole. Microneme proteins are the main adhesins involved in the attachment to the host-cell surface by an apicomplexan. The repertoire of membrane-spanning microneme proteins includes adhesins such as MIC2 and escorters such as MIC6. They are anticipated to bridge host-cell receptors to the parasite membrane during invasion (Soldati *et al.*, 2001). Dense granule proteins (GRA1 to 9) are rapidly and massively released in the parasitophorous vacuole after cell invasion by the parasite. They are of paramount importance in the developmental strategy of *T. gondii* in the parasitophorous vacuole (Mercier *et al.*, 2005). Beside these proteins, others such as proteases play a role in the intracellular processing events. Lipids are present in the membrane of *T. gondii*. (Coppens, 2006). Three categories can be described: lipids scavenged by the parasite from the host cell, lipids synthesized in large amounts by the parasite independently from the host cell, and lipids produced *de novo* by the parasite. Further studies are needed to know the exact role of these lipids in the pathogenesis of the disease.

11.2 Dissemination into the host

After infection, *T. gondii* actively crosses nonpermissive biological barriers such as the intestine, the blood-brain barrier and the placenta, thereby gaining access to tissues where it causes severe pathology depending largely on the host immunological status and the parasite virulence. Enhanced migration has been found to be associated with mouse virulent strains (Barragan and Sibley, 2003). More globally, apicomplexans are obligate intracellular protozoan parasites that rely on gliding motility for their migration across biological barriers and for host-cell invasion and egress (Keeley and Soldati, 2004). This unusual form of substrate-dependent motility is powered by the 'glideosome', a macromolecular complex consisting of adhesin proteins that are released apically and translocated to the posterior pole of the parasite by the action of an actomyosin system anchored in the inner membrane complex of the parasite. In *T. gondii*, MIC2 has an essential role during invasion. Remarkably, *T. gondii* is capable of infecting all types of nucleated cells and of crossing several important cellular barriers within the vertebrate host. Parasites rapidly disseminate following oral infection in mice and rat models, and are found in distant sites, such as the lung and the heart, within several hours of ingestion. Notably, many of the parasites are extracellular during this migration, which suggests that they actively travel through the lymphatic and blood vessels. Even if mouse-virulent (type-I) strains consistently exhibit a superior migratory capacity than the nonvirulent type-II and -III strains, the genetic determinants of the migratory phenotype remain to be elucidated.

11.3 Immune response

The immune response to *T. gondii* is very complex and, obviously, depends on the host immune status. The individual variations can be explained by the high level of heterogeneity in the genetic background, the differences in specific immune response depending on the tissue compartment (mucosal surface, CNS, placenta, etc.) and the differences between parasitic strains (virulence, stage conversion). Data in the literature are often obtained from animal models, and are sometimes hard to transpose to humans (Filisetti and Candolfi, 2004). In general, the immune response of an immunocompetent host developed in the course of the infection leads to the acquisition of a protective immunity, preventing re-infection. However, the parasite persists in the bradyzoite form inside the cysts. Cellular immunity is the key component of the host's immune reaction. The macrophages, T lymphocytes (TL), 'natural killer' (NK) cells and the cytokines are the major elements involved in the immune response. The balance of cytokine subversion and stimulation during infection probably results from the parasite's need to simultaneously avoid immune elimination and trigger immunity to prevent host death (Denkers, 2003).

T. gondii is capable of triggering the nonspecific activation of macrophages and NK cells along with other hematopoietic and non-hematopoietic cells. This nonspecific response begins immediately following the first contact between the parasite and the host. The activation of macrophages by IFN-γ is necessary to trigger the cytotoxic activity of the macrophages against *T. gondii*. The stimulation of IFN-γ production is dependent on a myeloid differentiation factor-88 signaling pathway used by Toll-like receptors (Yarovinsky and Sher, 2006). The destruction of the parasite, or the control of its replication, is the result of various mechanisms: oxidative mechanisms; nonoxidative mechanisms represented mainly by the production of nitrogen monoxide (NO) by macrophages activated by IFN-γ; and nonoxygen-dependant mechanisms, such as induction by IFN-γ of indoleamine 2,3-dioxygenase which degrades the tryptophan required for growth of the parasite. NK cells and macrophages exert a combined and synergetic action, activated by IFN-γ. Other cells, such as γ-δ TL, platelets, neutrophils, eosinophils, fibroblasts, endothelial cells, and so on are involved in this nonspecific resistance.

During the specific acquired immune response, the effector cells are stimulated by dendritic cells (antigen presenting cells) presenting the antigen. These effector cells, which are involved in resistance to *Toxoplasma* infection, then exert their function via a cytotoxic activity and/or the secretion of cytokines involved in the immune response. CD4+ (Th1 – production of IL-2 and IFN-γ- and Th2 – production of IL-4, IL-5, IL-6 and IL-10) and CD8+ TL are the effectors. CD4+ TL are required for the development of resistance during the early phase of the infection and for immunity during vaccination. This resistance is closely related to a type-1 response promoted by the IFN-γ and the IL-12 produced following activation of NK cells and macrophages. However, it has been shown that a synergistic action between CD4+ and CD8+ TL is necessary to control infections. CD8+ TL have a cytotoxic effect on tachyzoites or cells infected with the parasite. Cytokines produced during *T. gondii* infection can be divided in two groups: protective and regulatory cytokines, with a delicate balance between both groups. Protective cytokines are IFN-γ, IL-12, TNF-α, IL-6, IL-5, IL-15, IL-18 and IL-2. They play, globally, a major anti-*Toxoplasma* role. Regulatory cytokines (IL-4, IL-10 and TGF-β) regulate the immune response by an immunosuppressive action. Other molecules, such as chemokines, are also involved in the immune response to *T. gondii* (Brenier-Pinchart *et al.*, 2001). Their role is, roughly, to induce chemotaxis of NK cells, leukocytes and TL. Beside the direct host response, *T. gondii* is able to 'manipulate' and orientate the host cells. Intracellular infection can cause a blockade in the NFκB macrophage signaling pathway, correlating with reduced capacity for IL-12 and TNF-α production. The parasite also prevents STAT1 activity, resulting in decreased levels of IFN-γ-stimulated MHC surface antigen expression (Denkers, 2003). After infection, apoptosis is triggered in TL and other leukocytes, thereby leading to suppressed immune responses to the parasite (Lüder and Gross, 2005).

A humoral response is also involved in the pathogenesis of toxoplasmosis, while less important than cellular immunity. However, humoral immunity is the emerged part of the iceberg, since, in human medicine, the detection of antibodies is the most often used method to evaluate the immune status of the host. IgM antibodies first appear in serum, and then the IgG. IgG are involved in antibody-dependent cell cytotoxicity. IgA antibodies play a role in the mucosal immunity and are also present in the serum. IgE can be detected and used for diagnosis.

11.4 Relationship between toxoplasma and the central nervous system

Whatever the type of clinical toxoplasmosis (congenital or in immunocompromised patients), most of the clinical lesions due to the parasite are localized in the CNS, including the retina. So, beside the general mechanisms of the host-parasite interactions described *supra*, it is of paramount importance to discuss the relationship between *T. gondii* and the CNS tissues and cells (Hegab and Al Mutawa, 2003; Suzuki, 2002). *Toxoplasma* encephalitis (TE) in immunocompromised patients (mainly AIDS) is almost always due to reactivation of a chronic infection (cyst rupture and tachyzoite proliferation), while in the fetus the lesions are due to transplacental migration of tachyzoites invading immature tissues. IFN-γ plays a critical role in prevention of TE since the neutralization of its activity results in severe inflammation and development of areas of necrosis in mouse brains. It has been demonstrated that both CD4+ LT and CD8+ LT infiltrate the brain following infection in experimental models. However, microglia appears to be the major effector cells in the prevention of tachy-zoite proliferation in the brain. Microglia becomes activated *in vitro* to inhibit parasite proliferation following stimulation with IFN-γ. TNF-α and IL-6 are also involved in the activation of microglia, and NO mediates the inhibitory effect of activated microglia. Furthermore, these cells can produce TNF-α after *T. gondii* infection, this production being mediated by IFN-γ. IFN-γ-mediated activation of microglia in collaboration with autocrine TNF-α is probably one of the main resistance mechanisms of the brain against *T. gondii*. Granulocyte-macrophage colony-stimulating factor (GM-CSF) and transforming growth factor (TGF)-β are other cytokines that appear to be involved in the effector function of microglia. Astrocytes are also involved in the control of *T. gondii* infection of the brain and can inhibit parasite proliferation after stimulation by cytokines such as IFN-γ, IL-1β, IL-6 or TNF-α. This phenomenon is, at least in part, linked to the indoleamine 2,3-dioxygenase activity. Following infection with *T. gondii*, astrocytes become activated to produce IL-1, IL-6 and other cytokines and chemokines such as CCL2 (MCP-1). Furthermore, in humans, astrocytes are the host cells supporting most of the replication of *T. gondii* in the brain in reactivation, and support the cyst in latent infection. Neurons can also be infected by the parasite to a lesser extent, and have been shown to produce cytokines and chemokines after infection.

11.5 Conclusions about pathogenesis and immunology

Despite numerous and important works on this topic a lot remains to be done. In fact, the complexity of the mechanisms involved is so high that the challenge is, after the description of new results, to link their significance with previously described ones. Many aspects of the host-parasite relationship have been already described, but there is a lot of space for developments. Concerning the parasite itself, the strain characterization is to be improved in order to better understand the clinical relevance of the genotyping, and furthermore, it is of paramount importance to correlate (if possible) the genotype and the phenotype (virulence or not?) of each parasite isolate. This point is closely linked with the stage conversion mechanism, the control of which remains to be elucidated: parasite, host, genetic, epigenetic ... all of them?

12 Human Toxoplasmosis

12.1 Epidemiology

Seroprevalence

Serologic surveys demonstrate prevalence of infection from less than 10% to greater than 90% in various geographic locations. Infection depends on multiple factors including environmental conditions, animal fauna, cultural habits and socio-economic conditions (Tenter *et al.*, 2000). It is more prevalent in hot and wet areas than in dry and cold climates. Prevalence below 30% is observed in North America, Great Britain, Scandinavia and South East Asia. Prevalence higher than 60% is observed in moist tropical areas of Latin America and sub-Saharan Africa, where cats are abundant and the climate favors survival of oocysts. In developed countries, prevalence has fallen over the last 40 years. For instance, in France, where seroprevalence was reported to be 82% in 1960, a study performed in 2003 indicates a prevalence of 44% (AFSSA, 2005).

In a single country, prevalence may vary according to socio-economic factors. Lower socio-economic classes are usually more infected than upper classes, and acquire infection earlier.

Risk factors

Humans acquire toxoplasmosis mostly by ingestion of tissue cysts in infected meat or by ingestion of oocysts in food or water. Transmission can also occur through accidental inoculation, or organ transplants or through congenital infection during pregnancy. Risk factors of oocyst-transmitted toxoplasmosis vary according to climate-enhancing oocyst survival, food habits and other factors. Proximity of cats to human homes and smaller space for deposition of cat feces in urban areas may increase the possibility of contamination (Bahia-Oliveira *et al.*, 2003). However, owning cats may not be associated with toxoplasmosis demonstrating the key role of soil and water as sources of infection (Dumètre and Dardé, 2003). Infection can be strongly associated with soil contact, contributing up to 17% of infections among pregnant women in Europe (Cook *et al.*, 2000). Ingesting oocysts is responsible for infections in vegetarians (Dubey, 2004). Drinking unfiltered surface or underground waters has been associated with a high level of endemic toxoplasmosis in Brazil (Bahia-Oliveira *et al.*, 2003) and Poland (Sroka *et al.*, 2006). To date, few outbreaks of toxoplasmosis by ingesting oocyst from soil or water have been reported (Table 3). Oocysts have been rarely identified in the suspected sources of infection, in part due to the lack of specific and sensitive methods of detection. Public water supplies have been implicated in outbreaks in Canada (Bowie *et al.*, 1997) and Brazil (de Moura *et al.*, 2006). Drinking unfiltered surface water has been responsible for an outbreak in the Panamanian jungle (Benenson *et al.*, 1982) and for 11 cases of acute toxoplasmosis in a village in the Surinam-French Guiana border.

Tissue cysts of *T. gondii* contained in meat are another important source of infection for humans. Consummation of frozen imported meat, where tissue cysts are killed by freezing, is not important in the epidemiology of *T. gondii* (Kotula *et al.*, 1991). Meat processed by salting, curing, freezing or heating procedures are not the likely source of human exposure. The difference in cooking and eating habits might account in part for lower prevalence rates of *T. gondii* infection in Africa and Asia than in America

Table 3 Outbreaks of toxoplasmosis due to ingesting oocysts

Country	Year	Source of Contamination	No. of Cases	Oocysts Isolated
USA	1976	Soil	10 (the same family; 7 symptomatic)	No
USA	1977	Soil	37 (35 symptomatic)	No
USA	1979	Contact with infected kittens	9 (the same family; 6 symptomatic)	No
Panama	1979	Surface water in jungle	35 (32 symptomatic)	No
Brazil	1982	Soil	ND[a]	Yes[b]
Canada	1995	Chlorinated drinking water	100 (63 symptomatic)	No
Brazil	2002	Chlorinated drinking water	294 (155 symptomatic)	Yes[c]
Surinam	2005	Water	11 (the same family; 9 symptomatic)	No

[a] ND, not determined.
[b] Oocysts isolated by mouse bioassay.
[c] Oocysts isolated by PCR.

and Europe. Fresh meat probably represents the major risk to consumers in industrialized countries. Epidemiological studies in Europe showed that between 30 and 63% of *Toxoplasma* infections could be attributed to meat consumption. Risk factors that most strongly predicted acute infection in pregnant women were eating raw or undercooked lamb, beef or other meat (including venison, horse, rabbit and game birds; Cook *et al.*, 2000). The association between eating raw or undercooked meat and acute *Toxoplasma* infection was underlined in previous studies even if the types of meat could vary. Undercooked lamb and pork, but not beef, were identified in a Norwegian study (Kapperud *et al.*, 1996) whereas beef and lamb, but not pork, were identified risk factors in a French study (Baril *et al.*, 1999). An epidemiological study of risk factors for recent *Toxoplasma* infection in pregnant women in Southern Italy found a strong association with eating cured pork and raw meat (Buffolano *et al.*, 1996). To date, only a few outbreaks of foodborne toxoplasmosis by ingesting cysts from meat have been reported.

Although the principal source of *T. gondii* contamination is attributed to cattle, sheep or domestics pigs, consumption of wildlife meat can lead to *Toxoplasma* infection. Several acute toxoplasmosis cases were reported in hunters that have consumed uncooked infected meat from wild pigs and from cervids. Moreover, evisceration and handling of game may represent risks for human infection.

12.2 Clinical aspects of human toxoplasmosis

Acquired toxoplasmosis in immunocompetent patients

In immunocompetent adults, the infection route is in almost all cases oral. Very rare cases of accidental laboratory punctures can occur. Following ingestion of tissue cysts in meat or of oocysts, tachyzoites spread in the organism, via blood and lymphatic vessels. It is commonly accepted that, a few days after, tachyzoites convert into bradyzoites and cysts appear in CNS and in muscles. The infection is in most cases unapparent (80% of cases) and the clinical signs are restricted to lymph nodes, asthenia and mild fever. The immunity acquired is definitive, stimulated by the cysts and protects against further re-infections. The exact long-term behavior of the cysts in the CNS is still unknown. Do they remain dormant throughout life, or do they periodically rupture?

More rarely, patients develop severe clinical pictures with high and prolonged fever, and multivisceral failure (respiratory distress, hepatic cytolysis, renal failure) that may lead to death. Other presentations of these unusual severe cases are described: myositis, myocarditis, retinochoroiditis, cutaneous rash. Acquired ocular toxoplasmosis is more frequently observed in certain regions of South Brazil. Ocular toxoplasmosis was also described in 20% of symptomatic cases observed in a large outbreak in Canada. In all these cases, the infecting strain seems to play a major role as evidenced by the atypical strains found in severe cases acquired in French Guiana (Carme *et al.*, 2002), or in ocular toxoplasmosis in Brazil (Khan *et al.*, 2006b). But the possible roles of the inoculum size and of the genetic background of the host must not be forgotten.

Congenital toxoplasmosis

The pathogenesis of congenital toxoplasmosis is complex. The first step is the maternal infection occurring during pregnancy. A latent chronic infection cannot be transmitted to the fetus in an immunocompetent mother. During the dissemination phase, parasites can localize in the placenta and cross the placental barrier. It is not clear when exactly the tachyzoites cross the placental barrier, and this is a crucial point to determine the risk of clinical lesions in the fetus. Thirty to 40% of maternal infections are transmitted to the fetus in France, for example. But, the transmission rate and the severity of the fetal infection depend on several mechanisms, such as placenta leaks to allow transmission, or fetal immunological maturity, both linked to the time of pregnancy. Maternal infections acquired early during pregnancy are rarely transmitted but often cause severe fetal injuries (brain and ocular damage, multiorgan lesions, possible *in utero* death), while late maternal infections are highly transmitted (up to 80% of transmission at the end of the third trimester) but cause mainly light clinical lesions in the fetus and the newborn (asymptomatic congenital toxoplasmosis or clinical lesions restricted to retinochoroiditis) (Desmonts and Couvreur, 1974;

Dunn *et al.*, 1999; Hayde and Pollack, 2000). The parasite load in amniotic fluid can be, to some extent, linked to the severity of the infection. There is no proof that a particular genotype is more adapted to congenital transmission. *Toxoplasma* genotypes found associated with congenital toxoplasmosis in a country may simply reflect the genotype circulating in meat-producing animals and the environment of the same geographic areas. For instance, in France, the great majority of congenital toxoplasmosis is due to infection with type-II *Toxoplasma* strains (Ajzenberg *et al.*, 2002b), whereas they are due to atypical or recombinant genotypes in Brazil. Studies are ongoing to detect a relationship between genotype and congenital toxoplasmosis outcome.

In the fetus, the newborn and the infant, clinical lesions are mainly present in the CNS and the eye (encephalitis, hydrocephaly, and retinochoroiditis). However, in the case of very severe infection, other organs can be involved combining foci of necrosis and inflammation. After birth, the main target of the parasite, in case of mild infection, is the retina (part of the CNS) with risk of retinochoroiditis from birth to adult age. Ocular lesions have an unpredictable potential of development during a lifetime and in the case of bilateral and macular relapses, a patient can be blinded.

Now, most of the congenital toxoplasmosis cases are asymptomatic at birth, notably in countries where prevention and management protocols are used (serological control in pregnant women, antenatal diagnosis with ultrasound follow-up during pregnancy in case of seroconversion, neonatal diagnosis in newborns suspected of congenital toxoplasmosis, ante- and neonatal treatment of cases). However, the risk of retinochoroiditis, although diminished, is still present in these cases. In absence of treatment, up to 80% of infected children had ocular lesions when 20 years old (Koppe *et al.*, 1986). To date, it seems that in 30% of newborns with congenital toxoplasmosis treated during their first year of life, a lesion often appeared in the first 12 years (Wallon *et al.*, 2004).

Toxoplasmosis in immunodeficient patients

Chronic infections persist for years and can reactivate, causing severe disease in immunocompromised patients. Immunosuppression due to HIV is the first cause of reactivation. In AIDS patients, the reactivation of the disease from the dormant cysts is the basic mechanism underlying the clinical features. Newly acquired infection in these patients exists, but is rare. The decrease of the protective immunity is mainly due to the lack of CD4+ LT and the decrease of IFN-γ levels. This is indirectly confirmed by the facts that, in AIDS patients, *Toxoplasma* reactivation usually appears when the CD4+ T-cell count is less than 100 μl^{-1} of blood, and the increase in the number of CD4+ LT following HAART allows cessation of the prophylaxis with anti-*Toxoplasma* drugs without any relapse. Thus cysts are ruptured, bradyzoites convert into tachyzoites, and cerebral lesions occur after multiplication of virulent tachyzoites. The question that is, to date, unanswered is whether clinical lesions of the CNS are only due to local CNS cysts or if reactivation of peripheral cysts can cause TE after dissemination of parasites in the blood, causing a parasitemia. Cerebral toxoplasmosis is the first manifestation in AIDS patients (Luft *et al.*, 1993) and clinical symptoms are usually similar to cerebral abscess. Computerized Tomography (CT) or Magnetic Resonance Imaging (MRI) showing characteristic lesions are very helpful for diagnosis. Ocular lesions are often observed (second most frequent symptom) and associated with cerebral lesions in 10–20% (Holland, 2004). Pneumopathy or other clinical manifestations are less frequently encountered.

In immunocompromised patients other than AIDS patients, the pathogenesis of toxoplasmosis is more complex and two groups of patients can be identified. First, the mechanism of reactivation can be similar to the mechanism observed in AIDS patients. Dormant cysts reactivate because of immunosuppression. This can occur mainly in patients with hematological malignancies or undergoing immunosuppressive therapies (such as graft patients). Cerebral (and ocular) lesions are also seen in these patients, but disseminated toxoplasmosis with involvement of peripheral organs (such as lung or heart) and parasitemia is more often described than in AIDS patients. In patients with positive *Toxoplasma* serology before the immunosuppression, prophylaxis with anti-*Toxoplasma* drugs can be prescribed. The treatment of the underlying disease is of paramount importance to control *Toxoplasma* pathogenesis. Second, *Toxoplasma* can be transmitted from a chronically (or rarely acutely) infected donor to a non-infected recipient of a graft, the recipient developing an acute toxoplasmosis the severity of which depends on the level of immunosuppression. This is what is called a mismatch, donor+/recipient−. This kind of transmission, which occurs mainly in the case of heart transplantation (more rarely in case of kidney, bone marrow or other transplantations) can be avoided by a careful analysis of the serological profiles of the donor and the recipient and, if necessary, by a specific prophylaxis in the recipient. In this case, toxoplasmosis is generally acute and disseminated.

Ocular toxoplasmosis

Ocular involvement may occur in congenital toxoplasmosis, in acquired toxoplasmosis in immunocompetent patients or in immunosuppressed patients. It was widely accepted that almost all cases of ocular toxoplasmosis were the result of congenital infection. Over the past decade, the role of postnatally acquired *T. gondii* infection in the pathogenesis of ocular toxoplasmosis has been reassessed and it is now generally accepted that a substantial proportion of patients with toxoplasmic retinochoroiditis did acquire infection postnatally.

Although there is no precise data for the prevalence of ocular toxoplasmosis, it seems to vary according to epidemiological factors. It was estimated at 2% of *T. gondii*-infected individuals in the United States, whereas 17.7% of individuals in southern Brazil have retinal findings consistent with *T. gondii* infection. Higher rates of ocular toxoplasmosis may also be found in some parts of Africa. Examination of 97 individuals infected in the 1995 Victoria epidemic revealed ocular toxoplasmosis in 20 individuals (21%). The genotype of the infecting parasite, which is now known to present geographical variations, may be an important determinant in the ocular involvement as atypical genotypes were found associated with acquired ocular toxoplasmosis in Brazil and USA.

Initial retinal infection may be subclinical, with development of retinal lesions months or years later (Holland, 2004). Tissue destruction is probably attributable to proliferation of *T. gondii* and to inflammatory reactions, both leading to tissue destruction, but the relative importance of each factor may vary between hosts. It has been suggested that autoimmune reactions (antibodies against S-antigen or other retinal antigens) may also play a role in the inflammatory response associated with toxoplasmic retinochoroiditis, but the contribution of such nonspecific reactions to the various presentations of ocular toxoplasmosis has not been determined. The mechanism of recurrences, which are one of the main characteristics of *Toxoplasma* retinochoroiditis, in congenital as well as in acquired toxoplasmosis, is not well understood. Their frequency appears to decrease over time.

Clinical aspects vary from large destructive lesions to punctate lesions depending on the duration of active retinal infection and the characteristics of the accompanying inflammatory reactions. Complications due to intense inflammatory reactions can be observed (retinal detachment, cataract, cystoid macular oedema, glaucoma). Active inflammatory disease resolves without treatment, leaving hyperpigmented scars, and recurrences develop as 'satellite' lesions. Disease presentation is often more severe in immunocompromised patients. In addition to a higher prevalence of ocular involvement, older patients may also have ocular disease of greater severity. This is in favor of the role of immune dysfunction in the pathogenesis of ocular toxoplasmosis.

13 Diagnosis

Depending on the clinical context, the biological diagnosis of toxoplasmosis is achieved by serology and/or the demonstration of the parasite presence or the parasitic DNA in sample.

13.1 Serological diagnosis

Principal techniques for diagnosis are based on detection of specific antibodies in sera or other fluids (CSF or intraocular fluid) in the case of diagnosis of toxoplasmosis in a specific organ. The more widely used techniques are ELISA with crude *Toxoplasma* lysate as antigens or tests using complete formalized tachyzoites, such as direct agglutination or immunofluorescence. Some authors have developed ELISA tests with recombinant antigens. The sensitivity and specificity are variable depending on techniques and results are expressed according to a threshold. In order to conclude what is the serological status of a patient, different isotypes must be determined: IgM antibodies appear in the first 2 weeks after infection and disappear a few months later (depending on the technique); IgG antibodies appear later, rise during the first 3–4 months and will persist all life long, making them good markers of chronic infection. The detection of IgA and IgE antibodies is less widely used; they may be interesting for datation of infection as they usually disappear a few months after infection, usually earlier than IgM, but they are also occasionally detected in cases of reactivation or in cases of unusual immunological reactivity. Datation of infection is a crucial point during pregnancy to determine if a pregnant woman has a chronic infection (i.e. without risk for transmission to her fetus) or if she acquired infection during pregnancy, with a risk for materno-fetal transmission. Comparative results of two sera sampled at 3-week intervals and tested simultaneously allow the infection to be dated. More recently, tests based on avidity of IgG antibodies have been developed to facilitate datation. In congenital toxoplasmosis, neosynthetized antibodies present in newborns are detected using western-blot or ELIFA methods (Pinon *et al.*, 2001). Generally, these analytical techniques are the most sensitive in congenital toxoplasmosis diagnosis. In immunosuppressed patients, serology does not contribute to diagnosis, but a positive serology identifies patients with a potential risk of reactivation.

13.2 Parasite detection

Detection of the parasite may be made by direct examination of pathological products after Giemsa staining or immunofluorescence, or histopathology and immunochemistry. These last techniques allow description of the tissue lesions, but they have a low sensitivity. The reference method for *Toxoplasma* detection is mouse inoculation of samples (amniotic fluid and placenta in diagnosis of congenital toxoplasmosis; cerebral spinal fluid, blood, or organs depending of clinical context). Inoculated mouse presents a seroconversion if parasite was present in the sample, allowing diagnosis of infection. In positive mice, microscopic examination of mouse brain confirms the serological result and leads to isolation of the *Toxoplasma* strain responsible for infection. Although this technique is quite sensitive and highly specific, the major drawback is the delay in response (3–4 weeks). Furthermore, it requires animal accommodation and is performed only by a few reference laboratories. An alternative way to demonstrate parasite presence is the detection of parasitic DNA in the sample. This type of diagnosis has been used in reference laboratories for the last 15 years. There is no commercial kit available and various housekeeping methods have been developed, based on amplification of sequence repeated in the *Toxoplasma* genome (B1 gene, rDNA or a 200–300 repeated sequence of 529 bp). Sensitivity of PCR is generally equivalent or better than mouse inoculation but DNA detected in the sample is not proof of living parasite. PCR is recommended for antenatal diagnosis, for ocular toxoplasmosis and in immuno-suppressed patients in order to give a rapid result.

14 Treatment

14.1 Anti-Toxoplasma *drugs*

The drugs that are most currently used to treat toxoplasmosis fall into two main groups: folic-acid-synthesis inhibitors (dihydrofolate reductase and sulfonamides) and macrolides. Both drugs act only on the tachyzoites and leave the cysts unaffected. Atovaquone is the only drug that has some activity against the cystic form of the parasite. Extensive information and a bibliography on the *in vitro* and *in vivo* activities of anti-*Toxoplasma* drugs is available in the review published by the Eurotoxo European Consensus initiative on Prevention of Congenital Toxoplasmosis (http://eurotoxo.isped.u-bordeaux2.fr/WWW_PUBLIC/DOC/Pharmacodynamics_v26082005.pdf).

Folic-acid-synthesis inhibitors

Folate compounds are essential factors for several processes in *Toxoplasma* metabolism including nucleoside biosynthesis and thereby nucleic acid formation. Unlike mammalian cells, *T. gondii* does not possess a carrier system to take up performed folate and must synthesize it *de novo*. Dihydropteroate synthetase (DHPS) is one of the enzymes responsible for this synthesis, while dihydrofolate reductase (DHFR) maintains the folate in a reduced state.

Dihydrofolate reductase inhibitors

Since DHPS is absent in mammalian cells, it represents a unique drug target with a potentially high therapeutic index. By contrast, DHRF is represented in both mammalian and protozoan cells, and the therapeutic index of DHFR inhibitors is limited by the potential toxicity of these drugs in mammalian cells. This represents the main drawback of this family of drugs which may induce a folate deficiency, possibly responsible for severe hematological side effects and embryopathy.

Pyrimethamine. This 2,4-diamino pyrimidine has been known to be active against *T. gondii* for 50 years. *In vitro* culture methods showed that pyrimethamine is inhibitory for tachyzoite growth at a concentration less than 0.1–0.2 mg l^{-1}. Its parasiticidal effect can readily be evidenced, with marked alterations of nucleic and cell division in treated parasites. This high inhibitory activity was also confirmed using purified DHFR, pyrimethamine being approximately 1000-fold more active against the parasite than the human enzyme. Pyrimethamine was found ineffective on intracystic parasites in an *ex-vivo* model. Very few data are available on the susceptibility of various *T. gondii* strains to pyrimethamine. *In vitro* study performed on a few strains revealed a marked intrinsic heterogeneity in pyrimethamine sensitivity according to the strain genotype. This heterogeneity was not confirmed for 14 strains belonging to various genotypes, all being in the same range of susceptibility as the reference RH strain (type I; Derouin, unpublished). The *in vivo* activity of pyrimethamine for treatment of acute toxoplasmosis is well established in various experimental models (mainly murine models). Oral administration of pyrimethamine is protective but not curative and relapses are observed after cessation of treatment. Follow-up of parasitic burdens in infected tissues showed that this activity is delayed after initiation of treatment. There is no information on the efficacy of long-term treatment with pyrimethamine alone on cyst burden of chronically infected mice. However, several arguments are in favor of the absence of activity of pyrimethamine on cysts: *in vitro*, long-term incubation of pyrimethamine with purified cysts has no effect on their infectivity; *in vivo*, long-term administration of pyrimethamine plus sulfonamide to chronically infected mice does alter the microscopic aspect of brain cysts.

Trimethoprim.   *In vitro*, trimethoprim is inhibitory and parasiticidal for tachyzoites for concentrations greater than or equal to $2\,\mathrm{mg\,l^{-1}}$, that is a concentration 50 times higher than pyrimethamine in the same culture conditions. In animal models, several studies showed that the administration of trimethoprim alone was not protective. Because of this limited efficacy, trimethoprim is not used alone but always administered in combination with sulfamethoxazole (see below).

Sulfonamides and sulfone

As for DHFR inhibitors, the activity of sulfonamides and sulfone was demonstrated more than 50 years ago in murine models of infection and confirmed for various derivatives.

Despite some discrepancies, *in vitro* studies performed in tissue culture or on purified DHPS have shown that sulfonamides exhibit a broad range of inhibitory potency. Sulfadiazine and sulfamethoxazole are amongst the most effective compounds, whereas sulfadoxine is inhibitory at 10- to 20-fold higher concentrations. This marked inhibitory effect is not associated with morphological alterations of the parasite in tissue culture. A wide variety of sulfonamide compounds have been investigated in murine models of acute toxoplasmosis. In a large comparative-dose response study performed with 17 sulfonamides administered at 50 to $800\,\mathrm{mg\,kg^{-1}}$ to acutely infected mice, sulfadiazine was found to be the most effective sulfonamide with a 50% protective dose of $60\,\mathrm{mg\,kg^{-1}}$; it was $150\,\mathrm{mg\,kg^{-1}}$ for sulfadoxine and $700\,\mathrm{mg\,kg^{-1}}$ for sulfamethoxazole. Relapses are constantly observed after cessation of therapy.

As for pyrimethamine, there is no information on the efficacy of long-term treatment with sulfonamides alone on brain cyst burden of chronically infected mice. However, the lack of *in vitro* activity of sulfadiazine on purified cysts, as well as the poor efficacy of pyrimethamine/sulfonamides combinations on brain cyst burdens, are strong arguments for the absence of activity of sulfonamides on cysts.

Combination of DHFR inhibitors and sulfonamides
A remarkable synergy between these two families of drugs has been evidenced *in vitro* and *in vivo* for a variety of drug combinations.

Pyrimethamine and suppadiazine. Since the 1950s, the combination of pyrimethamine and sulfadiazine has been considered the most effective drug combination. This combination proved remarkably efficient in the treatment of severe forms of toxoplasmosis, both in experimental models and in humans but is ineffective on the cyst form of the parasite. Sequential follow-up of parasite burdens in acutely infected mice showed that, when treatment is initiated several days after IP infection, it does not eradicate the parasite from the tissues, since relapses are observed after cessation of therapy.

Pyrimethamine and sulfadoxine (Fansidar). Very little information is available on the *in vitro* and *in vivo* activities of this combination. Pyrimethamine and sulfadoxine are synergistic *in vitro* at various ratios and, as Fansidar, proved efficient for treatment of acute toxoplasmosis in mice models but failed to eradicate the parasite in chronically infected mice.

Trimethoprim and sulfamethoxazole. Cotrimoxazole combines trimethoprim and sulfamethoxazole at a 1/5 ratio. This ratio was chosen in order to provide, in tissues (mainly lung), the optimum synergistic ratio against bacteria (i.e. 1/20). Several *in vitro* and *in vivo* studies suggest that this ratio may not be optimal for *T. gondii*, due to the poor efficacy of trimethoprim as compared to pyrimethamine. *In vivo*, experimental studies agree that high doses of cotrimoxazole are needed to cure experimental acute infection in mice and that relapses are observed after cessation of therapy. In chronically infected mice, cotrimoxazole is ineffective on parasitic brain cysts. Administration of cotrimoxazole in pregnant mice infected with *T. gondii* indicated a better *in-utero* control of congenital toxoplasmosis by cotrimoxazole than by spiramycin.

Other folate inhibitors
Epiroprim. Epiroprim is a trimethoprim analogue. Despite a potent activity *in vitro*, this drug is ineffective in mice acutely infected with the RH strain of *T. gondii*. However, a marked synergistic effect is observed when it is combined with sulfadiazine or dapsone.

Trimetrexate and piritrexim. *In vitro*, these drugs are 10- to 100-fold more active that pyrimethamine, with a 50% inhibitory concentration (IC_{50}) at 0.0002 mg l^{-1} for trimetrexate. In mouse models, administration of piritrexim or trimetrexate alone were not curative and only prolonged survival of treated mice, compared to controls.

Triazine and PS-15. Recent data suggest that antimalarial triazine WR99210 is strongly inhibitory *in vitro* and in a murine model of acute toxoplasmosis. This is in agreement with the *in vivo* activity of its pro-drug (PS-15) for prophylactic of disseminated toxoplasmosis, especially when it is combined with dapsone.

Dapsone. Dapsone is inhibitory for *T. gondii* at a lower concentration than sulfadiazine with an IC_{50} of 0.5 mg l^{-1} in culture and 0.008 mg l^{-1} in a DPS inhibition assay. Despite this excellent activity *in vitro*, dapsone is poorly efficient or toxic in a murine model of acute toxoplasmosis when administered alone, but is efficient when administered for prophylaxis to pigs. When dapsone is combined with pyrimethamine at a noncurative dosage, a marked protection is obtained, with parasite clearance from blood, lung and brain; however, relapses are observed after cessation of therapy. This combination has no effect on brain cysts of chronically infected mice. Dapsone is also synergistic when combined with epiroprim.

Miscellaneous. Other highly potent and specific DHFR inhibitors have been identified either *in silico* or *in vitro* against purified DHFR, but these activities were not assessed in animal models. Data from these papers were not extracted for this review.

Macrolides and related compounds

This important family of antibiotics include 'classical' macrolides, such as spiramycin, roxithromycin, clarithromycin, lincosamides (clindamycin), azalides (azithromycin), and the recently developed ketolides, telithromycin (HMR 3647) and HMR 3004. All these drugs have some activity against *T. gondii in vitro* although this has only been demonstrated using sensitive culture assays and long-term incubations. The delayed mode of action of macrolides has been well documented for clindamycin. It was related to the targeting of protein synthesis in the cytoplasm and in the apicoplast but this effect is only detrimental for the parasite in the second parasite generation.

The inhibitory effect of macrolides progressively increases with increasing concentration of the drug in the culture and a complete inhibition of parasitic growth is only observed for high concentrations of drug. However, clindamycin seems unique since its 50% inhibitory concentration is 1000-fold lower than other macrolides and it is parasiticidal at 0.006 mg l^{-1}.

Among macrolides, azithromycin was effective against cysts *in vitro* at a high concentration (100 mg l^{-1}) and after long-term incubation. In the same assay, clindamycin was not effective. No data is available for other macrolides and for ketolides. In murine models of acute toxoplasmosis, macrolides are only partially protective even when administered at very high dosages. Azithromycin administered alone seems to be the most effective macrolide, but its efficacy on cerebral toxoplasmosis is limited by its poor brain penetration. Despite its excellent *in vitro* activity, clindamycin alone is poorly protective in murine models of acute toxoplasmosis. Its efficacy in animal models of ocular toxoplasmosis is controversial. In murine models of chronic infection, long-term administration of spiramycin, clarithromycin or azithromycin, but not of clindamycin, have resulted in a significant reduction of brain cyst burdens. Due to the moderate activity of the macrolides alone on *T. gondii*, the efficacy of their combination with other antiparasitic drugs has been evaluated experimentally. *In vitro*, a simple additive effect was observed when azithromycin, roxithromycin or clarithromycin were combined with sulfadiazine or pyrimethamine. By contrast, several *in vivo* studies demonstrated synergistic activities of various combinations of macrolides (azithromycin, clarithromycin, roxithromycin) and ketolides with other anti-*Toxoplasma* drugs like pyrimethamine, sulfadiazine, minocycline or atovaquone. This synergistic effect was remarkable with the combination of azithromycin and pyrimethamine: combination regimens consistently resulted in a marked and prolonged reduction of the parasitic burdens in blood, lung and brain, compared to those in mice treated with any of the agents alone. One possible hypothesis to explain this synergy is the complementary pharmacokinetics of the compounds on tissue infection.

Surprisingly, there is no relevant information on the *in vitro* and *in vivo* efficacy of drug combinations which comprised clindamycin, and particularly the combination of clindamycin with pyrimethamine, which is presently considered as an alternative to the combination of pyrimethamine plus sulfadiazine for treatment of cerebral toxoplasmosis (see below).

Atovaquone

This broad-spectrum antiparasitic drug is inhibitory for *T. gondii* by inhibition of mitochondrial electron transport processes, competing with the biological electron carrier ubiquinone. *In vitro*, a complete inhibition of the growth of *T. gondii* tachyzoites can be observed at very low concentrations (IC$_{50}$ < 0.1 mg l^{-1}) Of major interest is the activity of atovaquone on cysts; *in vitro* treatment of cysts isolated from brain of chronically infected mice resulted in loss of viability and infectivity of intracystic parasites. This effect was confirmed *in vivo* as prolonged oral treatment of chronically infected mice progressively reduced the mean number of brain cysts compared

to untreated mice. In addition, the cysts recovered from treated mice presented morphological abnormalities with an increase of degenerated intracystic bradyzoites.

In murine models of acute toxoplasmosis, administration of atovaquone alone resulted in a significant protection of infected mice which correlates with a marked reduction of parasitic burdens in tissues.

Several drug combinations including atovaquone have been assessed experimentally. *In vitro*, an additive effect was observed when atovaquone was combined with sulfadiazine, clarithromycin or minocycline but a mild significant antagonistic effect was noted between atovaquone and pyrimethamine. However, *in vivo*, the combinations of atovaquone with pyrimethamine, macrolides, and sulfonamides were more efficient than each drug administered alone but none of these combinations can be considered as efficient as the reference pyrimethamine-sulfadiazine therapy.

Other drugs
Antibiotics
Fluoroquinolones. Most fluoroquinolones are effective *in vitro*, probably by targeting the apicoplast. The most potent are trovafloxacin and gatifloxacin. In a mouse model of acute toxoplasmosis, trovafloxacin was also found to be the most effective fluoroquinolone and a synergistic activity was observed when it was combined with several anti-*T. gondii* drugs. Gatifloxacin was also found to act synergistically with pyrimethamine or with gamma interferon.

Whether quinolones have a place in the therapy of congenital toxoplasmosis is debatable, but these drugs are effective *in vitro* and *in vivo* against *T. gondii*.

Cyclines. Minocycline and doxycycline were found to be effective against *T. gondii in vitro* and *in vivo*. Minocycline was considered a promising agent for the treatment of toxoplasmic encephalitis, due to its high liposolubility, which favors CSF diffusion and brain penetration, and its synergistic activity when used in combination with clarithromycin.

Rifabutin/rifapentin. *In vitro*, rifabutin and rifapentine, rifamycin derivatives, are inhibitory for *T. gondii* and rifabutin is synergistic *in vitro* when combined with atovaquone. In murine models of toxoplasmosis, rifabutin and rifapentine are effective when administered at a high dosage (100 mg kg^{-1} day^{-1}). A significant synergy is observed when rifabutin is administered in combination with atovaquone, clindamycin, sulfadiazine, clarithromycin or pyrimethamine.

Other antiprotozoan drugs
Anticoccidial drugs such as monensin are active at very low concentration. *In vivo*, iclazuril and ponazuril are effective in a murine model of acute

toxoplasmosis, either when administered alone or combined with pyrimethamine. Arpinocid-N-oxide (Amprolium) was also proved to be active against cysts *in vitro*. Among antimalarial drugs, qinghaoshu and its derivatives, as well as trioxanes derivates, were found inhibitory for *T. gondii in vitro*.

Miscellaneous drugs

Anti-HIV1 protease inhibitors, ritonavir and nelfinavir, were found inhibitory for *T. gondii in vitro* at concentrations that can be achieved in human serum.

Several plant extracts and other compounds, such as neuroleptics, bisphosphonates and drugs affecting lipid synthesis, were found inhibitory *in vitro*; interestingly, azasterol also markedly enhanced the inhibitory effect of pyrimethamine and sulfadiazine.

Several reports have also shown that the administration of interferon gamma markedly potentiates the *in vivo* activity of several anti-*Toxoplasma* drugs in murine models of acute toxoplasmosis or toxoplasmic encephalitis. This was evidenced for roxithromycin, clindamycin, sulfadiazine, gatifloxacin and, to a lesser extent, with azithromycin. In the same way, administration of IL-12 potentiates the activity of clindamycin and atovaquone.

14.2. Drug Resistance of Toxoplasma gondii

Treatment failures have been reported for most drug regimens used for treatment of TE or congenital toxoplasmosis. These failures can be related to host factors (drug intolerance, malabsorption) and/or to the development of drug-resistant parasites or a lower susceptibility of the parasite strain. Experimentally, mutants resistant to most anti-*Toxoplasma* drugs can be selected. Drug efflux mechanisms can also be suspected in *T. gondii* by the existence of a transmembrane transport mediated by a Pgp homologue located on the parasite membrane complex and the presence of ABC-transporter coding genes. Whether these genes or the gene mutation are related to drug resistance need to be investigated.

14.3. Conclusions about anti-Toxoplasma drugs

An ideal drug would be preferentially parasiticidal against the different parasitic stages, prevent stage differentiation from tachyzoite to bradyzoite, would penetrate into cysts and be well distributed in the main sites of infection. Based on *in vitro* and *in vivo* data recorded over the last 35 years, this review shows:

(i) no such a drug exists today;
(ii) no marketed new compound can be expected for several years, despite the recent identification of new potential drug targets (apicoplast, parasite penetration).

From this review, it can be assumed that among drugs that can be used safely for treatment of congenital toxoplasmosis, no drug or drug regimen is as effective as the combination of pyrimethamine plus sulfadiazine. Among the 'new' macrolides, azithromycin seems to be the most efficient *in vivo*, and remarkably synergizes with pyrimethamine. Some new fluroquinolones are also active *in vivo* and *in vitro* but their potential toxicity is a major concern for treatment of congenital toxoplasmosis. Atovaquone remains of interest for reduction of parasite burden in a long-term treatment strategy. The combination of trimethoprim and sulfamethoxazole remains a possible therapy, although much less efficient that pyrimethamine and sulfadiazine.

It should be noted that experimental data are dramatically lacking for some major drug combinations (pyrimethamine plus clindamycin for example), for treatment of ocular toxoplasmosis and on the efficacy on brain cyst burden for long-term treatment of chronic infection. This review also points out major lacks of information on the susceptibility of *Toxoplasma* strains according to their genotypes. Differences of susceptibility have been occasionally reported but need to be confirmed on the basis of the new knowledge on strain genotyping. Similarly, the possible existence of natural or drug-selected resistance and their mechanisms need to be investigated.

14.4. Therapeutics

Toxoplasmosis in immunocompetent patients

In the absence of clinical symptoms, immunocompetent patients are not treated. In those presenting mild clinical symptoms, the possible benefit of a treatment with spiramycin or cotrimoxazole on the duration of symptoms is not demonstrated and treatment with the combination of pyrimethamine plus sulfadiazine (or clindamycin) is not recommended. Indication for treatment with this combination is required in the case of marked symptoms, especially for patients who acquired toxoplasmosis in countries where recombinant or atypical strains are highly prevalent (Carme *et al.*, 2002).

There is no clear recommendation for treatment of asymptomatic toxoplasmosis occurring in HIV-infected patients with CD4 counts greater than $200\,mm^{-3}$. In this setting, the risk for developing severe toxoplasmosis seems low, and as a primary approach, treatment would not be proposed. However, one may consider that a course of pyrimethamine and sulfadiazine might be given to reduce parasite spread and cyst formation and therefore reduce the delayed risk of reactivation.

Maternal toxoplasmosis and congenital toxoplasmosis

In pregnant women who presumably acquired toxoplasmosis during pregnancy, a treatment is indicated with the aim to reduce the risk of materno-fetal transmission of *T. gondii* and potential fetal lesions.

Treatment should be initiated as early as possible after diagnosis. Spiramycin is usually proposed until a prenatal diagnosis is performed on amniotic fluid. When PCR and mouse inoculation are negative, spiramycin is usually maintained until delivery. When parasite or parasite DNA is evidenced in amniotic fluid, spiramycin is switched to the combination of pyrimethamine plus sulfadiazine, administered with folinic acid to reduce bone-marrow suppression due to pyrimethamine. This therapeutic scheme is for women infected in the first or second trimester of pregnancy. In case of contamination occurring late in pregnancy (third trimester), fetal contamination is likely to occur (80–90% of cases) and can justify the administration of pyrimethamine plus sulfadiazine or sulfadoxine instead of spiramycin without undergoing amniocentesis.

In the last 30 years, several studies have shown the benefits of treatments, mainly on children's sequelea (Couvreur *et al.*, 1991; Foulon *et al.*, 1999). However, recent European multicentric studies performed on large cohorts of patients could not provide evidence of efficacy, neither on parasite transmission nor on fetal symptoms (SYROCOT, 2007). Because of these uncertainties, mainly due to the design of the studies that have been conducted up to now, most investigators consider that only a large randomized controlled clinical trial would provide valid evidence of the treatment benefit on materno-fetal transmission and fetal lesions (Petersen and Schmidt, 2003; SYROCOT, 2006).

Postnatal treatment of congenital toxoplasmosis relies on administration of pyrimethamine plus sulfonamides (sulfadiazine or sulfadoxine), either continuously or alternating with spiramycin. The duration of treatment markedly varies according to countries and centers, 3 months in Denmark to 1 or 2 years in USA or France (Petersen and Schmidt, 2003). As for prenatal treatment, several studies, including some from large cohorts of patients, clearly indicate a treatment benefit (McLeod *et al.*, 2006; Petersen and Schmidt, 2003) but controlled clinical trials comparing treatment regimens are still lacking.

Ocular toxoplasmosis

Despite the severity of chorioretinitis and the abundance of reports on treatment of ocular toxoplasmosis in the literature, there is still no consensus regarding the best treatment regimen (Holland, 2004). The combination of pyrimethamine with sulfadiazine is the most currently used and is often combined with corticosteroid according to the degree of inflammation. This combination is even associated with clindamycin. Several other drug regimens using single drug treatment, such as atovaquone or clindamycin, or drug combinations (macrolides or clindamycin with pyrmethamine, cotrimoxazole) have been proposed (Holland and Lewis, 2002). Despite an agreement on the need to treat acute toxoplasmic chorioretinitis, there is no clear definitive evidence of the beneficial effect of a short-term treatment on

duration of inflammation and permanent visual acuity (Standford *et al.*, 2003). On the other hand, long-term prophylactic treatment with cotrimoxazole was found efficient for prevention of relapses (Silveira *et al.*, 2002). As for prenatal treatment of congenital toxoplasmosis there is a crucial need for randomized controlled trials to determine optimal therapy and management of acute toxoplasmic chorioretinitis.

Toxoplasmosis in immunocompromised patients

In immunocompromised patients with CNS, ocular, or systemic toxoplasmosis, the reference treatment regimen for toxoplasmic encephalitis is the combination of pyrimethamine, at a loading dose of 100–200 mg day^{-1} for 2–3 days followed by 50–75 mg day^{-1}, with sulfadiazine, 4–6 g day^{-1}, for at least 4 weeks. Folinic acid is systematically added to prevent myelosuppression. Longer courses of treatment might be required in cases of extensive disease or poor response after 4 weeks. The rate of adverse effects is approximately 50% and they require discontinuation of one or both drugs in 20–25% of cases (Couvreur and Leport, 1999). The most frequent are cytopenia and rashes, attributed to pyrimethamine or to sulfadiazine. The primary alternative for sulfadiazine in patients who develop sulfonamide hypersensitivity is clindamycin (2.4 g j^{-1}), administered with pyrimethamine and leucovorin (AI). Clindamycin can be associated with fever, rash, and gastrointestinal symptoms (Danneman *et al.*, 1992; Katlama *et al.*, 1996). For patients who cannot take drugs by the oral route, the use of intravenous cotrimoxazole has been proposed and found not inferior to the combination of pyrimethamine plus sulfadiazine in one randomized study (Torre *et al.*, 1998). Because of its low cost and availability, cotrimoxazole is proposed as a possible alternative in countries where standard regimens are not available (Dedicoat and Livesley, 2006). Several other drug-combination regimens have been proposed as salvage therapy, such as atovaquone (3 g day^{-1}) administered orally combined with pyrimethamine or with sulfadiazine, and azithromycin or clarithromycin combined with pyrimethamine. Single drug therapies with pyrimethamine, sulfadiazine or macrolides should not be used. A possible exception could be the use of atovaquone for patients intolerant to both pyrimethamine and sulfadiazine, but physicians should be aware of the relationship between atovaquone plasma concentrations and response to therapy, requiring plasma level monitoring (Torres *et al.*, 1997).

After curative treatment of the acute phase, a maintenance therapy (secondary prophylaxis) is indicated to prevent recurrence as long as immunosuppression exists. It is usually based on the same combination of drugs used for curative treatment but administered at half doses. In AIDS patients receiving highly active antiviral therapy (HAART), maintenance treatment can be replaced by cotrimoxazole (Duval *et al.*, 2004). In those with sustained immunological restoration assessed by a CD4 count greater than

200 mm^{-3} for at least 3 months, maintenance therapy can be safely discontinued (Bertschy *et al.*, 2006; Miro *et al.*, 2006).

Primary chemoprophylaxis

Primary prophylaxis is strongly recommended for patients at risk for reactivation of a previously acquired toxoplasma infection, that is those who are seropositive for *T. gondii* and are profoundly immunocompromised. This mainly includes HIV-infected patients who have a CD4 count of less than 100 mm^{-3}, bone-marrow transplant recipients, and to a lesser extent organ transplant recipients and patients with severe T-cell lymphoma.

In HIV-infected patients, cotrimoxazole is the drug of choice (Couvreur *et al.*, 1999, Kaplan *et al.*, 2002) and is safe for HIV-infected pregnant women (Forna *et al.*, 2006; Walter *et al.*, 2006). Atovaquone alone or combined with pyrimethamine, or the combination of dapsone and pyrimethamine can be considered. Other single drug regimens are not recommended. In organ transplant recipients cotrimoxazole or the combination of pyrimethamine plus sulfadoxine have also been used safely (Anonymous, 2000; Baden *et al.*, 2003; Foot *et al.*, 1994). Primary prophylaxis should be given as long as the risk of reactivation persists and can be discontinued safely in HIV-infected patients who have responded to HAART with an increase in CD4 count to greater than 200 μl^{-1} for at least 3 months (Miro *et al.*, 2006). This is also supported by the fact that *T. gondii*-specific cellular immune response, including *T. gondii*-induced IFN-γ production was restored in these patients at a similar level to HIV-negative controls (Fournier *et al.*, 2001). Primary prophylaxis should be re-started if the CD4 count decreases to less than 100–200 mm^{-3}.

Primary prophylaxis should also be considered for organ transplant recipients who are seronegative for *T. gondii* and whose donor is seropositive for *T. gondii*, because of the potential risk of transmission of *Toxoplasma* cysts with the graft (Baden *et al.*, 2003).

15 Conclusions

Toxoplasma gondii was discovered one hundred years ago. However, its place in a book on emerging protozoan pathogens is justified due to the large amount of new data acquired during the last 20 years on the parasite itself and on the pathogenesis of *Toxoplasma* infection.

Toxoplasma has been considered as the best model system to study the biology of the Apicomplexa because it is readily amenable to genetic and biologic manipulation in the laboratory (Kim and Weiss, 2004). Major advances have been made in the understanding of host-cell invasion, of the role of secreted proteins, and on the host immune response in experimental models. High-throughput screening protocols have detected many interactions between the parasite and different host cells. Its genome has been sequenced. Studies are ongoing on *Toxoplasma* genetic diversity and description of the geographical structure of the parasite population. This, together with better environmental detection, should bring new insights into the global epidemiology of *Toxoplasma*. However, these major scientific advancements still do not have practical consequences for our understanding of disease pathogenesis in humans. There is an urgent need for research that can bring new tools to prevent, diagnose and treat this disease in infected humans, particularly fetuses, newborns and immunosuppressed patients. No new drug against *T. gondii* of real use in patients has been described for decades despite the work induced by the AIDS epidemic. Concerning the biological diagnosis of toxoplasmosis, the most recent tool is PCR that appeared in the 1980s.

16 Acknowledgments

I would like to thank Prof. C. Leport for her expert review of treatments of toxoplasmosis and Rodolphe Thiébault for his assistance in preparing the review of anti-*Toxoplasma* drugs.

17 References

AFSSA (2005) *Toxoplasmose: état des connaissances et évaluation du risque lié à l'alimentation – rapport du groupe de travail 'Toxoplasma gondii' de l'Afssa.* http://www.afssa.fr/ftp/afssa/34487-34488.pdf.

Ajzenberg, D., Banuls, A.L., Tibayrenc, M. and Dardé, M.L. (2002a) Microsatellite analysis of *Toxoplasma gondii* population shows a high polymorphism structured into two main clonal groups. *Int. J. Parasitol.* **32**: 7–38.

Ajzenberg, D., Cogné, N., Paris, L., Bessières, M.H., Thulliez, P., Candolfi, E., Pelloux, H., Marty, P. and Dardé, M.L. (2002b) Genotype of 86 *Toxoplasma gondii* isolates associated with human congenital toxoplasmosis and correlation with clinical findings. *J. Infect. Dis.* **186**: 684–689.

Ajzenberg, D., Bañuls, A.L., Su, C., Dumètre, A., Demar, M., Carme B. and Dardé, M.L. (2004) Genetic diversity, clonality and sexuality in *Toxoplasma gondii.* *Int. J. Parasitol.* **34**: 1185–1196.

Ajzenberg, D., Dumètre, A. and Dardé, M.L. (2005) Multiplex PCR for typing strains of *Toxoplasma gondii. J. Clin. Microbiol.* **43**: 1940–1943.

Ambroise-Thomas, P. and Pelloux, H. (1993) Toxoplasmosis-congenital and in immunocompromised patients: a parallel. *Parasitol. Today* **9**: 61–63.

Anonymous (2000) Centers for Disease Control and Prevention; Infectious Disease Society of America; American Society of Blood and Marrow Transplantation Guidelines for preventing opportunistic infections among hematopoietic stem cell transplant recipients. *MMWR Recomm. Rep.* **49**: 1–125, CE1–7.

Aspinall, T.V., Guy, E.C., Roberts, K.E., Joynson, D.H., Hyde, J.E. and Sims, P.F. (2003) Molecular evidence for multiple *Toxoplasma gondii* infections in individual patients in England and Wales: public health implications. *Int. J. Parasitol.* **33**: 97–103.

Baden, L.R., Katz, J.T., Franck, L., Tsang, S., Hall, M., Rubin, R.H. and Jarcho, J. (2003) Successful toxoplasmosis prophylaxis after orthoptic cardiac transplantation with trimethoprim-sulfamethoxazole. *Transplantation.* **75**: 339–343.

Bahia-Oliveira, L.M., Jones, J.L., Azevedo-Silva, J., Alves, C.C., Orefice, F. and Addiss, D.G. (2003) Highly endemic, waterborne toxoplasmosis in North Rio de Janeiro State, Brazil. *Emerg. Infect. Dis.* **9**: 55–62.

Baril, L., Ancelle, T., Goulet, V., Thulliez, P., Tirard-Fleury, V. and Carme, B. (1999) Risk factors for *Toxoplasma* infection in pregnancy: a case-control study in France. *Scand. J. Infect. Dis.* **31**: 305–309.

Barragan, A. and Sibley, L.D. (2003) Migration of *Toxoplasma gondii* across biological barriers. *Trends Microbiol.* **11**: 426–430.

Belli, S.I., Smith, N.C. and Ferguson, D.J.P. (2006) The coccidian oocyst: a tough nut to crack. *Trends Parasitol.* **22**: 416–423.

Benenson, M.W., Takafuji, E.T., Lemon, S.M., Greenup, R.L. and Sulzer, A.J. (1982) Oocyst-transmitted toxoplasmosis associated with ingestion of contaminated water. *N. Engl. J. Med.* **307**: 666–669.

Bertschy, S., Opravil, M., Cavassini, M., *et al.* (2006) Swiss HIV Cohort Study. Discontinuation of maintenance therapy against *Toxoplasma* encephalitis in AIDS patients with sustained response to anti-retroviral therapy. *Clin. Microbiol. Infect.* **12**: 666–671.

Black, M.W. and Boothroyd, J.C. (2000) Lytic cycle of *Toxoplasma gondii*. *Microbiol. Mol. Biol. Rev.* **64**: 607–623.

Bowie, W.R., King, A.S., Werker, D.H., Isaac-Renton, J.L., Bell, A., Eng, S.B. and Marion, S.A. (1997) Outbreak of toxoplasmosis associated with municipal drinking water. The BC Toxoplasma Investigation Team. *Lancet* **350**: 173–177.

Boyle, J.P., Rajasekar, B., Saeij, J.P.J., *et al.* (2006) Just one cross appears capable of dramatically altering the population biology of a eukaryotic pathogen like *Toxoplasma gondii*. *Proc. Natl Acad. Sci. USA* **103**: 10514–10519.

Bradley, P.J., Ward, C., Cheng, S.J., *et al.* (2005) Proteomic analysis of rhoptry organelles reveals many novel constituents for host-parasite interactions in *Toxoplasma gondii. J. Biol. Chem.* **280**: 34245–34258.

Brenier-Pinchart, M.P., Pelloux, H., Derouich-Guergour, D. and Ambroise-Thomas, P. (2001) Chemokines in host-protozoan parasite interactions. *Trends Parasitol.* **17**: 292–296.

Buffolano, W., Gilbert, R.E., Holland, F.J., Fratta, D., Palumbo, F. and Ades, A.E. (1996) Risk factors for recent *Toxoplasma* infection in pregnant women in Naples. *Epidemiol. Infect.* **116**: 347–351.

Buxton, D. (1990) Ovine toxoplasmosis: a review. *J. R. Soc. Med.* **83**: 509–511.

Carme, B., Bissuel, F., Ajzenberg, D., *et al.* (2002) Severe acquired toxoplasmosis in immunocompetent adult patients in French Guiana. *J. Clin. Microbiol.* **40**: 4037–4044.

Conrad, P.A., Miller, M.A., Kreuder, C., James, E.R., Mazet, J., Dabritz, H., Jessup, D.A., Gulland, F. and Grigg M.E. (2005) Transmission of *Toxoplasma*: clues from the study of sea otters as sentinels of *Toxoplasma gondii* flow into the marine environment. *Int. J. Parasitol.* **35**: 1155–1168.

Cook, A.J., Gilbert, R.E., Buffolano, W., Zufferey, J., Petersen, E., Jenum, P.A., Foulon, W., Semprini, A.E. and Dunn, D.T. (2000) Sources of *Toxoplasma* infection in pregnant women: a European multicentre case-control study. *Br. Med. J.* **321**: 142–147.

Coppens, I. (2006) Contribution of host lipids to *Toxoplasma* pathogenesis. *Cell Microbiol.* **8**: 1–9.

Couvreur, J. and Leport, C. (1999) *Toxoplasma gondii*. In: *Antimicrobial Therapy and Vaccines* (eds V.L. Yu, T.C. Merigan and S.L. Barriere).Willians & Wilkins, Baltimore, MD, pp. 600–612.

Couvreur, J., Thulliez, P., Daffos, F., Aufrant, C., Bompard, Y., Gesquierre, A. and Desmonts, G. (1991) Foetopathie toxoplasmique: traitement *in utero* par l'association pyriméthamine-sulfadiazine. *Arch. Fr. Pediatr.* **48**: 397–403.

Dannemann, B., McCutchan, J.A., Israelski, D., *et al.* (1992) Treatment of toxoplasmic encephalitis in patients with AIDS. A randomized trial comparing pyrimethamine plus clindamycin to pyrimethamine plus sulfadiazine. The California Collaborative Treatment Group. *Ann. Intern. Med.* **116**: 33–43.

Dardé, M.L. (1992) Isoenzyme analysis of 35 *Toxoplasma gondii* isolates and the biological and epidemiological implications. *J. Parasitol.* **78**: 786–794.

Dardé, M.L. (2004) Genetic analysis of the diversity in *Toxoplasma gondii. Ann. Ist. Super. Sanita.* **40**: 57–63.

de Moura, L., Bahia-Oliveira, L.M.G., Wada, M.Y., *et al.* (2006) Waterborne toxoplasmosis, Brazil: from field to gene. *Emerg. Infect. Dis.* **12**: 326–329.

Dedicoat, M. and Livesley, N. (2006) Management of toxoplasmic encephalitis in HIV-infected adults (with an emphasis on resource-poor settings). *Cochrane Database Syst. Rev.* **3**: CD005420.

Denkers, E. Y. (2003) From cells to signaling cascades: manipulation of innate immunity by *Toxoplasma gondii. FEMS Immunol. Med. Microbiol.* **39**: 193–203.

Desmonts, G. and Couvreur, J. (1974) Toxoplasmosis. In: *Current Diagnosis*, Vol. 7 (ed. R.B. Conn). WB Saunders Company, Philadelphia, PA, pp. 274–297.

Dubey, J.P. (2000) The scientific basis for prevention of *Toxoplasma gondii* infection: studies on tissues cyst survival, risk factor and hygiene measures. In: *Congenital Toxoplasmosis: Scientific Background, Clinical Management and Control* (eds. P. Ambroise-Thomas and E. Petersen). Springer-Verlag, Paris, pp. 271–275.

Dubey, J.P. (2004) Toxoplasmosis – a waterborne zoonosis. *Vet. Parasitol.* **126**: 57–72.

Dubey, J.P. and Beattie, C.P. (1988) *Toxoplasmosis of Animals and Humans*. CRC Press, Boca Raton, FL.

Dubey, J.P., Miller, N.L. and Frenkel, J.K. (1970) The *Toxoplasma gondii* oocyst from cat feces. *J. Exp. Med.* **132:** 636–662.

Dubey, J.P., Lindsay, D.S. and Speer, C.A. (1998) Structures of *Toxoplasma gondii* tachyzoites, bradyzoites, and sporozoites and biology and development of tissue cysts. *Clin. Microbiol. Rev.* **11**: 267–299.

Dubey, J.P., Gamble, H.R., Hill, D., Sreekumar, C., Romand, S. and Thuilliez, P. (2002) High prevalence of viable *Toxoplasma gondii* infection in market weight pigs from a farm in Massachusetts. *J. Parasitol.* **88**: 1234–1238.

Dubey, J.P., Hill, D.E., Jones, J.L., *et al.* (2005) Prevalence of viable *Toxoplasma gondii* in beef, chicken, and pork from retail meat stores in the United States: risk assessment to consumers. *J. Parasitol.* **91**: 1082–1093.

Dubey, J.P., Sundar N., Gennari S.M., *et al.* (2007) Biologic and genetic comparison of *Toxoplasma gondii* isolates in free-range chickens from the northern Para state and the southern state Rio Grande do Sul, Brazil revealed highly diverse and distinct parasite populations. *Vet. Parasitol.* **143**:182–8.

Dumètre, A. and Dardé, M.L. (2003) How to detect *Toxoplasma gondii* oocysts in environmental samples? *FEMS Microbiol. Rev.* **27**: 651–661.

Dumètre, A. and Dardé, M.L. (2005) Immunomagnetic separation of *Toxoplasma gondii* oocysts using a monoclonal antibody directed against the oocyst wall. *J. Microbiol. Methods* **61**: 209–217.

Dunn, D., Wallon, M., Peyron, F., Pertersen, E., Peckham, C. and Gilbert, R. (1999) Mother-to-child transmission of toxoplasmosis: risk estimates for clinical counselling. *Lancet* **353**: 1829–1833.

Duval, X., Pajot, O., Le Moing, V., Longuet, P., Ecobichon, J.L., Mentre, F., Leport, C. and Vilde, J.L. (2004) Maintenance therapy with cotrimoxazole for toxoplasmic encephalitis in the era of highly active antiretroviral therapy. *AIDS* **18**: 1342–1344.

Esseiva, A.C., Naguleswaran, A., Hemphill, A. and Schneider, A. (2004) Mitochondrial tRNA import in *Toxoplasma gondii*. *J. Biol. Chem.* **279**: 42363–42368.

Fayer, R., Dubey, J.P. and Lindsay, D.S. (2004) Zoonotic protozoa: from land to sea. *Trends Parasitol.* **20**: 531–536.

Ferguson, D.J.P. (2002) *Toxoplasma gondii* and sex: essential or optional extra. *Trends Parasitol.* **18**: 355–359.

Ferguson, D.J.P. (2004) Use of molecular and ultrastructural markers to evaluate stage conversion of *Toxoplasma gondii* in both the intermediate and definitive host. *Int. J. Parasitol.* **34**: 347–360.

Ferguson, D.J. and Hutchison, W.M. (1987) An ultrastructural study of the early development and tissue cyst formation of *Toxoplasma gondii* in the brains of mice. *Parasitol. Res.* **73**: 483–491.

Ferguson, D.J., Henriquez, F.L., Kirisits, M.J., Muench, S.P., Prigge, S.T., Rice, D.W., Roberts, C.W. and McLeod, R.L. (2005) Maternal inheritance and stage-specific variation of the apicoplast in *Toxoplasma gondii* during development in the intermediate and definitive host. *Eukaryot. Cell.* **4**: 814–826.

Filisetti, D. and Candolfi, E. (2004) Immune response to *Toxoplasma gondii. Ann. Ist. Super. Sanità* **40**: 71–80.

Foot, A.B., Garin, Y.J., Ribaud, P., Devergie, A., Derouin, F. and Gluckman, E. (1994) Prophylaxis of toxoplasmosis infection with pyrimethamine/sulfadoxine (Fansidar) in bone marrow transplant recipients. *Bone Marrow Transplant.* **14**: 241–245.

Forna, F., McConnell, M., Kitabire, F.N., Homsy, J., Brooks, J.T., Mermin, J. and Weidle, P.J. (2006) Systematic review of the safety of trimethoprim-sulfamethoxazole for prophylaxis in HIV-infected pregnant women: implications for resource-limited settings. *AIDS Rev.* **8**: 24–36.

Foulon, W., Villena, I., Stray-Pedersen, B., Decoster, A., Lappalainen, M., Pinon, J.M., Jenum, P.A., Hedman, K. and Naessens, A. (1999) Treatment of toxoplasmosis during pregnancy: a multicenter study of impact on fetal transmission and children's sequelae at age 1 year. *Am. J. Obstet. Gynecol.* **180**: 410–415.

Fournier, S., Rabian, C., Alberti, C., Carmagnat, M.V., Garin, J.F., Charron, D., Derouin, F. and Molina, J.M. (2001) Immune recovery under highly active anti-retroviral therapy is associated with restoration of lymphocyte proliferation and interferon-gamma production in the presence of *Toxoplasma gondii* antigens. *J. Infect. Dis.* **183**: 1586–1591.

Guerardel, Y., Leleu, D., Coppin, A., Lienard, L., Slomianny, C., Strecker, G., Ball, S. and Tomavo, S. (2005) Amylopectin biogenesis and characterization in the proto-zoan parasite *Toxoplasma gondii*, the intracellular development of which is restricted in the HepG2 cell line. *Microbes Infect.* **7**: 41–48.

Hayde, M. and Pollack, A. F. (2000) Clinical picture. Neonatal signs and symptoms. In: *Congenital Toxoplasmosis* (eds. P. Ambroise-Thomas E. and Petersen). Springer-Verlag, Paris, pp. 153–164.

Hegab, S.M. and Al-Mutawa, S.A. (2003) Immunopathogenesis of toxoplasmosis. *Clin. Exp. Med.* **3**: 84–105.

Hill, D.E., Benedetto, S.M., Coss, C., McCrary, J.L., Fournet, V.L. and Dubey, J.P. (2006) Effects of time and temperature on the viability of *Toxoplasma gondii* tissue cysts in enhanced pork loin. *J. Food Prot.* **69**: 1961–1965.

Holland G.N. (2004) Ocular toxoplasmosis: a global reassessment. Part II: Disease manifestations and management. *Am. J. Ophthalmol.* **137**: 1–17.

Holland, G.N. and Lewis, K.G. (2002) An update on current practices in the management of ocular toxoplasmosis. *Am. J. Ophthalmol.* **134**: 102–114.

Howe, D.K. and Sibley, L.D. (1995) *Toxoplasma gondii* comprises three clonal lineages: correlation of parasite genotype with human disease. *J. Infect. Dis.* **172**: 1561–1566.

Jung, C., Lee, C.Y. and Grigg, M.E. (2004) The SRS superfamily of *Toxoplasma* surface proteins. *Int. J. Parasitol.* **34**: 285–296.

Kaplan, J.E., Masur, H. and Holmes, K.K.; USPHS; Infectious Disease Society of America. (2002) Guidelines for preventing opportunistic infections among HIV-infected persons – 2002. Recommendations of the U.S. Public Health Service and the Infectious Diseases Society of America. *MMWR Recomm. Rep.* **51**: 1–52.

Kapperud, G., Jenum, P.A., Stray-Pedersen, B., Melby, K.K., Eskild, A. and Eng, J. (1996) Risk factors for *Toxoplasma gondii* infection in pregnancy. Results of a prospective case-control study in Norway. *Am. J. Epidemiol.* **144**: 405–412.

Katlama, C., De Wit, S., O'Doherty, E., Van Glabeke, M. and Clumeck, N. (1996) Pyrimethamine-clindamycin vs. pyrimethamine-sulfadiazine as acute and long-term therapy for toxoplasmic encephalitis in patients with AIDS. *Clin. Infect. Dis.* **22**: 268–275.

Keeley, A. and Soldati, D. (2004) The glideosome: a molecular machine powering motility and host-cell invasion by *Apicomplexa. Trends Cell Biol.* **14**: 528–532.

Khan, A., Taylor, S., Su, C., *et al.* (2005a) Composite genome map and recombination parameters derived from three archetypal lineages of *Toxoplasma gondii. Nucleic Acids Res.* **33**: 2980–2992.

Khan A., Su C., German M., Storch G.A., Clifford DB, Sibley L.D. (2005b). Genotyping of Toxoplasma gondii strains from immunocompromised patients reveals high prevalence of type I strains. *J Clin Microbiol.* **43(12):** 5881–7.

Khan, A., Bohme, U., Kelly, K.A., *et al.* (2006a) Common inheritance of chromosome Ia associated with clonal expansion of *Toxoplasma gondii. Genome Res.* **16**: 1119–1125.

Khan, A., Jordan, C., Muccioli, C., Vallochi, A.L., Rizzo, L.V., Belfort Jr, R., Vitor, R.W., Silveira, C. and Sibley, L.D. (2006b) Genetic divergence of *Toxoplasma gondii* strains associated with ocular toxoplasmosis, Brazil. *Emerg. Infect. Dis.* **12**: 942–949.

Kim, K. and Weiss, L.M. (2004) *Toxoplasma gondii*: the model apicomplexan. *Int. J. Parasitol.* **34**: 423–432.

Kissinger, J.C., Gajria, B., Li, L., Paulsen, I.T. and Roos D.S. (2003) ToxoDB: accessing the *Toxoplasma gondii* genome. *Nucleic Acids Res.* **31**: 234–236.

Kong, J.T., Grigg, M.E., Uyetake, L., Parmley, S. and Boothroyd, J.C. (2003) Serotyping of *Toxoplasma gondii* infections in humans using synthetic peptides. *J. Infect. Dis.* **187**: 1484–1495.

Koppe, J.G., Loewer-Sieger, D.H. and de Roever-Bonnet, H. (1986) Results of 20-year follow-up of congenital toxoplasmosis. *Lancet* **1**: 254–256.

Kotula, A.W., Dubey, J.P., Sharar, A.K., Andrews, C.D., Shen, S.K. and Lindsay, D.S. (1991) Effect of freezing on infectivity of *Toxoplasma gondii* tissue cysts in pork. *J. Food Prot.* **54**: 687–690.

Kuticic, V. and Wikerhauser, T. (1996) Studies of the effect of various treatments on the viability of *Toxoplasma gondii* tissue cysts and oocysts. *Curr. Top. Microbiol. Immunol.* **219**: 261–265.

Lehmann, T., Graham, D.H., Dahl, E., Sreekumar, C., Launer, F., Corn, J.L., Gamble, H.R. and Dubey, J.P. (2003) Transmission dynamics of *Toxoplasma gondii* on a pig farm. *Infect. Genet. Evol.* **3**: 135–141.

Lehmann, T., Graham, D.H., Dahl, E.R., Bahia-Oliveira, L.M.G., Gennari, S.M. and Dubey, J.P. (2004) Variation in the structure of *Toxoplasma gondii* and the roles of selfing, drift, and epistatic selection in maintaining linkage disequilibria. *Infect. Genet. Evol.* **4**: 107–114.

Lehmann, T., Marcet, P.L., Graham, D.H., Dahl, E.R. and Dubey, J.P. (2006) Globalization and the population structure of *Toxoplasma gondii*. *Proc. Natl Acad. Sci. USA* **103**: 11423–11428.

Lindsay, D.S., Collins, M.V., Jordan, C.N., Flick, G.J. and Dubey, J.P. (2005) Effects of high pressure processing on infectivity of *Toxoplasma gondii* oocysts for mice. *J. Parasitol.* **91**: 699–701.

Lindsay, D.S., Collins, M.V., Holliman, D., Flick, G.J. and Dubey, J.P. (2006) Effects of high-pressure processing on *Toxoplasma gondii* tissue cysts in ground pork. *J. Parasitol.* **92**: 195–196.

Lüder, C.G.K. and Gross, U. (2005) Apoptosis and its modulation during infection with *Toxoplasma gondii*: molecular mechanisms and role in pathogenesis. *Curr. Top. Microbiol. Immunol.* **289**: 219–238.

Luft, B.J., Hafner, R., Korzun, A.H., *et al.* (1993) Toxoplasmic encephalitis in patients with the acquired immunodeficiency syndrome. Members of the ACTG 077p/ANRS 009 Study Team. *N. Engl. J. Med.* **329**: 995–1000.

Lunden, A. and Uggla, A. (1992) Infectivity of *Toxoplasma gondii* in mutton following curing, smoking, freezing or microwave cooking. *Int. J. Food Microbiol.* **15**: 357–363.

Lyons, R.E., McLeod, R. and Roberts, C.W. (2002) *Toxoplasma gondii* tachyzoite-bradyzoite interconversion. *Trends Parasitol.* **18**: 198–201.

Maréchal, E. and Cesbron-Delauw, M.F. (2001) The apicoplast: a new member of the plastid family. *Trends Plant Sci.* **6**: 200–205.

Mazumdar, J., Wilson, E.H., Masek, K., Hunter, C.A. and Striepen, B. (2006) Apicoplast fatty acid synthesis is essential for organelle biogenesis and parasite survival in *Toxoplasma gondii*. *Proc. Natl Acad. Sci. USA.* **103**: 13192–13197.

McLeod, R., Boyer, K., Karrison, T., *et al.* (2006) Outcome of treatment for congenital toxoplasmosis, 1981–2004: the National Collaborative Chicago-Based, Congenital Toxoplasmosis Study. *Clin. Infect. Dis.* **15**: 1383–1394.

Mercier, C., Adjogbleb, K.D.Z., Daubenerb, W. and Cesbron-Delauw, M.F. (2005) Dense granules: are they key organelles to help understand the parasitophorous vacuole of all apicomplexa parasites? *Int. J. Parasitol.* **35**: 829–849.

Miro, J.M., Lopez, J.C., Podzamczer, D., *et al.* (2006) Discontinuation of primary and secondary *Toxoplasma gondii* prophylaxis is safe in HIV-infected patients after immunological restoration with highly active antiretroviral therapy: results of an open, randomized, multicenter clinical trial. *Clin. Infect. Dis.* **43**: 79–89.

Nichols, B.A. and Chiappino, M.L. (1987) Cytoskeleton of *Toxoplasma gondii.* *J. Protozool.* **34**: 217–226.

Nichols, B.A., Chiappino, M.L. and O'Connor, G.R. (1983) Secretion from the rhoptries of *Toxoplasma gondii* during host cell invasion. *J. Ultrastruct. Res.* **83**: 85–98.

Petersen, E. and Schmidt, D.R. (2003) Sulfadiazine and pyrimethamine in the postnatal treatment of congenital toxoplasmosis: what are the options? *Expert Rev. Anti. Infect. Ther.* **1**: 175–182.

Pinon, J.M., Dumon H., Chemla, C., *et al.* (2001) Strategy for diagnosis of congenital toxoplasmosis: evaluation of methods comparing mothers and newborns and standard methods for postnatal detection of immunoglobulin G, M and A antibodies. *J. Clin. Microbiol.* **39**: 2267–2271.

Poupel, O., Boleti, H., Axisa, S., Couture-Tosi, E. and Tardieux, I. (2000) Toxofilin, a novel actin-binding protein from *Toxoplasma gondii,* sequesters actin monomers and caps actin filaments. *Mol. Biol. Cell* **11**: 355–368.

Roos, D.S., Crawford, M.J., Donald, R.G.K., Kissinger, J.C., Klimczack, L.J. and Striepen, B. (1999) Origin, targeting, and function of the apicomplexan plastid. *Curr. Opin. Microbiol.* **2**: 426–432.

Saeij, J.P.J., Boyle, J.P. and Boothroyd, J.C. (2005) Differences among the three major strains of *Toxoplasma gondii* and their specific interactions with the infected host. *Trends Parasitol.* **21**: 476–481.

Schwartzman, J. D. and Pfefferkorn, E. R. (1983) Immunofluorescent localization of myosin at the anterior pole of the coccidian, *Toxoplasma gondii. J. Protozool.* **30**: 657–661.

Silveira, C., Belfort, R. Jr., Muccioli, C., Holland, G.N., Victora, C.G., Horta, B.L., Yu, F. and Nussenblatt, R.B. (2002) The effect of long-term intermittent trimethoprim/sulfamethoxazole treatment on recurrences of toxoplasmic retinochoroiditis. *Am. J. Ophthalmol.* **134**: 41–46.

Soldati, D., Dubremetz, J.F. and Lebrun, M. (2001) Microneme proteins: structural and functional requirements to promote adhesion and invasion by the apicomplexan parasite *Toxoplasma gondii. Int. J. Parasitol.* **31**: 1293–1302.

Speer, C.A., Clark, S. and Dubey, J.P. (1998) Ultrastructure of the oocysts, sporocysts, and sporozoites of *Toxoplasma gondii. J. Parasitol.* **84**: 505–512.

Sroka, J., Wojcik-Fatla, A. and Dutkiewicz, J. (2006) Occurrence of *Toxoplasma gondii* in water from wells located on farms. *Ann. Agric. Environ. Med.* **13**: 169–175.

Stanford, M.R., See, S.E., Jones, L.V. and Gilbert, R.E. (2003) Antibiotics for toxoplasmic retinochoroiditis: an evidence-based systematic review. *Ophthalmology* **110**(5): 926–931.

Su, C., Evans, D., Cole, R.H., Kissinger, J.C., Ajioka, J.W. and Sibley, L.D. (2003) Recent expansion of *Toxoplasma* through enhanced oral transmission. *Science* **299**: 414–416.

Su, C., Zhang, X. and Dubey, J.P. (2006) Genotyping of *Toxoplasma gondii* by multilocus PCR-RFLP markers: a high resolution and simple method for identification of parasites. *Int. J. Parasitol.* **36**: 841–848.

Suzuki, Y. (2002) Host resistance in the brain against *Toxoplasma gondii*. *J. Infect. Dis.* **185**(suppl 1): 58–65.

SYROCOT (Systematic Review on Congenital Toxoplasmosis) study group, Thiebaut, R., Leproust, S., Chene, G. and Gilbert, R. (2007) Effectiveness of prenatal treatment for congenital toxoplasmosis: a meta-analysis of individual patients' data. *Lancet* **369**: 115–122.

Tenter, A.M., Heckeroth, A.R. and Weiss, L.M. (2000) *Toxoplasma gondii*: from animals to humans. *Int. J. Parasitol.* **30**: 1217–1258.

Torre, D., Casari, S., Speranza, F., et al. (1998) Randomized trial of trimethoprim-sulfamethoxazole versus pyrimethamine-sulfadiazine for therapy of toxoplasmic encephalitis in patients with AIDS. Italian Collaborative Study Group. *Antimicrob. Agents Chemother.* **42**: 1346–1349.

Torres, R.A., Weinberg, W., Stansell, J., Leoung, G., Kovacs, J., Rogers, M. and Scott, J. (1997) Atovaquone for salvage treatment and suppression of toxoplasmic encephalitis in patients with AIDS. Atovaquone/Toxoplasmic Encephalitis Study Group. *Clin. Infect. Dis.* **24**: 422–429.

Villena, I., Aubert, D., Gomis, P., et al. (2004) Evaluation of a strategy for *Toxoplasma gondii* oocysts detection in water. *Appl. Environ. Microbiol.* **70**: 4035–4039.

Wallon, M., Kodjikian, L., Binquet, C., Garweg, J., Fleury, J., Quantin, C. and Peyron, F. (2004) Long-term ocular prognosis in 327 children with congenital toxoplasmosis. *Pediatrics* **113**: 1567–1572.

Walter, J., Mwiya, M., Scott, N., et al. (2006) Reduction in preterm delivery and neonatal mortality after the introduction of antenatal cotrimoxazole prophylaxis among HIV-infected women with low CD4 cell counts. *J. Infect. Dis .***194**: 1508–1510.

Warnekulasuriya, M.R., Johnson, J.D., and Holliman, R.E. (1998) Detection of *Toxoplasma gondii* in cured meats. *Int. J. Food Microbiol.* **45**: 211–215.

Weiss, L.M. and Kim, K. (2000) The development and biology of bradyzoites of *Toxoplasma gondii*. *Front. Biosci.* **5**: 391–405.

Wolf, A., Cowen, D. and Paige, B. (1939) Human toxoplasmosis: occurrence in infants as an encephalomyelitis verification by transmission to animals. *Science* **89**: 226–227.

Yarovinsky, F. and Sher, A. (2006) Toll-like receptor recognition of *Toxoplasma gondii*. *Int. J. Parasitol.* **36**: 255–259.

B3 *Isospora (Cystoisospora) belli*

Lynne S. Garcia

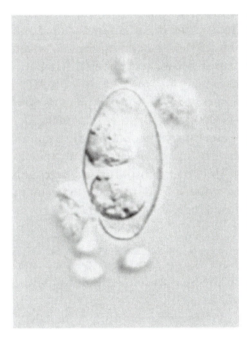 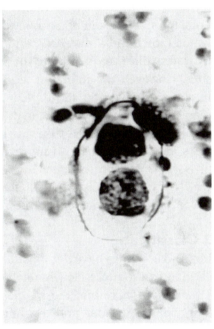

1 Introduction

Isospora belli was first described by Virchow in 1860 but not named until 1923. Human isosporiasis was first described in British troops with dysentery who were returning from Turkey during the First World War. Although isosporiasis has been found in many parts of the world, certain tropical areas in the Western Hemisphere contain some well-defined locations of endemic infections. These parasites can infect both adults and children, and intestinal involvement and symptoms are generally self-limiting unless the patient is immunocompromised. *I. belli* has also been implicated in traveler's diarrhea (Beaver *et al.*, 1984). However, unlike *Cryptosporidium* spp. and *C. cayetanensis*, large outbreaks of isosporiasis have not been reported. Only a few hundred cases of isosporiasis were described prior to recognition that this was an opportunistic infection in immunocompromised patients, particularly those infected with HIV. Transmission is via infective oocysts transmitted in contaminated food and water. Infections with *I. belli* tend to be self-limited and are characterized by watery diarrhea, similar to that caused by other coccidian human parasites. However, the infection tends to be chronic in patients with AIDS. In some tropical and subtropical areas, up to 20% of AIDS patients with diarrhea have been diagnosed with isosporiasis (Pape *et al.*, 1989). The organism is characterized by having mature oocysts containing two sporocysts, each containing four sporozoites. Currently, *I. belli* is the only species of *Isospora* known to infect humans. Unlike cryptosporidiosis, infection with *I. belli* is easily treated; however, the infection tends to recur or become chronic in immunocompromised individuals.

2 Classification

Molecular studies have demonstrated that *Isospora* spp. from primates and carnivores are more closely related to the Sarcocystiidae rather than the Eimeriidae. Therefore, the species will probably be transferred into the family Sarcocystiidae and the genus *Cystoisospora* (Franzen *et al.*, 2000). However, in this chapter, the organism will be referred to as *I. belli* and the disease as isosporiasis. In written publications, transition to the new genus name will most likely occur over the next few years. However, currently, the organism is in the family Eimeriidae.

3 Life Cycle and Morphology

Schizogonic and sporogonic stages in the life cycle of *I. belli* have been found in human intestinal mucosal biopsy specimens. Development in the intestine usually occurs within the epithelial cells of the distal duodenum and proximal jejunum (Figure 1). Stages of both asexual (trophozoite, schizont and merozoite) and sexual (macrogametocyte) phases of the life cycle of the parasite occur in the epithelium, always enclosed within a parasitophorous vacuole. Eventually, oocysts are passed in the stool; they

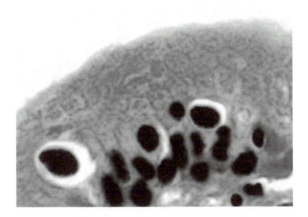

Figure 1 *Isospora belli* jejunal biopsy specimen containing a large, densely staining oval body in the apical cytoplasm of an enterocyte. The clear space around the developing organism is due to shrinkage that often occurs during tissue processing.

are long and oval and measure 20–33 μm by 10–19 μm. Usually the oocyst contains only one immature sporont, but two may be present (Figures 2 and 3). Continued development occurs outside the body with the development of two mature sporocysts, each containing four sporozoites, which can be recovered from the fecal specimen. The sporulated oocyst is the infective stage that excysts in the small intestine, releasing the sporozoites, which penetrate the mucosal cells and initiate the life cycle. The lifecycle stages – schizonts, merozoites, gametocytes, gametes, and oocysts – are structurally similar to those seen in the other coccidia.

Isospora cysts containing undeveloped sporozoites or merozoites have been seen in extraintestinal locations, which include lymph nodes, liver, and spleen (Michiels *et al.*, 1994). Zoites have also been found in lymphatics, which may provide a means of transport from the intestine to other body sites. It is possible that these stages may be responsible for reactivation of infection in AIDS patients, even following effective therapy.

The patent period is not known but may be as short as 15 days. Chronic infections develop in some patients, and oocysts can be shed for several months to years. In one particular case, an immunocompetent individual had symptoms for 26 years and *I. belli* was recovered in stool a number of times over a 10-year period.

4 Clinical Disease

Clinical symptoms include diarrhea, which may last for long periods (months to years), weight loss, abdominal colic, and fever; diarrhea is the main symptom (Cranendonk *et al.*, 2003; Guk *et al.*, 2005). Bowel movements

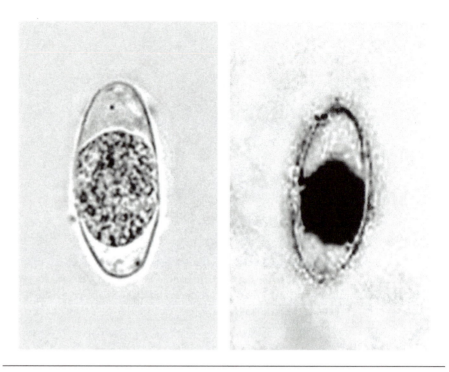

Figure 2 *Isospora belli* oocysts containing a single sporoblast; these oocysts are normally seen in patients with diarrhea and are not infective when passed. (*Left*) Wet mount from stool sedimentation concentration; (*Right*) stained with modified acid-fast stain.

(usually 6 to 10 per day) are watery to soft, foamy, and offensive smelling, suggesting a malabsorption process. Eosinophilia is found in many patients, recurrences are quite common, and the disease is more severe in infants and young children. The diarrhea cannot be differentiated from that caused by other coccidia and the microsporidia.

Patients who are immunosuppressed, particularly those with AIDS, often present with profuse diarrhea associated with weakness, anorexia, and weight loss. In a series of AIDS cases, cryptosporidiosis was more prevalent in the group of patients who had diarrhea for two or more weeks than those who had diarrhea for less than two weeks (22.82% vs. 10.29%). Other frequent parasitic infections were isosporiasis (10.6%), giardiasis (8.3%) and strongyloidiasis (6.9%). In this particular group, the most common opportunistic pathogens were *Cryptosporidium* spp. and *I. belli* and the most frequent nonopportunistic pathogens were *Giardia lamblia* and the nematode *Strongyloides stercoralis* (Garcia *et al.*, 2006).

To assess the prevalence of intestinal parasites in a cohort of HIV-infected adults in Cameroon, a cross-sectional study was conducted (Sarfati *et al.*, 2006). Detection of parasites was performed in 181 stool samples from 154 HIV-infected patients with a mean CD4 cell count of 238 cells mm^{-3}.

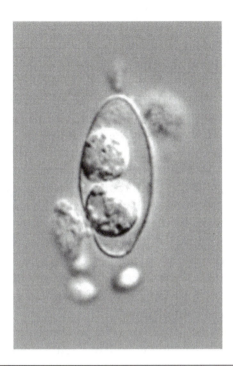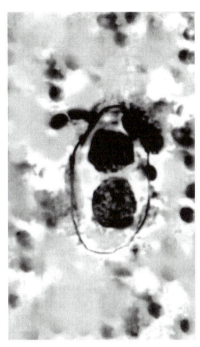

Figure 3 *Isospora belli* oocysts containing two sporoblasts; these oocysts are normally seen in patients with more normal stools and are not infective when passed. (*Left*) Wet mount from stool sedimentation concentration; (*Right*) stained with modified acid-fast stain.

Only 35 patients (22%) were receiving antiretroviral therapy at the time of stool sampling, and 46 (29%) had diarrhea. Opportunistic protozoa were found in 15 patients (9.7%), 8 of whom (53%) had diarrhea. *Enterocytozoon bieneusi* was found in eight patients, *C. parvum* in six patients, and *I. belli* in three patients. The prevalence of opportunistic protozoa among patients with CD4 cell counts less than 50 mm^{-3} was 32%.

In one patient with a well-documented case of isosporiasis of long standing, a series of biopsies showed a markedly abnormal mucosa with short villi, hypertrophied crypts, and infiltration of the lamina propria with eosinophils, neutrophils, and round cells. It has been recommended that physicians consider *I. belli* in AIDS patients with diarrhea who have immigrated from or traveled to Latin America, are Hispanics born in the United States, are young adults, or have not received prophylaxis with trimethoprim/sulfamethoxazole (TMP-SMX) for *P. jiroveci*. It has also been recommended that AIDS patients traveling to Latin America and other developing countries be advised of the waterborne and foodborne transmission of *I. belli* and that they may want to consider chemoprophylaxis.

Extraintestinal infections in AIDS patients have been reported. One patient was a 38-year-old white man who presented with progressive dyspnea and fever; he also complained of dysphagia, nausea, vomiting, and brown watery diarrhea (eight or nine episodes daily) and had lost 20 lb in 2 months. He was diagnosed with *P. jiroveci* pneumonia and oropharyngeal candidiasis and was treated with TMP-SMX and pentamidine. He improved and was discharged, but he was readmitted complaining of nausea, vomiting and diarrhea; he was diagnosed with giardiasis and treated. Five months later he was diagnosed as having *I. belli* and *Entamoeba histolytica* infection and was treated. Three months later he presented with dyspnea, fever, diarrhea, and generalized wasting; he was diagnosed with cytomegalovirus pneumonia and died 2 weeks later. At autopsy, microscopic findings associated with *I. belli* infection were seen in the lymph nodes and walls of both the small and large intestines. Intracellular zoites were seen in the cytoplasm of histiocytes. Each organism was surrounded by a thick eosinophilic cyst wall in routine histologic preparations with hematoxylin and eosin, and the cyst wall was PAS positive. Examination of the intestinal tissues revealed intraepithelial asexual and sexual stages of *I. belli*, as well as some merozoites that appeared to be in cells of the lamina propria.

The second case involved a 30-year-old black woman who was living in France but was originally from Burkina Faso. Initially she was symptomatic with fever, diarrhea, and weight loss. She was diagnosed with esophageal candidiasis and *I. belli* infection. She was treated with TMP-SMX and placed on maintenance therapy but suffered episodes of recurrent infection over the next 3 years. Examination of samples collected at autopsy revealed stages of *I. belli* in the intestine, mesenteric and mediastinal lymph nodes, liver, and spleen. The extraintestinal stages were always observed as single organisms that did not stain well with acid-fast stains. Massive infection with plasmacytosis and some eosinophils, but no granulomatous reaction, was observed.

In a third AIDS patient who presented with watery, nonbloody diarrhea and fever, examination of small intestine biopsy specimens revealed merozoites in the intestinal lumen, lamina propria, and lymphatic channels. Confirmation of the merozoites within the lymphatic channels documents a means of dissemination to lymph nodes and other tissues. Additional studies of extraintestinal tissue cysts have identified early tissue cysts that lack a developed cyst wall, demonstrating that more than one tissue cyst can occupy a host cell, describing the distribution of micronemes and the shedding of zoite membranes, and identifying tubular structures in the inner tissue cyst wall and inner compartment.

Charcot-Leyden crystals derived from eosinophils have also been found in the stools of patients with *I. belli* infection. The diarrhea and other symptoms may continue in compromised patients, even those on immunosuppressive therapy, when the regimen of therapy is discontinued. This infection has been found in homosexual men, all of whom were immunosuppressed and had several months of diarrhea.

In patients with AIDS, eosinophilia has been strongly associated with isosporiasis, particularly in those patients without weight loss, but with low CD4+ cell counts (<100–200 cells mm^{-3}; Certad *et al.*, 2003). The finding of unizoic cysts of *I. belli* in lymphoid tissue of a patient with AIDS may be responsible for drug resistance and/or relapses (Frenkel *et al.*, 2003). Also, in those patients with severe malabsorption, failure to reach therapeutic levels of nitazoxanide in plasma and bile may be responsible for treatment failure.

In a recent study in Brazil, 100 HIV/AIDS patients (Group 1) and 85 clinically healthy individuals (Group 2) were submitted to coproparasitological examination. Intestinal parasites were detected in 27% of patients from Group 1 and in 17.6% from Group 2. In Group 1 the most frequent parasites were *Strongyloides stercoralis* (12%), with two cases of hyperinfection; *Isospora belli*, 7%; *Cryptosporidium* spp., 4%; with 1 asymptomatic case and hookworm, 4%. Of the infected patients from Group 1 who reported to be chronic alcoholics, 64.3% had strongyloidiasis. Only 6 of the 27 infected patients from Group 1 were on highly antiretroviral therapy (HAART). In Group 2 the most frequent parasites were *S. stercoralis*, 7.1%; hookworm, 7.1% and *Giardia lamblia*, 3.5%. In conclusion, diagnosing intestinal parasites in HIV/AIDS patients is necessary especially in those who report to be chronic alcoholics or are not on antiretroviral treatment (Silva *et al.*, 2005).

Opportunistic intestinal parasite infection should be suspected in any HIV-infected patient with advanced disease presenting with diarrhea. Also, the importance of tropical epidemic nonopportunistic intestinal parasite infections among HIV-infected patients should not be neglected.

Previous studies from African countries where HIV-1 infection is prevalent have shown that infections with *C. parvum*, *I. belli* and microsporidia are frequently associated with chronic diarrhea in AIDS patients. However, information about the occurrence of these parasites in HIV-2-associated AIDS cases with chronic diarrhea is limited. In Guinea-Bissau, the country with the highest prevalence of HIV-2 in the world, a study of stool parasites in patients was undertaken (Lebbad *et al.*, 2001). Stool specimens were screened for parasitic infections from 52 adult patients with chronic diarrhea; 37 were HIV-positive and fulfilled the clinical criteria of AIDS (five HIV-1, 28 HIV-2, and four dually infected with HIV-1 and HIV-2). Of the HIV-2-positive patients, 25% were infected with *C. parvum*, 11% with *I. belli* and 11% with microsporidia. The three patients with microsporidiosis, all HIV-2-infected, are to our knowledge the first cases reported from Guinea-Bissau. Other stool parasites such as *Blastocystis hominis*, hookworm and *S. stercoralis* were observed both among HIV-positive and HIV-negative patients.

It is now well-established that *I. belli* infection is frequent in patients with acquired immunodeficiency syndrome in tropical areas. It has also been reported in other immunodepressive diseases, such as lymphoblastic leukemia, adult T-cell leukemia, Hodgkin's disease, and non-Hodgkin's lymphoma in non-HIV-infected patients (Resiere *et al.*, 2003). In nontropical areas of the world, *I. belli* can cause severe chronic diarrhea in patients with malignancies whose country of origin is in an endemic area.

5 Pathogenesis

Abnormal mucosa has been observed with moderate to severe villous atrophy, hypertrophied crypts, and infiltration of the lamina propria with eosinophils, neutrophils, and round cells. Collagen deposition in the lamina propria has also been seen.

In a case of granulomatous endometritis caused by coccidiosis in an immunologically uncompromised 63-year-old patient, the glandular epithelium of the endometrium contained numerous intracytoplasmic cysts, corresponding to periodic acid-Schiff-positive and methenamine-silver-negative sporoblasts. The endometrial glands revealed reactive phenomena, such as eosinophilic and squamous glandular metaplasia and intraluminal desquamation. Although parasites were absent, nonnecrotizing epithelioid granulomata were present in the stroma. These findings were not confirmed by stool examination, but the organisms were probably *I. belli*. There was no evidence of other foci of the disease (De Otazu *et al.*, 2004).

Endoscopic biopsy evaluation of patients with AIDS who had unexplained chronic diarrhea revealed an occult infection in half of the cases. However, villus and crypt changes in advanced HIV infection were independent of diarrhea or enteric infection and did not correlate with AIDS enteropathy. Subnormal epithelial proliferation in response to injury could be a factor, but the underlying cause of the changes may be due to multiple factors (Greenson *et al.*, 1991).

I. belli has also been documented as causing chronic diarrhea and acalculous cholecystitis in patients with AIDS. Although no structural means of differentiating *Isospora* to the species level are available at the light or electron microscopy levels, the unizoite cysts seen are probably part of the cycle of *I. belli*, rather than other species that could be pathogenic in immunocompromised individuals (Velasquez *et al.*, 2001).

6 Diagnosis

Examination of a fecal specimen for the oocysts is recommended. However, wet preparation examination of fresh material either as a direct smear or as concentrated material is recommended rather than the permanent stained smear (Garcia, 2007; Garcia *et al.*, 2003; National Committee for Clinical Laboratory Standards, 1997; Rigo and Franco, 2002). The oocysts are very pale and transparent and can easily be overlooked. They can also be very difficult to see if the concentration sediment is from polyvinyl-alcohol-preserved stool. The light level should be reduced, and additional contrast should be obtained with the microscope for optimal examination conditions. It is also quite possible to have a positive biopsy specimen but not recover the oocysts in the stool because of the small numbers of organisms present. These organisms are modified acid-fast positive and can also be demonstrated by using auramine-rhodamine stains. Organisms tentatively identified by using auramine-rhodamine stains should be confirmed by wet smear examination or modified acid-fast stains, particularly if the stool contains other cells or excess artifact material (more normal stool consistency).

In a comparison of the modified Ziehl-Neelsen stain (MZN) with the newer method, acid-fast-trichrome (AFT), the authors concluded that the sensitivity and specificity of the MZN was superior to that of the AFT. However, with minor changes, the AFT method would be appropriate for use in the diagnosis of intestinal coccidia, as well as the possible detection of microsporidia.

Like *Cyclospora cayetanensis, I. belli* unstained oocysts will autofluoresce; they appear blue/violet under ultraviolet light and green under violet or blue/violet light (Figure 4). However, often the oocysts are seen in the concentration sediment wet preparation, and additional testing is not required for the diagnosis of isosporiasis.

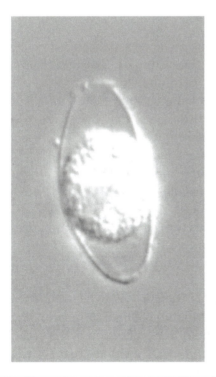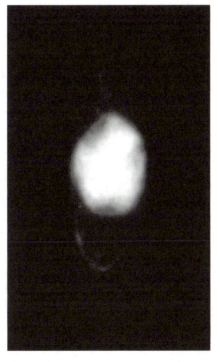

Figure 4 *Isospora belli* oocysts containing a single sporoblast, stained using optical brightening agent stain (*Calcofluor white*).

The development of *I. belli* has been studied in different cell lines. Merozoites were observed in all kinds of cells, whereas sporogony was demonstrated only in Hct-8. This implied that not only the human cell line can be infected, but also some animal cell lines. Unizoites could be found in Vero cells. The merozoites were transferred to a new culture cell for three passages and maintained for 2 weeks, but no oocyst production was observed in any culture cells during cultivation. Eventually, this approach may provide additional methods of isolation and identification (Siripanth *et al.*, 2004).

Intracellular development of *I. belli* has also been demonstrated in four different mammalian cell lines. Human ileocecal adenocarcinoma (HCT-8), epithelial carcinoma of lung (A549), Madin-Darby bovine kidney (MDBK), and African green monkey kidney (VERO) were exposed *in vitro* to *I. belli* sporozoites, which had been isolated from the feces of HIV-AIDS patients. Parasites invaded all the cellular types after exposure and multiplication was demonstrated after 24 h. More merozoites formed in VERO cells, followed by HCT-8. In the MDBK and HCT-8 cells, the parasitophorous vacuole was less evident and immobile merozoites were observed in the cytoplasm. In VERO cells, parasitophorous vacuoles contained up to 16 motile sporozoites. However, no oocysts were found in any of the cell types used (Oliveira-Silva *et al.*, 2006).

6.1 Key points – laboratory diagnosis

1. The oocysts are more easily recovered and identified when wet preparations are examined (direct smear, concentration smear), rather than the permanent stained smear (trichrome, iron-hematoxylin).
2. Oocysts recovered in a concentrated sediment from polyvinyl-alcohol-preserved stool may be very difficult to detect (the oocyst wall is very difficult to see), while oocysts concentrated from formalin-preserved specimens are much easier to see microscopically.
3. A biopsy can be positive while no organisms are seen in the stool. This is not necessarily due to poor-quality laboratory work but may reflect normal findings related to the life cycle.
4. Although these organisms can be stained with auramine-rhodamine stains, they should be confirmed by wet smears or modified acid-fast stains.

7 Treatment

Effective eradication of the parasites has been achieved with cotrimoxazole, trimethoprim-sulfamethoxazole, pyrimethamine-sulfadiazine, primaquine phosphate-nitrofurantoin, and primaquine phosphate-chloroquine phosphate (Abramowicz, 1995; Beaver *et al.*, 1984). Other drugs proven to be ineffective include dithiazanine, tetracycline, metronidazole, phanquone, and quinacrine hydrochloride. The drug of choice is trimethoprim-sulfamethoxazole, which is classified as an investigational drug for treatment of this infection. TMP (160 mg) -SMX (800 mg) is given every 6 h for 10 days and then twice a day for 3 weeks. In patients allergic to sulfonamides, pyrimethamine alone (50–75 mg daily) has cured infections. In immuno-suppressed patients with recurrent or persistent infection, therapy must be continued indefinitely.

A 60-year-old immunocompetent patient with chronic biliary isospori-asis failed to respond to orally administered cotrimoxazole prophylaxis and orally administered treatment with nitazoxanide, a 5-nitrothiazole benzamide compound. Severe malabsorption was thought to be responsi-ble for the subtherapeutic levels of nitazoxanide in plasma and bile, result-ing in treatment failure (Bialek *et al.*, 2001). Intravenously administered cotrimoxazole stopped the shedding of *I. belli* oocysts in bile within 5 days, thus excluding suspected resistance to cotrimoxazole. Patients with malabsorption and cholangitis due to coccidia such as *I. belli* and *Cryptosporidium* spp. or due to microsporidial infections often fail to respond to therapy.

In a 2-year-old child, who was a known case of systemic vasculitis receiv-ing prolonged corticosteroids therapy, a case of severe debilitating diarrhea due to isosporiasis was reported. The infection was refractory to treatment with dihydrofolate reductase inhibitor combined with sulfonamide such as cotrimoxazole. This is another clinical case where an immunocompro-mised patient failed to respond to therapy to which isosporiasis usually responds well (Malik *et al.*, 2005).

8 Epidemiology and Prevention

Although *I. belli* tends to be more common in tropical areas of the world, isosporiasis is emerging as a major infection in immunocompromised patients, particularly those infected with HIV. However, the incidence in AIDS patients with chronic diarrhea varies tremendously. In striking con-trast to *Cryptosporidium* spp. finding *I. belli* in children with AIDS is rare.

I. belli is thought to be the only species of *Isospora* that infects humans, and no other reservoir hosts are recognized for this infection. Transmission is through ingestion of water or food contaminated with mature, sporu-lated oocysts. Sexual transmission by direct oral contact with the anus or perineum has also been postulated, although this mode of transmission is probably much less common.

The oocysts are very resistant to environmental conditions and may remain viable for months if kept cool and moist; oocysts usually mature within 48 h following stool evacuation and are then infectious. It has been speculated that diagnostic methods for laboratory examinations may tend to miss the organisms when they are present. Since transmission is via the infective oocysts, prevention centers on improved personal hygiene meas-ures and sanitary conditions to eliminate possible fecal-oral transmission from contaminated food, water, and possibly environmental surfaces.

9 References

Abramowicz, M. (ed) (1995) Drugs for parasitic infections. *Med. Lett.* **37**: 99–108.

Beaver, P.C., Jung, R.C. and Cupp, E.W. (1984) *Clinical Parasitology*, 9th Edn. Lea & Febiger, Philadelphia, PA.

Bialek, R., Overkamp, D., Rettig, I. and Knobloch, J. (2001) Case report: nitazoxanide treatment failure in chronic isosporiasis. *Am. J. Trop. Med. Hyg.* **65**: 94–95.

Certad, G., *et al.* (2003) Isosporiasis in Venezuelan adults infected with human immunodeficiency virus: clinical characterization. *Am. J. Trop. Med. Hyg.* **69**: 217–222.

Cranendonk, R.J., *et al.* (2003) *Cryptosporidium parvum* and *Isospora belli* infections among patients with and without diarrhea. *East Afr. Med. J.* **80**: 398–401.

De Otazu, R.D., *et al.* (2004) Endometrial coccidiosis. *J. Clin. Pathol.* **57**: 1104–1105.

Franzen, C., *et al.* (2000) Taxonomic position of the human intestinal protozoan parasite *Isospora belli* as based on ribosomal RNA sequences. *Parasitol. Res.* **86**: 669–676.

Frenkel, J.K., *et al.* (2003) *Isospora belli* infection: observation of unicellular cysts in mesenteric lymphoid tissues of a Brazilian patient with AIDS and animal inoculation. *J. Eukaryot. Microbiol.* **50** (Suppl): 682–684.

Garcia, C., *et al.* (2006) Intestinal parasitosis in patients with HIV-AIDS. *Rev. Gastroenterol. Peru* **26**: 21–24.

Garcia, L.S. (2007) *Diagnostic Medical Parasitology*, 5th Edn. ASM Press, Washington, DC.

Garcia, L.S., Shimizu, R.Y. and Deplazes, P. (2003) Specimen collection, transport, and processing: parasitology. In: *Manual of Clinical Microbiology*, 8th Edn (eds P.R. Murray, E.J. Baron, J.H. Jorgensen, M.P. Pfaller and R.H. Yolken). ASM Press, Washington, DC.

Greenson, J.K., *et al.* (1991) AIDS enteropathy: occult enteric infections and duodenal mucosal alterations in chronic diarrhea. *Ann. Intern. Med.* **114**: 366–372.

Guk, S.M., *et al.* (2005) Parasitic infections in HIV-infected patients who visited Seoul National University Hospital during the period 1995–2003. *Korean J. Parasitol.* **43**: 1–5.

Lebbad, M., *et al.* (2001) Intestinal parasites in HIV-2 associated AIDS cases with chronic diarrhoea in Guinea Bissau. *Acta Trop.* **80**: 45–49.

Malik, S., *et al.* (2005) Refractory isosporiasis. *Indian J. Pediatr.* **72**: 437–439.

Michiels, J.F., *et al.* (1994) Intestinal and extraintestinal *Isospora belli* infection in an AIDS patient. *Pathol. Res. Pract.* **190**: 1089–1093.

National Committee for Clinical Laboratory Standards (1997) *Procedures for the Recovery and Identification of Parasites from the Intestinal Tract, Approved*

Guideline, M28-A. National Committee for Clinical Laboratory Standards, Villanova, PA.

Oliveira-Silva, M.B., *et al.* (2006) *Cystoisospora belli*: in vitro multiplication in mammalian cells. *Exp. Parasitol.* May 2: 114(3):189–92.

Pape, J.W., Verdier, R.I. and Johnson, W.D. (1989) Treatment and prophylaxis of *Isospora belli* infection with the acquired immunodeficiency syndrome. *N. Engl. J. Med.* **320**: 1044–1047.

Resiere, D., Vantelon, J.M., Bouree, P., Chachaty, E., Nitenberg, G. and Blot, F. (2003) *Isospora belli* infection in a patient with non-Hodgkin's lymphoma. *Clin. Microbiol. Infect.* **10**: 1065–1067.

Rigo, C.R. and Franco, R.M. (2002) Comparison between the modified Ziehl-Neelsen and acid-fast-trichrome methods for fecal screening of *Cryptosporidium parvum* and *Isospora belli. Rev. Soc. Bras. Med. Trop.* **35**: 209–214.

Sarfati, C., *et al.* (2006) Prevalence of intestinal parasites including microsporidia in human immunodeficiency virus-infected adults in Cameroon: a cross-sectional study. *Am. J. Trop. Med. Hyg.* **74**: 162–164.

Silva, C.V., *et al.* (2005) Intestinal parasitic infections in HIV/AIDS patients: experience at a teaching hospital in central Brazil. *Scand J Infect Dis.* **37**(3): 211–215.

Siripanth, C., *et al.* (2004) Development of *Isospora belli* in Hct-8, Hep-2, human fibroblast, BEK and Vero culture cells. *Southeast Asian J. Trop. Med. Public Health* **35**: 796–800.

Velasquez, J.N., *et al.* (2001) Isosporosis and unizoite tissue cysts in patients with acquired immunodeficiency syndrome. *Hum. Pathol.* **32**: 500–505.

B4 *Babesia microti*

Jeremy S. Gray and
Louis M. Weiss

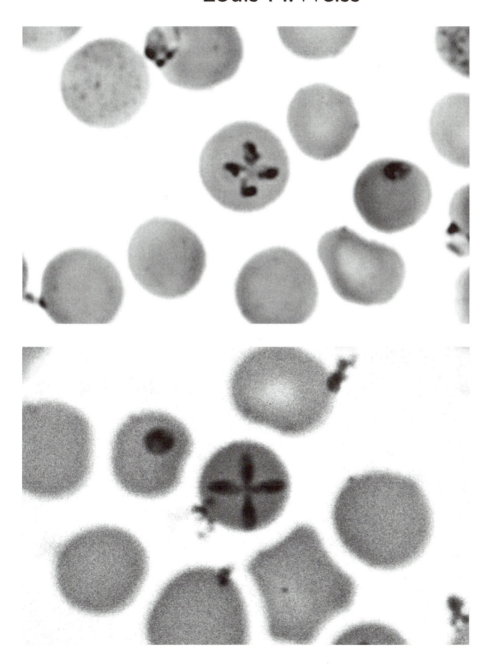

1 Introduction

Babesia microti was first described in an African mongoose by Franca (1908) under the generic name *Smithia*. Franca was subsequently credited as the authority in 1910 and the organism assigned to the genus *Babesia*, which was defined by Starcovici in 1893. In the following years similar organisms were observed and described in a range of hosts, particularly rodents, under a range of names. Killick-Kendrick (1974) recommended that all similar parasites of rodents be assigned to the genus *Babesia* until more data become available for reclassification of the group. In time most rodent intraerythrocytic piroplasms came to be referred to as *B. microti* and they are found in a large range of host species throughout the world. Scientific interest in these parasites increased enormously when the first human case caused by *B. microti* was reported (Western *et al.*, 1970).

The morphology of the parasites in stained thin blood smears is variable, but typically they appear as ring forms that may be round, oval or amoeboid, with a small nucleus and conspicuous vacuole, usually with chromatin extending round the margins, except when the parasite is degenerating when it has a punctate appearance. Less frequent budding forms may be seen and also tetrads resembling a Maltese cross, which have been said to be characteristic of *B. microti* (Figure 1A), though they also occur in *B. equi* in horses, *B. gibsoni* in dogs and also in other species such as *B. divergens* when in abnormal hosts such as humans or gerbils (Figure 1B). In early stages of the infection most erythrocytes contain one parasite, but as the parasitemia nears its peak, multiinfected erythrocytes become more common, and budding and Maltese-cross forms are also seen more frequently. In most strains the parasites occupy a central position in the erythrocyte, but some occur on the margins (accolé position). This difference may sometimes be due to differences in host erythrocytes but is also seen in erythrocytes of the same laboratory host (Tsuji *et al.*, 2001) and even when the two strains are very closely related as in the case of GI and HK strains in gerbils (*Meriones unguiculatus*; Figure 2).

B. microti is transmitted by various species of ixodid ticks and so far all have been found to belong to the single genus *Ixodes*. The ticks acquire the infection by ingesting the parasites with their blood meal, which then multiply in the gut wall as the tick digests the blood and develops to the next stage. The parasites then invade other organs in the developing ticks and are found in the salivary glands of the newly molted tick, ready for transmission to a new host.

2 Taxonomy

Babesia parasites were first recognized in Romanian cattle in 1888 by Babes in which they were observed to cause hemoglobinuria and fever. They were subsequently classified as apicomplexan parasites and assigned to the suborder Piroplasmidea and family Babesiidae on the basis of their exclusive

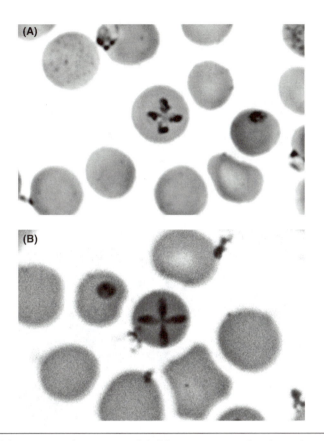

Figure 1 Maltese-cross forms in gerbil (*Meriones unguiculatus*) erythrocytes: (A) *Babesia microti*, (B) *Babesia divergens*.

invasion of erythrocytes, multiplication by budding rather than schizogony, lack of hemozoin and transmission by ticks. In 1893, Smith and Kilbourne established the principle of transmission of arthropod-borne infections to vertebrates when they reported that the cause of Texas cattle fever, a parasite they named *Pyrosoma*, later changed to *Babesia bigemina*, was tick-transmitted. Over 100 species have now been identified, infecting many mammalian and some avian species. Traditionally, babesias were grouped on the basis of their morphology, host and vector specificity, and susceptibility to drugs, but in recent years molecular phylogenetic analyses have proved useful in clarifying a confused situation, sometimes resulting in the emergence of new groups. One of the outcomes of such analyses has been confirmation of the previously expressed view that *B. microti* is only distantly related to the 'true' babesias such as *B. bigemina*, *B. bovis* and *B. divergens*, that are best known as parasites of domestic livestock (Figure 3). In a seminal study on *B. microti* by Goethert and Telford (2003), DNA analyses were performed on fragments of the 18S rRNA and β-tubulin genes of parasites collected in several countries (USA, Switzerland, Spain, Russia),

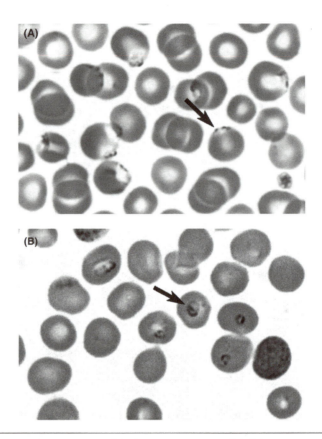

Figure 2 Morphology of related *Babesia microti* in gerbil (*Meriones unguiculatus*) erythrocytes: (A) American GI, (B) European HK. Central and accolé-positioned parasites are indicated by arrows.

from a variety of vertebrate hosts (humans, voles, woodmice, shrews, foxes, skunks, raccoons, dogs) and from ticks. Their analyses resulted in identification of three separate clades and they concluded that *B. microti*, long regarded as a single species, consists of a genetically diverse species complex. Clade 1 contained mostly rodent parasites and also the majority of strains thought to be zoonotic. Clade 2 contained carnivore parasites, and Clade 3 contained rodent parasites that are probably not zoonotic (Figure 4). A separate comparison (Gray, 2006) of 18SrDNA sequences deposited in GenBank demonstrated that a European rodent isolate, 'Munich' (GenBank AB071177) is clearly distinct from any of the Goethert and Telford rodent isolates (98.6% homology versus GI) and more recent studies have shown that the Munich strain is identical to isolates from ticks (Pieniazek *et al.*, 2006) and *Microtus* spp. (Sinski *et al.*, 2006) from Poland. In Japan, the first strain to be characterized (Kobe) was isolated from a human case in the central region of the country (Saito-Ito *et al.*, 2000). Another Japanese strain (Hobetsu) was found to be much more widespread than the Kobe strain.

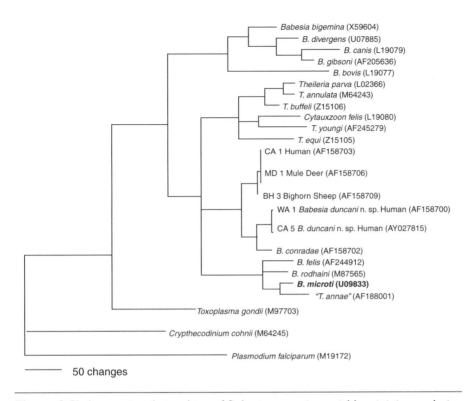

Figure 3 Phylogenetic relationships of *Babesia* spp. using neighbor-joining analysis of the 18SrRNA gene (modified from Kjemtrup *et al.*, 2006).

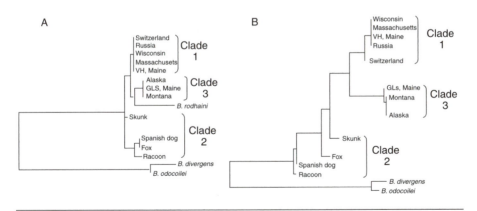

Figure 4 Phylogenetic relationships of *Babesia microti* strains using maximum-likelihood analysis of: (A) 18SrRNA gene, (B) β-tubulin gene (modified from Goethert and Telford, 2003).

Neither was closely related to the American zoonotic strains and both seemed to be especially associated with the large Japanese field mouse, *Apodemus speciosus* (Tsuji *et al.*, 2001). A third Japanese strain, referred to as a 'US-type', is closely related to the American zoonotic strains, according to β-tubulin gene homology, and has been found in rodents in a limited region in Hokkaido, northern Japan (Zamoto *et al.*, 2004a). This strain also occurs in South Korea, Vladivostok in Russia, and Xinjiang in China, and appears to utilize a wider range of hosts, including insectivores, than the Hobetsu and Kobe strains (Zamoto *et al.*, 2004b). The taxonomic positions of zoonotic strains of *B. microti* are considered in later sections (sections 6 and 11.5).

The rather loosely constituted Clade 2 group of Goethert and Telford's 2003 study are all parasites of carnivores, and the recently described *B. microti*-like parasite of dogs in Spain (GenBank AY534602, *Theileria annae?*) clearly belongs to this group since it shows 100% 18S rDNA homology to a Clade 2 parasite (GenBank AY144702) from a fox in Cape Cod, MA, USA. The assigning of this Spanish dog parasite to the genus *Theileria* by Camacho *et al.* (2001) emphasizes the lack of certainty in the classification of *B. microti*-like parasites.

3 Life Cycle

Theileria spp., also tick-transmitted, are distinguished from *Babesia* spp. by their development in host lymphocytes in the initial phase of infection and in view of the taxonomic proximity of *B. microti* to the theilerias there has been some speculation that *B. microti* also undergoes intralymphocytic development, as has been demonstrated for the horse babesia *B. equi*, now classified as a theileria by some authorities. Although there is one report of lymphocyte invasion by *B. microti* (Mehlhorn and Schein, 1987) this unpublished observation was never verified or repeated. It is therefore assumed that when sporozoites are inoculated into the host by the feeding tick they invade the erythrocytes within a matter of hours. Penetration of erythrocytes is an active process consisting of contact with the erythrocyte membrane, orientation of the parasite's apical pole so that apical organelles come into apposition with the cell membrane, membrane fusion between parasite and erythrocyte, invagination of the erythrocyte membrane, and entry of the parasite. The erythrocyte membrane is not disrupted by this process and initially encloses the parasite but disappears soon afterwards, so that the parasite is in direct contact with the erythrocyte cytoplasm (Mehlhorn and Schein, 1987). The parasite is now described as a trophozoite and multiplies by a process of budding resembling binary fission, each time producing two or four merozoites that invade new erythrocytes, resulting in a rapidly increasing parasitemia. The next stage in the life cycle is infection of the vector tick by these blood stages.

Several studies have been conducted on babesia development in the tick vector and *B. microti* was the subject of one of the most detailed, which was also the first to provide firm evidence of the sexuality of babesia parasites (Rudzinska *et al.*, 1983). Electron microscopy revealed that there are two forms of intraerythrocytic parasite in circulating erythrocytes. In one, representing the great majority, differentiation of the cytoplasm occurs resulting in the formation of new organelles, whereas in the other the cytoplasm remains undifferentiated, though the parasite grows larger and becomes coiled and twisted (Rudzinska *et al.*, 1979). These organisms are believed to be the gametocytes and no development takes place until they reach the gut of the tick. About 10 h after the commencement of larval feeding, erythrocytes containing parasites can be found in the tick gut. The majority of the parasites degenerate and disappear but some differentiate and form a cytosome and microtubules that extend to the outside of the main body. By the time the ticks are replete and have detached, these developing organisms have a structure resembling an arrowhead at one end. These are the 'StrahlenKorper', first described by Koch (1906) and suspected of being either gametes or degenerating forms. Fusion (syngamy) takes place between apparently identical organisms resulting in formation of the zygote, which penetrates the peritrophic membrane and becomes internalized by cells of the gut epithelium; the arrowhead disappears and the zygote becomes spherical. Elongate primary kinetes are produced in these gut cells and secondary kinetes then invade the hemolymph and penetrate other organs, including cells of the salivary glands where the parasites form a sporoblast meshwork occupying the whole cell. Unlike most other babesias, *B. microti* does not invade the ovaries, so that vertical transmission is absent. No further development in the salivary glands takes place until the nymphal stage begins feeding on a new host, whereupon large numbers of sporozoites, stimulated by the increase in temperature, are formed by budding and are inoculated into the host with the saliva (Karakashian *et al.*, 1986). The life cycle of *B. microti* is summarised in Figure 5.

4 Vector Relations

The vectors of all the variants of *B. microti* studied so far are hard ticks (Ixodidae) of the genus *Ixodes* (Table 1). The majority of *Ixodes* spp. are endophilic (nidicolous) and inhabit the burrows or nests of small animals or birds where the unfed stages wait for the host. They engorge on the host for a few days and then drop off and, after locating a suitable microenvironment with a humidity of approximately 80%, develop to the next stage, or in the case of females, lay eggs. There are three instars: the larvae that emerge from the eggs, the nymphs that develop from larvae and finally the adult females and males that develop from nymphs (Figure 6). The males rarely feed and never engorge. In members of the *I. ricinus* species complex, such as *I. persulcatus*, *I. pacificus*, *I. ricinus* and *I. scapularis*, the ticks adopt an exophilic (campestral) life style in which the unfed stages occur in the open and ascend the vegetation to ambush passing hosts. These are the species that are important as vectors of disease because they have ready access to humans and livestock. The larvae tend to feed on small animals

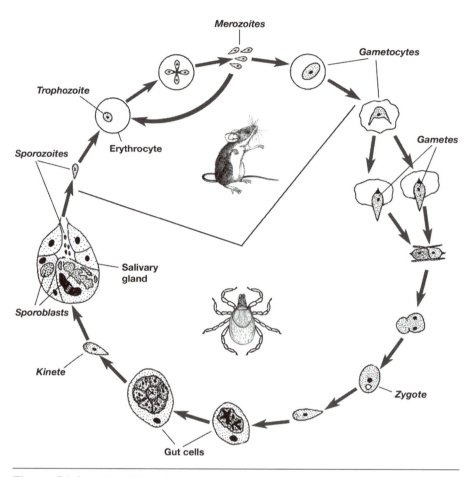

Figure 5 Life cycle of *Babesia microti* (mouse image courtesy of Canterbury Environmental Education Centre, Kent, UK).

such as mice and voles, the nymphs on medium-sized animals such as squirrels and birds, and the adult females on large animals such as deer, cattle and sheep (Figure 7). The vectors of *B. microti* include both endophilic species such as *I. trianguliceps*, that have no significance in transmission of disease to humans, but may be important in maintaining infection in reservoir host populations, and exophilic species such as *I. scapularis*, that are the major vectors of human babesiosis. It has been speculated that an endophilic species, *I. muris*, may have maintained *B. microti* in New England, USA, in the recent past, but has now been displaced by *I. scapularis* (Spielman *et al.*, 1984), largely as a result of a rapidly expanding deer population, deer being the maintenance hosts for *I. scapularis*.

Ixodes scapularis, the deer or black-legged tick, is the vector of the zoonotic American strain of *B. microti*. *I. scapularis* was formerly known as *I. dammini* and is still referred to by that name by some authorities. It was

Table I *Ixodes* **spp. vectors and principal rodent reservoir hosts of** *Babesia microti*

Tick Species	Main Reservoir Host	Location	Original Reference
I. angustus	*Microtus oeconomus* (Alaskan vole)	Alaska, USA	Fay and Rausch, 1969
I. eastoni	*Microtus montanus* (montane vole)	Wyoming, USA	Watkins *et al.*, 1991
I. ricinus	*Microtus agrestis* (meadow vole)	Europe	Walter, 1984
I. ovatus	*Apodemus speciosus* (Japanese field mouse)	Japan	Tsuji *et al.*, 2001
(*I. persulcatus*	Unknown (ticks only)	Russia	Alekseev *et al.*, 2003)
I. scapularis (dammini)	*Peromyscus leucopus* (white-footed mouse)	New England, mid-west, USA	Spielman, 1976
I. spinipalpis	*Microtus ochrogaster* (prairie vole)	Colorado, USA	Burkot *et al.*, 2000
I. trianguliceps	*Clethrionomys glareolus* (bank vole)	Europe	Hussein, 1980

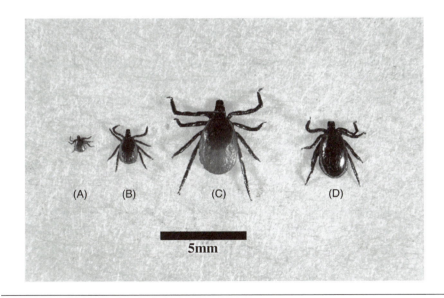

Figure 6 *Ixodes ricinus* L., a vector of *Babesia microti* in Europe – unfed stages: (A) larva, (B) nymph, (C) adult female, (D) adult male.

identified as the vector by Spielman (1976) as a result of transmission experiments with the natural reservoir *Peromyscus leucopus* (white-footed mouse), and with a human isolate of the parasite.

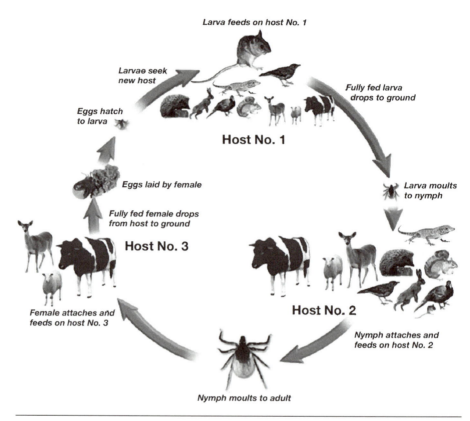

Figure 7 Life cycle of *Ixodes ricinus* complex ticks, with size of host indicating relative importance for each tick stage. Several other *Ixodes* spp. vectors of *B. microti* are restricted entirely to rodent hosts.

Telford (1998) made a strong epidemiological argument for the retention of the specific name *dammini* for the northern populations of *I. scapularis* and this may be lent some support by the apparent absence of any records of *B. microti* in southern populations of *I. scapularis*. There are also no records of any babesias being transmitted naturally by *I. pacificus*, a close relative of *I. scapularis* from the west coast of the USA, and the vector of *Borrelia burgdorferi* and possibly *Anaplasma phagocytophilum* in that region (Fritz *et al.*, 2005; Lane *et al.*, 2001). *I. pacificus* appears to be a competent vector of *B. microti* in the laboratory (Oliveira and Kreier, 1979), as is *Dermacentor andersoni* though sparingly (Genga and Kreier, 1976). However, the vectors of western USA human babesiosis, including those cases caused by *B. duncani* are unknown. The observations of Wilson and Chowning (1904) (cited by Kjemtrup and Conrad, 2000) on *Babesia* spp. in the blood of Rocky Mountain spotted fever (RMSF) patients, and their claims of tick transmission, were not followed up once the cause of RMSF was found to be a rickettsia.

Another American tick that has been firmly identified as a vector of *B. microti* is *I. spinipalpis* in Colorado (Burkot *et al.*, 2000), where the reservoir host is the prairie vole, *Microtus ochrogaster*. This tick species does not bite man and there are no indications that the strain of *B. microti* involved is zoonotic. Other American species incriminated as vectors include *I. angustus* and *I. muris* (Goethert *et al.*, 2003). *I. angustus*, which belongs to the subgenus *Pholeoixodes*, was reported to transmit a *B. microti*-like organism to Alaskan voles (Fay and Rausch, 1969). It was shown recently that this parasite, though morphologically indistinguishable, is genotypically distinct from the American zoonotic strains (Goethert *et al.*, 2006).

It appears that European *B. microti* is transmitted by two very different tick species. The first is *I. trianguliceps*, which like *I. angustus*, also a member of the *Pholeoixodes* subgenus, is a specialist rodent tick that rarely if ever bites man and is probably responsible for the transmission of *B. microti* throughout Europe, including Great Britain, Poland, Sweden, and Russia. The bank vole, *Clethrionomys glareolus*, is probably the main reservoir host. The second, *I. ricinus*, which belongs to the *Ixodes* subgenus, as does *I. scapularis*, was identified as a vector of *B. microti* in Germany by Walter (1984). The field vole, *Microtus agrestis*, was identified as the natural reservoir, though *C. glareolus* could also be infected in the laboratory. *B. microti* transmission by *I. ricinus* using Mongolian gerbils was confirmed by Gray *et al.* (2002) and the parasite has been detected in *I. ricinus* specimens collected from vegetation in Slovenia (Duh *et al.*, 2001), Poland (Pieniazek *et al.*, 2006; Sinski *et al.*, 2006; Skotarczak and Sawczuk, 2003), Switzerland (Casati *et al.*, 2006; Foppa *et al.*, 2002), Hungary (Kalman *et al.*, 2003), southern Germany (Hartelt *et al.*, 2004), and the Czech Republic (Rudolf *et al.*, 2005). In most cases the investigators focused on nymphs, which is the principle instar parasitizing humans (Robertson *et al.*, 2000), but adults were also found to be infected. Overall infection rates (nymphs plus adults) in these studies varied from 0.9% in Hungary (Kalman *et al.*, 2003) to 13% in Poland (Skotarczak and Sawczuk, 2003), which may be due to differences in laboratory procedures but are more likely to indicate differing strains or genotypes of the parasite. At least two different strains seem to be transmitted by *I. ricinus*. Identical strains to the HK strain, that was the subject of transmission experiments (Gray *et al.*, 2002; Walter, 1984), were found in ticks from Berlin and Switzerland, and a distinctly different strain was detected in *I. ricinus* in Poland (Pieniazek *et al.*, 2006; Sinski *et al.*, 2006) that is identical to a Munich mouse isolate. The natural reservoirs of this latter parasite in Poland appear to be *Microtus* spp. (Sinski *et al.*, 2006).

Ixodes persulcatus of Eurasia and temperate Asia shares the capability with *I. scapularis* and *I. ricinus* of transmitting Lyme borreliosis, but there is very little information on its role as a vector of babesiosis. One PCR study found 7 out of 738 adult ticks collected in the St Petersburg area in Russia to be infected with *B. microti*, in each case accompanied by another pathogen such as *Borrelia* spp., *Ehrlichia* spp. or tick-borne encephalitis virus (Alekseev *et al.*, 2003). Another study in Western Siberia did not detect *B. microti* (Rar *et al.*, 2005). The low prevalence may be due in part to the fact that only adults were analysed. Nymphs are more likely to be infected than adults, but since *I. persulcatus* nymphs are endophilic, they are difficult to collect by blanket-dragging.

There has been considerable interest in the biology of *B. microti* in Japan since the report of an autochthonous human case transmitted by blood transfusion (Saito-Ito *et al.*, 2000), but the only vector implicated in transmission of *B. microti* in Japan is *I. ovatus*, specimens of which were found to have infected salivary glands (Saito-Ito *et al.*, 2004; Yano *et al.*, 2005), although neither of the two types of *B. microti* that were detected matched the zoonotic (Kobe) strain.

Transmission studies that were conducted on *I. scapularis* (Piesman *et al.*, 1986; Spielman *et al.*, 1984), *I. ricinus* (Gray *et al.*, 2002; Walter and Weber, 1981) and *I. trianguliceps* (Randolph, 1995) established that *B. microti* is generally acquired by larval ticks and transmitted by nymphs. These laboratory studies are supported by epidemiological data from New England where Piesman *et al.* (1987) established that the nymphal stage is responsible for most transmission to humans. Infections acquired by larvae do not persist through to the adult instar, but adults may be infected as a result of nymphal acquisition of infection. Although this has epidemiological significance for zoonotic babesiosis, it is not relevant to the maintenance of the infection in nature, since adults do not feed on rodents and large animals such as deer are not susceptible to *B. microti* (Piesman *et al.*, 1979). Transovarial transmission does not occur, so unfed larvae are not infected (Gray *et al.*, 2002; Spielman *et al.*, 1984; Walter and Weber, 1981). Oral transmission has been documented twice (Konopka and Sinski, 1996; Malagon and Tapia, 1994) and although apparently a minor feature of transmission, may contribute to the force of infection of small rodents through cannibalism or ingestion of engorging nymphal ticks groomed from accompanying animals.

Two studies have investigated the effects of the parasite on the tick itself. No pathogenic effects were found but increased larval weights and molting success were reported for *I. trianguliceps* (Randolph, 1991). Hu *et al.* (1997) reported similar effects on *I. scapularis* and suggested that a mutualistic relationship between the parasite and the vector exists, mediated by immunosuppressive effects on the host and resulting in enhanced survival of both species.

5 Host Relations

5.1 Course of infection

The course of infection of *B. microti* has been well documented in laboratory animals, particularly mice, hamsters and gerbils. Tick-induced infections generally take 7–14 days to become patent followed by a period of exponential parasite multiplication, at the end of which parasitemias often exceed 10%. A slower growth period then ensues as the supply of erythrocytes dwindles and host resistance mechanisms attempt to bring the infection under control. Parasitemia may climb to as high as 80%, with many erythrocytes containing more than one parasite and exoerythrocytic parasites can be seen. At this point in the infection animals will be severely anemic and complications such as disseminated intravascular coagulation may occur, often resulting in death. The course of infection and outcome will vary according to the strain of parasite, and species and strain of host. Hamsters and gerbils tend to be much more susceptible than laboratory mice, and intraperitoneal injections of infected erythrocytes often cause acute and fatal infections. In early passages in a new host *B. microti* may have difficulty in establishing, but in subsequent infections sometimes goes through a highly virulent phase before adapting to the host to the point of permitting survival following high parasitemias (Cullen and Levine, 1987; Gray and Pudney *et al.*, 1999; Ike *et al.*, 2005; Lykins *et al.*, 1975; Oz and Hughes, 1996; Wozniak *et al.*, 1996). The tendency for infections of laboratory mice to resolve more readily than those of hamsters and gerbils has meant that mice have been much used for studies on immunological mechanisms, whereas hamsters and gerbils are regarded as more suitable for drug-efficacy studies. In natural hosts such as white-footed mice (*Peromyscus leucopus*; Piesman and Spielman, 1982), meadow voles (*Microtus* spp.; Van Peenen and Healy, 1970; Walter, 1984; Watkins *et al.*, 1991) and bank voles (*Clethrionomys glareolus*; Randolph, 1995; Walter, 1984), acute infections also occur especially if induced by intraperitoneal injections, but in nature most infected animals have low or inapparent parasitemias, having gone through a relatively low-grade acute phase following tick infection. The resulting carrier-phase may last several months, if not for the life of the animal, and even those with no detectable parasitemias in thin blood smears may be infectious for ticks (Telford and Spielman, 1993). This carrier state is responsible for persistence of the pathogen in the environment.

5.2 Resistance mechanisms

In the naive host infecting sporozoites from the tick salivary gland are free in the bloodstream for a short time and are probably not affected by nonspecific resistance mechanisms. Once they have invaded erythrocytes and the parasitemia starts to rise, macrophages and natural killer (NK) cells (i.e., cytotoxic lymphocytes that do not require specific activation) are thought to have a role in limiting the growth of the parasite population (Homer and Persing, 2005). Macrophages produce soluble factors such as tumour necrosis factor alpha (TNF-α), nitric oxide (NO) and reactive oxygen species (ROS). These factors are thought to diffuse into the erythrocyte to have their effect and probably cause intraerythrocytic degeneration of parasites producing the so-called crisis forms that appear after the parasitemia peak, though by this stage of the infection specific immunological mechanisms have also come into play. Macrophages may also control infection by the phagocytosis of opsonized erythrocytes and parasites and NK cells produce IFN-γ: their activity has been correlated with resistance to *B. microti* in inbred strains of mice (Eugui and Allison, 1980). Results from another study suggest that IFN-γ may not be important in resistance early in the infection, since IFN-γ-deficient mice developed a less severe parasitemia and were still able to clear the infection (Clawson *et al.*, 2002), but IFN-γ is critical in acquired immunity (late response) as IFN-γ knockout mice are highly susceptible to re-challenge with *B. microti* (Igarashi *et al.*, 1999) and mice with IFN-γ-receptor deficiency have a higher parasitemia and greater mortality when infected with the WA1 babesia (*B. duncani*; Aguilar-Delfin *et al.*, 2003). Macrophages are also a source of the important regulatory cytokine, interleukin-12 (IL-12), which has been shown to stimulate NK cells to produce IFN-γ in bovine splenic cell cultures following exposure to *B. bovis* antigens (Goff *et al.*, 2006). IL-12 also stimulates the growth and development of T-cells and therefore provides an important link to the immunological mechanisms that come into play in the later stages of a primary infection.

The spleen has been shown in many studies to have an essential role in resisting primary infections of *Babesia* species (reviewed by Zwart and Brocklesby, 1979) and this lymphoid organ is packed with macrophages and NK cells, as well as T- and B-lymphocytes. It has an important role in specific immunity to babesia infections and several studies have demonstrated the protective effect of transferring splenocytes from immune to naive animals (Homer *et al.*, 2000). Other studies have demonstrated that the spleen is not essential for the development of immunity, though in splenectomised mice that have received cells from *B. microti*-immune donors, erythrocytes containing degenerated parasites continued to circulate for much longer than in intact animals. This suggests that a major function of the spleen is to specifically remove infected cells from circulation, probably through a combination of the spleen microcirculation and stimulated phagocytic cell activity (de Vos *et al.*, 1987).

The effective specific immune response to babesia infections appears to be primarily cell-mediated. T-cells have been shown to be essential by experiments that demonstrated fulminating infections in congenitally athymic mice in contrast to the transient parasitemias observed in normal mice (Ruebush and Hanson, 1980a), resistance to infection following adoptive T-cell transfer to severe combined immunodeficient (SCID) mice (Matsubara *et al.*, 1993), and immunity in naive mice following T-cell transfer (Ruebush and Hanson, 1980b). It appears that the CD4+ T-helper cells are mainly involved, since mice that are depleted of these cells by

treatment with anti-CD4 monoclonal antibodies became relatively susceptible to infection with *B. microti* (Igarashi *et al.*, 1999). This study also found
that treatment of immune mice with anti-IFN-γ sera partially reduced protection, which contrasts with the apparent lack of effect of IFN-γ early in
infection (Clawson *et al.*, 2002). However, anti-IL-2, IL-4 and TNF-α monoclonal antibodies did not interfere with immunity (Igarashi *et al.*, 1999)
even though TNF-α is thought to have a role in intraerythrocyte degeneration in primary infections. Mice deficient in CD4+ helper cells experience
a longer duration of parasitemia (Hemmer *et al.*, 2000) and mice depleted
of CD8+ T cells have an increased resistance to babesia infection (Igarashi
et al., 1999).

Antibodies appear to have little direct role in immunity to *B. microti*,
though invading sporozoites are probably vulnerable to antibodies for a
short period before they penetrate erythrocytes. Immune serum transfer to
naive mice conferred a small degree of immunity to the closely related
B. rodhaini (Abdalla *et al.*, 1978), but suppression of B-lymphocyte antibody
production by irradiation had little effect (Zivkovic *et al.*, 1984).

5.3 Co-infection and immunomodulation

Some of the vectors of *B. microti* transmit several other pathogens and
there has been considerable interest in the effects that these co-infections
have on pathogenesis and diagnosis (reviewed by Belongia, 2002; Swanson
et al., 2006). *B. microti* has been shown to depress the ability of mice to
respond immunologically to sheep erythrocytes (Purvis, 1977), and
Randolph (1991) demonstrated that infection of bank voles (*Clethrionomys
glareolus*) with *B. microti* interfered with their development of immunity to
ticks. *B. microti* immunosuppressant effects are thought to be responsible
for the exacerbation of Lyme borreliosis pathology and one of the few
instances of Lyme borreliosis possibly resulting in fatality through spirochetal pancarditis may have been due to concurrent infection with
B. microti (Marcus *et al.*, 1985). Krause *et al.* (1996) presented evidence for
a synergistic pathological effect in a study on patients suffering from concurrent Lyme borreliosis and babesiosis. These patients experienced an
illness with a mixture of symptoms associated with the two diseases that
was longer and more severe than the illness seen in patients with either
Lyme borreliosis or babesiosis alone. Circulating spirochetal DNA was
detected more than three times as often in these co-infected patients.
Experimental evidence for the exacerbation of *B. burgdorferi*-arthritis in
mice through co-infection with *B. microti* was reported by Moro *et al.*
(2002). So far no experiments have been reported suggesting that infection
with *Borrelia burgdorferi* affects *B. microti* infections. However, another
pathogen, *Anaplasma phagocytophilum*, transmitted by *Ixodes* spp. ticks, is
well known to cause immunosuppression in ruminants in Europe
(Woldehiwet, 2006). The strains of *A. phagocytophilum* that infect humans

in the USA have not been associated with immunosuppression in clinical cases, but Holden *et al.* (2005) found that in *B. burgdorferi* infections spirochete numbers increased and antibody levels decreased in mice co-infected with *A. phagocytophilum*. As yet there is no evidence for any interaction of *A. phagocytophilum* with *B. microti* infections, though a co-infection model should be possible since both pathogens can infect laboratory rodents. *Ehrlichia chaffensis*, which is closely related to *A. phagocytophilum* but transmitted by a different tick species, has been reported in a fatal co-infection with *B. microti* in an elderly man, but the nature of the interaction between the two pathogens was not clear (Javed *et al.*, 2001).

The mechanisms involved in *B. microti* immunodepression are not clear, but presumably various regulatory cytokines are important components. It has been suggested (Homer and Persing, 2005) that the balance between Th-1 and Th-2 responses may be fundamental to pathogen immunological interactions. It is known that *B. burgdorferi* infections in mice are controlled by a Th-2 CD4+ T-cell response (Matyniak and Reiner, 1995) and if infection with *B. microti* skews the immune response towards Th-1 T-cell development, susceptibility to *B. burgdorferi* would probably increase. Some support for this hypothesis is available from the finding that *B. microti* infection can slow the rejection of *Trichuris muris* worms (Phillips and Wakelin, 1976), but the converse interaction, where worm infections (*Heligosomoides polygyrus*) were superimposed on *B. microti* infections, did not result in prolonged or heightened *B. microti* parasitemias (Behnke *et al.*, 1999). The reported shift in T-cell development caused by HIV infection may be partially responsible for the chronicity of *B. microti* in HIV-infected patients.

5.4 Persistence of infection

A chronic carrier state in natural hosts of *Babesia* spp. has been recognised for many years (Homer and Persing, 2005). In experimental animals, *B. microti* parasitemias may be maintained for two or more years (Lykins *et al.*, 1975). Natural reservoirs such as the white-footed mouse (*P. leucopus*) can be parasitemic for at least 4 months and probably remain infected for life (Spielman *et al.*, 1981). Less is known about persistence in humans but the fact that more than 50 cases have occurred as a result of blood transfusion transmission (Leiby, 2006) is evidence that asymptomatic persistent infection occurs and PCR analysis of seropositive blood donor samples showed that 10 out of 19 contained *B. microti* DNA (Leiby *et al.*, 2005). Krause *et al.* (1998) investigated parasite persistence over a 5-year period in 24 patients who did not receive antibabesial therapy following acute babesiosis and showed that parasite DNA persisted in blood for a mean of 82 days. One patient had recrudescent disease after 2 years of asymptomatic carriage. It is now generally assumed that asymptomatic individuals can be infected for several years.

The mechanisms that enable *B. microti* to persist in their hosts are not known, but Allred (2003) identified four possible ones: antigenic variation, cytoadhesion/sequestration, host protein binding, and immunosuppression induction through control of host cytokine production. Antigenic variation has only been demonstrated among the babesias in *B. bovis* and *B. rodhaini*, but might be involved in *B. microti* since it has been shown (Lodes *et al.*, 2000) that different alleles of an immunodominant antigen are expressed by *B. microti* infecting patients from different areas and that the gene family involved probably undergoes frequent recombination. Cytoadhesion is probably an important mechanism in *B. bovis*, especially when acting in concert with antigenic variation, but does not seem to occur in *B. microti* (Clarke *et al.*, 2006).

6 Zoonotic Babesiosis

Following the initial record of human babesia infections in RMSF patients in Montana, USA in 1904 (Kjemtrup and Conrad, 2000), it was not until the 1960s that American human babesiosis re-emerged on the west coast of the USA, this time in California (Scholtens *et al.*, 1968). Soon after this the first east coast American case was reported and *B. microti* was incriminated as the causal agent (Western *et al.*, 1970). As additional cases of *B. microti* babesiosis emerged in Wisconsin and Minnesota this parasite became acknowledged as the commonest cause of human babesiosis worldwide with several hundred cases documented, the majority of them in patients with intact spleens (Kjemtrup and Conrad, 2000).

Other species involved in human disease include *B. divergens*, the main cause of European babesiosis, and *B. duncani* (previously called WA-1), the recently identified agent of infections occurring on the west coast of the USA. The first well-documented case of human babesiosis was probably caused by *B. divergens*, though referred to at the time as *B. bovis*, which occurred in 1956 in a splenectomized 33-year-old Yugoslav, who grazed his babesia-infected cattle on tick-infested pastures (Skrabalo and Deanovic, 1957). *B. divergens* is the primary cause of babesiosis in European cattle and approximately 30 human cases have now been reported from France, Britain, Ireland, Spain, Sweden, Switzerland, former Yugoslavia and the former USSR, all in splenectomized individuals and all presenting as acute cases with manifest hemoglobinuria. Early cases were invariably fatal, but mortality rates have been reduced to approximately 40% as a result of increased awareness and the use of more appropriate drugs. *I. ricinus* is regarded as the tick species responsible for the transmission of *B. divergens* to humans, and human isolates of *B. divergens* have been transmitted by *I. ricinus* (Zintl *et al.*, 2003). Recently a related but distinct species has been reported to cause a less acute disease in Europe (Herwaldt *et al.*, 2003a), probably also transmitted by *I. ricinus* and most likely a natural parasite of deer (Duh *et al.*, 2005). Some isolated *B. divergens*-like cases have occurred outside Europe, including the USA, but although very closely related to *B. divergens*, these parasites seem to be distinct (Gray, 2006).

Cases of babesiosis occurring in the western USA in 1968 and 1981, and a further seven cases reported from 1991 to 1994, were recognized as distinct from *B. microti* (Kjemtrup and Conrad, 2000). One of these, which occurred in northern California, was designated CA1, and a second from Washington State as WA1. So far 11 cases caused by similar or identical organisms have been reported, five of them in asplenic patients, one of whom died. The CA1 isolate was found to be unrelated antigenically to *B. microti*, but sera from infected patients were reactive to a small canine babesia from southern California, then thought to be *B. gibsoni* but now renamed *B. conradae* (Kjemtrup *et al.*, 2006). Recently, isolates WA1 and CA5 were characterized in detail and the parasite named *B. duncani* (Conrad *et al.*, 2006). *B. duncani* shows only subtle morphological differences from *B. microti*, but analysis of the 18SrRNA gene clearly demonstrated that, together with isolates from dogs and wildlife, it belongs to a separate phylogenetic group (Figure 3) and was furthermore indistinguishable from two other human isolates, CA6 and WA2-clone 1, though distinguishable from the dog isolate *B. conradae*. The clinical manifestations of the *B. duncani* disease are very similar to those caused by *B. microti* (see below). However, unlike *B. microti*, the reservoir hosts and vectors of *B. duncani* are unknown at present.

Uncharacterized *B. microti*-like babesias in human patients have been recorded in Brazil, China, Egypt, Mexico, South Africa and Taiwan (Gorenflot *et al.*, 1998), and most recently in India (Marathe *et al.*, 2005). Although *B. microti* is widespread in Europe, very few human cases have been confirmed to date. Apart from isolated cases in Taiwan and Japan (Saito-Ito *et al.*, 2000; Shih *et al.*, 1997), the majority of human *B. microti*-babesiosis cases have been reported from the east coast and mid-west of the USA and an extensive literature is now available on clinical manifestations, diagnosis, treatment and epidemiology.

7 Clinical Manifestations

Patients with babesiosis caused by *B. microti* show a wide range of signs and symptoms. About 25% adults and 50% children are asymptomatic or only show very mild 'flu-like' symptoms in cases that may not result in medical consultations and are therefore rarely diagnosed (Krause, 2002). At the other end of the spectrum very severe manifestations may occur in patients who have been splenectomised, are receiving immunosuppressive therapy, or are elderly. These cases may show high fever, chills, night sweats, myalgia, hemolytic anemia, hemoglobinuria and jaundice. Hypotension, hepatomegaly and splenomegaly may also be present. Early diagnosis is essential for such at-risk patients, as highlighted by Herwaldt *et al.* (1995), who reported 3 fatal cases out of 10 in Wisconsin, all of whom were diagnosed incidentally.

A review of 139 hospitalized cases, of which 9 died, in New York State (White *et al.*, 1998), identified the most common symptoms in severe cases as malaise and fever (both 91%), shaking chills (77%), diaphoresis (69%) and nausea (57%). Dark urine indicating hemoglobinuria or jaundice was observed in only 12%, though practically all patients were anemic to some extent. Only 37% were aware of a tick-bite 30 days prior to admission. Most of the patients (73%) were over 55 years of age, half of them (52%) had a history of chronic disease and 12% had been splenectomised. Another 12% had a history of Lyme borreliosis, which can be attributed to the fact that the same tick vector is involved, but increased severity of disease in such co-infections may also be involved (Sweeney *et al.*, 1998). Complications were observed in 43% in which congestive heart failure (25%) and acute respiratory syndrome (18%) were the most frequent.

In another reported series of 34 hospitalized patients with an average age of 53 years in Long Island, New York, life-threatening complications included acute respiratory failure (21%), disseminated intravascular coagulation (18%), congestive heart failure (12%), coma (9%) and renal failure (6%; Hatcher *et al.*, 2001). Death occurred in 9% of these patients. Laboratory findings in acute cases include parasitemias in excess of 10% (sometimes as high as 80% in immunocompromised patients), normocytic normochromic anemia, thrombocytopenia, occasional leucopenia, occasionally atypical lymphocytes, elevated serum transaminases, alkaline phosphatases, lactic dehydrogenase and unconjugated bilirubin, and protein and hemoglobin in the urine. Hemoglobinuria is sometimes frank and urine color may range from pink to black depending on the severity and stage of the hemolytic anemia.

In most respects severe babesiosis resembles fulminant malaria due to *Plasmodium falciparum*, but without any cerebral symptoms that are sometimes characteristic of malaria. A recent study of a severe human babesiosis case confirmed that multiorgan failure in babesiosis was not caused by erythrocyte sequestration, which is a feature of falciparum malaria (Clarke *et al.*, 2006).

8 Pathogenesis

In heavy infections the primary pathological event seems to be hemolysis of erythrocytes by multiplying parasites and this can lead to wholesale destruction of erythrocytes, resulting in hemolytic anemia and jaundice, which has toxic effects on many organs especially the liver and kidneys. The relatively few studies conducted on the pathogenesis of *B. microti* infections in rodent models, show that when high parasitemias occur (i.e. more than 50%) the main pathological changes are a leucocytosis following the parasitemic peak, intravascular hemolysis, glomerulonephritis, hepatitis, pneumonitis, myocarditis and splenomegaly (Cullen and Levine, 1987; Ike *et al.*, 2005; Lykins *et al.*, 1975; Oz and Hughes, 1996). Leucopenia and thrombocytopenia may develop when infections become chronic. Parasitemias do not always relate directly to the degree of anemia, especially in long-standing infections, suggesting that erythrocyte destruction is not only due to hemolysis and the removal of infected cells by splenic and liver macrophages. When host and parasite are well adapted, few symptoms are seen even when parasitemias rise to more than 50% before falling again (Wozniak *et al.*, 1996). Factors affecting susceptibility have been addressed in mouse models and Vannier *et al.* (2004) have shown that age-related adaptive immunity to *B. microti* is genetically determined, which may be relevant to the known higher risk of disease in some elderly patients.

In humans, symptoms occur at parasitemias that are less than 1% and there is now good evidence to suggest that many of these, such as fever, myalgia, renal insufficiency, coagulopathy and hypotension, may be caused by excessive production of pro-inflammatory cytokines (Clark and Jacobson, 1998). Highly elevated levels of cytokines, including TNF-α, IFN-γ, IL-2 and IL-6, have been observed in babesiosis (Shaio and Lin, 1998) and excessive cytokine production is thought to contribute to acute respiratory failure (Boustani *et al.*, 1994). Experimental treatment of human volunteers with TNF-α (Jabukowski *et al.*, 1989) resulted in a remarkably similar range of symptoms to those seen in malaria and human babesiosis. So far no comparative studies have been conducted in experimental animal models on the pathological effects produced by TNF-α, administration and *B. microti*-infection, though experiments have been carried out on the effects of various cytokines, including TNF-α, on immunity to *B. microti* infections in mice (Igarashi *et al.*, 1999).

9 Diagnostic Features

Patient residence in or travel to endemic areas is an important component of accurate diagnosis. Though few patients remember a tick bite, incubation periods have been estimated to be between 1 and 6 weeks (Homer and Persing, 2005); however, the period between infection and appearance of disease may be much longer than this. A long-standing asymptomatic infection may manifest itself following an immunosuppressive event such as splenectomy. Recrudescence following immunosuppression has been demonstrated in animal models (Corrier and Wagner, 1988; Gray and Pudney, 1999; Ruebush *et al.*, 1981) and in one human case symptoms appeared 2 years after presumed infection (Krause *et al.*, 1998).

Acute babesiosis often presents as a medical emergency requiring rapid diagnosis and treatment. The profound changes in the blood picture and in serum enzyme levels, together with general symptoms, are now well documented so that in endemic areas rapid diagnosis of such cases is possible. However, the majority of *B. microti*-babesiosis cases occur in spleen-intact and otherwise healthy patients, and these constitute more of a diagnostic challenge. They are characterised by relatively vague 'flu-like' symptoms including slight fever, myalgia, night sweats, nausea and weight loss, and their symptoms may have been continuing for some months. Unlike other tick-borne infections such as rickettsiosis, Lyme borreliosis and human anaplasmosis, skin rashes are not seen in babesiosis cases, but diagnosis can be complicated by co-infection with other pathogens transmitted by the same tick vectors, resulting in a confusing complex of symptoms (Swanson *et al.*, 2006).

Specific diagnosis is made by detection of parasites by light microscopy in Giemsa-stained blood smears, but this may be difficult because symptoms generally commence before patency and patients rarely have parasitemias of more than 1%. Positive smears usually show ring forms that resemble the trophozoites of *Plasmodium falciparum*, though babesia parasites do not show pigment deposition (hemozoin). *Babesia* species also lack the synchronous stages seen in other *Plasmodium* species. Maltese-cross forms (Figure 1A) are of limited diagnostic value since they are rarely seen in human patients and even in rodent models do not constitute more than 1% of circulating parasites (Yokoyama *et al.*, 2003). Reliable detection of parasites in blood smears requires training and experience. For example, platelets are easily mistaken for parasites and sometimes give rise to the view that extracellular forms occur frequently, which is rarely the case except in high parasitemias. Hemolysis-induced Pappenheimer bodies may also be mistaken for parasites (Carr *et al.*, 1991).

PCR based on the primers developed by Persing *et al.*, (1992) is rapidly becoming the test of choice for confirmation of on-going infection in which *B. microti* has been incriminated by serology. Stringent cleanliness and control procedures are required to avoid false-positives resulting from amplicon contamination, and it is further recommended that PCR products be sequenced for confirmation. The persistence of DNA detectable by PCR correlates with the presence of symptoms and positive serological tests (Krause *et al.*, 1998). An alternative or adjunct confirmatory approach is to inoculate laboratory rodents (usually golden hamsters or gerbils) intraperitoneally with blood samples from patients. The erythrocytes can be concentrated by centrifugation and the rodents may be treated with corticosteroid immunosuppressants in order to reduce the prepatent period, which is generally between 2 and 4 weeks.

Serological diagnosis is appropriate in chronic cases and the main test in use is the indirect immunofluorescence (IFA: IgG) assay. Although interpretation of the assay is inherently subjective and requires a trained microscopist, this assay provides high levels of specificity and sensitivity in the right hands (Krause *et al.*, 1994). Titers greater than 1:64 are considered diagnostic for infection and a titer of more than 1:1024 indicates active or recent infection, though titers may decline to less than 1:64 12 weeks after infection. Immunoblot provides even greater specificity (99%), but is more time consuming and relatively expensive (Ryan *et al.*, 2001). A reliable ELISA would be useful for epidemiological studies involving the handling of a large number of samples or for blood screening, but early tests were limited by nonspecific reactions. A rodent model has recently been used to develop an ELISA with an acceptable level of specificity (95%; Loa *et al.*, 2004) and attempts have been made to identify immunodominant, secreted antigens for the development of diagnostic tests using defined antigens (Homer *et al.*, 2003).

10 Treatment with Antimicrobials

B. microti has long been utilized as a model for livestock babesiosis, primarily in relation to host immunity, and has occasionally been used for drug screening, although this role was usually reserved for the closely related *B. rodhaini* (Beveridge, 1953). However, these parasites are not as susceptible as cattle babesias to standard veterinary babesicides, which lends weight to the molecular data suggesting *B. microti* is taxonomically separate from livestock *Babesia* spp.

When *B. microti* became a drug target as a zoonotic pathogen, clinicians were faced with the difficulty of the relatively poor activity of established babesicides, but also the absence of appropriate drugs licensed for use in humans. Antimalarials such as chloroquine were used unsuccessfully in early cases (Miller *et al.*, 1978) and subsequent drug screening experiments in rodent models (Marley *et al.*, 1997; Wittner *et al.*, 1996) have confirmed that even more recently developed antimalarials, such as mefloquine and artemisinin, have nothing to offer in this regard. The outcomes of attempts to treat human cases with drugs that have some babesicidal activity, such as diminazene or pentamidine, also proved unsatisfactory, partly because of the side effects caused by the doses required to produce clinical improvement, but also because of failure to clear parasites (Francioli *et al.*, 1981; Ruebush *et al.*, 1979). Shortly after these early treatment attempts, quinine, in combination with clindamycin, was used successfully in a transfusion case (Wittner *et al.*, 1982) and the effectiveness of this regimen was confirmed subsequently in experiments in hamsters (Rowin *et al.*, 1982). This drug combination became the treatment of choice over the next 18 years, despite the fact that both drugs, especially quinine, can cause significant side effects at therapeutic levels (Krause, 2002; Weiss, 2002). It also became apparent that the use of this drug regimen is sometimes inadequate for patients with severe babesiosis, especially if they are immunocompromised through splenectomy, immunosuppressant therapy or HIV infection (Falagas and Klempner, 1996; Herwaldt *et al.*, 1995; Ong *et al.*, 1990).

Azithromycin was investigated as a replacement for clindamycin and was found to have promising activity as a monotherapy in hamsters, and efficacy was improved by adding quinine to the regimen (Weiss *et al.*,

1993). In two subsequent human cases, where the use of quinine and clin-damycin had failed to produce a reduction in parasitemia, addition of azithromycin resulted in a satisfactory response (Shaio and Yang, 1997; Shih and Wang, 1998).

Gupta *et al.* (1995) reported that treatment of an immunosuppressed patient with quinine and clindamycin failed to prevent a relapse, but that the parasitemia was brought under control by atovaquone. The antibabesial activity of atovaquone against *Babesia microti* was subse-quently confirmed in hamsters (Hughes and Oz, 1995; Wittner *et al.*, 1996) and gerbils (Gray and Pudney, 1999). Parasitaemias were rapidly reduced by monotherapy, but it proved difficult to clear parasites from the circula-tion and surviving parasites appeared to develop resistance (Gray and Pudney, 1999; Wittner *et al.*, 1996). The addition of azithromycin (Wittner *et al.*, 1996) or clindamycin (Gray and Pudney, 1999) resulted in clearance of parasites from the blood as determined by blood-smear examination, but subsequent immunosuppression with dexamethasone caused recrudescences in atovaquone/clindamycin-treated gerbils.

Since atovaquone/azithomycin in rodent models appeared to provide the best efficacy without accompanying resistance dangers, the perform-ance of this drug combination in humans was compared with that of quinine/clindamycin in a prospective randomized trial involving 58 sub-jects with nonlife-threatening babesiosis in New England, USA (Krause *et al.*, 2000). In both regimens no parasites were detectable in blood smears after 3 weeks and parasite DNA was no longer detectable after 12 weeks. The atovaquone/azithromycin combination showed slightly shorter per-sistence of parasite DNA (11 weeks) than quinine/clindamycin, but this difference was not statistically significant. The symptoms of patients that received atovaquone/azithromycin resolved at least as rapidly as those who received quinine/clindamycin, and whereas four in the quinine/clin-damycin group required hospitalization in order to receive intravenous quinine, none in the atovaquone/azithromycin group were hospitalized. Clearer differences between the two treatment groups were evident when side effects were examined. In the quinine/clindamycin group 72% of sub-jects reported adverse side effects, mostly associated with quinine, whereas only 15% of the subjects in the atovaquone/azithromycin group were affected. Furthermore, side effects in the quinine/clindamycin group were generally more severe.

The combination of azithromycin and atovaquone appears to provide the best treatment option in terms of reduction in parasitemias, minimal side effects and ease of administration without inducing resistance, and is now recommended as the regimen of choice for *B. microti* infection despite relatively high cost (Weiss, 2002; Wormser *et al.*, 2006). Children and neonates with babesiosis can also be treated with atovaquone and azithromycin as an alternative to quinine and clindamycin.

The advantages of atovaquone/azithromycin therapy were clearly demonstrated in two cases involving immunocompromised patients, in which initial quinine/clindamycin therapy failed to control parasitemias (Bonoan *et al.*, 1998; Lux *et al.*, 2003). Supportive treatment for patients with severe babesiosis may include apheresis or erythrocyte exchange transfusion (Bonoan *et al.*, 1998), but it is not known whether it reduces the frequency of complications (Weiss, 2002).

Consideration should also be given to the development of regimens for treatment of *B. microti* patients infected with other tick-borne infections, such as *B. burgdorferi* or *A. phagocytophilum*. Treatment with doxycyline is appropriate if either co-infecting pathogen is found, since this drug is effective against both *B. burgdorferi* and *A. phagocytophilum*, and has also been shown to have a prophylactic effect against *B. canis* (Vercammen *et al.*, 1996).

There are few prospects for the development of new drugs for human babesiosis since it affects a relatively small number of people and there appear to be few other pathogens with similar drug susceptibilities. However, 8-aminoquinoline has shown promise (Marley *et al.*, 1997) and recent *in vitro* studies (Bork *et al.*, 2003) have identified some antifungal agents as potent growth inhibitors of the horse parasite *B. (Theileria) equi*, which is more closely related to *B. microti* than any of the other large animal babesias. Table 2 summarises efficacy data for a range of antimicrobials against *B. microti* in hamsters and gerbils. A variety of procedures and assessment parameters were used with parasites of varying virulence and hosts of varying susceptibility, so the data are presented for consideration within, rather than between, studies. Amongst other things they demonstrate that antimalarials such as chloroquine, mefloquine and artesunate are relatively ineffective; that the babesicide of choice for veterinary use, imidocarb, is less effective than atovaquone; that combined atovaquone and azithromycin eliminates parasites and prevents recrudescence; and that there are compounds that perform well in the laboratory, such as 8-aminoquinoline, but which are unlikely to be developed for use against this relatively rare disease unless other applications can be found.

11 Epidemiology

11.1 Emergence of the pathogen

B. microti first emerged as a human pathogen on the east coast of the USA in 1969 (Western *et al.*, 1970) and was initially known as Nantucket fever after the island of that name off the Massachusetts coast where the first 14 cases occurred. The parasite was identified as *B. microti* from its morphological appearance in blood smears and was subsequently isolated in hamsters and designated the 'Gray' strain (Gleason *et al.*, 1970). Over the next decade more than 40 infections, acquired on off-coast Massachussets islands, were recorded. Although all the initial Nantucket cases occurred in

Table 2 Activity of antimicrobials against *Babesia microti* in laboratory rodents

Compounds	mg kg⁻¹ day⁻¹	Host	Effect	Author
		Gerbil	% Suppression of parasitemia day 3 & 5 pt	Ruebush *et al.*, 1980
4-methyl primaquine	100		99.2/99.9	
Diminazene	25		91.7/99.6	
Diamidine	100		92.0, 93.9/99.3, 99.5	
Pentamidine	10		17.9/63.0	
Chloroquine	100		23.3/0.0	
Tetracycline	300		23.2/2.7	
Clindamycin	300		0.0/9.4	
		Hamster	% Suppression of peak parasitemia	Rowin *et al.*, 1982
Clindamycin	150		49.1	
Quinine/clindamycin	125/150		55.8	
		Hamster	% Suppression of parasitemia day 7 pt	Weiss *et al.*, 1993
Azithromycin	150		98.4	
Azithromycin/quinine	150/250		99.1	
		Hamster	% Suppression of parasitemia day 4 pt	Hughes and Oz, 1995
Atovaquone	80		69.1	
	150		98.5	
	300		99.99	
		Hamster	% Suppression of parasitemia day 8 pi	Wittner *et al.*, 1996
Azithromycin	150		38.6	
Atovaquone	100		99.8	
Atovaquone/ azithromycin	100/150		100.0; no recrudescence	
		Gerbil	Days to clear parasites	Gray and Pudney, 1999
Imidocarb	12 × 1		15.2	
Atovaquone	50 × 1		8.4	
Atovaquone/ clindamycin	50/150		4.0; recrudescence after immunosuppression	

Continued

Table 2 *Continued*

Compounds	mg kg⁻¹ day⁻¹	Host	Effect	Author
		Hamster	% Suppression of parasitemia day 3 pt	Marley *et al.*, 1997
8-aminoquinoline	12.5		100	
Clindamycin/quinine	150/250		22	
Artesunate	20		20	
Mefloquine	250 × 1		21	
Pentamidine	40		0.0	

pt = post-treatment
pi = post-infection

spleen-intact patients, about half these subsequent patients were asplenic. Epidemiological investigations on Nanucket Island revealed that the minimum age of patients with parasitologically confirmed infections was 49, that most cases occurred between July and August, and 5 of 21 patients recorded a tick bite 1–4 weeks before illness (Ruebush *et al.*, 1981b). The black-legged or deer tick *Ixodes dammini* (subsequently reclassified as *I. scapularis*) was incriminated as the vector of *B. microti* by experimental transmission by nymphal ticks to hamsters (Spielman, 1976) and to rhesus monkeys (Ruebush *et al.*, 1981a). Detection of the parasite in tick salivary glands indicated that about 5% of ticks were infected (Piesman and Spielman, 1980). Reservoir hosts were identified as voles (*Microtus pennsylvanicus*) and white-footed mice (*Peromyscus leucopus*; Healy *et al.*, 1976). *P. leucopus* was the most abundant species and its central role as a reservoir of this pathogen was confirmed by Spielman *et al.* (1981) who found that it was more frequently parasitized than other animals on the island and that parasitemias were sustained for as long as 4 months. In a later quantitative experimental study, captured *P. leucopus* were exposed to laboratory-reared larval ticks and the resulting nymphs examined for salivary gland infection. About 30% of the ticks became infected and, although only 25% of the mice were parasitemic by blood smear, virtually all of them were infectious for ticks (Telford and Spielman, 1993). It was soon apparent that an endemic focus of *B. microti* infection occurred along the eastern seaboard, including Massachusetts, Connecticut and Long Island, New York. In 1983 a case was reported from an inland location when an asplenic resident of Wisconsin was diagnosed (Steketee *et al.*, 1985) and cases have also been recorded from Minnesota (Herwaldt *et al.*, 2002; Setty *et al.*, 2003). There is evidence that the pathogen is extending its range southwards with at least 40 diagnoses made in New Jersey over an 8-year period, which strongly suggests that the disease is now endemic in that state (Herwaldt *et al.*, 2003b).

11.2 Prevalence

Several hundred cases have now been recorded in the USA but this is probably an underestimate since babesiosis is not reportable and symptoms may be mild and transient, with a large proportion of infections thought to be asymptomatic. A serosurvey during a single transmission season on Shelter Island, New York showed point prevalence values of 4.4% for June and 6.9% for October, whereas 300 samples from the metropolitan area of New York showed insignificant titers (Filstein *et al.*, 1980). Another serosurvey found that in endemic areas prevalence rates ranged from 2.5 to 10.2%, with higher seroprevalence in subjects that were positive for Lyme borreliosis. Children were found to be infected as often as adults, though they present with clinical symptoms less frequently (Krause *et al.*, 1991). However, a serosurvey for a range of tick-borne infections in 671 individuals at high risk of exposure to ticks, mainly on eastern Long Island, New York found antibodies in only seven people (1%), whereas 13% had antibodies to one or more tick-borne pathogens, including Lyme borreliosis, ehrlichiosis and Rocky Mountain spotted fever (Hilton *et al.*, 1999). A 10-year serosurvey and case-finding study on Block Island, Rhode Island, involving 1487 participants, who were initially seronegative (Krause *et al.*, 2003), demonstrated that approximately 10% seroconverted over the study period and the case incidence was 1.6%, which was not significantly different from the case incidence of Lyme borreliosis (2.1%).

11.3 Neonatal babesiosis

Several cases of babesiosis infection have occurred among neonates (preterm to 1 month old) and eight of them were probably acquired by blood transfusion. This seems a relatively small number considering that approximately 40 adult infections (probably more) have been transmitted by blood tranfusion (Leiby, 2006). It is possible that this is due to incomplete recognition of the disease in the neonatal population (Fox *et al.*, 2006) or to lower susceptibility in this apparently vulnerable population compared with some adult cases. In at least one infected neonate no symptoms occurred and infection was only recognized because this patient was part of a transfusion-transmitted cluster. Another two neonates who received the same infected blood did not develop infections (Dobroszycki *et al.*, 1999). All but one of the neonatal transfusion cases have been caused by *B. microti*, the exception being due to a WA1-type (*B. duncani*) infection (Kjemtrup *et al.*, 2002).

Significant though variable clinical symptoms, including fever, hepatosplenomegaly, and anemia, occurred in most of these infected neonates, but they made uncomplicated recoveries. The maximum parasitemia was 8% and all but one received antibabesial treatment, usually quinine and clindamycin, but also azithromycin and atovaquone in one case and quinine/clindamycin followed by atovaquone in another (Dobroszycki *et al.*, 1999). No adverse effects of treatment were reported.

Two cases of neonatal infection with *B. microti* are presumed to have been caused by tick-bite. One of these had a tick removed 2 weeks before diagnosis and both showed evidence of infection with *B. burgdorferi*, a reliable marker for tick-bite. These infants had relatively low parasitemias (6 and 1% respectively) and made good recoveries following treatment with antibabesials (Fox *et al.*, 2006). A further two neonatal cases are presumed to have been acquired by vertical transmission (Esernio-Jennsen *et al.*, 1987; New *et al.*, 1997). Neither had been exposed to ticks or received blood transfusions and in both cases the mothers had experienced tick-bite and were seropositive for *B. microti*. These babies were born at term and diagnosis was made at 4 and 5 weeks of age when clinical symptoms including anemia and hepatosplenomegaly were seen and parasitemias of 5% and 4% were detected. Both cases responded satisfactorily to antibabesial therapy (quinine/clindamycin in one case and quinine/clindamycin followed by atovaquone in the other).

There are questions about the phagocytic activity of the spleen in neonates and increased susceptibility to other intracellular pathogens has prompted an aggressive therapeutic and prophylactic approach in the management of babesiosis in these patients (Dobroszycki *et al.*, 1999). Exchange transfusion has not been utilised in any published accounts of neonatal babesiosis but there is at least one unpublished report of the use of this therapy (Fox *et al.*, 2006).

11.4 Geographical distribution

The vast majority of *B. microti* cases have been recorded in limited areas in the US (coastal New England and upper Midwestern States). Cases have also occurred in Japan (Saito-Ito *et al.*, 2000; Wei *et al.*, 2001), Taiwan (Shih *et al.*, 1997) and Europe (Humiczewska and Kuzńa-Grygiel, 1997; Nohynkova *et al.*, 2003). Both these European cases appear to have been imported, the first from Brazil into Poland and the second from the USA into the Czech Republic. Although a Swiss case was reported as being autochthonous (Meer-Scherrer *et al.* 2004), doubts have been expressed about the diagnosis (Gray, 2006; Telford and Maguire, 2006). A more recent case with strong supporting laboratory evidence was reported in an immunosuppressed German patient (Hildebrandt *et al.* 2007) and this appears to be the first confirmed case contracted in Europe.

There is good evidence that *B. microti* human infections, if not clinical cases, occur relatively frequently in Europe. Krampitz *et al.* (1986) was the first to demonstrate *B. microti* antibodies in human blood samples in Germany and Foppa *et al.* (2002) reported a seroprevalence of 1.5% in 369 blood donors residing in an area in Switzerland where ticks were abundant. Hunfeld *et al.* (2002), using an IFA with an estimated specificity of 97.5%, detected antibodies to *B. microti* in 12 blood samples collected from 144 individuals (8.3%) residing in the Rhein-Main area of Germany who were clinically or serologically positive for Lyme borreliosis. In another 81 individuals, who had a history of tick-bite, a further 9 (11.1%) positive samples were found, whereas only 2 of 120 blood donors were seropositive (1.7%). The overall seroprevalence of all those with evidence of exposure to ticks was 11.5%.

In many parts of Europe, including the UK, the main vector of *B. microti* appears to be *Ixodes trianguliceps*, a specialist rodent parasite that is not known to bite humans. However, at least one German strain (HK) originally isolated from a field vole, *Microtus agrestis*, is efficiently transmitted by *I. ricinus* in the laboratory (Gray *et al.*, 2002; Walter, 1984) and *B. microti* has now been detected in *I. ricinus* in several European countries including the Czech Republic, Hungary, Slovenia and Switzerland (see section 4).

Since *I. ricinus* is also the vector of Lyme borreliosis, cases of this disease with unusual symptoms are worth investigating for concurrent *B. microti* infection (Gray, 2006; Meer-Scherrer *et al.*, 2004). Similar strategies were adopted by Baumann *et al.* (2003) and Arnez *et al.* (2003), who investigated patients that developed febrile illnesses following tick bites, though no evidence of *B. microti* infections were found in either study.

11.5 Molecular epidemiology

Goethert and Telford (2003) found that, according to 18S rDNA and β-tubulin gene homology, the majority of *B. microti* strains thought to be zoonotic, belong to a single clade (Clade1). This clade included Swiss and Russian isolates, supporting the possibility of the occurrence of zoonotic strains in Europe. It is notable in this context that *I. ricinus* can transmit an American zoonotic strain (GI) to gerbils as readily as it transmits a German strain (HK; Gray *et al.*, 2002). Comparison of 18S rRNA gene sequences of various strains with the GI strain (Table 3) shows that the GI strain is identical to Goethert and Telford's Clade-1 Nantucket strain and that the Swiss and Russian strains are almost identical. Further Swiss strains from ticks (Casati *et al.*, 2006) proved to be identical to Goethert and Telford's Swiss strain and these, together with one from Berlin, were found to be the same as the HK strain. The Russian strain isolated from a vole differed from these European strains by only two base pairs. It is thus apparent that there are several strains in Europe that are very closely related to American zoonotic strains and most of these seem to be transmitted by a known vector of tick-borne zoonoses, *I. ricinus*.

The zoonotic Japanese Kobe strain, although of rodent origin, does not appear to be closely related to the American zoonotic reference strain (GI; Table 3). A Japanese strain (Hobetsu) is more widespread, but is not closely related to the Kobe strain (98.8% 18SrRDNA homology). So far it has only been isolated from rodents, but two isolates were shown to be infective for human erythrocytes in hu-RBC-SCID mice (Tsuji *et al.*, 2001) indicating zoonotic potential. Another rodent parasite showing 100% β-tubulin gene homology with the zoonotic US types, including the GI strain, has recently been identified in northern Japan (Zamoto *et al.*, 2004a) and similar geno-types to this potentially zoonotic Clade-1 parasite have also been found on mainland northeastern Eurasia (Zamoto *et al.*, 2004b).

11.6 Risk factors for infection

The primary risk factor for infection with *B. microti* is a tick bite, particu-larly that of the nymphal stage of *Ixodes scapularis*. Infected *I. scapularis* nymphs are most abundant in the environment in late spring-early sum-mer so the greatest risk of infection occurs when appropriate tick habitat, such as deciduous woodland (or well-vegetated peri-domestic sites), is vis-ited in May or June (Piesman *et al.*, 1987). Nymphs require 36–48 h for the parasites to mature to infectious sporozoites in their salivary glands, so prompt removal is a good preventive measure. However, an engorging nymph only measures 2 or 3 mm in diameter and is difficult to detect. The majority of babesiosis patients do not seem to recall a tick bite in the pre-ceding few weeks (White *et al.*, 1998). Adult ticks are often infected but do not transmit *B. microti* efficiently for approximately 48 h of feeding, so the engorging females are usually large enough to detect and remove before transmission occurs. Other explanations for the relatively low involvement of adults in transmission to humans in the USA are that they are much less numerous than nymphs in the habitat and also that they are active later in the year when the public spend less time in prime tick habitat.

The two other means of transmission of *B. microti* to humans besides tick-bite are transplacental transmission and blood transfusion. Transplacental transmission is rare and only two cases have been reported in the literature (Ersenio-Jenssen *et al.*, 1987; New *et al.*, 1997). There are, however, over 40 reports of transmission by blood transfusion, and in view of the subclinical status of many infected individuals and the persistence of the parasite in blood, this route of infection is a cause for concern (Leiby, 2006). Exact numbers of cases are difficult to determine because the dis-ease is not reportable and furthermore it is likely that new cases are not considered sufficiently noteworthy to publish details. The patients in the published cases range in age from neonates to 79 years of age and the vast majority have occurred in the USA. While transfusion-transmitted cases are becoming more common, few estimates of risk have been made. Five transfusion cases in 1997 led to the identification of three donors, repre-senting three units of blood out of a total of 500 000 units donated at that time. This corresponds to an incidence of 6 infected units per million (Kjemtrup and Conrad., 2000). Two studies in an intense focus of infection in Connecticut estimated the risk of acquiring infection from a unit of ery-throcytes to be 1 in 601 (Gerber *et al.*, 1994) and 1 in 1,800 (Cable and Leiby, 2003). The numbers of infected blood units donated for transfusion are probably increasing and since there are no recognized screening tests, only

Table 3 Homology of selected isolates of *Babesia microti* of zoonotic interest (EMBL-EBI ClustalW analysis of 18S rRNA gene), with reference to the GI strain (GenBank accession number AF231348)

Strain	Accession No.	Base Pairs Compared	% Homology	Source	Author
Nantucket (USA)	AY144722	1255	100.0	Human, tick	Goethert and Telford, 2003
Switzerland	AY144692	1255	99.9	Tick	Goethert and Telford, 2003
HK, Hannover, (Germany)	AB085191	1705	99.9	Vole	Tsuji et al., 2002[a]
Russia	AY144693	1254	99.9	Vole	Goethert and Telford, 2003
Berlin (Germany)	AF231349	1705	99.9	Tick	Zahler et al., 2000[a]
Hobetsu (Japan)	AB050732	1705	99.2	Mouse	Tsuji et al., 2001
Kobe (Japan)	AB032434	1705	99.1	Human, mouse	Saito-Ito et al., 2000

[a]Unpublished GenBank deposition.

experimental procedures to inactivate parasites (Zavizion *et al.*, 2004) and no consensus on developing and licensing such tests and procedures, the feasibility of preventing increased contamination of the blood supply seems uncertain (Leiby, 2006). It has been suggested that people from endemic areas should be prevented from donating blood in the summer months but this would not exclude chronic infections, and disallowing all donations from endemic areas would have an unacceptable impact on the blood supply. At present the only measure in place seems to be to ask intending donors if they have had babesiosis, but since most infections are asymptomatic this cannot prevent infected blood from entering the system (Kjemtrup *et al.*, 2000).

11.7 Risk factors for disease

Among hospitalized patients the case fatality rate seems to be about 5% (Meldrum *et al.*, 1992; White *et al.*, 1998). Relatively advanced age is one of the risk factors for clinical babesiosis, as shown in a study on 139 hospitalized patients, 73% of whom were over 55 years old (White *et al.*, 1998). In another study patients with severe disease were found to be an average of 30 years older than those with mild symptoms or asymptomatic infections (Krause *et al.*, 1998) and in a 10-year cohort study on 1487 subjects, older people (>50 years) were more likely to be hospitalized, though the frequency and duration of symptoms was similar to those in other age groups (Krause *et al.*, 2003). Age has also been identified as a factor in the susceptibility of animals to babesiosis. Inverse age resistance is well known in cattle babesiosis and it has been suggested that in young animals immune responses to the parasite occur earlier and are concentrated in the spleen, a vital organ for efficient control of babesiosis, whereas in adults they are delayed and more systemic, perhaps contributing more to pathology than to resistance (Zintl *et al.*, 2005). Whether similar mechanisms operate against *B. microti* is not known, but juvenile *P. leucopus* were more frequently parasitemic than adults at the end of the transmission season (Spielman *et al.*, 1981) and Habicht *et al.* (1983) reported that older laboratory mice had difficulty clearing parasites compared with juveniles, though peak parasitemias were less. Vannier *et al.* (2004) demonstrated that age-related resistance appeared to be genetically determined in laboratory mice and this may have relevance to human babesiosis.

Epidemiological data do not suggest that there are significant gender differences in frequency of infection, but White *et al.* (1998) reported that backward stepwise logistic regression of data for 139 hospitalized cases suggested that male patients were four times more likely to develop severe babesiosis. Some animal experiments also support the hypothesis of greater male susceptibility. Hughes and Randolph (2001) showed that testosterone in wood mice (*Apodemus sylvaticus*) and bank voles (*Clethrionomys glareolus*) decreased resistance to *B. microti*, and Barnard *et al.*, (1994) showed that in high-ranking male house mice (*Mus musculi*) increased susceptibility to *B. microti* was associated with high levels of serum testosterone and corticosteroids.

Asplenia is certainly a risk factor for severe babesiosis (White *et al.*, 1998), as are other immunocompromising conditions such as HIV infection (Falagas and Klempner, 1996; Froberg *et al.*, 2004) and immunosuppressive therapy (Gupta *et al.*, 1995; Herwaldt *et al.*, 1995; Lux *et al.*, 2003; Perdrizet *et al.*, 2000). Unlike some other zoonotic babesias, however, the majority of *B. microti* infections, many of which are asymptomatic, occur in spleen-intact individuals. Some of these cases can be very severe (Dacey *et al.*, 2001; Dorman *et al.*, 2000) and presumably age-susceptibility is the main predisposing factor in the absence of other health issues.

Co-infection as a risk factor has come more into focus as knowledge of the complex of pathogens transmitted by the vector ticks has increased. The first clinical studies of babesiosis attributed erythema chronicum migrans (ECM) to *B. microti* infection, but ECM (or EM as it is now called) has since been acknowledged to be pathognomonic for Lyme borreliosis, which is transmitted by the same tick species. Meldrum *et al.* (1992) reported that 31 of 136 patients (23%) in the New York area had evidence of concurrent Lyme borreliosis and White *et al.* (1998) found that 12% of 139 patients hospitalized with babesiosis had a history of Lyme borreliosis. Since there is no firm evidence that Lyme borreliosis exacerbates babesiosis in human patients, Lyme borreliosis infection should be a viewed as a marker for possible *B. microti* infection, rather than as a risk factor for clinical babesiosis. There is however, evidence for exacerbation of Lyme borreliosis by concurrent babesiosis infection, presumably through the immunosuppressant effects of *B. microti* (Krause *et al.*, 1996; Marcus *et al.*, 1985; Moro *et al.*, 2002), as discussed in section 5.

12 Prevention and Control

B. microti was the first human pathogen to be associated with the black-legged tick, *I. scapularis*, and was followed in the early 1980s by the causal agent of Lyme borreliosis, *B. burgdorferi*, and then by *Anaplasma phagocytophilum*, the cause of human granular ehrlichiosis. Prevention methods have therefore been developed for a complex of zoonotic pathogens transmitted by the same vector, with the focus mainly on Lyme borreliosis, which has by far the highest incidence. Despite the fact that relatively few reservoir host species occur, with the white-footed mouse being the most important, it has not proved any easier to suppress *B. microti* babesiosis than Lyme borreliosis. The tick maintenance hosts and spirochete reservoir hosts are wild animals and targeting their ticks clearly presents major difficulties. Success depends largely on the development of self-medication systems and studies have been conducted on the provision of acaricide-treated cotton waste for use as nesting material by small mammals, the use of self-applicators impregnated with acaricide for deer, the feeding of systemic acaricides, such as the avermectins (e.g., ivermectin) and anti-chitin growth regulators (e.g., benzoylphenylurea derivatives), and on the application of novel agents such as pathogenic fungi (e.g., *Metarhizium anisopliae*).

Measures focused more on the deer themselves, the maintenance host for ticks, include exclusion by fencing, and depopulation by culling or removal (or perhaps contraception in the future). Several studies have been conducted in North America and, although effects have been demonstrated, there is a widespread view that such measures would not be sufficiently practical for large-scale use, though effective methods of controlling ticks on wildlife could have an important impact in peri-domestic situations and in parks.

The most effective prophylactic measures aimed at ticks consist of preventing them from feeding on humans by the wearing of appropriate clothing, the application of various repellents such as N,N-diethyl-toluamide (DEET) and synthetic pyrethroids (e.g., permethrin) to clothing, and most importantly the dissemination of information and advice, including the need for inspection and prompt removal of any ticks (Stafford and Kitron, 2002).

An increasing number of cases are being transmitted by blood transfusion and at present there are no specific prevention methods in place. An efficient automatable screening system is desirable, as are methods to routinely inactivate such pathogens in donated blood products. However, the low incidence of the disease except in local areas of intense transmission militates against any commercial interest in developing such technologies.

Vaccine development is limited for the same reason, but additionally by the enormous technical difficulties entailed in developing an effective vaccine against a parasite that seems able to evade the immune system efficiently and persist for long periods. The only effective vaccines against cattle babesiosis are still based on live organisms, despite considerable interest and research in the development of molecular vaccines (de Waal and Combrink, 2006). Antibiotic prophylaxis following a tick bite, using a single dose of doxycyline, has been extensively studied in Lyme borreliosis (Nadelman *et al.*, 2001), but similar studies do not exist for the prevention of babesiosis.

13 Conclusions

The full impact of human infection with *B. microti* is difficult to assess since the disease is not notifiable and the pathogen has been shown to interact subtly with other diseases. However, it is likely that, as in the case of Lyme borreliosis, infection is increasing as deer and tick populations continue to expand. The emergence of *B. microti* babesiosis in countries other than the USA is also likely to continue as awareness increases and human populations come into increased contact with wildlife. Considerable recent progress has been made in unraveling the taxonomy of these parasites and defining the characteristics of zoonotic forms, which should make the task of detecting the emergence of babesiosis in new areas easier.

14 References

Abdalla, H.S., Hussein, H.S. and Kreier, J.P. (1978) *Babesia rodhaini*: passive protection of mice with immune serum. *Tropenmed. Parasitol.* **29**: 295–306.

Aguilar-Delfin, I., Wettstein, P.J. and Persing, D.H. (2003) Resistance to acute babesiosis is associated with interleukin-12- and gamma interferon-mediated responses and requires macrophages and natural killer cells. *Infect. Immun.* **71**: 2002–2008.

Alekseev, A.N., Semenov, A.V. and Dubinina, H.V. (2003) Evidence of *Babesia microti* infection in multi-infected *Ixodes persulcatus* ticks in Russia. *Exp. Appl. Acarol.* **29**: 345–353.

Allred, D.R. (2003). Babesiosis: persistence in the face of adversity. *Trends Parasitol.* **9**: 51–55.

Arnez, M., Luznik-Bufon, T., Avsic-Zupanc, T., Ruzic-Sabljic, E., Petrovec, M., Lotric-Furlan, S. and Strle, F. (2003) Causes of febrile illnesses after a tick bite in Slovenian children. *Pediatr. Infect. Dis. J.* **22**: 1078–1083.

Baumann, D., Pusterla, N., Peter, O., Grimm, F., Fournier, P.E., Schar, G., Bossart, W., Lutz, H. and Weber, R. (2003) [Fever after a tick bite: clinical manifestations and diagnosis of acute tick bite-associated infections in northeastern Switzerland]. *Dtsch Med. Wochenschr.* **128**: 1042–1047.

Barnard, C.J., Behnke, J.M. and Sewell, J. (1994) Social behaviour and susceptibility to infection in house mice (*Mus musculus*): effects of group size, aggressive behaviour and status-related hormonal responses prior to infection on resistance to *Babesia microti*. *Parasitology* **108**: 487–496.

Behnke, J.M., Sinski, E. and Wakelin, D. (1999) Primary infections with *Babesia microti* are not prolonged by concurrent *Heligmosomoides polygyrus*. *Parasitol. Int.* **48**: 183–187.

Belongia, E.A. (2002) Epidemiology and impact of coinfections acquired from Ixodes ticks. *Vector Borne Zoonot. Dis.* **2**: 265–273.

Beveridge, E. (1953) *Babesia rodhaini*; a useful organism for the testing of drugs designed for the treatment of piroplasmosis. *Ann. Trop. Med. Parasitol.* **47**: 134–138.

Bonoan, J.T., Johnson, D.H. and Cunha, B.A. (1998) Life-threatening babesiosis in an asplenic patient treated with exchange transfusion, azithromycin, and atovaquone. *Heart Lung* **27**: 424–428.

Bork, S., Yokoyama, N., Matsuo, T., Claveria, F.G., Fujisaki, K. and Igarashi, I. (2003) Clotrimazole, ketoconazole, and clodinafop-propargyl as potent growth inhibitors of equine *Babesia* parasites during in vitro culture. *J. Parasitol.* **8**: 604–606.

Boustani, M.R., Lepore, T.J., Gelfand, J.A. and Lazarus, D.S. (1994) Acute respiratory failure in patients treated for babesiosis. *Am. J. Respir. Crit. Care. Med.* **149**:1689–1691.

Burkot, T.R., Schneider, B.S., Pieniazek, N.J., Happ, C.M., Rutherford, J.S., Slemenda, S.B., Hoffmeister, E., Maupin, G.O. and Zeidner, N.S. (2000) *Babesia microti* and *Borrelia bissettii* transmission by *Ixodes spinipalpis* ticks among prairie voles, *Microtus ochrogaster*, in Colorado. *Parasitology* **121**: 595–599.

Cable, R.G. and Leiby, D.A. (2003) Risk and prevention of transfusion-transmitted babesiosis and other tick-borne diseases. *Curr. Opin. Hematol.* **10**: 405–411.

Camacho, A.T., Pallas, E., Gestal, J.J., Guitian, F.J., Olmeda, A.S., Goethert, H.K. and Telford, S.R. (2001) Infection of dogs in north-west Spain with a *Babesia microti*-like agent. *Vet. Rec.* **149**: 552–555.

Carr, J.M., Emery, S., Stone, B.F. and Tulin, L. (1991) Babesiosis. Diagnostic pitfalls. *Am. J. Clin. Pathol.* **95**: 774–777.

Casati, S., Sager, H., Gern, L. and Piffaretti, J.C. (2006) Presence of potentially pathogenic *Babesia* sp. for human in *Ixodes ricinus* in Switzerland. *Ann. Agric. Environ. Med.* **13**: 65–70.

Clark, I.A. and Jacobson, L.S. (1998) Do babesiosis and malaria share a common disease process? *Ann. Trop. Med. Parasitol.* **92**: 483–488.

Clark, I.A., Budd, A.C., Hsue, G., Haymore, B.R., Joyce, A.J., Thorner, R. and Krause, P.J. (2006) Absence of erythrocyte sequestration in a case of babesiosis in a splenectomized human patient. *Malaria J.* **5**: 69.

Clawson, M.L., Paciorkowski, N., Rajan, T.V., La Vake, C., Pope, C., La Vake, M., Wikel, S.K., Krause, P.J. and Radolf, J.D. (2002) Cellular immunity, but not gamma interferon, is essential for resolution of *Babesia microti* infection in BALB/c mice. *Infect. Immun.* **70**: 5304–5306.

Conrad, P.A., Kjemtrup, A.M., Carreno, R.A., Thomford, J., Wainwright, K., Eberhard, M., Quick, R., Telford 3rd, S.R. and Herwaldt, B.L. (2006) Description of *Babesia duncani* n.sp. (Apicomplexa: Babesiidae) from humans and its differentiation from other piroplasms. *Int. J. Parasitol.* **36**: 779–789.

Corrier, D.E. and Wagner, G.G. (1988) Comparison of the effect of T-2 toxin with that of dexamethasone or cyclophosphamide on resistance to *Babesia microti* infection in mice. *Am. J. Vet. Res.* **49**: 2000–2003.

Cullen, J.M. and Levine, J.F. (1987) Pathology of experimental *Babesia microti* infection in the Syrian hamster. *Lab. Anim. Sci.* **37**: 640–643.

Dacey, M.J., Martinez, H., Raimondo, T., Brown, C. and Brady, J. (2001) Septic shock due to babesiosis. *Clin. Infect. Dis.* **33**: 37–38.

De Vos, A.J., Dalgliesh, R.J. and Callow, L.L. (1987) Babesia. In: *Immune Responses in Parasitic Infections*, Vol. III. Protozoa (ed. E.J.L. Soulsby). CRC Press, Boca Raton, FL.

de Waal, D.T. and Combrink, M.P. (2006) Live vaccines against bovine babesiosis. *Vet. Parasitol.* **138**: 88–96.

Dobroszycki, J., Herwaldt, B.L., Boctor, F., *et al.* (1999) A cluster of transfusion-associated babesiosis cases traced to a single asymptomatic donor. *J. Am. Med. Assoc.* **281**: 927–930.

Dorman, S.E., Cannon, M.E., Telford 3rd, S.R., Frank, K.M. and Churchill, W.H. (2000) Fulminant babesiosis treated with clindamycin, quinine, and whole-blood exchange transfusion. *Transfusion* **40**: 375–380.

Duh, D., Petrovec, M. and Avsic-Zupanc, T. (2001) Diversity of *Babesia* infecting European sheep ticks (*Ixodes ricinus*). *J. Clin. Microbiol.* **39**: 3395–3397.

Duh, D., Petrovec, M., Bidovec, A. and Avsic-Zupanc, T. (2005) Cervids as babesiae hosts, Slovenia. *Emerg. Infect. Dis.* **11**: 1121–1123.

Esernio-Jenssen, D., Scimeca, P.G., Benach, J.L. and Tenenbaum, M.J. (1987) Transplacental/perinatal babesiosis. *J. Pediatr.* **110**: 570–572.

Eugui, E.M. and Allison, A.C. (1980) Differences in susceptibility of various mouse strains to haemoprotozoan infections: possible correlation with natural killer activity. *Parasite Immunol.* **2**: 277–292.

Falagas, M.E. and Klempner, M.S. (1996) Babesiosis in patients with AIDS: a chronic infection presenting as fever of unknown origin. *Clin. Infect. Dis.* **22**: 809–812.

Fay, F.G. and Rausch, R.L. (1969) Parasitic organisms in the blood of arvicoline rodents in Alaska. *J. Parasitol.* **55**: 1258–1265.

Filstein, M.R., Benach, J.L., White, D.J., Brody, B.A., Goldman, W.D., Bakal, C.W. and Schwartz, R.S. (1980) Serosurvey for human babesiosis in New York. *J. Infect. Dis.* **141**: 518–521.

Foppa, I.M., Krause, P.J., Spielman, A., Goethert, H., Gern, L., Brand, B. and Telford 3rd, S.R. (2002) Entomologic and serologic evidence of zoonotic transmission of *Babesia microti*, eastern Switzerland. *Emerg. Infect. Dis.* **8**: 722–726.

Fox, L.M., Wingerter, S., Ahmed, A., Arnold, A., Chou, J., Rhein, L. and Levy, O. (2006) Neonatal babesiosis: case report and review of the literature. *Pediatr. Infect. Dis. J.* **25**: 169–173.

Franca, C. (1908) Sur une piroplasme nouvelle chez une mangouste. *Bull. Soc. Path. Exot.* **1**: 410–412.

Francioli, P.B., Keithly, J.S., Jones, T.C., Brandstetter, R.D. and Wolf, D.J. (1981) Response of babesiosis to pentamidine therapy. *Ann. Intern. Med.* **94**: 326–330.

Fritz, C.L., Bronson, L.R., Smith, C.R., Crawford-Miksza, L., Yeh, E. and Schnurr, D. (2005) Clinical, epidemiologic, and environmental surveillance for ehrlichiosis and anaplasmosis in an endemic area of northern California. *J. Vector Ecol.* **30**: 4–10.

Froberg, M.K., Dannen, D. and Bakken, J.S. (2004) Babesiosis and HIV. *Lancet* **363**: 704.

Genga, U.E. and Kreier, J.P. (1976) Brief note: transmission experiments with *Babesia microti* (Gray strain) using *Dermacentor andersoni* Stiles as a vector. *Ohio J. Sci.* **76**: 188–189.

Gerber, M.A., Shapiro, E.D., Krause, P.J., Cable, R.G., Badon, S.J. and Ryan, R.W. (1994) The risk of acquiring Lyme disease or babesiosis from a blood transfusion. *J. Infect. Dis.* **170**: 231–234.

Gleason, N.N., Healy, G.R., Western, K.A., Benson, G.D. and Schultz, M.G. (1970) The "Gray" strain of *Babesia microti* from a human case established in laboratory animals. *J. Parasitol.* **56**: 1256–1257.

Goethert, H.K. and Telford 3rd, S.R. (2003) What is *Babesia microti? Parasitology* **127**: 301–309.

Goethert, H.K., Lubelcyzk, C., LaCombe, E., Holman, M., Rand, P., Smith Jr, R.P. and Telford 3rd, S.R. (2003) Enzootic *Babesia microti* in Maine. *J. Parasitol.* **89**: 1069–1071.

Goethert, H.K., Cook, J.A., Lance, E.W. and Telford, S.R. (2006) Fay and Rausch 1969 revisited: *Babesia microti* in Alaskan small mammals. *J. Parasitol.* **92**: 826–831.

Goff, W.L., Storset, A.K., Johnson, W.C. and Brown, W.C. (2006) Bovine splenic NK cells synthesize IFN-gamma in response to IL-12-containing supernatants from *Babesia bovis*-exposed monocyte cultures. *Parasite Immunol.* **28**: 221–228.

Gorenflot, A., Moubri, K., Precigout, E., Carcy, B. and Schetters, T.P. (1998) Human babesiosis. *Ann. Trop. Med. Parasitol.* **92**:489–501.

Gray, J.S. (2006) Identity of the causal agents of human babesiosis in Europe. *Int. J. Med. Microbiol.* **296**(Suppl 40): 131–136.

Gray, J.S. and Pudney, M. (1999) Activity of atovaquone against *Babesia microti* in the Mongolian gerbil, *Meriones unguiculatus. J. Parasitol.* **85**: 723–728.

Gray, J., von Stedingk, L.V., Gurtelschmid, M. and Granstrom, M. (2002) Transmission studies of *Babesia microti* in *Ixodes ricinus* ticks and gerbils. *J. Clin. Microbiol.* **40**: 1259–1263.

Gupta, P., Hurley, R.W., Helseth, P.H., Goodman, J.L. and Hammerschmidt, D.E. (1995) Pancytopenia due to hemophagocytic syndrome as the presenting manifestation of babesiosis. *Am. J. Hematol.* **50**: 60–62.

Habicht, G.S., Benach, J.L., Leichtling, K.D., Gocinski, B.L. and Coleman, J.L. (1983) The effect of age on the infection and immunoresponsiveness of mice to *Babesia microti. Mech. Age. Dev.* **23**: 357–369.

Hartelt, K., Oehme, R., Frank, H., Brockmann, S.O., Hassler, D. and Kimmig, P. (2004) Pathogens and symbionts in ticks: prevalence of *Anaplasma phagocytophilum* (*Ehrlichia* sp.), *Wolbachia* sp., *Rickettsia* sp., and *Babesia* sp. in Southern Germany. *Int. J. Med. Microbiol.* **293**(Suppl 37): 86–92.

Hatcher, J.C., Greenberg, P.D., Antique, J. and Jimenez-Lucho, V.E. (2001) Severe babesiosis in Long Island: review of 34 cases and their complications. *Clin. Infect. Dis.* **32**: 1117–1125.

Healy, G.R., Spielman, A. and Gleason, N. (1976) Human babesiosis: reservoir in infection on Nantucket Island. *Science.* **192**: 479–480.

Hemmer, R.M., Ferrick, D.A. and Conrad, P.A. (2000) Up-regulation of tumor necrosis factor-alpha and interferon-gamma expression in the spleen and lungs of mice infected with the human *Babesia* isolate WA1. *Parasitol. Res.* **86**: 121–128.

Herwaldt, B.L., Springs, F.E., Roberts, P.P., Eberhard, M.L., Case, K., Persing, D.H. and Agger, W.A. (1995) Babesiosis in Wisconsin: a potentially fatal disease. *Am. J. Trop. Med. Hyg.* **53**: 146–151.

Herwaldt, B.L., Neitzel, D.F., Gorlin, J.B., *et al.* (2002) Transmission of *Babesia microti* in Minnesota through four blood donations from the same donor over a 6-month period. *Transfusion* **42**: 1154–1158.

Herwaldt, B.L., Caccio, S., Gherlinzoni, F., *et al.* (2003a) Molecular characterization of a non-*Babesia divergens* organism causing zoonotic babesiosis in Europe. *Emerg. Infect. Dis.* **9**: 942–948.

Herwaldt, B.L., McGovern, P.C., Gerwel, M.P., Easton, R.M. and MacGregor, R.R. (2003b) Endemic babesiosis in another eastern state: New Jersey. *Emerg. Infect. Dis.* **9**: 184–188.

Hildebrandt, A., Hunfeld, K.P., Baier, M., Krumbholz, A., Sachse, S., Lorenzen, T., Kiehntopf, M., Fricke, H.J. and Straube, E. (2007) First confirmed autochthonous case of human *Babesia microti* infection in Europe. *Eur. J. Clin. Microbiol. Infect. Dis.* **26**: 595–601.

Hilton, E., DeVoti, J., Benach, J.L., Halluska, M.L., White, D.J., Paxton, H. and Dumler, J.S. (1999) Seroprevalence and seroconversion for tick-borne diseases in a high-risk population in the northeast United States. *Am. J. Med.* **106**: 404–409.

Holden, K., Hodzic, E., Feng, S., Freet, K.J., Lefebvre, R.B. and Barthold, S.W. (2005) Coinfection with *Anaplasma phagocytophilum* alters *Borrelia burgdorferi* population distribution in C3H/HeN mice. *Infect. Immun.* **73**: 3444.

Homer, M.J. and Persing, D.H. (2005) Babesiosis. In: *Tick-borne Diseases of Humans* (eds J.L. Goodman, D.T. Dennis and D.E. Sonenshine). ASM Press, Washington, DC.

Homer, M.J., Aguilar-Delfin, I., Telford 3rd, S.R., Krause, P.J. and Persing, D.H. (2000) Babesiosis. *Clin. Microbiol. Rev.* **13**: 451–469.

Homer, M.J., Lodes, M.J., Reynolds, L.D., Zhang, Y., Douglass, J.F., McNeill, P.D., Houghton, R.L. and Persing, D.H. (2003) Identification and characterization of putative secreted antigens from *Babesia microti*. *J. Clin. Microbiol.* **41**: 723–729.

Hu, R., Hyland, K.E. and Markowski, D. (1997) Effects of *Babesia microti* infection on feeding pattern, engorged body weight, and molting rate of immature *Ixodes scapularis* (Acari: Ixodidae). *J. Med. Entomol.* **34**: 559–564.

Hughes, W.T. and Oz, H.S. (1995) Successful prevention and treatment of babesiosis with atovaquone. *J. Infect. Dis.* **172**: 1042–1046.

Hughes, V.L. and Randolph, S.E. (2001) Testosterone increases the transmission potential of tick-borne parasites. *Parasitology* **123**: 365–371.

Humiczewska, M. and Kuzńa-Grygiel, W. (1997) A case of imported human babesiosis in Poland. *Wiad. Parazytol.* **43**: 227–229.

Hunfeld, K.P., Lambert, A., Kampen, H., Albert, S., Epe, C., Brade, V. and Tenter, A.M. (2002) Seroprevalence of *Babesia* infections in humans exposed to ticks in midwestern Germany. *J. Clin. Microbiol.* **40**: 2431–2436.

Hussein, H.S. (1980) *Ixodes trianguliceps:* seasonal abundance and role in the epidemiology of *Babesia microti* infection in north-western England. *Ann. Trop. Med. Parasitol.* **74**: 531–539.

Igarashi, I., Suzuki, R., Waki, S., *et al.* (1999) Roles of CD4(+) T cells and gamma interferon in protective immunity against *Babesia microti* infection in mice. *Infect. Immun.* **67**: 4143–4148.

Ike, K., Komatsu, T., Murakami, T., Kato, Y., Takahashi, M., Uchida, Y. and Imai, S. (2005) High susceptibility of Djungarian hamsters (*Phodopus sungorus*) to the infection with *Babesia microti* supported by hemodynamics. *J. Vet. Med. Sci.* **67**: 515–520.

Jakubowski, A.A., Casper, E.S., Gabrilove, J.L., Templeton, M.A., Sherwin, S.A. and Oettgen, H.F. (1989) Phase I trial of intramuscularly administered tumor necrosis factor in patients with advanced cancer. *J. Clin. Oncol.* **7**: 298–303.

Javed, M.Z., Srivastava, M., Zhang, S. and Kandathil, M. (2001) Concurrent babesiosis and ehrlichiosis in an elderly host. *Mayo Clin. Proc.* **76**: 563–565.

Kalman, D., Sreter, T., Szell, Z. and Egyed, L. (2003) *Babesia microti* infection of anthropophilic ticks (*Ixodes ricinus*) in Hungary. *Ann. Trop. Med. Parasitol.* **97**: 317–319.

Karakshian, S.J., Rudzinska, M.A., Spielman, A., Lewengrub, S. and Campbell, J. (1986) Primary and secondary ookinetes of *Babesia microti* in the larval stages of the tick *Ixodes dammini. Can. J. Zool.* **64**: 328–339.

Killick-Kendrick, R. (1974) Parasitic protozoa of the blood of rodents. II. Haemogregarines, malaria parasites and piroplasms of rodents: an annotated checklist and host index. *Acta Trop.* **31**: 28–69.

Kjemtrup, A.M. and Conrad, P.A. (2000) Human babesiosis: an emerging tick-borne disease. *Int. J. Parasitol.* **30**: 1323–1337.

Kjemtrup, A.M., Lee, B., Fritz, C.L., Evans, C., Chervenak, M. and Conrad, P.A. (2002) Investigation of transfusion transmission of a WA1-type babesial parasite to a premature infant in California. *Transfusion* **42**: 1482–1487.

Kjemtrup, A.M., Wainwright, K., Miller, M., Penzhorn, B.L. and Carreno, R.A. (2006) *Babesia conradae*, sp. Nov., a small canine babesia identified in California. *Vet. Parasitol.* **138**: 103–111.

Koch, R. (1906) Beitrage zur Entwicklungsgeschichte der Piroplasmen. *Z. Hyg. Infektionskr.* **54**: 1–9.

Konopka, E. and Sinski, E. (1996) [Experimental infection of mice with *Babesia microti*: characterization of parasitemia] *Wiad. Parazytol.* **42**: 395–406.

Krampitz, H.E., Buschmann, H. and Münchoff, P. (1986) Gibt es latente Babesien-infektionen beim Menschen in Süddeutschland? *Mitt. Österr. Ges. Tropenmed. Parasitol.* **8**: 233–243.

Krause, P.J. (2002) Tick-borne diseases: babesiosis. *Med. Clin. North Am.* **86**: 361–373.

Krause, P.J., Telford 3rd, S.R., Ryan, R., Hurta, A.B., Kwasnik, I., Luger, S., Niederman, J., Gerber, M. and Spielman, A. (1991) Geographical and temporal distribution of babesial infection in Connecticut. *J. Clin. Microbiol.* **29**: 1–4.

Krause, P.J., Telford 3rd, S.R., Ryan, R., Conrad, P.A., Wilson, M., Thomford, J.W. and Spielman, A. (1994) Diagnosis of babesiosis: evaluation of a serologic test for the detection of *Babesia microti* antibody. *J. Infect. Dis.* **169**: 923–926.

Krause, P.J., Telford 3rd, S.R., Spielman, A., *et al.* (1996) Concurrent Lyme disease and babesiosis. Evidence for increased severity and duration of illness. *J. Am. Med. Assoc.* **275**: 1657–1660.

Krause, P.J., Spielman, A., Telford 3rd, S.R., *et al.* (1998) Persistent parasitemia after acute babesiosis. *N. Engl. J. Med.* **339**: 160–165.

Krause, P.J., Lepore, T., Sikand, V.K., *et al.* (2000) Atovaquone and azithromycin for the treatment of babesiosis. *N. Engl. J. Med.* **343**: 1454–1458.

Krause, P.J., McKay, K., Gadbaw, J., *et al.* (2003) Tick-Borne Infection Study Group. Increasing health burden of human babesiosis in endemic sites. *Am. J. Trop. Med. Hyg.* **68**: 431–436.

Lane, R.S., Foley, J.E., Eisen, L., Lennette, E.T. and Peot, M.A. (2001) Acarologic risk of exposure to emerging tick-borne bacterial pathogens in a semirural community in northern California. *Vector Borne Zoonot. Dis.* **1**: 197–210.

Leiby, D.A. (2006) Babesiosis and blood transfusion: flying under the radar. *Vox Sang.* **90**: 157–165.

Leiby, D.A., Chung, A.P., Gill, J.E., Houghton, R.L., Persing, D.H., Badon, S. and Cable, R.G. (2005) Demonstrable parasitemia among Connecticut blood donors with antibodies to *Babesia microti*. *Transfusion* **45**: 1804–1810.

Loa, C.C., Adelson, M.E., Mordechai, E., Raphaelli, I. and Tilton, R.C. (2004) Serological diagnosis of human babesiosis by IgG enzyme-linked immunosorbent assay. *Curr. Microbiol.* **49**: 385–389.

Lodes, M.J., Houghton, R.L., Bruinsma, E.S., Mohamath, R., Reynolds, L.D., Benson, D.R., Krause, P.J., Reed, S.G. and Persing, D.H. (2000) Serological expression cloning of novel immunoreactive antigens of *Babesia microti*. *Infect. Immun.* **68**: 2783–2790.

Lux, J.Z., Weiss, D., Linden, J.V., Kessler, D., Herwaldt, B.L., Wong, S.J., Keithly, J., Della-Latta, P. and Scully, B.E. (2003) Transfusion-associated babesiosis after heart transplant. *Emerg. Infect. Dis.* **9**: 116–119.

Lykins, J.D., Ristic, M. and Weisiger, R.M. (1975) *Babesia microti:* pathogenesis of parasite of human origin in the hamster. *Exp. Parasitol.* **37**: 388–397.

Malagon, F. and Tapia, J.L. (1994) Experimental transmission of *Babesia microti* infection by the oral route. *Parasitol. Res.* **80**: 645–648.

Marathe, A., Tripathi, J., Handa, V. and Date, V. (2005) Human babesiosis – a case report. *Indian J. Med. Microbiol.* **23**: 267–269.

Marcus, L.C., Steere, A.C., Duray, P.H., Anderson, A.E. and Mahoney, E.B. (1985) Fatal pancarditis in a patient with coexistent Lyme disease and babesiosis. Demonstration of spirochetes in the myocardium. *Ann. Intern. Med.* **103**: 374–376.

Marley, S.E., Eberhard, M.L., Steurer, F.J., Ellis, W.L., McGreevy, P.B. and Ruebush 2nd, T.K. (1997) Evaluation of selected antiprotozoal drugs in the *Babesia microti*-hamster model. *Antimicrob. Agents Chemother.* **41**: 91–94.

Matsubara, J., Koura, M. and Kamiyama, T. (1993) Infection of immunodeficient mice with a mouse-adapted substrain of the gray strain of *Babesia microti*. *J. Parasitol.* **79**: 783–786.

Matyniak, J.E. and Reiner, S.L. (1995) T helper phenotype and genetic susceptibility in experimental Lyme disease. *J. Exp. Med.* **181**: 1251–1254.

Meer-Scherrer, L., Adelson, M., Mordechai, E., Lottaz, B. and Tilton, R. (2004) *Babesia microti* infection in Europe. *Curr. Microbiol.* **48**: 435–437.

Mehlhorn, H. and Schein, E. (1987) The piroplasms: lifecycle and sexual stages. *Adv. Parasitol.* **23**: 37–103.

Meldrum, S.C., Birkhead, G.S., White, D.J., Benach, J.L. and Morse, D.L. (1992) Human babesiosis in New York State: an epidemiological description of 136 cases. *Clin. Infect. Dis.* **15**: 1019–1023.

Miller, L.H., Neva, F.A. and Gill, F. (1978) Failure of chloroquine in human babesiosis (*Babesia microti*): case report and chemotherapeutic trials in hamsters. *Ann. Intern. Med.* **88**: 200–202.

Moro, M.H., Zegarra-Moro, O.L., Bjornsson, J., Hofmeister, E.K., Bruinsma, E., Germer, J.J. and Persing, D.H. (2002) Increased arthritis severity in mice coinfected with *Borrelia burgdorferi* and *Babesia microti*. *J. Infect. Dis.* **186**: 428–431.

Nadelman, R.B., Nowakowski, J., Fish, D., *et al.* (2001) Prophylaxis with single-dose doxycycline for the prevention of Lyme disease after an *Ixodes scapularis* tick bite. *N. Engl. J. Med.* **345**: 79–84.

New, D.L., Quinn, J.B., Qureshi, M.Z. and Sigler, S.J. (1997) Vertically transmitted babesiosis. *J. Pediatr.* **131**: 163–164.

Nohynkova, E., Kubek, J., Mest'ankova, O., Chalupa, P. and Hubalek, Z. (2003) [A case of *Babesia microti* imported into the Czech Republic from the USA] *Cas. Lek. Cesk.* **142**: 377–381.

Oliveira, M.R. and Kreier, J.P. (1979) Transmission of *Babesia microti* using various species of ticks as vectors. *J. Parasitol.* **65**: 816–817.

Ong, K.R., Stavropoulos, C. and Inada, Y. (1990). Babesiosis, asplenia, and AIDS. *Lancet* **336**: 112.

Oz, H.S. and Hughes, W.T. (1996) Acute fulminating babesiosis in hamsters infected with *Babesia microti*. *Int. J. Parasitol.* **26**: 667–670.

Perdrizet, G.A., Olson, N.H., Krause, P.J., Banever, G.T., Spielman, A. and Cable, R.G. (2000) Babesiosis in a renal transplant recipient acquired through blood transfusion. *Transplantation* **70**: 205–208.

Persing, D.H., Mathiesen, D., Marshall, W.F., Telford, S.R., Spielman, A., Thomford, J.W. and Conrad, P.A. (1992) Detection of *Babesia microti* by polymerase chain reaction. *J. Clin. Microbiol.* **30**: 2097–2103.

Phillips, R.S. and Wakelin, D. (1976) Suppression of immunity in mice in the nematode *Trichuris muris* by concurrent infection with rodent piroplasms. *Trans R. Soc. Trop. Med. Hyg.* **68**: 276.

Pieniazek, N., Sawczuk, M. and Skotarczak, B. (2006) Molecular identification of *Babesia* parasites isolated from *Ixodes ricinus* ticks collected in northwestern Poland. *J. Parasitol.* **92**: 32–35.

Piesman, J. and Spielman, A. (1980) Human babesiosis on Nantucket Island: prevalence of *Babesia microti* in ticks. *Am. J. Trop. Med. Hyg.* **29**: 742–746.

Piesman, J. and Spielman, A. (1982) *Babesia microti*: infectivity of parasites from ticks for hamsters and white-footed mice. *Exp. Parasitol.* **53**: 242–248.

Piesman, J., Spielman, A., Etkind, P., Ruebush 2nd, T.K. and Juranek, D.D. (1979) Role of deer in the epizootiology of *Babesia microti* in Massachusetts, USA. *J. Med. Entomol.* **15**: 537–540.

Piesman, J., Karakashian, S.J., Lewengrub, S., Rudzinska, M.A. and Spielman, A. (1986) Development of *Babesia microti* sporozoites in adult *Ixodes dammini*. *Int. J. Parasitol.* **16**: 381–385.

Piesman, J., Mather, T.N., Dammin, G.J., Telford 3rd, S.R., Lastavica, C.C. and Spielman, A. (1987) Seasonal variation of transmission risk of Lyme disease and human babesiosis. *Am. J. Epidemiol.* **126**: 1187–1189.

Purvis, A.C. (1977) Immunodepression in *Babesia microti* infections. *Parasitology* **75**: 197–205.

Randolph, S.E. (1991) The effect of *Babesia microti* on feeding and survival in its tick vector, *Ixodes trianguliceps*. *Parasitology* **102**: 9–16.

Randolph, S.E. (1995) Quantifying parameters in the transmission of *Babesia microti* by the tick *Ixodes trianguliceps* amongst voles (*Clethrionomys glareolus*). *Parasitology* **110**: 287–295.

Rar, V.A., Fomenko, N.V., Dobrotvorsky, A.K., Livanova, N.N., Rudakova, S.A., Fedorov, E.G., Astanin, V.B. and Morozova, O.V. (2005) Tickborne pathogen detection, Western Siberia, Russia. *Emerg. Infect. Dis.* **11**: 1708–1715.

Robertson, J. N., Gray, J. S. and Stewart, P. (2000) Tick bite and Lyme borreliosis risk at a recreational site in England. *Eur. J. Epidemiol.* **16**: 647–652.

Rowin, K.S., Tanowitz, H.B. and Wittner, M. (1982) Therapy of experimental babesiosis. *Ann. Intern. Med.* **97**: 556–558.

Rudolf, I., Golovchenko, M., Sikutova, S., Rudenko, N., Grubhoffer, L. and Hubalek, Z. (2005) *Babesia microti* (Piroplasmida: Babesiidae) in nymphal *Ixodes ricinus* (Acari: Ixodidae) in the Czech Republic. *Folia Parasitol. (Praha)* **52**: 274–276.

Rudzinska, M.A., Spielman, A., Lewengrub, S., Trager, W. and Piesman, J. (1983) Sexuality in piroplasms as revealed by electron microscopy in *Babesia microti*. *Proc. Natl. Acad. Sci. U S A.* **80**: 2966–2970.

Rudzinska, M.A., Spielman, A., Riek, R.F., Lewengrub, S.J. and Piesman, J. (1979) Intraerythrocytic 'gametocytes' of *Babesia microti* and their maturation in ticks. *Can. J. Zool.* **57**: 424–434.

Ruebush, M.J. and Hanson, W.L. (1980a) Thymus dependence of resistance to infection with *Babesia microti* of human origin in mice. *Am. J. Trop. Med. Hyg.* **29**: 507–515.

Ruebush, M.J. and Hanson, W.L. (1980b) Transfer of immunity to *Babesia microti* of human origin using T lymphocytes in mice. *Cell Immunol.* **52**: 255–265.

Ruebush, T.K. 2nd, Contacos, P.G. and Steck, E.A. (1980). Chemotherapy of *Babesia microti* infections in Mongolian Jirds. *Antimicrob. Agents. Chemother.* **18**: 289–291.

Ruebush 2nd, T.K., Rubin, R.H., Wolpow, E.R., Cassaday, P.B. and Schultz, M.G. (1979) Neurologic complications following the treatment of human *Babesia microti* infection with diminazene aceturate. *Am. J. Trop. Med. Hyg.* **28**: 184–189.

Ruebush 2nd, T.K., Collins, W.E. and Warren, M. (1981a) Experimental *Babesia microti* infections in *Macaca mulatta*: recurrent parasitemia before and after splenectomy. *Am. J. Trop. Med. Hyg.* **30**: 304–307.

Ruebush, 2nd, T.K., Juranek, D.D., Spielman, A., Piesman, J. and Healy, G.R. (1981b) Epidemiology of human babesiosis on Nantucket Island. *Am. J. Trop. Med. Hyg.* **30**: 937–941.

Ryan, R., Krause, P.J., Radolf, J., Freeman, K., Spielman, A., Lenz, R. and Levin, A. (2001) Diagnosis of babesiosis using an immunoblot serologic test. *Clin. Diagn. Lab. Immunol.* **8**: 1177–1180.

Saito-Ito, A., Tsuji, M., Wei, Q., *et al.* (2000) Transfusion-acquired, autochthonous human babesiosis in Japan: isolation of *Babesia microti*-like parasites with hu-RBC-SCID mice. *J. Clin. Microbiol.* **38**: 4511–4516.

Saito-Ito, A., Yano, Y., Dantrakool, A., Hashimoto, T. and Takada, N. (2004) Survey of rodents and ticks in human babesiosis emergence area in Japan: first detection of *Babesia microti*-like parasites in *Ixodes ovatus*. *J. Clin. Microbiol.* **42**: 2268–2270.

Scholtens, R.G., Braff, E.H., Healy, G.R., and Gleason, N. (1968) A case of babesiosis in man in the United States. *Am. J. Trop. Med. Hyg.* **17**: 810–813.

Setty, S., Khalil, Z., Schori, P., Azar, M. and Ferrieri, P. (2003) Babesiosis. Two atypical cases from Minnesota and a review. *Am. J. Clin. Pathol.* **120**: 554–559.

Shaio, M.F. and Lin, P.R. (1998) A case study of cytokine profiles in acute human babesiosis. *Am. J. Trop. Med. Hyg.* **58**: 335–337.

Shaio, M.F. and Yang, K.D. (1997) Response of babesiosis to a combined regimen of quinine and azithromycin. *Trans R. Soc. Trop. Med. Hyg.* **91**: 214–215.

Shih, C.M. and Wang, C.C. (1998) Ability of azithromycin in combination with quinine for the elimination of babesial infection in humans. *Am. J. Trop. Med. Hyg.* **59**: 509–512.

Shih, C.M., Liu, L.P., Chung, W.C., Ong, S.J. and Wang, C.C. (1997) Human babesiosis in Taiwan: asymptomatic infection with a *Babesia microti*-like organism in a Taiwanese woman. *J. Clin. Microbiol.* **35**: 450–454.

Sinski, E., Bajer, A., Welc, R., Pawelczyk, A., Ogrzewalska, M. and Behnke, J.M. (2006) *Babesia microti*: prevalence in wild rodents and *Ixodes ricinus* ticks from the Mazury Lakes District of North-Eastern Poland. *Int. J. Med. Microbiol.* **296**(Suppl 40): 137–143.

Skotarczak, B. and Sawczuk, M. (2003) [Occurrence of *Babesia microti* in ticks *Ixodes ricinus* on selected areas of western Pomerania]. *Wiad. Parazytol.* **49**: 273–280.

Skrabalo, Z. and Deanovic, Z. (1957) Piroplasmosis in man; report of a case. *Doc. Med. Geogr. Trop.* **9**: 11–16.

Spielman, A. (1976) Human babesiosis on Nantucket Island: transmission by nymphal *Ixodes* ticks. *Am. J. Trop. Med. Hyg.* **25**: 784–787.

Spielman, A., Etkind, P., Piesman, J., Ruebush, T.K. 2nd, Juranek, D.D. and Jacobs, M.S. (1981) Reservoir hosts of human babesiosis on Nantucket Island. *Am. J. Trop. Med. Hyg.* **30**: 560–565.

Spielman, A., Levine, J.F. and Wilson, M.L. (1984) Vectorial capacity of North American *Ixodes* ticks. *Yale J. Biol. Med.* **57**: 507–513.

Stafford, K. III and Kitron, U. (2002). Environmental Management for Lyme Borreliosis Control In: Gray J, Kahl O, Lane R & Stanek G (Eds) Lyme Borreliosis: Biology, Epidemiology and Control, pp. 301–334. CAB International.

Steketee, R.W., Eckman, M.R., Burgess, E.C., Kuritsky, J.N., Dickerson, J., Schell, W.L., Godsey Jr, M.S. and Davis, J.P. (1985) Babesiosis in Wisconsin. A new focus of disease transmission. *J. Am. Med. Assoc.* **253**: 2675–2678.

Swanson, S.J., Neitzel, D., Reed, K.D. and Belongia, E.A. (2006) Coinfections acquired from ixodes ticks. *Clin. Microbiol. Rev.* **19**: 708–727.

Sweeney, C.J., Ghassemi, M., Agger, W.A. and Persing, D.H. (1998) Coinfection with *Babesia microti* and *Borrelia burgdorferi* in a western Wisconsin resident. *Mayo Clin. Proc.* **73**: 338–341.

Telford 3rd., S.R. (1998) The name *Ixodes dammini* epidemiologically justified. *Emerg. Infect. Dis.* **4**: 132–134.

Telford, S. and Maguire, J.H. (2006) Babesiosis. In: *Tropical Infectious Diseases: Principles, Pathogens, and Practice*, 2nd Edn (eds R.L. Guerrant, D.H. Walker and P.F. Weller). Churchill Livingstone, London, pp. 1063–1071.

Telford 3rd, S.R. and Spielman, A. (1993) Reservoir competence of white-footed mice for *Babesia microti*. *J. Med. Entomol.* **30**: 223–227.

Tsuji, M., Wei, Q., Zamoto, A., *et al.* (2001) Human babesiosis in Japan: epizootiologic survey of rodent reservoir and isolation of new type of *Babesia microti*-like parasite. *J. Clin. Microbiol.* **39**: 4316–4322.

Vannier, E., Borggraefe, I., Telford 3rd, S.R., Menon, S., Brauns, T., Spielman, A., Gelfand, J.A. and Wortis, H.H. (2004) Age-associated decline in resistance to *Babesia microti* is genetically determined. *J. Infect. Dis.* **189**: 1721–1728.

Van Peenen, P.F. and Healy, G.R. (1970) Infection of *Microtus ochrogaster* with piroplasms isolated from man. *J. Parasitol.* **56**: 1029–1031.

Vercammen, F., De Deken, R. and Maes, L. (1996) Prophylactic treatment of experimental canine babesiosis (*Babesia canis*) with doxycycline. *Vet. Parasitol.* **66**: 251–255.

Walter, G. (1984) [Transmission and course of parasitemia of *Babesia microti* (Hannover I strain) in the bank vole (*Clethrionomys glareolus*) and field vole (*Microtus agrestis*)]. *Acta Trop.* **41**: 259–264.

Walter, G. and Weber, G. (1981) [A study on the transmission (transstadial, transovarial) of *Babesia microti*, strain 'Hannover i', in its tick vector, *Ixodes ricinus*]. *Tropenmed. Parasitol.* **32**: 228–230.

Watkins, R.A., Moshier, S.E., O'Dell, W.D. and Pinter, A.J. (1991) Splenomegaly and reticulocytosis caused by *Babesia microti* infections in natural populations of the montane vole, *Microtus montanus. J. Protozool.* **38**: 573–576.

Wei, Q., Tsuji, M., Zamoto, A., Kohsaki, M., Matsui, T., Shiota, T., Telford 3rd, S.R. and Ishihara, C. (2001) Human babesiosis in Japan: isolation of *Babesia microti*-like parasites from an asymptomatic transfusion donor and from a rodent from an area where babesiosis is endemic. *J. Clin. Microbiol.* **39**: 2178–2183.

Weiss, L.M., Wittner, M., Wasserman, S., Oz, H.S., Retsema, J., Tanowitz, H.B. (1993) Efficacy of azithromycin for treating *Babesia microti* infection in the hamster model. *J. Infect. Dis.* **168**: 1289–1292.

Weiss, L.M. (2002) Babesiosis in humans: a treatment review. *Expert Opin. Pharmacother.* **3**: 1109–1115.

Western, K.A., Benson, G.D., Gleason, N.N., Healy, G.R. and Schultz, M.G. (1970) Babesiosis in a Massachusetts resident. *N. Engl. J. Med.* **283**: 854–856.

White, D.J., Talarico, J., Chang, H.G., Birkhead, G.S., Heimberger, T. and Morse, D.L. (1998) Human babesiosis in New York State: Review of 139 hospitalized cases and analysis of prognostic factors. *Arch. Intern. Med.* **158**: 2149–2154.

Wittner, M., Rowin, K.S., Tanowitz, H.B., Hobbs, J.F., Saltzman, S., Wenz, B., Hirsch, R., Chisholm, E. and Healy, G.R. (1982) Successful chemotherapy of transfusion babesiosis. *Ann. Intern. Med.* **96**: 601–604.

Wittner, M., Lederman, J., Tanowitz, H.B., Rosenbaum, G.S. and Weiss, L.M. (1996) Atovaquone in the treatment of *Babesia microti* infections in hamsters. *Am. J. Trop. Med. Hyg.* **55**: 219–222.

Woldehiwet, Z. (2006) *Anaplasma phagocytophilum* in ruminants in Europe. In: Century of Rickettsiology: Emerging, Reemerging Rickettsioses, Molecular Diagnostics, and Emerging Veterinary Rickettsioses. *Ann. N.Y. Acad. Sci.* **1078**: 446–460.

Wormser, G.P., Dattwyler, R.J., Shapiro, E.D., *et al.* (2006) The clinical assessment, treatment, and prevention of Lyme disease, human granulocytic anaplasmosis, and babesiosis: clinical practice guidelines by the Infectious Diseases Society of America. *Clin. Infect. Dis.* **43**: 1089–1134.

Wozniak, E.J., Lowenstine, L.J., Hemmer, R., Robinson, T. and Conrad, P.A. (1996) Comparative pathogenesis of human WA1 and *Babesia microti* isolates in a Syrian hamster model. *Lab. Anim. Sci.* **46**: 507–515.

Yano, Y., Saito-Ito, A., Anchalee, D. and Takada, N. (2005) Japanese *Babesia microti* cytologically detected in salivary glands of naturally infected tick *Ixodes ovatus*. *Microbiol. Immunol.* **49**: 891–897.

Yokoyama, N., Bork, S., Nishisaka, M., *et al.* (2003) Roles of the Maltese cross form in the development of parasitemia and protection against *Babesia microti* infection in mice. *Infect. Immun.* **71**: 411–417.

Zamoto, A., Tsuji, M., Kawabuchi, T., Wei, Q., Asakawa, M. and Ishihara, C. (2004a) U.S.-type *Babesia microti* isolated from small wild mammals in Eastern Hokkaido, Japan. *J. Vet. Med. Sci.* **66**: 919–926.

Zamoto, A., Tsuji, M., Wei, Q., *et al.* (2004b) Epizootiologic survey for *Babesia microti* among small wild mammals in northeastern Eurasia and a geographic diversity in the beta-tubulin gene sequences Japan. *J. Vet. Med. Sci.* **66**: 785–792.

Zavizion, B., Pereira, M., de Melo, J.M., *et al.* (2004) Inactivation of protozoan parasites in red blood cells using INACTINE PEN110 chemistry. *Transfusion* **44**: 731–738.

Zintl, A., Gray, J.S., Skerrett, H.E. and Mulcahy, G. (2005) Possible mechanisms underlying age-related resistance to bovine babesiosis. *Parasite Immunol.* **27**: 115–120.

Zintl, A., Mulcahy, G., Skerrett, H.E., Taylor, S.M. and Gray, J.S. (2003) *Babesia divergens*, a bovine blood parasite of veterinary and zoonotic importance. *Clin. Microbiol. Rev.* **16**: 622–636.

Zivkovic, D., Seinen, W., Kuil, H., Albers-van Bemmel, C.M. and Speksnijder, J.E. (1984) Immunity to Babesia in mice. I. Adoptive transfer of immunity to *Babesia rodhaini* with immune spleen cells and the effect of irradiation on the protection of immune mice. *Vet. Immunol. Immunopathol.* **5**: 343–357.

Zwart, D. and Brocklesby, D.W. (1979) Babesiosis: non-specific resistance, immunological factors and pathogenesis. *Adv. Parasitol.* **17**: 49–113.

C Ciliates

C1 *Balantidium coli*

Lynne S. Garcia

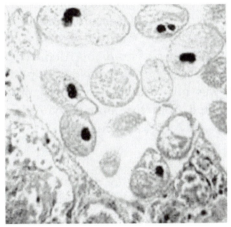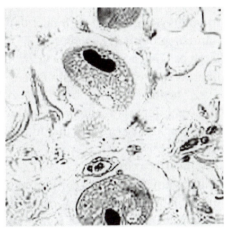

1 Introduction

Balantidium coli was first described by Malmsten in 1857 from the dysenteric stools of two patients and was soon also observed by Leuckart in 1861 and Stein in 1862, who transferred the species to the genus *Balantidium* (prior genera were *Paramoecium*, *Leukophyra*, and *Holophyra*; Beaver *et al.*, 1984).

B. *coli* is widely distributed in hogs, particularly in warm and temperate climates, and in monkeys in the tropics. Human infection is found in warmer climates, sporadically in cooler areas, and in institutionalized groups with low levels of personal hygiene. It is rarely recovered in clinical specimens within the United States (Garcia, 1999, 2007).

In a Japanese study, a total of 375 fecal samples of 56 mammalian species belonging to 17 families of 4 orders were examined for the detection of *B. coli*. As a result, *B. coli* was found in 6 species belonging to 4 families of 2 orders (Primates and Artiodactyla) of the host animals examined. White-handed gibbon (*Hylobates lar*), squirrel monkey (*Saimiri sciurea*) and Japanese macaque (*Macaca fuscata*) were new hosts for *B. coli*. All the wild boar (*Sus scrofa*) and chimpanzees (*Pan troglodytes*) examined were positive. The highest number of *B. coli* was obtained from a chimpanzee (1230 g^{-1} feces). No *B. coli* was detected from the animals of orders Rodentia and Carnivora, including dogs and cats (Nakauchi, 1999).

2 Classification

The phylum Ciliophora contains ciliated protozoa that are characterized by patterns of cilia distribution that allows them to be classified using morphological criteria. *B. coli* is the only ciliate that infects humans. Although classification is not difficult, it is interesting to note that the ciliates may be more closely related to the sporozoans than was previously believed.

3 Life Cycle and Morphology

B. coli is the only pathogenic ciliate and is the largest of the protozoa that parasitize humans (Figures 1 and 2). Both the trophozoite and cyst forms are found. The trophozoite is quite large, oval, and covered with short cilia; it measures approximately 50–100 μm in length and 40–70 μm in width. The organism can easily be seen in a wet preparation on lower power. The anterior end is somewhat pointed and has a cytostome present; in contrast, the posterior end is broadly rounded. The cytoplasm contains many vacuoles with ingested bacteria and debris. There are two nuclei within the trophozoite, one very large bean-shaped macronucleus and the smaller round micronucleus. The organisms normally live in the large intestine.

The cyst is formed as the trophozoite moves down the intestine. Nuclear division does not occur in the cyst; therefore, only two nuclei are present, the macronucleus and the micronucleus. The cysts measure from 50 to 70 μm in diameter.

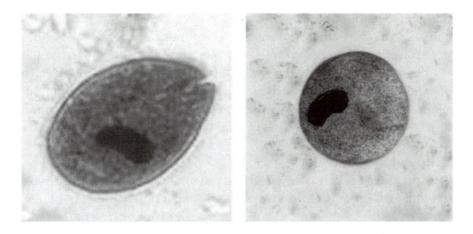

Figure 1 *Balantidium coli* saline/iodine wet mount. (Left) *B. coli* trophozoite – note the large bean-shaped macronucleus and the cilia ('fuzz') around the periphery; (right) *B. coli* cyst – note the more rounded shape, cilia are more difficult to see, and the presence of the macronucleus.

The localization and activity of D glucose-6-phosphatase (G-6-Pase) and alkaline phosphatase (AlP) in the trophozoites of *B. coli* isolated from pig intestine content have been investigated using ultrastructural and cytochemical methods. The activity of G-6-Pase was demonstrated on the membranes of the endoplasmic reticulum, particularly in the cortical part of the trophozoites. In addition, the product of the reaction to G-6-Pase was concentrated in the vesicular structures, which were distributed along the reticular membranes. These structures were described as vesicles similar to

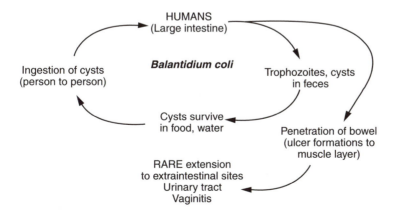

Figure 2 Life cycle of *Balantidium coli* demonstrating the infective stage, and the development of the trophozoites and cysts. Note that tissue invasion and rare dissemination to extraintestinal sites are included in the life cycle. (Courtesy of L.S. Garcia, 2007 *Diagnostic Medical Parasitology*, Edn 5. ASM Press, Washington, DC.)

glycosomes, containing enzymes of glycogenolysis. It is very likely that hydrolases in *B. coli* are formed on the rough reticular membranes without the involvement of cisterns of the Golgi complex. The ultrastructural deposits of the reaction to G-6-Pase and AlP in the trophozoites of *B. coli* indicate that some membranes of the rough endoplasmic reticulum and small vacuoles with a strong reaction to these enzymes can play a similar role to the Golgi complex (Skotarczak and Kolodziejczyk, 2005).

4 Clinical Disease

Some individuals with *B. coli* infections are totally asymptomatic, whereas others have symptoms of severe dysentery similar to those seen in patients with amoebiasis. Symptoms usually include diarrhea or dysentery, tenesmus, nausea, vomiting, anorexia, and headache. Insomnia, muscular weakness, and weight loss have also been reported. The diarrhea may persist for weeks to months prior to the development of dysentery. There may be tremendous fluid loss, with a type of diarrhea similar to that seen in cholera or in some coccidial or microsporidial infections.

   B. coli has the potential to invade tissue. On contact with the mucosa, *B. coli* may penetrate the mucosa with cellular infiltration in the area of the developing ulcer (Figure 3). Some of the abscess formations may extend to the muscular layer. The ulcers may vary in shape, and the ulcer bed may be full of pus and necrotic debris. Although the number of cases is small, extraintestinal disease has been reported (peritonitis, urinary tract, inflammatory vaginitis).

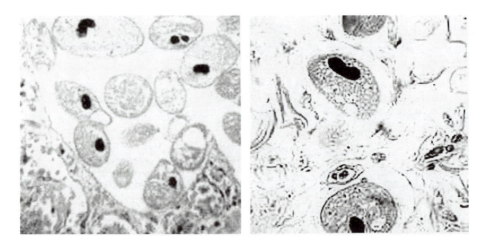

Figure 3 Two images of *Balantidium coli* trophozoites that have penetrated intestinal tissue. In most of the organisms, the macronucleus is visible. In those organisms in which the macronucleus is not visible, the tissue cut was made through the organism cytoplasm, but missed the macronucleus.

Confirmation of two cases of dysentery caused by *B. coli* with numerous colonic ulcers was documented by colonoscopy and diagnosed by endoscopic biopsies. Unfortunately, the good response to tetracyclines in the younger patient was not repeated in the older patient, who died in septic shock despite antibiotic therapy (Castro *et al.*,1983).

One case report involves an alcoholic pork-butcher who presented with severe colitis with peritonitis, caused by *B. coli*. This parasite is common in a variety of domestic and wild mammals, mainly pigs; however, its prevalence rate in humans is very low – particularly in industrialized, northern countries, including France. The infection is most frequently acquired by ingesting food or water contaminated by pig feces, and it may be asymptomatic or may cause acute diarrhea. It is important to consider the risk of this parasitic disease in susceptible patients presenting with bloody diarrhea (Ferry *et al.*, 2004). The infection has also been recovered in a patient with suspected Crohn's disease (Agapov, 2006).

A urinary bladder infection has been reported as a co-infection with *B. coli* and *Trichomonas vaginalis*. Although this presentation is apparently very uncommon, infection may be linked to ectopic *B. coli* in the bladder (Maleky, 1998). In another case, a patient with non-Hodgkin's lymphoma from Turkey was diagnosed as having diarrhea caused by *B. coli*. This case emphasizes the fact that *B. coli* should also be considered as a possible pathogen in immunocompromised patients with diarrhea (Yazar *et al.*, 2004).

A 59-year-old woman suffering from chronic lymphocytic leukemia developed pulmonary lesions, and bronchoalveolar lavage was performed for possible confirmation of a systemic fungal infection. However, direct microscopic analysis revealed ciliated protozoa identified as *B. coli*. On very rare occasions this parasite may invade extraintestinal organs, and in this case the lungs of an immunocompromised patient were involved. This case is unusual as balantidiasis is rare in Europe, the patient had no obvious contact with pigs, and there was no history of diarrhea prior to pulmonary colonization. Metronidazole was rapidly administered, and the condition improved after 24–48 h. Certainly, this case points up the importance of considering this infection in immunocompromised patients, even if they do not have diarrhea (Anargyrou *et al.*, 2003).

In another case of pulmonary disease, a fatal case was reported of *B. coli* pneumonia in a 71-year-old woman suffering from anal cancer. The diagnosis was made by the discovery of motile trophozoites in a wet mount from bronchial secretions. While the usual habitat of the parasite is the colon, lung balantidiasis is very rare, but has been reported (Vasilakopoulou *et al.*, 2003).

A unique case of chronic balantidiasis has been described, in which the patient presented with chronic colitis and inflammatory polyposis of the rectum and sigmoid colon and an intrapulmonary mass. Histology of the colonic polyps showed *B. coli*, and both *Aspergillus* and *B. coli* were found in the aspirate of the pulmonary mass. The patient was treated with doxycycline HCl 100 mg day^{-1} for 10 days with complete clinical recovery and marked improvement of the endoscopic appearance of the colonic mucosa (Ladas *et al.*, 1989).

Infection with *B. coli* has also been reported in HIV-infected patients with chronic diarrhea (Cermeno *et al.*, 2003). A case of balantidial dysentery has also been reported in a patient positive for HIV in French Guiana. Apparently, this case was the first described in medical literature of co-infection with HIV and *B. coli*. The patient also presented with disseminated histoplasmosis. Immunosuppression probably played a role in the evolution of asymptomatic *B. coli* carriage to clinical dysentery. This clinical case did not present any complications; however, treatment with doxycycline had to be carried out for 20 days in order to obtain a clinical and parasitological cure (Clyti *et al.*, 1998).

In a patient who was surgically treated for acute appendicitis, many *B. coli* trophozoites were found in the sections of cecal appendix (Gonzalez Sanchez, 1978). In another case, *B. coli* has also been confirmed in cervico-vaginal cytology smears (Rivasi and Giannotti, 1983), and *B. coli* were detached in a routine cytologic smear of ascitic fluid. The morphologic features were clearly discernible in the Papanicolaou-stained smears (Lahiri *et al.*, 1977).

5 Pathogenesis

The infection fundamentally affects the colon and causes variable clinic pictures, from asymptomatic to serious dysenteric forms. In one study ten piglets and four monkeys free from *Balantidium* were dosed with human fecal homogenate which contained 1.2×10^4–4.8×10^4 *B. coli* cysts. The infection resulted in severe diarrhea in piglets 1–6 and hydrocortisone-treated monkeys 1–2, moderate diarrhea in piglets 7–10 and a subclinical infection in monkeys 3–4. In piglets 1–3 and monkeys 1–2, heavy infection of the intestinal mucosa extended from the terminal ileum to the rectum and the mucosa was severely damaged. In piglets 4–10, infection was heavy in the large intestine with moderate mucosal damage (Yang *et al.*, 1995).

Using ultrastructural and cytochemical techniques, peroxisomes of the trophozoites of *B. coli* isolated from pig fecal material were investigated. The peroxisomes of *B. coli* trophozoites from pigs with subclinical balantidiasis were less than 0.8 mm in diameter whereas those from pigs with acute balantidiasis were greater than 0.8 μm in diameter. These findings may have some relevance to pathogenesis in pigs, as well as in humans (Skotarczak, 1997).

Cytoenzymatic assays on *B. coli* with the use of a reaction-detecting membrane-coupled hydrolase, that is, ATP-ase, permitted identification of the mucocysts. The shape, size, and location of mucocysts in *B. coli* trophozoites were found to correspond to descriptions of these structures in other ciliates. The mucocysts were more numerous in *B. coli* trophozoites isolated from symptomatic balantidiasis-affected pigs (Group I), and the product of reaction to ATP-ase was more copious than in Group II trophozoites. However, not all the bubble-like structures with similar morphological features reacted positively to the enzyme. The discrepancy was explained by the cytoenzymatic reaction to β-GR. The reaction product was visible in the vesicular structures, situated above the plasmolemma, although some of them contained no reaction product. Thus the presence of two types of secretory structure can be inferred: the mucocysts, with ATP-ase in their membranes, and other extrusomes containing active β-GR (Skotarczak, 1999).

Adult white rats were immunized by numerous subcutaneous injections of antigens obtained from cultures of *B. coli* and *B. suis*. After the rats were sensitized they were infected with cultural forms of *Balantidium*. In the infected rats, 75% were found to have ciliates in the lumen of the large intestine. In the tissues of the intestinal wall, up to the muscular layer observed, changes included hyperemia, edema, hemorrhage and ulcers. Using the macrophage migration test, it was established that in rats during their immunization and following infection, lymphocytes appear, which are sensitized in relation to the *Balantidium* antigens and strongly suggest the development of a cellular immune response (Karapetian *et al.*, 1978).

6 Diagnosis

Routine stool examinations, particularly wet preparation examinations of fresh and concentrated material, will demonstrate the organisms (Figure 4;

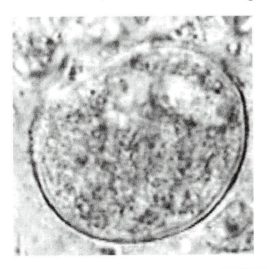

Figure 4 Saline wet mount of *Balantidium coli* without iodine. Note that the organism characteristics are somewhat difficult to see; however, the macronucleus is still visible.

Garcia, et al., 2003; National Committee for Clinical Laboratory Standards). Organism recognition and identification on a permanent stained smear may be very difficult. These protozoa are so large that they tend to stain very darkly, thus obscuring any internal morphology. *B. coli* organisms may even be confused with helminth eggs because of their size, particularly when the cilia are not visible. The recovery of *B. coli* from specimens within the United States is rare. Nonetheless, laboratories are expected to be able to identify these organisms and are called upon to do so with proficiency testing specimens.

In a study of parasites from swine feces examined for autofluorescence, cysts of *B. coli* emitted light after excitation with UV light. Although not a common screening approach, this method may be used for the examination of wet mounts of fecal specimens (Daugschies *et al.*, 2001).

Various methods and media are available for the establishment and maintenance of the following organisms in culture: *Entamoeba histolytica, Giardia lamblia, Trichomonas vaginalis, Dientamoeba fragilis, Blastocystis hominis,* and *Balantidium coli*. However, while culture methods are of limited importance in the diagnostic laboratory, they are essential to most research laboratories (Clark and Diamond, 2002).

Direct immunofluorescence assays are used in the clinical virology laboratory for the rapid detection of viruses. An assessment of the cellularity of specimens submitted for DFA is necessary for the most effective use of this assay. This assessment ensures that an adequate number of the appropriate cells are present for examination. During this assessment, clinical virologists may encounter unfamiliar cellular elements or cellular fragments. One of these elements, ciliocytophthoria, has been misinterpreted as a parasite in specimens submitted for cytologic testing. The authors describe a similar case in which a technologist thought that ciliocytophthoria possibly represented a ciliated parasite in a nasopharyngeal specimen sent for respiratory syncytial virus DFA. After a thorough morphologic examination, the staff dismissed the possibility of a ciliated parasite. We confirmed this entity as ciliocytophthoria using morphologic criteria and the Diff-Quik stain. This near misidentification of ciliocytophthoria as a ciliated parasite, *B. coli*, provides the opportunity to raise the awareness about ciliocytophthoria and to emphasize the importance of differentiating ciliated human cells from the parasitic ciliate (Hadziyannis *et al.*, 2000).

6.1 Key Points – Laboratory Diagnosis

1. Since the organisms are so large, they can frequently be seen under low power (100 ×), particularly in a concentration sediment wet mount.
2. If wet mounts are examined using high dry power (400 ×), the organisms may be confused with vegetable cells (both cells and parasites will be larger than other intestinal protozoa). The cilia tend to be short and can be missed on microscopic examination.

3. Although this infection is very rare within the United States, all laboratories may receive positive specimens for proficiency testing.
4. These organisms do not stain well (too large and thick) on the permanent stained smear (trichrome, iron-hematoxylin) and can be confused with fecal debris (including helminth eggs), hence the need to make the diagnosis by using wet smears (from direct mounts or concentrate sediment; Figure 4).

7 Treatment

For treatment of *B. coli*, tetracycline is the drug of choice, although it is considered investigational for this infection. Iodoquinol or metronidazole may be used as alternatives (Abramowicz, 2004).

In an earlier study, therapeutic activity of metronidazole against *B. coli* was evaluated by using two drug regimens in 20 human cases of balantidiasis. All patients also harbored *Trichuris trichiura*. Children received a total amount of drug that varied from 2.5 g over 5 days to 7.5 g in a period of 10 days. Adults were given 5 g in 5 days or 12.5 g during a period of 10 days. In all patients *B. coli* disappeared after the fourth day of treatment. Frequent posttreatment stool examination remained negative. All patients tolerated the drug well and there were no side effects (Garcia-Laverde and de Bonilla, 1975).

8 Epidemiology and Prevention

Balantidium coli is the etiologic agent of balantidiasis, an infrequent zoonosis of worldwide distribution. In a study of the clinical and epidemiological aspects of balantidiasis in a rural community in the Bolivar State in Venezuela, 50 people and 12 pigs were evaluated. Fecal samples were analyzed by direct examination and by concentration. The rate of intestinal infection detected was 88.0% for the human population and 83.3% for the pigs. The prevalence of human and porcine balantidiasis was 12.0% (6/50) and 33.3% (4/12), respectively. The disease was only detected in children, all of them with multiple parasites and with clinical manifestations. Deficient environmental sanitation, absence of basic services in the dwellings, low socioeconomic level, and the presence of pigs infected with *B. coli* are the factors that explain and maintain the conditions favorable to the transmission of balantidiasis in the population studied (Devera *et al.*, 1999).

B. coli infection was studied in 2124 Aymara children 5–19 years of age from the schools of 22 communities of the northern Bolivian Altiplano over a 5-year period. Infection with *B. coli* was found in 11 of the communities surveyed, with prevalences of 1.0–5.3% (overall prevalence 1.2%). The prevalences observed are some of the highest reported and did not differ significantly among the various age groups or between boys and girls. These prevalences, the apparent absence of symptoms or signs of illness due to this parasite in the schoolchildren surveyed at the time of stool sampling, and the consistency of stool samples of the infected students suggest that they are apparently asymptomatic carriers. Infection with *B. coli* must be considered as an endemic anthropozoonosis in the area studied. A relationship between *B. coli* infection and Altiplanic pigs is suggested (Esteban *et al.*, 1998).

Although *B. coli* has been reported from many different simian hosts, domestic hogs probably serve as the most important reservoir host for human infection. In areas where pigs are the main domestic animal, the incidence of human infection can be quite high. Particularly susceptible to infection are persons working as pig farmers or in slaughterhouses (28% infection in New Guinea). Human infection is fairly rare in temperate areas, although once the infection is established it can develop into an epidemic, particularly where poor environmental sanitation and personal hygiene are found. This situation has been seen in mental hospitals in the United States. Preventive measures involve increased attention to personal hygiene and sanitation measures, since the mode of transmission is ingestion of infective cysts through contaminated food or water (Muriuki *et al.*, 1998; Owen, 2005).

Entamoeba histolytica and *B. coli* are parasitic protozoa that cause amoebic dysentery and balantidiasis, respectively. Both intestinal infections are spread via a fecal-oral route, with cysts as the infective stage. Nonhuman primates and swine are reservoirs for *E. histolytica* and *B. coli*, and the diseases they cause are acquired from cysts, usually in sewage-contaminated water. Amoebic dysentery and balantidiasis are examples of zoonotic waterborne infections, though human-to-human transmission can also occur (Schuster and Visvesvara, 2004).

In a study of some gastrointestinal parasites commonly observed in six different Kenyan nonhuman primates on the basis of their health implications for humans, both helminths and protozoan parasites were detected in varying rates in all primate species. *Trichuris* sp. was the most frequent helminth followed by *Strongyloides fuelleborni*, *Strongyles* sp. and *Schistosoma mansoni*, in that order. *Entamoeba coli* was the most common protozoan followed, respectively, by *B. coli* and *E. histolytica*. All primate species examined but one were infected with all the parasites listed.

Cryptosporidium spp. was found in both clinically normal and diarrheic baboons and vervets. Most taxa of parasites observed could impact human welfare directly through infection and cause of illness, and indirectly through increased cost of livestock production and decreased availability of animal proteins. The potential of some of the agents to cause opportunistic infections in immunocompromised persons was suggested as a likely threat. This would warrant such person's exemption from high-risk operations at primate and other animal facilities in developing countries.

Specific studies are ongoing to clarify the epidemiology, socio-economic impact and pathogenicity of the primate parasites to other species of animals and humans (Muriuki *et al.*, 1998).

In a study designed to determine the prevalence of intestinal parasitic infections in the residents of four Italian psychiatric institutions, the authors examined stool specimens collected in triplicate from 238 residents, enrolled between May 1995 and May 1996. *Trichuris trichiura* ova, *G. lamblia* cysts and trophozoites, *C. parvum* oocysts, and *B. coli* cysts were found in the fecal samples from 22 residents (9.2%). Statistical analyses revealed that the presence of pathogenic parasites in fecal specimens was significantly associated with diarrhea, nausea, vomiting, abdominal pain, fever, behavioral aberrations and nonpathogenic protozoa, but did not demonstrate any other significant associations between these parasites and the other variables, such as pruritus, mucus or blood in the stools and presence of fecal leukocytes. Data analyses revealed that both pathogenic and nonpathogenic parasites were significantly more common in institutionalized patients than in controls. The rare presence of clinical signs and symptoms in colonized patients represents an important public health problem, since the presence of asymptomatic carriers among residents with low hygienic conditions, raises concern of transmission of parasitic infections to professional staff and other residents. Since the eradication of parasitic colonization in residential facilities is hard to reach, an effective prevention is the only measure to deal with this public health problem (Giacometti *et al.*, 1997).

Protozoa of nose, mouth, and pharynx of 30 randomly chosen female caries patients at an odontological clinic of the National Autonomous University of Mexico, were surveyed by culture from swabs. Culture tubes of swabs from each patient were observed every other day over 5 weeks. Pathogenic protozoa found included *E. histolytica*, *Naegleria fowleri*, *Acanthamoeba castellanii*, *A. culbertsoni*, and *B. coli*. This isolation of pathogens suggests that healthy patients may be healthy carriers of cysts of protozoa, mainly amoebae, responsible for several diseases, including primary amoebic meningoencephalitis. Small pathogenic free-living amoebae have not been isolated before from females in Mexico; however, in this study many species of free-living protozoa were also cultured from swabs from the patients (Rivera *et al.*, 1984).

Most parasites of humans in New Guinea are cosmopolitan species, widely distributed and highly prevalent in the island. This impoverished fauna includes anthroponotic species normally occurring at low prevalences, for example *Isospora belli*, *Dientamoeba fragilis*, *Trichomonas hominis* and a few zoonotic forms with pigs as reservoir hosts, notably *B. coli* and *Entamoeba polecki* (Barnish and Ashford, 1989).

An area for prevention that is often over-looked is mechanical transmission by cockroaches. These arthropods represent an important reservoir for infectious pathogens, including *B. coli*; therefore, control of cockroaches will substantially minimize the spread of infectious diseases in our environment (Tatfeng *et al.*, 2005).

9 References

Abramowicz, M. (ed.) (2004) Drugs for parasitic infections. *Med. Lett.* **46**: 1–12.

Agapov, M. (2006) Balantidiasis in a patient with suspected Crohn's disease. *Endoscopy* **38**: 655.

Anargyrou, K., Petrikkos, G.L., Suller, M.T., Skiada, A., Siakantaris, M.P., Osuntoyinbo, R.T., Pangalis, G. and Vaiopoulos, G. (2003) Pulmonary *Balantidium coli* infection in a leukemic patient. *Am. J. Hematol.* **73**: 180–183.

Barnish, G. and Ashford, R.W. (1989) Occasional parasitic infections of man in Papua New Guinea and Irian Jaya (New Guinea). *Ann. Trop. Med. Parasitol.* **83**: 121–135.

Beaver, P.C., Jung, R.C. and Cupp, E.W. (1984) *Clinical Parasitology*, 9th Edn. Lea & Febiger, Philadelphia, PA.

Castro, J., Vazquez-Iglesias, J.L. and Arnal-Monreal, F. (1983) Dysentery caused by *Balantidium coli* – report of two cases. *Endoscopy* **15**: 272–274.

Cermeno, J.R., Hernandez De Cuesta, I., Uzcategui, O., Paez, J., Rivera, M. and Baliachi, N. (2003) *Balantidium coli* in an HIV-infected patient with chronic diarrhoea. *AIDS* **17**: 941–942.

Clark, C.G. and Diamond, L.S. (2002) Methods for cultivation of luminal protists of clinical importance. *Clin. Microbiol. Rev.* **15**: 329–341.

Clyti, E., Aznar, C., Couppie, P., el Guedj, M., Carme, B. and Pradinaud, R. (1998) A case of coinfection by *Balantidium coli* and HIV in French Guiana. *Bull. Soc. Pathol. Exot.* **91**: 309–311.

Daugschies, A., Bialek, R., Joachin, A. and Mundt, M.C. (2001) Autofluorescence microscopy for the detection of nematode eggs and protozoa, in particular *Isospora suis*, in swine faeces. *Parasitol. Res.* **87**: 409–412.

Devera, R., Requena, I., Velasquez, V., Castillo, H., Guevara, R., De Sousa, M., Marin, C. and Silva, M. (1999) Balantidiasis in a rural community from Bolivar State, Venezuela. *Bol. Chil. Parasitol.* **54**: 7–12.

Esteban, J.G., Aguirre, C., Angles, R., Ash, L.R. and Mas-Coma, S. (1998) Balantidiasis in Aymara children from the northern Bolivian Altiplano. *Am. J. Trop. Med. Hyg.* **59**: 922–927.

Ferry, T., Bouhour, D., De Monbrison, F., Laurent, F., Dumouchel-Champagne, H., Picot, S., Piens, M.A. and Granier, P. (2004) Severe peritonitis due to *Balantidium coli* acquired in France. *Eur. J. Clin. Microbiol. Infect. Dis.* **23**: 393–395.

Garcia, L.S. (1999) Flagellates and ciliates. *Clin. Lab. Med.* **19**: 621–638.

Garcia, L.S. (2007) *Diagnostic Medical Parasitology*, 5th Edn. ASM Press, Washington, DC.

Garcia, L.S., Shimizu, R.Y. and Deplazes, P. (2003) Specimen collection, transport, and processing: parasitology. In: *Manual of Clinical Microbiology*, 8th Edn (eds P.R. Murray, E.J. Baron, J.H. Jorgensen, M.P. Pfaller and R.H. Yolken). ASM Press, Washington, DC.

Garcia-Laverde, A. and de Bonilla, L. (1975) Clinical trials with metronidazole in human balantidiasis. *Am. J. Trop. Med. Hyg.* **24**: 781–783.

Giacometti, A., Cirioni, O., Balducci, M., et al. (1997) Epidemiologic features of intestinal parasitic infections in Italian mental institutions. *Eur. J. Epidemiol.* **13**: 825–830.

Gonzalez Sanchez, O. (1978) Acute appendicitis caused by *Balantidium coli. Rev. Cubana Med. Trop.* **30**: 9–13.

Hadziyannis, E., Yen-Lieberman, B., Hall, G. and Procop, G.W. (2000) Ciliocytophthoria in clinical virology. *Arch. Pathol. Lab. Med.* **124**: 1220–1223.

Karapetian, A.E., Isaakian, Z.S. and Zavgorodniaia, A.M. (1978) Importance of cellular immunity factors in the pathogenesis of experimental balantidiasis. *Parazitologiia* **12**: 323–326.

Ladas, S.D., Savva, S., Frydas, A., Kaloviduris, A., Hatzioannou, J. and Raptis, S. (1989) Invasive balantidiasis presented as chronic colitis and lung involvement. *Dig. Dis. Sci.* **34**: 1621–1623.

Lahiri, V.L., Elhence, B.R. and Agarwal, B.M. (1977) *Balantidium* peritonitis diagnosed on cytologic material. *Acta Cytol.* **21**: 123–124.

Maleky, F. (1998) Case report of *Balantidium coli* in human from south of Tehran, Iran. *Indian J. Med. Sci.* **52**: 201–202.

Muriuki, S.M., Murugu, R.K., Munene, E., Karere, G.M. and Chai, D.C. (1998) Some gastro-intestinal parasites of zoonotic (public health) importance commonly observed in old world hon-human primates in Kenya. *Acta Trop.* **71**: 73–82.

Nakauchi, K. (1999) The prevalence of *Balantidium coli* infection in fifty-six mammalian species. *J. Vet. Med. Sci.* **61**: 63–65.

National Committee for Clinical Laboratory Standards (1997) *Procedures for the Recovery and Identification of Parasites from the Intestinal Tract, Approved Guideline, M28-A.* National Committee for Clinical Laboratory Standards, Villanova, PA.

Owen, I.L. (2005) Parasitic zoonoses in Papua New Guinea. *J. Helminthol.* **79**: 1–14.

Rivasi, F. and Giannotti, T. (1983) *Balantidium coli* in cervico-vaginal cytology. a case report. *Pathologica* **75**: 439–442.

Rivera, F., Medina, F., Ramirez, P., Alcocer, J., Vilaclara, G. and Robles, E. (1984) Pathogenic and free-living protozoa cultured from the nasopharyngeal and oral regions of dental patients. *Environ. Res.* **33**: 428–440.

Schuster, F.L. and Visvesvara, G.S. (2004) Amebae and ciliated protozoa as causal agents of waterborne zoonotic disease. *Vet. Parasitol.* **126**: 91–120.

Skotarczak, B. (1997) Ultrastructural and cytochemical identification of peroxisomes in *Balantidium coli*, Ciliophora. *Folia Biol. (Krakow)* **45**: 117–120.

Skotarczak, B. (1999) Cytochemical identification of mucocysts in *Balantidium coli* trophozoites. *Folia Biol. (Krakow)* **47**: 61–65.

Skotarczak, B. and Kolodziejczyk, L. (2005) An electron microscopic study of the phosphatases in the ciliate *Balantidium coli. Folia Morphol. (Warsz.)* **64**: 282–286.

Tatfeng, Y.M., Usuanlele, M.U., Orukpe, A., Digban, A.K., Okodua, M., Oviasogie, F. and Turay, A.A. (2005) Mechanical transmission of pathogenic organisms: the role of cockroaches. *J. Vector Borne Dis.* **42**: 129–134.

Vasilakopoulou, A., Dimarongona, K., Samakovli, A., Papadimitris, K. and Aviami, A. (2003) *Balantidium coli* pneumonia in an immunocompromised patient. *Scand. J. Infect. Dis.* **35**: 144–146.

Yang, Y., Zeng, L., Li, M. and Zhou, J. (1995) Diarrhoea in piglets and monkeys experimentally infected with *Balantidium coli* isolated from human faeces. *J. Trop. Med. Hyg.* **98**: 69–72.

Yazar, S., Altuntas, F., Sahin, I. and Atambay, M. (2004) Dysentery caused by *Balantidium coli* in a patient with non-Hodgkin's lymphoma from Turkey. *World J. Gastroenterol.* **10**: 458–459.

D Flagellates

D1 Diplomonadida – *Giardia* spp.

Edward L. Jarroll

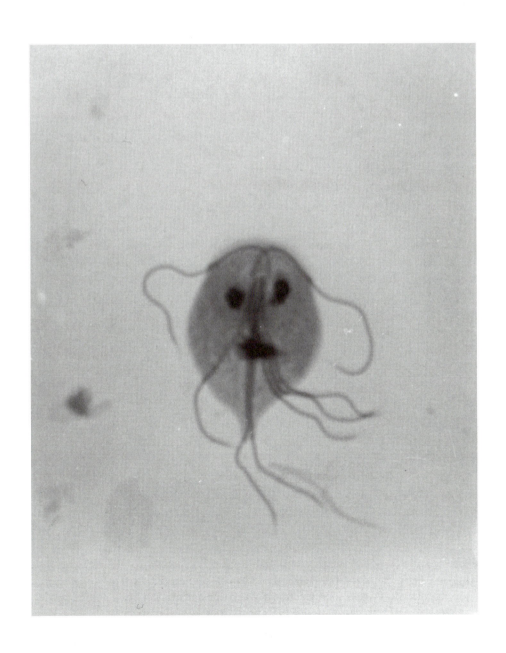

1 Introduction

Parasitic protozoa in the genus *Giardia* are members of the order Diplomonadida (literally two animals), which derives its name in part from the fact that *Giardia* trophozoites have two morphologically identical nuclei and vaguely resemble two cells stuck together. This intestinal parasite (Figure 1) has eight flagella, a ventral adhesive disk, a microtubular median body, mitosomes, and lysosome-like peripheral vesicles, but lacks typical Golgi complexes, functional mitochondria and aerobic respiration.

Giardia enters the vertebrate host when cysts are ingested in food, in water or on fomites contaminated with feces from infected hosts. Trophozoites, encased within the cyst wall, establish infection in the upper

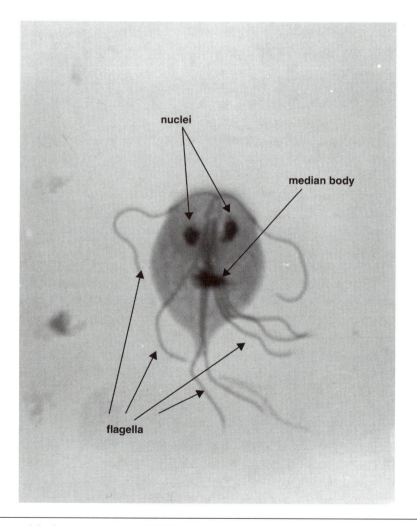

Figure 1 Light micrograph (~1600×) of Giemsa-stained *Giardia* trophozoite showing morphologically identical nuclei, microtubular median body and flagella (courtesy of D. Feely, University of Nebraska College of Dentistry).

small intestine after they excyst. *Giardia* exits the host's intestine with the feces once trophozoites re-form cysts in the lower small and upper large intestine.

2 History of Discovery

The protozoan, now called *Giardia*, was first described by van Leeuwenhoek in 1681 from his own stool (Dobell, 1920); Lambl (1859) had described this same organism as *Cercomonas intestinalis*. The first use of the name *Giardia* for the genus was when Grassi (1879) described *G. agilis* from a tadpole.

Giardia from humans has been called *G. lamblia* as well as *Lamblia intestinalis* (the latter was especially used in Eastern Europe), but Kulda and Nohýnková (1996) determined that *G. intestinalis* is its more appropriate name. Numerous names for what we now call *Giardia* appear in the literature undoubtedly because of the lack of effective communication in earlier times.

3 Taxonomy

The importance of understanding *Giardia* taxonomy becomes clearer as one observes that, for the most part, many known *Giardia* species (especially the cyst forms) share almost indistinguishable morphological features. Prior to the late 1980s, some believed that there were at least 40 species of *Giardia* each colonizing a different host species (Kulda and Nohýnková, 1978; Meyer and Jarroll, 1980; Thompson, 2002). These original 40 'species', except for the human parasite *G. lamblia*, apparently were named for the host from which they were described with little or no regard for morphological differences in the parasite.

In the late 1980s, many investigators began following the morphology-based *Giardia* nomenclature proposed by Filice (1952); namely, that there are only three *Giardia* species based on median body morphology (see section 6). A claw hammer-shaped median body is typical of *G. duodenalis*, a round type is typical of *G. muris*, and a long teardrop-shaped median body typifies *G. agilis*. The *G. duodenalis* type includes many of the original 40 species previously described, including *G. lamblia*, and that are considered the same type but from many different hosts. *G. muris* is found in rodents and *G. agilis* is from tadpoles.

Besides the three types in Filice's (1952) taxonomic scheme, three new species have been suggested in more recent times based on sound morphological and molecular criteria: *G. psittaci* from parakeets (Erlandsen and Bemrick, 1987), *G. ardeae* from great blue herons (Erlandsen *et al.*, 1990a and b), and *G. microti* from muskrats and voles (Feely, 1988).

Attempts to separate *Giardia*, especially the *G. duodenalis* type, into molecularly defined taxonomic groups began with an isoenzyme comparison by Bertram *et al.* (1983). Two years later, Nash *et al.* (1985) used restriction endonuclease analysis of DNA (RFLP) to distinguish 15 *G. lamblia* (syn. *intestinalis* or *duodenalis*) isolates. Electrophoretic karyotyping of *Giardia* showed that some *G. duodenalis*-type isolates from humans have closely related chromosomes and chromosomal patterns (Adam *et al.*, 1988a and b; Campbell *et al.*, 1990), but that *Giardia* from herons and mice had very different chromosomal patterns (Campbell *et al.*, 1990) from those of *duodenalis* type *Giardia* from human and cat isolates. For a complete review of such analyses see Adam (2001).

Today, genotyping (reviewed by Thompson, 2002) has helped sort out some of the earlier confusion in the *G. duodenalis* group identifying at least seven genotypes (and hosts from which they have been isolated): Assemblage A (human, livestock, cat, dog, beaver, guinea pig, slow loris); Assemblage B (human, slow loris, chinchilla, dog, beaver, rat, siamang); Dog (dog); Cat (cat); Livestock (alpaca, cattle, goat, pig, sheep); Rat (domestic rats); and Muskrats/voles (wild rodents). Subgroups have been identified within some of these assemblages. Since some *Giardia* infecting lower mammals (including domesticated ones) may also infect humans, giardiasis represents a potential zoonosis (a disease of animals that can be transmitted to humans), and molecular typing information can be useful when studying the epidemiology of giardiasis in a variety of situations including outbreaks.

4 Ecological Distribution

Giardia is ubiquitous and is transmitted by the fecal-oral route: cysts can survive in cold water for several months but may only last a few days at 37°C (Bingham *et al.*, 1979). Cysts, however, do not survive freezing, desiccation or artificial seawater for even 24 hours (Bingham *et al.*, 1979; Meyer and Jarroll, 1980).

Giardia infection (giardiasis) is truly worldwide from arctic to tropical climates (Meyer and Jarroll, 1980). Members of the genus have been reported in hosts literally from fish to mammals (Thompson, 2002).

5 Life Cycle

Giardia exhibits a simple and direct life cycle (meaning that no intermediate host is required in the life cycle; Meyer and Jarroll, 1980). The trophozoite (Figure 2A, B), a vegetative (actively feeding and metabolizing) stage, is pear-shaped, measuring from 7 to 10 μm at its widest part, and 10–15 μm long. The cyst (Figure 2C, D; trophozoites encased in a multilayered cyst wall) is oval, measuring from 10 to 15 μm in length. Cysts are excreted in the host's feces and are transmitted to the next host when cysts contaminate food, water or fomites. Ingested, viable cysts excyst after passage through the stomach and exposure to an acid level of hydrogen ion concentration (and possibly HCO_3^-; Bingham *et al.*, 1979; Feely, 1986). After excystation, trophozoites colonize (Figure 3) and reproduce by binary

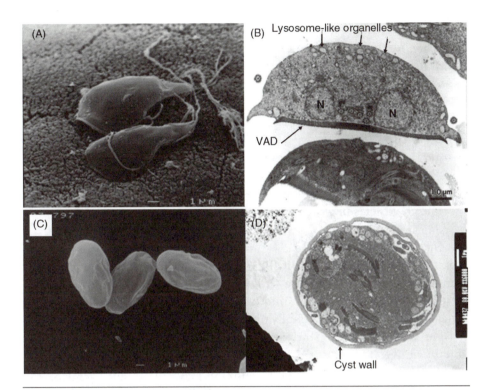

Figure 2 *Giardia* life cycle. (A) Scanning electron micrograph of trophozoites attached to intestinal epithelial microvilli (courtesy B. Koudela, University of Brno Veterinary School); (B) transmission electron micrograph of transversely sectioned trophozoites showing two nuclei, ventral adhesive disk (VAD), and lysosome-like organelles (courtesy D. Feely, University of Nebraska College of Dentistry); (C) scanning electron micrograph of cysts (courtesy B. Koudela, University of Brno Veterinary School); (D) transmission electron micrograph of cyst showing the cyst-wall filaments (courtesy of B. Koudela, University of Brno Veterinary School). VAD, ventral adhesive disk.

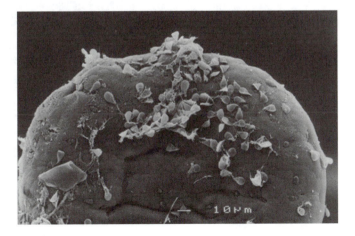

Figure 3 *Giardia* trophozoites attached to an intestinal villus. Micrograph courtesy B. Koudela (University of Brno Veterinary School).

fission in the host's small intestine where, in the presence of bile, they form cysts. *G. microti* cysts appear structurally different internally from those of *G. intestinalis* (see below), and there have been no reports of cysts from *G. ardeae* or *G. agilis*.

6 Biology

6.1 Organelles

Giardia trophozoites exhibit two morphologically identical nuclei (four in cysts), microtubular median bodies, four pairs of bilaterally symmetrical flagella, lysosome-like vesicles, a ventral adhesive disk, mitosomes and specialized vesicles in encysting trophozoites.

6.2 Nuclei

Trophozoites have two genetically (Yu *et al.*, 2002) and morphologically identical nuclei per cell (Figure 4). Thus, *Giardia* nuclei differ from those of the ciliate macronucleus and micronucleus, which are not only morphologically distinct but functionally distinct as well.

Giardia's nuclei have nearly equal amounts of DNA, are both transcriptionally active (Kabnick and Peattie, 1990) and divide nearly simultaneously (Weisehahn *et al.*, 1984). Elmendorf *et al.* (2000) used stable transfection systems to demonstrate that both nuclei are able to import nuclear proteins and support the idea that both nuclei are active.

Microscopic analyses (Sagolla *et al.*, 2006) showed that *Giardia*'s nuclei have a semi-open mitosis in which two extranuclear spindles attach to the chromatin through openings at the poles of the nuclear membranes.

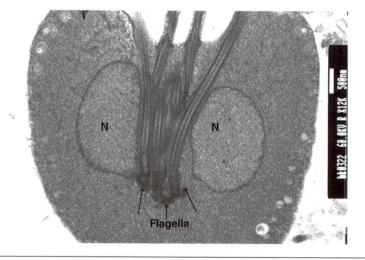

Figure 4 Transmission electron micrograph of a *Giardia* trophozoite showing the two morphologically identical nuclei and flagella (courtesy of B. Koudela, University of Brno Veterinary School).

Furthermore, these authors showed that the nuclei remain separate during mitosis and that the daughter cells inherit them with mirror-image symmetry.

6.3 Median body and flagella

The median body (Figure 5; mb – bundles of microtubules, mts) lies approximately dorsal to the *Giardia* trophozoite's caudal flagella and in the trophozoite's midline. Soltys and Gupta (1994) reported tubulins in the mts and Meng *et al.* (1996) suggested that the mbs are sites where mts assemble for incorporation into the ventral adhesive disk (see below).

Piva and Benchimol (2004) observed that mbs occur in roughly 80% of trophozoites and that they vary in shape, location and number. Mbs may be attached to the plasma membrane, caudal flagella, adhesive disk, or protrude to the trophozoite surface, but they apparently do not float freely in the cell. Mbs seemingly occur in cells regardless of whether or not they are in interphase or mitosis.

Centrin, a member of the EF-hand family of calcium-binding protein with calcium-sensitive contractility, has been localized by monoclonal antibodies to the mb, basal bodies, adhesive disk, funis (a short array of mts arising from the axonemes of the caudal flagella; Benchimol *et al.* (2004)) and the flagellar axonemes (Correa *et al.*, 2004).

Giardia's four pairs of flagella (Figure 4) appear to have the typical microtubular eukaryotic flagellar structure and arise from two sets of basal bodies at or near the midline of the trophozoite. *Giardia* trophozoites are weakly motile and use of their flagella produces their characteristic 'falling leaf' motility. The flagella do not appear to be used for attachment.

Nohýnková *et al.* (2006) reported that the eight parental flagella persist during trophozoite division, but they are transformed and distributed

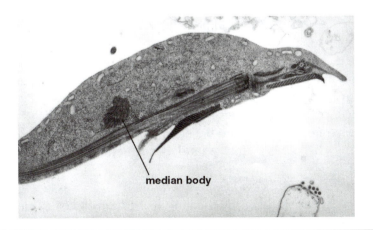

median body

Figure 5 Transmission electron micrograph of a *Giardia* trophozoite showing the microtubular median body (courtesy of D. Feely, University of Nebraska College of Dentistry).

between the daughter trophozoites in a semiconservative manner with each daughter trophozoite receiving a different set of four parental flagella. The posterolateral and ventral pairs of flagella develop *de novo* during mitosis in both daughter trophozoites. The first evidence of division was a gradual shift of points of emergence of the two posteriorly directed parental anterolateral flagella. From the original lateral positions, the flagella migrate anteriorly along the cell's periphery, meet at the midline, and exit at the subapical region of the dorsal surface. The flagellum's exit from the left side was ventral, while that of the right flagellum was dorsal. The flagella moved apart, migrating backward while pointing forward, eventually assuming a lateral position on both sides of the dividing trophozoite. Each of the parent anterolateral flagella emerged on the opposite side with respect to its interphase position. Both of them were directed laterally, and either to the right or left. Posterolateral, ventral, and caudal parental flagella retain their original interphase positions; newly formed flagella were not observed at this stage.

From the dorsal aspect, the left daughter trophozoite received an anterolateral, a caudal, and two posterolateral parental flagella; the right daughter trophozoite received an anterolateral, a caudal, and two ventral parent flagella. The pairs of parent posterolateral and ventral flagella transformed, by reorientation, to pairs of anterolateral flagella in the left and right daughter trophozoites. In each daughter, the parent's anterolateral flagellum became, by reorientation during division, a caudal flagellum and joined the parent caudal flagellum completing the daughter's pair. Pairs of ventral and posterolateral flagella were assembled *de novo* in each trophozoite and each basal body pair in the interphase daughter trophozoites contains a parental basal body along with a new one. Because of the flagellar transformation during division, the resulting interphase daughter trophozoites have flagella of different ages and functions. The newest flagella are those of the ventral and posterolateral pairs with the anterolateral pair of flagella being a generation older. The caudal pair has the oldest flagella. The flagellum arising from the transformed parent anterolateral flagellum is two generations old and the other was at least one generation older. Nohýnková *et al.* (2006) also show that basal body and flagellar maturation occurs over three successive cell cycles in *Giardia*. Thus in this scheme, each newly assembled flagellum goes through three transformations before it becomes a mature caudal flagellum.

6.4 Ventral adhesive disk and ventrolateral flange

The ventral adhesive disk (VAD; Figure 6) in *Giardia* covers up to three quarters of the ventral surface of the trophozoites depending upon the species. This concave microtubular structure is used to attach trophozoites to the microvillous brush border of the host's intestinal epithelium (Figure 7) leaving indentations of unknown duration after the trophozoites detach.

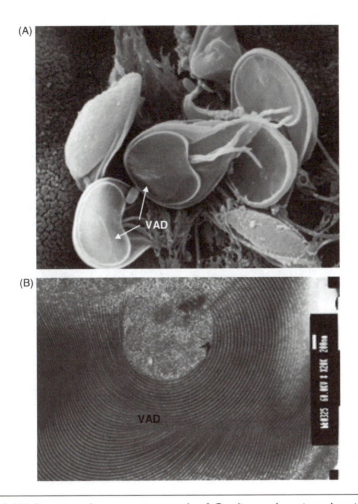

Figure 6 (A) Scanning electron micrograph of *Giardia* trophozoites showing the ventral adhesive disk (courtesy of D. Feely, University of Nebraska College of Dentistry). (B) Transmission electron micrograph showing microtubules of the ventral adhesive disk of a single *Giardia* trophozoite (courtesy of B. Koudela, University of Brno Veterinary School). VAD, ventral adhesive disk.

Apparently, the VAD is used as a holdfast to prevent the trophozoite from being expelled prematurely (before cyst formation) from the intestine. This attachment may fail during bouts of severe diarrhea when the stool contains more trophozoites than cysts.

The VAD is associated with the ventral membrane through the protofilaments, and there are groups of 29–38 kDa α-helical proteins (termed giardins) that form part of the microribbons attaching the mts to the membrane. There are at least 23 different forms of giardins including α1-, α2-, β- and γ-giardins (Aggarwal *et al.*, 1989; Alonso and Peattie, 1992; Holbertson *et al.*, 1988; Nohria *et al.*, 1992; Peattie *et al.*, 1989).

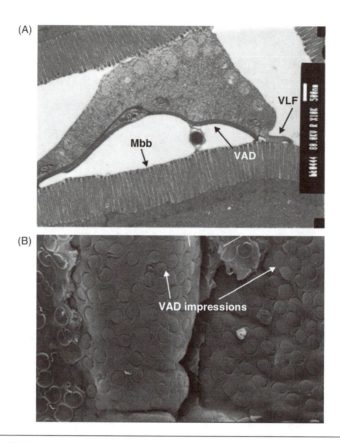

Figure 7 (A) Transmission electron micrograph showing a *Giardia* trophozoite attached to the epithelial microvillous brush border (mbb; VAD, ventral adhesive disc; VFL, ventrolateral flange). (B) Scanning electron micrograph of intestinal villus showing indentations caused by *Giardia*'s ventral adhesive disc. Micrographs courtesy of B. Koudela (University of Brno Veterinary School).

In some *Giardia* species, the VAD narrows to a lateral crest with a flexible ventrolateral flange (VLF; Figures 7 and 8). Using high resolution field emission scanning electron microscopy, Erlandsen *et al.* (2004) showed the attachment of trophozoites to the top of pillars in the microfabricated substrates. Their adhesion was apparently mediated by direct attachment of the VLF to the tops of the pillars. Attachment to the microfabricated surfaces was 16% of that observed for attachment mediated by the VAD. Their report provides at least some experimental evidence that the VLF may play a role in trophozoite adhesion. Interestingly, one of the morphological characteristics upon which the description of *G. psittaci* from parakeets rests is that this species lacks a VLF (Erlandsen and Bemrick, 1987).

6.5 Mitosomes
All living eukaryotes apparently are descended from an ancestor that had a mitochondrion and in anaerobic eukaryotes the organelle has been

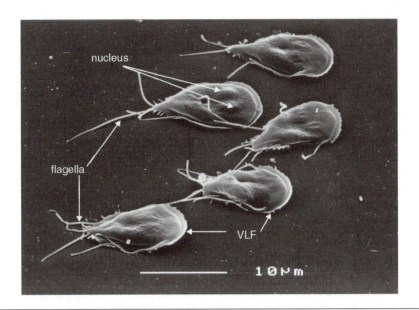

Figure 8 Scanning electron micrograph of *Giardia* trophozoites showing the ventrolateral flange (VLF; courtesy of B. Koudela, University of Brno Veterinary School).

modified into either hydrogenosomes, which generate energy and molecular hydrogen for the host cell, or mitosomes, which do not (Tovar *et al.*, 2003; van der Giezen *et al.*, 2005). Mitosomes (also found in *Entamoeba*) are reportedly double-membrane-bounded remnants of mitochondria that function in iron-sulphur (FeS) protein maturation. *Giardia* contains mitosomes (~25 to more than 100 per trophozoites measuring between 137–142 × 61–68 nm) and does not exhibit typical eukaryotic mitochondria or hydrogenosomes (Tovar *et al.*, 2003). Richards and van der Giezen (2006) showed that *Giardia* mitosomes have α-proteobacterium-derived, mitochondrial-type FeS cluster proteins supporting a single common ancestry for mitochondria and mitosomes. They proposed that their findings demonstrate that the α-proteobacterial endosymbiosis occurred before the last common ancestor of all eukaryotes.

The *Giardia* mitosome has Cpn60, mitochondrial Hsp70 and ferredoxin as well as vestiges of the independent import pathways (Regoes *et al.*, 2005). Mitosomes appear in *Giardia* either peripherally or centrally depending upon their cellular location. The division and segregation of the central mitosomes associated with the mts cytoskeleton appears to co-ordinate with the cell cycle but peripheral mitosomes are inherited randomly by daughter trophozoites. Dolezal *et al.* (2006) reported that in *Giardia* FeS center proteins, GiiscS, GiiscU and (2Fe2S) ferredoxin, were targeted to the mitosomes and exhibited targeting signals recognizable by mitosomal protein import machinery. Further, proteins resembling mitochondrial inner membrane translocase and peptidase were detected in the *Giardia* mitosome.

6.6 Lysosome-like vesicles (peripheral vesicles)

In the 1980s, Lindmark (1980, 1988) demonstrated that there are hydrolytic enzymes, notably acid phosphatase (a marker for lysosomes in other eukaryotic systems) in nonencysting *Giardia* trophozoites, and that their activities are associated with a particle population having a density in sucrose of about 1.18. These particles (Figure 9) were referred to as lysosome-like organelles. Additionally, Reiner *et al.* (1990) showed that there is regulated transport of cyst-wall proteins and a distinct constitutive lysosomal pathway. They localized acid phosphatase activity to the endoplasmic reticulum, atypical Golgi complex, and small constitutive peripheral vacuoles which function as lysosomes.

6.7 Genome

The *Giardia* genome has been sequenced and is available at www.mbl.edu/Giardia. It exhibits a haploid size between 10.6 and 12.3 Mb (Adam *et al.*, 1988b; Fan *et al.*, 1991), and telomere-flanked linear chromosomes

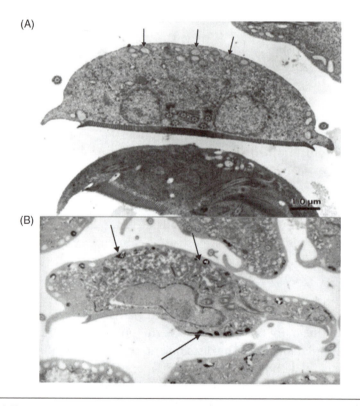

Figure 9 Transmission electron micrograph of *Giardia* trophozoites. (A) Transverse section showing unstained lysosome-like organelles. (B) Longitudinal section showing lysosome-like organelles stained for acid phosphatase (black precipitate). Arrows indicate lysosome-like organelles. Micrographs courtesy of D. Feely (University of Nebraska College of Dentistry).

and chromatin are formed by chromosomal DNA associating with four histones.

Five chromosome groups have been identified that range from 1.6 to 3.8 Mb and trophozoites appear to alternate between 4N (stationary phase) and 8N (Adam *et al.*, 1988b; Bernander *et al.*, 2001; Le Blancq and Adam, 1998). Yu *et al.* (2002) showed that the two nuclei have complete sets of each of the five chromosomes.

6.8 Transcription

Giardia transcription is eukaryotic: transcripts originate in the nucleus and are transported to the cytoplasm for translation (Adam, 2001). There is polyadenylation but there are short 5' untranslated regions suggesting that the promoters may be near the beginning of an open reading frame. Some transcripts appear to lack capped 5' ends. Only two introns have been reported in *Giardia* to date: the first a spliceosomal-type 35-bp intron in a gene encoding a putative (2Fe-2S) ferredoxin (Nixon *et al.*, 2002) and the second in the ribosomal gene Rp17a (Russell *et al.*, 2005).

Sun *et al.* (2006) reported that *Giardia* has four genes with putative GARP domains, a large family of DNA-binding proteins from plants that may have use in a number of cellular functions such as transcription, differentiation and phosphotransfer signaling. One of these genes, GARP-like protein 1, increased slightly during encystment and these investigators suggest that it may function as a transcriptional activator.

6.9 Giardiavirus

'So naturalists observe, a flea
Has smaller fleas that on him prey;
And these have smaller still to bite 'em
And so proceed *ad infinitum*.'

By *Jonathan Swift*

Giardia too has its smaller 'fleas' in the form of *G. lamblia* virus, a 33-nm double-stranded, nonenveloped, icosahedral virus with about 200 copies in the nuclei (De Jonckheere and Gordts, 1987; Wang and Wang, 1986; Wang *et al.*, 1993). A second 6.2-kb double-stranded RNA virus has been described from *Giardia* (Tai *et al.*, 1996); this virus exhibited a capsid protein of only 95 kDa.

7 Metabolism

7.1 Purines and pyrimidines

For many years, parasitic protozoa have been known to salvage rather than synthesize their purine requirements *de novo*. However, the first report that a parasitic protozoan does not synthesize its pyrimidines *de novo* came when Lindmark and Jarroll (1982) first showed this for *Giardia*. In 1983, they also reported the same finding for *Tritrichomonas foetus* (Jarroll and Lindmark, 1983). Aldritt *et al.* (1985) were the first to elucidate the details of *Giardia's* salvage pathways for its pyrimidine requirements.

Wang and Aldritt (1983) also were the first to show that *Giardia*, like other parasitic protozoa, do not synthesize purines *de novo*. In lieu of *de novo* purine and pyrimidine syntheses, *Giardia* uses salvage pathways to acquire its nucleic acid requirements. Purine nucleosides are salvaged and hydrolyzed to their respective bases (Figure 10). Deoxyribonucleotides, used in DNA synthesis, must be taken up directly since *Giardia* lacks ribonucleotide reductase activity and thus cannot convert ribonucleotides to deoxyribonucleotides.

Salvage of pyrimidine nucleosides serves as the source for nucleotide synthesis (Figure 10) in *Giardia*. Thymidine and uridine are likely transported by the same transporter with the uridine being converted to uracil by either a phosphorylase or hydrolase (Davey *et al.*, 1991; Ey *et al.*, 1992).

7.2 Amino acids
Arginine dihydrolase pathway
While glucose appears to be the only carbohydrate energy source for *Giardia* trophozoites, energy can be generated by metabolizing arginine via the arginine dihydrolase pathway (Figure 11). As with glucose catabolism,

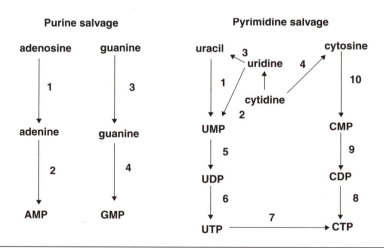

Figure 10 Purine and pyrimidine salvage pathways in *Giardia*. Purine salvage enzymes: 1, adenosine hydrolase; 2, adenine phosphoribosyltransferase; 3, guanosine hydrolase; 4, guanine phosphoribosyltransferase. Pyrimidine salvage enzymes: 1, uracil phosphoribosyltransferase; 2, uridine phosphotransferase (postulated); 3, uridine/thymine phosphorylase; 4, cytidine hydrolase; 5, UMP kinase; 6, UDP kinase; 7, CTP synthetase; 8, CDP kinase; 9, cytosine phosphoribosyltransferase; 10, cytidine deaminase.

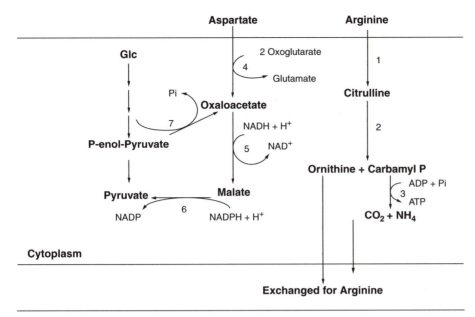

Figure 11 Arginine dihydrolase pathway and aspartate metabolism in *Giardia*. Enzymes: 1, arginine deaminase; 2, ornithine transcarbamylase; 3, carbamate kinase; 4, aspartate transaminase; 5, malate dehydrogenase; 6, malate dehydrogenase (decarboxylating); 7, phosphoenolpyruvate carboxyphosphotransferase.

arginine catabolism can produce ATP by substrate-level phosphorylation of ADP while degrading arginine to ornithine and ammonia (Edwards *et al.*, 1992; Schofield *et al.*, 1990, 1992).

Aspartate

Aspartate is converted to oxaloacetate and in turn to malate and eventually pyruvate (Figure 11). Additional energy generation may occur from this conversion via pyruvate (Mendis *et al.*, 1992).

Cysteine

Cysteine helps protect *Giardia* from the toxic effects of oxygen (Gillin and Diamond, 1981a, 1981b; Gillin and Reiner, 1982; Luján and Nash, 1994). This aerotolerant anaerobe requires cysteine, at about 16 mM, at least in axenic culture medium since *Giardia* does not synthesize cysteine *de novo* or make it from cysteine; it is acquired mainly by passive diffusion.

7.3 Lipids

Giardia trophozoites are lipid auxotrophs (Jarroll *et al.*, 1981), that is, they do not synthesize their lipid requirements *de novo*. While isoprenylation of proteins probably occurs in *Giardia*, the complete conversion of mevalonate to isoprenoids and then to cholesterol does not (Luján *et al.*, 1995c). Phospholipids and cholesterol are salvaged from the environment from lipoproteins, β-cyclodextrins and bile salts (Farthing *et al.*, 1985; Gillin *et al.*, 1986; Luján *et al.*, 1996b) in its external milieu.

7.4 Carbohydrate and partial tricarboxylic acid cycle components

Giardia trophozoites and cysts are metabolically active; cysts are active at a level about 20% that of trophozoites as assessed by oxygen uptake (Jarroll, 1991; Lindmark, 1980; Paget *et al.*, 1989, 1993a, 1993b). The oxygen uptake measurements are indirect assessments of metabolic activity because protozoa, such as *Giardia*, *Entamoeba*, *Trichomonas* and others that lack mitochondria, do not metabolize glucose aerobically and do not take up oxygen to use as the final electron acceptor in cytochrome-mediated electron transport. *Giardia* lives attached mainly to the mucosa of the small intestine and are subjected to small quantities of potentially toxic oxygen in the 0 to 60 μM range. Superoxide dismutase and NADH oxidases have been proposed as possible mechanisms by which *Giardia* take up and detoxify oxygen (Brown *et al.*, 1995, 1996; Lindmark, 1980).

Giardia metabolizes glucose, its only known carbohydrate energy source, and generates ATP by substrate-level phosphorylation (Lindmark, 1980; Figure 12). Glucose is converted to pyruvate via glycolysis and pyruvate enters intermediary metabolism to yield ethanol, acetate, alanine, and carbon dioxide as end products. Alanine (major) and acetate predominate over ethanol under anaerobic conditions, but ethanol predominates and alanine is inhibited even in low concentrations (ca. 0.25 μM–46 μM) of oxygen (Paget *et al.*, 1993b). Alanine also may be useful to *Giardia* in regulating osmotic variations in the environment.

8 Differentiation

Many protozoa, free-living and parasitic, have one feature in common: they differentiate from trophozoites to cysts (encystment) when triggered by some external factor(s) unfavorable to trophozoite survival, and one of these stimuli is nutrient deprivation. Ultimately, when conditions occur in which trophozoites can survive, trophozoites exit the cyst wall (excystation).

8.1 Encystment (encystation)

In vivo, *Giardia* encystment takes place in the lower small intestine with trophozoites rounding up, elaborating a cyst wall and forming intact cysts by the time they reach the colon (Meyer and Jarroll, 1980). *In vitro*, the ability to generate cysts axenically has only been possible since the late 1980s. Gillin *et al.* (1987) and Schupp *et al.* (1988) were the first to report methods for the *in vitro* production of cysts from axenically grown trophozoites; *in vitro* encystment requires conditions of pH 7.8 and bile salts. Originally, bile (or its derivatives) were believed to induce encystment directly but Luján *et al.* (1996a and b, 1997) demonstrated that trophozoites encyst

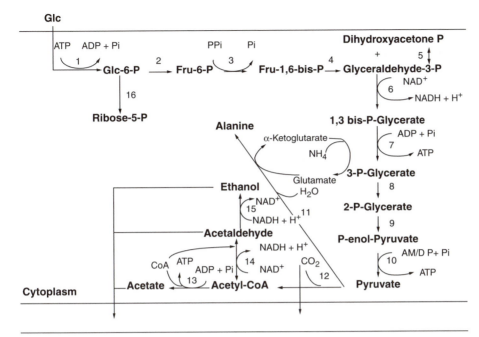

Figure 12 Glucose metabolism in *Giardia*. Enzymes: 1, hexokinase; 2, glucose phosphate isomerase (not confirmed); 3, pyrophosphate-dependent phosphofructokinase; 4, fructose bisphosphate aldolase; 5, triosephosphate isomerase; 6, glyceraldehyde-3-phosphate dehydrogenase; 7, phosphoglycerate kinase (not confirmed); 8, phosphoglyceromutase (not confirmed); 9, enolase (not confirmed); 10, PEP carboxykinase; 11, alanine aminotransferase; 12, GDH; 13, pyruvate:ferredoxin (Fd) oxidoreductase; 14, acetyl-CoA synthetase; 15, alcohol dehydrogenase E (also has acetaldehyde dehydrogenase activity); 16, pentose phosphate pathway.

because of bile's sequestration of cholesterol provided by the serum used in the growth and encystment media.

Encystment in *Giardia* entails a series of yet-to-be identified signaling events culminating in the differentiation of a feeding trophozoite into a tetranucleate cyst form resulting from karyokinesis in the trophozoite during encystment. The encysting trophozoite typically does not undergo cytokinesis (cellular division) or VAD organization until it excysts. *G. microti* cysts, apparently an exception to this rule, frequently contain two differentiated trophozoites with mature VADs pointing in opposite directions (Feely, 1988). Cysts resembling those reported for *G. intestinalis* and *G. muris* were in greater abundance in preparations made from intestinal contents and were interpreted as immature cysts.

A possible Golgi complex and encystation-specific vesicles

Reiner *et al.* (1990) demonstrated structures that resemble Golgi complexes in encysting but not in nonencysting trophozoites. While still not proven conclusively, newer evidence supporting the presence of a Golgi complex in encysting trophozoites comes from finding ADP-ribosylation factor (ARF), a GTP-binding protein required for clathrin, and COP1-coated Golgi vesicles (Lee *et al.*, 1992; Murtagh *et al.*, 1992). Luján *et al.* (1995a) showed that brefeldin A inhibited the anti-ARF and anti-β-COP labeling of the peripheral vesicles, further suggesting their Golgi origin.

The appearance of cytoplasmic encystation-specific vesicles (ESVs) is the earliest morphologic change detectable by light microscopy in encysting trophozoites (Reiner *et al.*, 1989). Cyst wall antigens were localized to these vesicles by immunocytochemistry, suggesting that the vesicles function in export of cyst-wall constituents, and acid phosphatase activity is not detectable in ESV (Reiner *et al.*, 1990).

Ultrastructural studies showed that regulated transport and secretion of cyst-wall antigens in ESV occurs during encystment (Reiner *et al.*, 1989). Encystment antigens were localized in simple Golgi membrane stacks in cisternae and localized within ESV. ESVs and these atypical Golgi complexes are not observed in nonencysting trophozoites or in cysts after completion of encystment. Lanfredi-Rangel *et al.* (2003) suggested from microscopic analyses that the endoplasmic reticulum cisternae dilate to form clefts that enlarge into ESVs, and Stefanic *et al.* (2006) reported that ESVs lack the morphological characteristics of Golgi cisternae and the sorting functions. They detected HSP70-BiP and some proteasome subunits, and the BiP was exported to the ESVs and later retrieved by the KDEL signal at its C-terminus. These observations lead to the suggestion that continuous ER-associated quality control activity and retrograde Golgi-to-ER transport are needed for cyst-wall material maturation.

Giardia homologues of ERK1 and ERK2, members of the mitogen-activated protein kinase (MAPK) pathway, were described by Ellis *et al.* (2003). Native ERK1 and ERK 2 had about equal phosphorylating efficiency (for myelin basic protein as substrate) but recombinant ERK2 (rERK2) was more efficient than rERK1. Both proteins exhibited decreased kinase activity during encystment. Except early in encystment (first 2 h) when ERK1 activity increased around 2.5-fold, the levels of phosphorylated ERK1 and 2 were reduced during encystment.

8.2 Cyst-wall structure

During encystment, trophozoites encase themselves in a wall composed of an inner membranous and an outer filamentous portion (Feely *et al.*, 1990). The outer filamentous portion is 0.3–0.5 μm thick, is composed of filaments measuring 7–20 nm in diameter and is arranged in a tightly packed meshwork (Erlandsen *et al.* 1989, 1990a, 1996). There are two layers to the inner membrane of *Giardia*'s cyst wall and those are composed of an inner and outer membrane. The outer membrane is apparently derived from the cell membrane of the original encysting trophozoite. The inner one arises from the new membranes of the cell undergoing encystment. An early external indication of encystment is the appearance of around 15-nm-diameter blebs on the encysting trophozoite's cell membrane about 10 h after it is induced to encyst. Formed cysts begin to appear at about 14 h post induction and increase by 16 h. Carbohydrate-specific tags indicated that the polysaccharide exposure on the trophozoite surface precedes fibril patch assembly, and that the pattern of filament assembly during *Giardia* encystment resembles that for microfibrils of the β (1, 3) glucan in *Candida albicans* (Arguello-Garcia *et al.*, 2002).

8.3 Cyst-wall synthesis

Cyst-wall synthesis (Figure 13) in *Giardia* requires the induction of pathways of protein and carbohydrate (polysaccharide) synthesis. The filamentous portion of the *Giardia* cyst wall is composed of around 63% of a unique polysaccharide and around 37% proteins (Gerwig *et al.*, 2002).

Cyst-wall proteins

There are at least three specific *Giardia* cyst-wall proteins (Cwps): Cwp 1 (26 kDa) and Cwp 2 (39 kDa; Luján *et al.*, 1996a and b, 1997) and Cwp 3 (27.3 kDa; Sun *et al.*, 2003). Each is encoded by single-copy genes and the transcripts for Cwp1 and 2 increase about 140-fold while that of Cwp 3 increases about 100-fold above those in nonencysting trophozoites. Cwp 1 and 2 have a 61% identity in their overlapping regions; Cwp 3 is 36% identical to Cwp 1 and 34% to Cwp 2 in the leucine-rich repeat (LRR) regions. There are five such tandem copies in Cwp 1 and 2; Cwp 3 has four with a possible fifth LRR. There is also a 14-residue cysteine-rich conserved region in all of the Cwps. These three proteins localize to ESVs as well as to the mature cyst wall (Luján and Touz, 2003; Luján *et al.*, 1995a,1995b, 1997; Sun *et al.*, 2003). The concentration of Cwp1/Cwp2 complex at specific regions of rough endoplasmic reticulum may initiate secretory granule biogenesis (Luján and Touz, 2003). Luján *et al.* (1997) reported up-regulation of chaperones, immunoglobulin heavy-chain-binding protein and protein disulfide isomerases apparently to assist with folding Cwps. Hehl *et al.* (2000) demonstrated that during encystment a green fluorescent protein (Gfp) chimera with Cwp 1 formed labeled dense-granule-like vesicles and showed subsequent incorporation of Cwp 1-Gfp into the cyst wall. Furthermore, they showed that the *N*-terminal domain of Cwp 1 was required for targeting Gfp to the secretory compartments while the central LRRs were required for chimera association with extracellular cyst-wall components (Hehl *et al.*, 2000).

Sun *et al.* (2003) showed that the LRRs and the *N*-terminal region were required for targeting AU1-tagged Cwp 3 into secretory compartments, while the C-terminus was necessary for its incorporation into the cyst wall. Cwp 1 was not seen in mature cysts causing speculation that there is a negative regulation of Cwp 1 synthesis late in encystment (Hehl *et al.*, 2000).

The three Cwps are exported as disulfide-bonded heterodimers in ESVs, demonstrated by the fact that dithiothreitol inhibited formation of ESVs (ca. 85%) and reduced the Cwps to monomers; dimers reformed when dithiothreitol was removed (Reiner *et al.*, 2001). Gottig *et al.* (2006) showed that Cwp2 aggregates Cwp1 and Cwp3 via its conserved LRR and by so doing regulates ESV formation.

Touz *et al.* (2002a) demonstrated a novel 54-kDa encystment protein they termed *Giardia* granule-specific protein, gGsp. gGsp is located solely in ESVs, exhibits calcium-binding characteristics, is induced during encystment and exhibits no homology to any other proteins in reference databases. However, when they applied antisense to the gGsp gene, Cwp 2 formed and was transported to granules as normally expected, but the granules failed to express their contents to the outside of the trophozoite. gGsp may also function together with chaperones to keep Cwp 1/Cwp 2 from assembling inside the ESVs prematurely (Luján and Touz, 2003; Touz *et al.* 2002a).

An around 45-kDa cysteine proteinase, induced during encystment, plays a role in proteolytic cleavage and release of Cwp 2 prior to cyst-wall formation. Cysteine proteinase inhibitors E64, ALLM, ALLN and bestatin inhibited cyst-wall production (Touz *et al.*, 2002b). Of course, other proteinases may also be involved but were not inhibited or identified.

Cyst-wall carbohydrates

Originally, the *Giardia* cyst-wall carbohydrate was identified as chitin, a β-1,4-homopolysaccharide of *N*-acetylglucosamine based on lectin binding and by finding chitin synthase activity (Gillin *et al.*, 1987; Ward *et al.*, 1985), but a study by Jarroll *et al.* (1989) showed that this was not the case. A thorough carbohydrate analysis of purified *Giardia* cyst-wall filaments revealed that about 63% is composed of a unique *N*-acetylgalactosamine (GalNAc) [homopolysaccharide D-GalNAc β (1,3) D-GalNAc (*N*-acetylgalactosamine)]$_n$, which is not deacetylated during its formation (Gerwig *et al.*, 2002). This polysaccharide is highly insoluble and its insolubility is most likely due to strong interchain interactions. The persistence length (PL) for the polysaccharide is about 35 Å (the smaller the PL the more flexible the polysaccharide) while that of cellulose is 90–125 Å. Potentially, a covalent linkage between the polysaccharide and some of the proteins in the cyst wall may play a role in the insolubility; the existence of such a linkage remains unproven (Gerwig *et al.*, 2002).

Synthesis of UDP-N-acetylgalactosamine

GalNAc is below the limits of detection in growing trophozoites, but is synthesized and demonstrable during encystment (Jarroll *et al.*, 1989). Extracellular formation of the unique GalNAc polysaccharide (see below) requires intracellular UDP-GalNAc synthesis *de novo* from glucose (Figure 13).

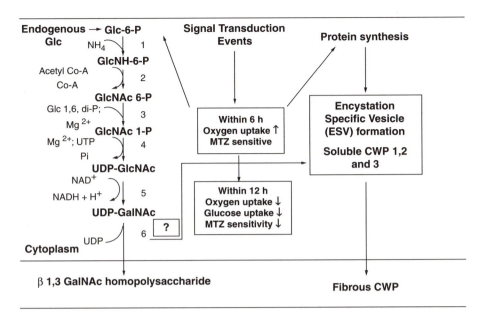

Figure 13 Cyst-wall synthesis. Enzymes: 1, glucosamine 6-P isomerase; 2, glucosamine 6-P N-acetylase; 3, phospho N-acetylglucosamine mutase; 4, uridine diphospho (UDP) N-acetylglucosamine pyrophosphorylase; 5, UDP-N-acetylglucosamine 4′-epimerase; 6, cyst-wall synthase.

Macechko *et al.* (1992) showed clearly that a pathway of enzymes was induced during encystment which converted glucose 6-P to UDP-GalNAc and that the activity of these enzymes localized to the cytoplasm of encysting trophozoites and includes: glucosamine 6-P isomerase (EC 5.3.1.10), glucosamine 6-P N-acetylase (EC 2.3.1.4), phospho N-acetylglucosamine mutase (EC 2.7.5.2), uridine diphospho (UDP)-N-acetylglucosamine pyrophosphorylase (EC 2.7.7.23) and UDP-N-acetylglucosamine 4′-epimerase (EC 5.1.3.7). Glucosamine 6-phosphate isomerase is transcriptionally regulated, and during encystment and in mature cysts this isomerase appears to be modified by ubiquitin attachment and might be targeted for destruction by a ubiquitin-mediated pathway (Lopez *et al.*, 2002). Lopez *et al.* (2003) demonstrated that the regulation of glucosamine 6-phosphate N-acetyltransferase, phosphoacetylglucosamine mutase, UDP-N-acetylglucosamine pyrophosphorylase, and UDP-N-acetylglucosamine 4-epimerase was at the level of transcription by Northern and Western blot analyses of mRNA and protein, respectively, and by nuclear run-on assays of these and the previously analyzed glucosamine 6-phosphate deaminase (glucosamine 6-P isomerase; Steimle *et al.*, 1997). In addition to the induced UDP-N-acetylglucosamine pyrophosphorylase (Lopez *et al.*, 2003; Mok and Edwards, 2005; Mok *et al.*, 2005), Bulik *et al.* (2000) reported a constitutive UDP-N-acetylglucosamine pyrophosphorylase activity which was stimulated in the anabolic direction toward UDP-GalNAc synthesis by

GlcN-6-P, the product of the glucosamine 6-P deaminase. Şener *et al.* (2004) demonstrated changes in the cytoplasmic level of the sugar phosphate intermediates generated by these enzymes during encystment in *Giardia*. The first and most notable absolute increase was in the level of GlcN-6-P (~fivefold above nonencysting levels) which reportedly stimulates the cUap of the two UDP-*N*-GlcNAc pyrophosphorylase activities. The largest relative increase in sugar phosphates, however, was in UDP-GlcNAc (~ninefold) above nonencysting levels.

Epimerase activity has been detected in a number of organisms such as humans, rats (Thoden *et al.*, 2001; Wohlers *et al.*, 1999) and bacteria such as *Bacillus* (Nishikawa *et al.*, 1986), *Pseudomonas aeruginosa* strain NP 250075 (Creuzenet *et al.*, 2000), and *Yersinia* (Bengoechea *et al.*, 2002). The epimerase from these organisms is able to catalyze two similar, but distinct, reversible reactions: UDP-GlcNAc to UDP-GalNAc and UDP-Glc to UDP-Gal (EC 5.1.3.2). The conversion of UDP-Gal to UDP-Glc is part of the Leloir pathway, which is necessary for the breakdown of lactose to glucose (Thoden *et al.*, 2001; Wohlers *et al.*, 1999). When, epimerase activity was measured in crude *Giardia* lysates, there was an increase in UDP-GlcNAc/GalNAc activity when trophozoites were induced to encyst, but there was no detectable UDP-Gal to Glc activity (Macechko *et al.*, 1992). This observation agreed with the finding that the Leloir pathway is absent in this organism since *Giardia* cannot metabolize galactose (Jarroll *et al.*, 1989). While the *Giardia* epimerase catalyzes the reversible epimerization of UDP-GlcNAc to UDP-GalNAc, the reverse reaction apparently is favored. Therefore, an excess of UDP-GlcNAc is apparently required to drive the reaction towards the synthesis of UDP-GalNAc, which is needed for cyst-wall formation (Lopez *et al.*, 2007). They also speculate that the activity of cyst-wall synthase (see below) is required to remove the UDP-GalNAc rapidly.

Synthesis of β-1,3-N-acetylgalactosamine

The UDP-GalNAc synthesized in the cytoplasm must then be incorporated into the extracellular β-1,3-GalNAc polysaccharide that will eventually become the polysaccharide portion of the filaments. This appears to be done by a previously undiscovered, inducible β-1,3-*N*-acetylgalactosaminyltransferase termed cyst-wall synthase (CWS; Karr and Jarroll, 2004). CWS has been partially purified but not cloned. It is very specific for UDP-GalNAc and does not incorporate UDP-UDP-GlcNAc or other UDP-sugars into the polysaccharide. CWS requires a divalent cation and prefers Ca^{2+} or Mg^{2+} over Zn^{2+}, Cu^{2+} or Mn^{2+}. Thus, it can be inhibited by chelating agents such as EDTA. Unlike the other inducible enzymes in this pathway, CWS is particle associated. The particles with which it is associated appear to be different from the lysosome-like organelles based on isodensity centrifugation (Karr and Jarroll, 2004). In fact, the vesicles are possibly the ESV formed during encystment or a subpopulation of them, but this remains to be demonstrated.

Cyst-wall synthesis regulation

Bazan-Tejeda *et al.* (2006) demonstrated that *Giardia* express protein kinase C (PKC) in beta, delta, epsilon, theta and zeta isoforms. PKC, a family of serine/threonine kinases, regulate several cellular processes including growth and differentiation in eukaryotes. These investigators employed specific polyclonal antibodies against mammalian PKC isoforms, which showed changes in the isoforms expression pattern during encystment induction. PKC inhibitors blocked encystment in a dose-dependent manner suggesting that these kinases may play important roles during this differentiation process.

Gibson *et al.* (2006) studied the structural and functional characteristics of the regulatory subunit of PKA (gPKAc) and its involvement in encystment. gPKAc is stimulated by intracellular cyclic AMP, and therefore links that important second messenger system to the equally important protein phosphorylation cascade for transducing biological signals. These authors noted increased gPKAc activity during encystment but no significant change in the protein concentration. gPKAc was localized to the anterior flagella, basal bodies and caudal flagella in trophozoites but these were absent in encysting cells at later stages. The punctuate distribution of gPKAc was mostly over the cell periphery. That possible enrichment of the active gPKAc during later encystment stages may have implications in completion of the encystment process or cyst priming for efficient excystation was suggested.

8.4 Excystation

Excystation (the process whereby a trophozoite exits the cyst wall) occurs in the upper small intestine of the vertebrate host when *Giardia* cysts are induced by the acidic conditions in the stomach to exit the fibrous cyst wall in which they are encased when they were ingested. Bingham and Meyer (1979) and Bingham *et al.* (1979) showed that it was the hydrogen ion concentration rather than the counter ion which was the important stimulus for excystation and that the entire process could occur within 15–30 min of induction in HCl at pH 2, followed by removal from that solution and the addition of growth medium used for axenic cultivation of the trophozoites. Excystation in *G. muris*, but not *G. intestinalis*, cysts was induced with bicarbonate in a pH 7.5 phosphate buffer solution (Feely *et al.*, 1991).

Trypsin inhibitor and DIDS (4-4′diisothiocyanatostilbene-2-2′-disulfonic acid) inhibit excystation, the latter suggesting that acidification of vacuoles is required for the process, but excystation is promoted by pancreatic proteases and Ward *et al.* (1997) described a cathepsin-B-family cysteine proteinase, which they called CP2, localized to lysosome-like vesicles and required for excystation. Cysteine proteinase inhibitors, which did not affect trophozoite growth, did inhibit excystation *in vitro*. Lysosomal acid phosphatase, localized to peripheral vesicles, apparently is also required for the dephosphorylation of cyst-wall proteins needed for excystation (Slavin *et al.*, 2002). Because the *Giardia* cyst wall is made, in large part, of a novel β-1,3-GalNAc homopolymer synthesized by a novel CWS, it is likely that there exists an enzyme which specifically degrades this polysaccharide during or after excystation even though such an enzyme has not yet been reported.

Niño and Wasserman (2003) reported that during excystation mRNA transcripts of some metabolic enzyme increased. Pyruvate: ferredoxin oxidoreductase (PFOR) and alcohol dehydrogenase transcripts appeared after the first induction phase, and acetyl CoA synthetase and glutamate dehydrogenase increased throughout excystation.

9 Methods of Isolation and Axenic Cultivation

The history of *Giardia* trophozoite cultivation was reviewed by Radulescu and Meyer (1990). *Giardia* trophozoites were first cultured by A. E. Karapetyan from the Soviet Union in 1960 (Karapetyan, 1960, 1962). Karapetyan's initial medium contained a tryptic digest of meat, serum, chick embryo extract and Earle's or Hank's solution, but was not axenic since it also had chick fibroblasts and *Candida guillermondi* in the culture along with *Giardia*. By 1962, Karapetyan had substituted *Saccharomyces* for *Candida* and omitted chick fibroblasts.

E. A. Meyer was the first person to successfully grow *Giardia in vitro* axenically (Meyer, 1970). Meyer's first axenized cultures (Meyer, 1970) contained only *Giardia* trophozoites because he had been able to substitute yeast extract for living yeast. His first medium was replaced in 1976 (Fortess and Meyer, 1976) by HSP (Hank's balanced salt solution: human serum, cysteine and Phytone peptone – soybean meal digested with papain) and he was able to grow human *Giardia* axenically. HSP was gradually replaced by TPS-1 (trypticase, panmede – an ox-liver digest – and bovine serum; Visvesvara, 1980) and eventually by Diamond's TYI-S-33 (trypticase, yeast extract, iron and serum; Diamond *et al.*, 1978) originally used for *Entamoeba histolytica* culture.

Today, most who study *Giardia* in culture use Keister's medium (Keister, 1983) a casein digest, cysteine and yeast extract supplemented with bovine serum and a small amount of bile, but bile is not required for *Giardia* growth *in vitro*. This medium is filter-sterilized not autoclaved, and normally the medium contains penicillin and streptomycin to prevent contamination. Cultures are normally grown in medium-filled, screw-capped flasks or tubes that prevent as much oxygen as possible from mixing with the medium. Tubes are usually tilted at a 45° angle and cell culture flasks are generally placed on one side.

Axenic culture can be initiated by excysting cysts *in vitro* (see section 8.4) but antibiotics are required to control bacterial overgrowths. Before *in vitro* excystation methods were developed, the cultures were initiated with trophozoites taken from the gut of sacrificed animals.

G. ardea have been cultured *in vitro* but require a slightly different medium than *G. intestinalis* (Erlandsen *et al.*, 1990a and b). Interestingly and despite diligent efforts to do so, *G. muris* trophozoites have never been cultured *in vitro*; *G. psittaci* also remains uncultured.

10 Storage of *Giardia*

Giardia cultures can be cryopreserved by standard methods using DMSO (dimethylsulfoxide) to prevent ice crystal formation and stored at −80°C or in a nitrogen refrigerator. To re-initiate culture growth, the tubes containing these frozen trophozoites should be placed in a water bath at 37°C until thawed and transferred to sterile culture medium at 37°C. Personal experience informs that it is often wise to increase the serum concentration from the 10% normally used in culturing to as high as 15–20% until these revived cells establish confluent growth.

Giardia cysts do not survive freezing in distilled water or saline at −20°C, but cysts apparently have been frozen successfully in either Keister's medium or physiological saline, containing 5% and 7.5% dimethylsulfoxide as the cryoprotectant (Dickerson *et al.*, 1991; Meyer and Jarroll, 1980). Dickerson *et al.* (1991) observed that the viability of cryopreserved cysts depended upon the number of cysts inoculated, the length of time cysts are held at 4°C before cryopreservation, and the cryopreserving medium. Infection was established in gerbils by inoculating them with cysts that had been cryopreserved for up to 67 weeks.

11 Giardiasis

Giardia, the etiological (causative) agent for giardiasis in a variety of animals including humans, is one of the most common intestinal parasites in the world causing an estimated 2.8×10^6 infections per year in humans (possibly an underestimate), and contributing to diarrhea and nutritional deficiencies in children in developing regions (Ali and Hill, 2003). *Giardia* is responsible for numerous waterborne outbreaks of diarrheal disease in humans, and is a frequent cause of traveler's diarrhea especially in developing countries. Huang and White (2006) have reviewed our current knowledge and practices regarding giardiasis.

11.1 Epidemiology

Giardia is the most commonly diagnosed intestinal parasite (Huang and White, 2006). Giardiasis is a typical fecal-oral disease with the cyst shed in the feces of the infected host contaminating food, water or fomites, ultimately leading to the ingestion of those cysts by the next host. Typically, giardiasis in humans occurs in areas where sanitation standards are low. Human out-breaks have been reported numerous times often associated with failure in a water-supply purification system, the lack of proper water purification, drinking from untreated/unfiltered surface water especially by 'outdoor'-type individuals, unsanitary practices in day care/nursery settings, and a few from foodborne transmission (Meyer and Jarroll, 1980). While *Giardia* and giardiasis are worldwide in distribution the incidence appears greatest in the tropics, in developing countries and in day-care situations. While infants and young children seem to have the highest incidence of infection, giardiasis occurs at all ages, and apparently without gender preference normally. Venereal transmission of giardiasis has been reported for male homosexuals who engage in oral-anal sexual activity (Meyer and Jarroll, 1980)

Epidemic giardiasis has usually been associated with waterborne outbreaks especially in cold to temperate climates and the possibility of cross-species transmission is certainly likely (Thompson, 2002; Wade *et al.*, 2002). The cold temperatures, in fact, are probably beneficial to cysts longevity (Bingham *et al.*, 1979).

11.2 Pathophysiology and clinical manifestations

Giardiasis range from asymptomatic to symptomatic with severe malab-sorption (Meyer and Jarroll, 1980) and can present with acute or chronic disease with diarrhea, weight loss, flatulence, dehydration, and abdominal discomfort. How this pathology is produced is still unclear but a thorough recent review of this topic is available (Buret *et al.*, 2002). Among the patho-physiologies observed are villous atrophy, decreased epithelial permeabil-ity, and decreased levels of epithelial disaccharidases (maltase, lactase, sucrase, etc.), lipases and proteases (trypsin and chymotrypsin notably). These changes may be due as much to factors of the host as to those of the parasite. Villous atrophy may be caused at least in part by T-cell activation and mast cell hyperplasia. What appears clear at least currently is that trophozoites do not invade the mucosa. There is some indication that *Giardia* may damage the enterocyte, especially the cytoskeletal α-actinin, F-actin, villin and ezrin. Reports of cytotoxic compounds from trophozoites causing *in vitro* cell culture damage are inconclusive and no compound(s) have been demonstrated to confirm this.

The signs and symptoms of acute giardiasis may resemble *Campylobacter* infection, viral enteritis bacillary dysentery, bacterial food poisoning, amoebiasis or toxigenic *Escherichia coli*-produced traveler's diarrhea. Of course, it is possible that more than one of these infections is going on simultaneously with giardiasis. Giardiasis can move to a chronic stage in which symptoms become milder but in which the original signs and symptoms may recur. Flatulence, mild diarrhea, soft stools, yellowish frothy stools or even intestinal cramps and distention are possible. Weight loss can occur along with constipation, headache, belching, heartburn and rarely chills, fever and vomiting (Buret *et al.*, 2002; Huang and White, 2006; Meyer and Jarroll, 1980).

11.3 Diagnosis

Traditional diagnostic tests for giardiasis include stool examinations. In cases of formed stools, one would expect only cysts, but in diarrheic stools, trophozoites might out-number cysts or even replace them (Meyer and Jarroll, 1980). Stool examinations follow standard protocols for the collection, concentration and microscopic examination. In cases of diarrhea, one can examine concentrated samples for trophozoites in a wet-mount slide. Giardiasis has a prepatent period, established by the first time cysts are detected in stool samples, of about 9 days (Rendtorff, 1954). Cyst production can be quite variable over the period of infection and thus three stool samples collected over three consecutive days is often used (Meyer and Jarroll, 1980), but Danciger and Lopez (1975) recommended a 4- to 5-week time frame before declaring samples negative for cysts. Stool samples should be declared negative before other more invasive tests are used.

A variety of tests are now available for detecting *Giardia* antigens by immunoassays, ELISA and direct fluorescence antibody assays. EnteroTest (HDC Corporation, Milpitas, CA), duodenal aspiration, string tests, and in rare cases duodenal biopsy are possibilities (Huang and White, 2006).

11.4 Treatment

The drug of choice for treating human giardiasis remains metronidazole (MTZ), a 5-nitroimidazole; other 5-nitroimidazoles (tinidazole – TNZ, secnidazole and ornidazole) are also effective but only MTZ and TNZ are approved currently for use in the US; for children it is nitazoxanide (Medical Letter, 2004). MTZ and TNZ are also used to treat other anaerobic protozoan infections such as trichomoniasis and amoebiasis (Jarroll, 1994). All of these anaerobic protists have an enzyme known as pyruvate: ferredoxin oxido reductase (PFOR). The MTZ is reduced by the low redox potential of PFOR and the reduction products apparently result in toxic compounds which kill the trophozoites. Because MTZ is reduced upon entry into the cell, there is no gradient formed to prevent additional MTZ from entering the cells. MTZ has no effect on encysting *Giardia* or cysts after about 12 hours into the encystment process (Paget *et al.*, 1989, 1993a, 1998).

A number of workers have stated that MTZ resistance may be beginning to appear *in vitro* and in clinical settings (reviewed by Reynoldson, 2002); the mechanism for this appears to be a decrease in the production of PFOR. When considering resistance, especially in clinical settings, it is important to distinguish between true resistance to the drug and the failure of patients to follow the course of treatment.

Other drugs have a history of use against *Giardia* including Atabrine and furazolidone. Atabrine (quinacrine and mepacrine) is no longer the drug of choice for giardiasis because of often-encountered side effects (nausea, vomiting, dizziness, headaches, and rarely toxic psychoses) and fura-zolidone (furoxone, a synthetic nitrofuran that also can cause vomiting, nausea, diarrhea and malaise) was prescribed to children because it is available as a liquid. Upcroft *et al.* (1996) reported that they had induced quinacrine resistance in laboratory stocks of *Giardia duodenalis* and that the quinacrine was actively excluded by resistant trophozoites.

Paromomycin, a minimally absorbed aminoglycoside with unknown mechanism against *Giardia*, remains an alternative drug. Benzimidazoles are tubulin inhibitors that have also been used against giardiasis but their use remains questionable (Reynoldson, 2002). Some benzimidazoles are quite toxic to mammals.

Paget *et al.* (2004) reported that menadione-generated radicals kill *Giardia intestinalis* trophozoites and cysts. The encysting *Giardia* exhibit reduced metabolic activity, reduced glucose uptake and MTZ resistance (Paget *et al.*, 1989, 1993a, 1998).

Jarroll and Şener (2003) suggested that since the trigger for encystment is usually depletion of a vital nutrient, and assuming encystment is irre-versible at some point, then inhibiting the process could cause the encyst-ing trophozoites to die. While this concept has yet to be tested experimentally, it does seem likely that there exist targets in the pathways for potential drug design. The two most obvious targets are the UDP-GlcNAc 4′-epimerase and CWS. The epimerase differs from that found in humans and the CWS, which is not found in humans, appears essential for *Giardia* cyst-wall filament formation.

11.5 Control of waterborne giardiasis

Numerous outbreaks of giardiasis have been associated with drinking water (Huang and White, 2006; Meyer and Jarroll, 1980). Many of the earliest outbreaks were often associated with failure of a chemical disinfec-tant in a public water system or with drinking from untreated water sources.

Water treatment criteria were based originally on the amount of disin-fectant required to kill selected bacterial indicators such as *Escherichia coli*. However, *Giardia* cysts are more resistant to chemical disinfectants than are typical coliform bacteria such as *E. coli* (Jarroll, 1999). Normally used chemical disinfectants include chlorine, chlorine dioxide, chloramines and ozone. The concentration of ozone required to inactivate *Giardia* cysts is much lower than that for the chlorine-based disinfectants. The action of all of these chemical disinfectants is affected by the amount of residual disin-fectant (free after the disinfectant demand of the solution and associated organics and inorganics that react with the halogen), temperature and often pH of the solution being treated, especially for halogen-based ones. In general, the higher the demand of the system for the disinfectant, the lower the water temperature and the higher the pH, the less effective is the disinfectant (Jarroll, 1999).

UV irradiation also kills *Giardia* cysts at relatively low dosages. However, there remain issues with protocols and reliability that need to be resolved before it can be used routinely (Jakubowski and Craun, 2002).

Today, the effective inactivation and removal of *Giardia* cysts from public drinking water can be accomplished using a multibarrier approach and a combination of disinfection, coagulation and filtration (Jakubowski and Craun, 2002). Relying on anyone of these alone increases the possibility that a breakdown in the single system could allow infectious cysts to survive the treatment.

11.6 Antigenic variation and variable surface proteins

A family of approximately 150 cysteine-rich surface antigens (variable surface proteins, VSPs) with a CXXC motif (Adam *et al.*, 1988), varying in size from about 50 to 200 kDa, are present on *Giardia in vitro* and *in vivo* (Adam *et al.*, 1991; Aggarwal and Nash, 1988; Kulkova *et al.*, 2006; Nash and Keister, 1985; Nash *et al.*, 1990); only one of these VSP antigens seems to be expressed at a given time (Kulkova *et al.*, 2006). Apparently, VSPs are not glycosylated (Nash *et al.*, 1983) since they do not bind lectins.

Evidence from *Giardia lamblia*-infected human volunteers (Nash *et al.*, 1990), athymic mice and SCID mice (Gottstein and Nash, 1991) suggests that VSPs change in response to humoral antibodies against the specific VSPs. It is possible that these VSPs may also aid the parasite in adapting to various intestinal environmental conditions (Nash *et al.*, 1991).

Kulkova *et al.* (2006) integrated a hemagglutinin (HA) epitope-tagged h7 VSP into the *Giardia* GS genome. It underwent antigenic variation in the same manner, but independent of, the native H7 VSP. When analyzed, clones that expressed or failed to express showed an absence of HA-tagged h7 gene rearrangements after switching and acetylation of histone lysine residues within the 167 nucleotides 5′ to the expression of HA-tagged h7 gene, suggesting epigenetic mechanisms in antigenic variation. Palmitoylation of HA-VSPH7 by palmitoyl acyl transferase to the cysteine in the conserved hydrophobic tail amino acids CRGKA appears to control VSP-mediated signaling and processing (Touz *et al.*, 2005).

11.7 Immune response and therapy

Whether or not *Giardia* elicits a true host immune response has been debated for many years. Evidence is mounting that there is a host immune response. Nash *et al.* (1987) and Daniels and Belosevic (1994) reported that serum and mucosal antibody levels rose in animal and human hosts with giardiasis. IgM specific for *Giardia* appeared in the serum and at the gut mucosa about 10 days after exposure; IgA and serum IgG levels rose within a week (Soliman *et al.*, 1998; Yanke *et al.*, 1998). However, not all experimental animals mount an IgA or IgG response. Antibody-mediated trophozoite lysis has been reported when serum or bile with anti-*Giardia* antibodies was used for *in vitro* assays. Whether this occurs *in vivo* is still in question. While an intact cellular immune system seems to be required it does not seem to play a major role itself in the immune response (reviewed by Olson *et al.*, 2002).

Experimental oral and injectable vaccines have been used in mice, dogs and cats (reviewed Olson *et al.*, 2002). In each instance, animals given the vaccines exhibited less signs and symptoms of giardiasis than those that did not. While vaccines are now available for dogs and cats, no human *Giardia* vaccine is approved.

12 Conclusions

Giardia, a binucleate protozoan parasite, remains a common intestinal infection of humans and other vertebrates. Human giardiasis is transmitted fecal-orally often in waterborne outbreaks but also in day-care settings and perhaps zoonotically. The infection may go undetected in asymptomatic cases, or in symptomatic cases it may be confused with a number of other gastrointestinal infections. Giardiasis is typically treated with metronidazole (or tinidazole) though other drugs are available; some reports of resistance to metronidazole and Atabrine have appeared but remain in question. Whether or not there is a protective immune response to giardiasis in humans is debatable but there is evidence that antibodies to *Giardia* occur in many hosts including humans. Currently there is no vaccine against human giardiasis although an animal vaccine maybe efficacious.

This protozoan is interesting from a number of biological perspectives not the least of which is its two morphologically and genetically identical nuclei in each trophozoite. *Giardia* is a lipid auxotroph, salvages, but does not synthesize *de novo*, purine or pyrimidine requirements, and does not synthesize *de novo* many of its amino acid requirements. This evolutionarily early diverging protozoan lacks mitochondria, typical Golgi, aerobic respiration and cytochrome-mediated electron transport, but it exhibits substrate-level phosphorylation from glucose and arginine for ATP production. *Giardia* does exhibit mitosomes (FeS maturation organelle evolutionarily derived from a mitochondrion), a microtubular median body, a microtubular adhesive disk, and specific encystation vesicles for transporting its cyst-wall material externally when encysting. The cyst wall is a combination of novel cyst-wall proteins (Cwp1, 2 and 3), and a novel cyst-wall polysaccharide (β-1,3-GalNAc homopolymer). The polysaccharide is synthesized, at least in part, by a novel membrane-associated transferase tentatively named cyst-wall synthase. Whether or not the proteins and polysaccharide of the cyst wall are covalently linked or are extruded by the same mechanisms remain unknown.

13 References

Adam, R. (2001) Biology of *Giardia. Clin. Rev. Microbiol.* **14**: 447–475.

Adam, R., Aggarwal, A., Lal, A., de la Cruz, V., McCutchan, T. and Nash, T. (1988a) Antigenic variation of a cysteine-rich protein in *Giardia lamblia. J. Exp. Med.* **167**: 109–118.

Adam, R., Nash, T. and Wellems, T. (1988b) The *Giardia lamblia* trophozoite contains sets of closely related chromosomes. *Nucleic Acids Res.* **16**: 4555–4567.

Adam, R., Nash, T. and Wellems, T. (1991) Telomeric location of *Giardia* rDNA genes. Mol. Cell Biol. **11**: 3326–3330.

Aggarwal, A. and Nash, T. (1988) Antigenic variation of *Giardia lamblia* in vivo. *Infect. Immun.* **56**: 1420–1423.

Aggarwal, A., Adam, R. and Nash, T. (1989) Characterization of a 29.4-kilodalton structural protein of *Giardia lamblia* and localization of the ventral disk. *Infect. Immun.* **57**: 1305–1310.

Aldritt, S., Tien, P. and Wang, C. (1985) Pyrimidine salvage in *Giardia lamblia. J. Exp. Med.* **161**: 437–445.

Ali, S. and Hill, D. (2003) *Giardia intestinalis. Curr. Opin. Infect. Dis.* **16**: 453–460.

Alonso, R. and Peattie, D. (1992) Nucleotide sequence of a second alpha giardin gene and molecular analysis of the alpha giardin genes and transcripts in *Giardia lamblia. Mol. Biochem. Parasitol.* **50**: 95–104.

Arguello-Garcia, R., Arguello-Lopez, C., Gonzales-Robles, A., Castillo-Figueroa A. and Ortega-Pierres, M. (2002) Sequential exposure and assembly of cyst wall filaments on the surface of encysting *Giardia duodenalis. Parasitology* **125**: 209–219.

Bazan-Tejeda, M., Arguello-Garcia, R., Bermudez-Cruz, R., Robles-Flores, M. and Ortega-Pierres, G. (2006) Protein kinase C isoforms from *Giardia duodenalis*: identification and functional characterization of a beta-like molecule during encystment. *Arch. Microbiol.* **187**: 55–66.

Benchimol, M., Piva, B., Campanati, L. and DeSouza, W. (2004) Visualization of the funis of *Giardia lamblia* by high-resolution field emission scanning electron microscopy – new insights. *J. Struct. Biol.* **147**: 102–115.

Bengoechea, J., Pinta, E., Salminen, T., Oertelt, C., Holst, O., Radziejewska-Lebrecht, J., Piotrowska-Seget, Z., Venho, R. and Skurnik, M. (2002) Functional characterization of Gne (UDP-*N*-acetylglucosamine-4′-epimerase), Wzz (chain length determinant), and Wzy (O-antigen polymerase) of *Yersinia enterocolitica* serotype O:8. *J. Bacteriol.* **184**: 4277–4287.

Bernander, R., Palm, J. and Svärd, S. (2001) Genome ploidy in different stages of *Giardia lamblia* life cycle. *Cell Microbiol.* **35**: 55–62.

Bertram, M., Meyer, E., Lile, J. and Morse, S.A. (1983) A comparison of isozymes of five axenic *Giardia* isolates. *J. Parasitol.* **69**: 793–801.

Bingham, A. and Meyer, E. (1979) *Giardia* excystation can be induced *in vitro* in acidic solutions. *Nature* **277**: 301–302.

Bingham, A., Jarroll, E., Meyer, E. and Radulescu, S. (1979) *Giardia* sp.: physical factors of excystation *in vitro* and excystation vs. eosin-exclusion as determinants of viability. *Exp. Parasitol.* **47**: 284–291.

Brown, D., Upcroft, J. and Upcroft, P. (1995) Free radical detoxification in *Giardia duodenalis. Mol. Biochem. Parasitol.* **72**: 47–56.

Brown, D., Upcroft, J. and Upcroft, P. (1996) A H_2O-producing NADH oxidase from the protozoan parasite *Giardia duodenalis. Eur. J. Biochem.* **241**: 155–161.

Bulik, D., van Ophem, P., Manning, J., Shen, Z., Newburg, D. and Jarroll, E. (2000) UDP-N-Acetylglucosamine pyrophosphorylase: a key enzyme in encysting *Giardia* is allosterically regulated. *J. Biol. Chem.* **275**: 14722–14728.

Buret, A., Scott, K. and Chin, A. (2002) Giardiasis: pathophysiology and pathogenesis. In: *Giardia*: the Cosmopolitan Parasite (eds B. Olson, M. Olson and P. Wallis). CABI, New York, NY, pp. 109–125.

Campbell, S., van Keulen, H., Erlandsen, S., Senturia, J. and Jarroll, E. (1990) *Giardia* sp.: comparison of electrophoretic karyotypes. *Exp. Parasitol.* **71**: 470–482.

Correa, G., Morgado-Diaz, J. and Benchimol, M. (2004) Centrin in *Giardia lamblia* – ultrastructural localization. *FEMS Microbiol. Lett.* **233**: 91–96.

Creuzenet, C., Belanger, M., Wakarchuk, W. and Lam, J. (2000) Expression, purification, and biochemical characterization of WbpP, a new UDP-GlcNAc C4 epimerase from *Pseudomonas aeruginosa* serotype O6. *J. Biol. Chem.* **275**: 19060–19067.

Danciger, M. and Lopez, M (1975) Numbers of *Giardia* in the feces of infected children. *Am. J. Trop. Med. Hyg.* **24**: 237–242.

Daniels, C. and Belosevic, M. (1994) Serum antibody responses by male and female C57BL/6 mice infected with *Giardia muris. Clin. Exp. Immunol.* **97**: 424–429.

Davey, R., Ey, P. and Mayrhofer, G. (1991) Characteristics of thymine transport in *Giardia intestinalis* trophozoites. *Mol. Biochem. Parasitol.* **48**: 163–171.

De Jonckheere, J. and Gordts, B. (1987) Occurrence and transfection of a *Giardia* virus. *Mol. Biochem. Parasitol.* **23**: 85–89.

Diamond, L., Harlow, D. and Cunnick, C. (1978) A new medium for the axenic cultivation of *Entamoeba histolytica* and other *Entamoeba. Trans R. Soc. Trop. Med. Hyg.* **72**: 431–432.

Dickerson, J., Visvesvara, G., Walker, E. and Feely, D. (1991) Infectivity of cryopreserved *Giardia* cysts for Mongolian gerbils (*Meriones unguiculatus*). *J. Parasitol.* **77**: 688–691.

Dobell, C. (1920) The discovery of the intestinal protozoa of man. *Proc. R. Soc. Med.* **13**: 1–15.

Dolezal, P., Smíd, O., Rada, P., Zubácová, Z., Bursać, D., Suták, R., Nebesárová, J., Lithgow, T. and Tachezy, J. (2006) *Giardia* mitosomes and trichomonad hydrogenosomes share a common mode of protein targeting. *Proc. Natl Acad. Sci. USA* **102**: 10924–10929.

Edwards, M., Schofield, P., O'Sullivan, W. and Costello, M. (1992) Arginine metabolism during culture of *Giardia intestinalis. Mol. Biochem. Parasitol.* **53**: 97–103.

Ellis, J., Davila, M. and Chakrabarti, R. (2003) Potential involvement of extracellular signal-regulated kinase 1 and 2 in encystation of a primitive eukaryote, *Giardia lamblia.* Stage-specific activation and intracellular localization. *J. Biol. Chem.* **278**:1936–1945.

Elmendorf, H., Singer, S. and Nash, T. (2000) Targeting of proteins to the nuclei of *Giardia lamblia. Mol. Biochem. Parasitol.* **106**: 315–319.

Erlandsen, S. and Bemrick, W. (1987) SEM evidence for a new species, *Giardia psittaci. J. Parasitol.* **73**: 623–629.

Erlandsen, S., Bemrick, W. and Pawley, J. (1989) High resolution electron microscopic evidence for the filamentous structure of the cyst wall in *Giardia muris* and *Giardia duodenalis. J. Parasitol.* **75**: 787–797.

Erlandsen, S. L., Bemrick, W., Schupp, D., Shields, J., Jarroll, E., Sauch, J. and Pawley, J. (1990a) High-resolution immunogold localization of *Giardia* cyst wall antigens using field emission SEM with secondary and backscatter electron imaging. *J. Histochem. Cytochem.* **38**: 625–632.

Erlandsen, S., Bemrick, W., Wells, C., Feely, D., Knudson, L., Campbell, S., van Keulen, H. and Jarroll, E. (1990b) Axenic culture and characterization of *Giardia ardeae* from the great blue heron (*Ardea herodias*). *J. Parasitol.* **76**: 717–724.

Erlandsen, S., Macechko, P.T., van Keulen, H. and Jarroll, E. (1996) Formation of the *Giardia* cyst wall: studies on extracellular assembly using immunogold labeling and high resolution field emission SEM. *J. Eukaryot. Microbiol.* **43**: 416–420.

Erlandsen, S., Russo, A. and Turner, J. (2004) Evidence for adhesive activity of the ventrolateral flange in *Giardia lamblia. J. Eukaryot. Microbiol.* **51**: 73–80.

Ey, P., Davey, R. and Duffield, G. (1992) A low-affinity nucleobase transporter in the tsa417-like subfamily of variant-specific surface protein genes in *Giardia intestinalis. Mol. Biochem. Parasitol.* **99**: 55–68.

Fan, J., Korman, S., Cantor, C. and Smith, C. (1991) *Giardia lamblia*: haploid genome size determined by pulse field gel electrophoresis is less than 12 MB. *Nucleic Acids Res.* **19**: 1905–1908.

Farthing, M., Keusch, G. and Carey, M. (1985) Effects of bile and bile salts on growth and membrane lipid uptake by *Giardia lamblia.* Possible implications for pathogenesis of intestinal disease. *J. Clin. Invest.* **76**: 1727–1732.

Feely, D. (1986) A simplified method for *in vitro* excystation of *Giardia muris. J. Parasitol.* **72**: 474–475.

Feely, D. (1988) Morphology of the cyst of *Giardia microti* by light and electron microscopy. *J. Protozool.* **35**: 52–54.

Feely, D., Holbertson, D. and Erlandsen, S. (1990) The biology of *Giardia.* In: *Giardiasis* (ed. E.A. Meyer). Elsevier, New York, NY, pp. 11–50.

Feely, D., Gardner, M. and Hardin, E. (1991) Excystation of *Giardia muris* induced by a phosphate-bicarbonate medium: localization of acid phosphatase. *J. Parasitol.* **77**: 441–448.

Fortess, E. and Meyer, E. (1976) Isolation and axenic cultivation of *Giardia* trophozoites from the guinea pig. *J. Parasitol.* **62**: 689.

Filice, F. (1952) Studies on the cytology and life history of a *Giardia* from the laboratory rat. *Univ. Calif. Publ. Zool.* **57**: 53–146.

Gerwig, G., van Kuik, J., Leeflang, B., Kamerling, J., Vliegenthart, J., Karr, C. and Jarroll, E. (2002) Conformational studies of the β (1–3)-*N*-acetyl-D-galactosamine polymer of the *Giardia lamblia* filamentous cyst wall. *Glycobiology* **12**: 1–7.

Gibson, C., Schanen, B., Chakrabarti, D. and Chakrabarti, R. (2006) Functional characterisation of the regulatory subunit of cyclic AMP-dependent protein kinase A homologue of *Giardia lamblia*: differential expression of the regulatory and catalytic subunits during encystation. *Int. J. Parasitol.* **36**: 791–799.

Gillin, F. and Diamond, L. (1981a) *Entamoeba histolytica* and *Giardia lamblia*: growth responses to reducing agents. *Exp. Parasitol.* **51**: 382–391.

Gillin, F. and Diamond, L. (1981b) *Entamoeba histolytica* and *Giardia lamblia*: effects of cysteine and oxygen tension on trophozoite attachment to glass and survival in culture media. *Exp. Parasitol.* **52**: 9–17.

Gillin, F. and Reiner, D. (1982) Attachment of the flagellate *Giardia lamblia*: role of reducing agents, serum, temperature, and ionic composition. *Mol. Cell Biol.* **2**: 369–377.

Gillin, F., Gault, M., Hofmann, A., Gurantz, D. and Sauch, J. (1986) Biliary lipids support serum-free growth of *Giardia lamblia*. *Infect. Immun.* **53**: 641–645.

Gillin, F., Reiner, D., Gault, M., Douglas, H., Das, S., Wunderlich, A. and Sauch, J. (1987) Encystation and expression of cyst antigens by *Giardia lamblia in vitro*. *Science* **235**: 1040–1043.

Gottig, N., Elías, E., Quiroga, R., Nores, M., Solari, A., Touz, M. and Luján, H. (2006) Active and passive mechanisms drive secretory granule biogenesis during differentiation of the intestinal parasite *Giardia lamblia*. *J. Biol. Chem.* **281**: 18156–18166.

Gottstein, B. and Nash, T. (1991) Antigenic variation in *Giardia lamblia*: infection of congenitally athymic nude and scid mice. *Parasite Immunol.* **13**: 649–659.

Grassi, B. (1879) Dei protozoi parassiti e specialmente di quelliche sono nell'uomo. *Gazz. Med. Ital. Lomb.* **39**: 445–448.

Hehl, A., Marti, M. and Kohler, P. (2000) Stage-specific expression and targeting of cyst wall protein-green fluorescent protein chimeras in *Giardia*. *Mol. Biol. Cell* **11**: 1789–1800.

Holbertson, D., Baker, D. and Marshall, J. (1988) Segmented alpha-helical coiled-coil structure of the protein giardin from the *Giardia* cytoskeleton. *J. Mol. Biol.* **204**: 789–795.

Huang, D. and White, A. (2006) An updated review on *Cryptosporidium* and *Giardia*. *Gastroenterol. Clin. North Am.* **35**: 291–314.

Jakubowski, W. and Craun, G. (2002) Towards a better understanding of host specificity and the transmission of *Giardia*: the impact of molecular epidemiology. In: Giardia: *the Cosmopolitan Parasite* (eds B. Olson, M. Olson and P. Wallis). CABI, New York, NY, pp. 217–238.

Jarroll, E. (1991) *Giardia* cysts: their biochemistry and metabolism. In: *Biochemical Protozoology as a Basis for Drug Design* (eds G. Coombs and M. North). Taylor and Francis, London, pp. 52–60.

Jarroll, E. (1994) Biochemical mechanisms of action of antigiardial drugs. In: Giardia: *from Molecules to Disease* (eds A. Thompson, J. Reynoldson and A. Lymbury). CAB International, Oxford, pp. 329–337.

Jarroll, E.L. (1999) Sensitivity of protozoa to disinfection. Intestinal protozoa. In: *Principles and Practice of Disinfection Preservation and Sterilization* (eds B. Hugo, G. Ayliffe and D. Russell). Blackwell Scientific Publications, Oxford, pp. 251–257.

Jarroll, E. and Lindmark, D. (1983) Pyrimidine metabolism in *Tritrichomonas foetus. J. Parasitol.* **69**: 846–849.

Jarroll, E. and Şener, K. (2003) Potential drug targets in cyst-wall biosynthesis by intestinal protozoa. *Drug Resist. Updates* **6**: 239–246.

Jarroll, E., Muller, P., Meyer, E. and Morse, S. (1981) Lipid and carbohydrate metabolism of *Giardia lamblia. Mol. Biochem. Parasitol.* **2**: 187–196.

Jarroll, E., Manning, P., Lindmark, D., Coggins, J. and Erlandsen, S. (1989) *Giardia* cyst wall-specific carbohydrate: evidence for the presence of galactosamine. *Mol. Biochem. Parasitol.* **32**: 121–132.

Kabnick, K. and Peattie, D. (1990) In situ analyses reveal that the two nuclei of *Giardia lamblia* are equivalent. *J. Cell. Sci.* **95**: 353–360.

Karapetyan, A. (1960) Methods of *Lamblia* cultivation. *Tsitologiya* **2**: 379–384.

Karapetyan, A. (1962) *In vitro* cultivation of *Giardia duodenalis. J. Parasitol.* **48**: 337–340.

Karr, C. and Jarroll, E. (2004) Cyst wall synthase: *N*-acetylgalactosaminyl-transferase activity is induced to form the novel GalNAc polysaccharide in the *Giardia* cyst wall. *Microbiology* **150**: 1237–1243.

Keister, D. (1983) Axenic cultivation of *Giardia lamblia* in TYI-S-33 supplemented with bile. *Trans R. Soc. Trop. Med. Hyg.* **77**: 487–488.

Kulda, J. and Nohýnková, E. (1978) Flagellates of the human intestinal tract and of intestine of other species. In: *Parasitic Protozoa* (ed. J.P. Kreier). Academic Press, New York, NY, pp. 1–138.

Kulda, J. and Nohýnková, E. (1996) *Giardia* in humans and animals. In: *Parasitic Protozoa* (ed. J.P. Kreier). Academic Press, New York, NY, pp. 225–422.

Kulkova, L., Singer, M. and Nash, T. (2006) Epigenetic mechanisms are involved in the control of *Giardia lamblia* antigenic variation. *Mol. Microbiol.* **61**: 1533–1542.

Lambl, W. (1859) Mikroskopische untersuchungen der Darmexcrete. *Vierteljahrsschr. Prakst. Heikd.* **61**: 1–58.

Lanfredi-Rangel, A., Attias, M., Reiner, D., Gillin, F. and DeSouza, W. (2003) Fine structure of the biogenesis of *Giardia lamblia* encystation secretory vesicles. *Struct. Biol.* **143**: 153–163.

Le Blancq, S. and Adam, R. (1998) Structural basis of karyotype heterogeneity in *Giardia lamblia. Mol. Biochem. Parasitol.* **97**: 199–208.

Lee, F., Moss, J. and Vaughan, M. (1992) Human and *Giardia* ADP-ribosylation factors (ARFs) complement ARF function in *Saccharomyces cerevisiae. J. Biol. Chem.* **267**: 24441–24445.

Lindmark, D. (1980) Energy metabolism of the anaerobic protozoon *Giardia lamblia. Mol. Biochem. Parasitol.* **1**: 1–12.

Lindmark, D. (1988) Giardia lamblia: localization of hydrolase activities in lysosome-like organelles of trophozoites. *Exp. Parasitol.* **65**: 141–147.

Lindmark, D. and Jarroll, E. (1982) Pyrimidine metabolism in *Giardia lamblia. Mol. Biochem. Parasitol.* **5**: 291–296.

Lopez, A., Hossain, M. and van Keulen, H. (2002) *Giardia intestinalis* glucosamine 6-phosphate isomerase: the key enzyme to encystment appears to be controlled by ubiquitin attachment. *J. Eukaryot. Microbiol.* **49**: 134–136.

Lopez, A., Şener, K. Jarroll, E. and van Keulen, H. (2003) Transcription regulation is demonstrated for five key enzymes in *Giardia intestinalis* cyst wall polysaccharide biosynthesis. *Mol. Biochem. Parasitol.* **128**: 51–57.

Lopez, A., Şener, K., Trosien, J., Jarroll, E. and van Keulen, H. (2007) UDP-*N*-acetylglucosamine 4'-epimerase from the intestinal protozoan *Giardia intestinalis* lacks UDP-glucose 4'-epimerase activity. *J. Eukaryot. Microbiol.* **54**: 154–160.

Luján, H. and Nash, T. (1994) The uptake and metabolism of cysteine by *Giardia lamblia* trophozoites. *J. Eukaryot. Microbiol.* **41**: 169–175.

Luján, H. and Touz, M. (2003) Protein trafficking in *Giardia lamblia. Cell Microbiol.* **5**: 427–434.

Luján, H., Marotta, A., Mowatt, M., Sciaky, N., Lippincott-Schwarts, J. and Nash, T. (1995a) Developmental induction of Golgi structure and function in the primitive eukaryote *Giardia lamblia. J. Biol. Chem.* **270**: 4612–4618.

Luján, H., Mowatt, M., Conrad, J., Bowers, B. and Nash, T. (1995b) Identification of a novel *Giardia lamblia* cyst wall protein with leucine-rich repeats. Implications for secretory granule formation and protein assembly into the cyst wall. *J. Biol. Chem.* **270**: 29307–29313.

Luján, H.D., Mowatt, M., Chen, G. and Nash, T. (1995c) Isoprenylation of proteins in the protozoan *Giardia lamblia. Mol. Biochem. Parasitol.* **72**: 121–127.

Luján, H., Mowatt, M., Byrd, L. and Nash, T. (1996a) Cholesterol starvation induces differentiation of the intestinal parasite *Giardia lamblia. Proc. Natl Acad. Sci. USA* **93**: 7628–7633.

Luján, H., Mowatt, M. and Nash, T. (1996b) Lipid requirements and lipid uptake by *Giardia lamblia* trophozoites in culture. *J. Eukaryot. Microbiol.* **43**: 237–242.

Luján, H., Mowatt, M. and Nash, T. (1997) Mechanisms of *Giardia lamblia* differentiation into cysts. *Microbiol. Mol. Biol. Rev.* **61**: 294–304.

Macechko, P.T., Steimle, P., Lindmark, D., Erlandsen, S. and Jarroll, E. (1992) Galactosamine synthesizing enzymes are induced when *Giardia* encyst. *Mol. Biochem. Parasitol.* **56**: 301–310.

Medical Letter (2004) Tinidazole (Tindamax) – a new antiprotozoal drug. *Med. Lett. Drugs Therap.* **46**: e1–e12.

Mendis, A., Thompson, R., Reynoldson, J., Armson, A., Meloni, B. and Gunsberg, S. (1992) The uptake and conversion of l-[U-^{14}C] aspartate and l-[U-^{14}C] alanine to $^{14}CO_2$ by intact trophozoites of *Giardia duodenalis*. *Comp. Biochem. Physiol. Ser. B* **102**: 235–239.

Meng, T., Aley, S., Svärd, S., Smith, M., Huang, B., Kim, J. and Gillin, F. (1996) Immunolocalization and sequence of caltractin/centrin from the early branching eukaryote *Giardia lamblia*. *Mol. Biochem. Parasitol.* **78**: 103–108.

Meyer, E. (1970) Isolation and axenic cultivation of *Giardia* trophozoites from the rabbit, chinchilla, and cat. *Exp. Parasitol.* **27**: 179–183.

Meyer, E. and Jarroll, E. (1980) Giardiasis. *Am. J. Epidemiol.* **11**: 1–12.

Mok, M. and Edwards, M. (2005) Kinetic and physical characterization of the inducible UDP-*N*-acetylglucosamine pyrophosphorylase from *Giardia intestinalis*. *J. Biol. Chem.* **280**: 39363–39372.

Mok, M., Tay, E., Sekyere, E., Glenn, W., Bagnara, A. and Edwards, M. (2005) *Giardia intestinalis*: molecular characterization of UDP-*N*-acetylglucosamine pyrophosphorylase. *Gene* **357**: 73–82.

Murtagh, J., Mowatt, M., Lee, C., Lee, F., Mishima, K., Nash, T., Moss, J. and Vaughan, M. (1992) Guanine nucleotide-binding proteins in the intestinal parasite *Giardia lamblia*. Isolation of a gene encoding an approximately 20-kDa ADP-ribosylation factor. *J. Biol. Chem.* **267**: 9654–9662.

Nash, T. and Keister, A. (1985) Differences in excretory-secretory products and surface antigens among 19 isolates of *Giardia*. *J. Infect. Dis.* **152**: 1166–1171.

Nash, T., Gillin, F. and Smith, P. (1983) Excretory-secretory products of *Giardia lamblia*. *J. Immunol.* **131**: 2004–2010.

Nash, T., McCutchan, T., Keister, D., Dame, J., Conrad, J. and Gillin, F. (1985) Restriction-endonuclease analysis of DNA from 15 *Giardia* isolates obtained from human and animals. *J. Infect. Dis.* **153**: 64–73.

Nash, T., Herrington, D., Losonsky, G. and Levine, M. (1987) Experimental human infections with *Giardia lamblia*. *J. Infect. Dis.* **156**: 974–984.

Nash, T., Herrington, D., Levine, M., Conrad, J. and Merritt, J. (1990) Antigenic variation of *Giardia lamblia* in experimental human infections. *J. Immunol.* **144**: 4362–4369.

Nash, T., Merritt, J. and Conrad, J. (1991) Isolate and epitope variability in susceptibility of *Giardia lamblia* to intestinal proteases. *Infect. Immun.* **59**: 1334–1340.

Niño, C. and Wasserman, M. (2003) Transcription of metabolic enzyme genes during the excystation of *Giardia lamblia. Parasitol. Intl.* **52**: 291–298.

Nishikawa, J., Iwawaki, H., Takubo, Y., Nishihara, T. and Kondo, M. (1986) Appearance of uridine 5′-diphospho-N-acetylglucosamine-4-epimerase during sporulation of *Bacillus megaterium. Microbiol. Immunol.* **30**: 1085–1093.

Nixon, J., Wang, A., Morrison, H., *et al.* (2002) A spliceosomal intron in *Giardia lamblia. Proc. Natl Acad. Sci. USA* **99**: 3701–3705.

Nohria, A., Alnso, A. and Peattie, D. (1992) Identification and characterization of gamma-giardin and the gamma-giardin gene from *Giardia lamblia. Mol. Biochem. Parasitol.* **56**: 27–37.

Nohýnková, E., Tůmová, P and Kulda, J. (2006) Cell division of *Giardia intestinalis*: flagellar developmental cycle involves transformation and exchange of flagella between mastigonts of a diplomonad cell. *Eukaryot. Cell* **5**: 753–761.

Olson, M.E., Hannigan, C.J., Gaviller, P.F., Fulton, L.A. (2002) The use of a Giardia vaccine as an immunotherapeutic agent in dogs. *Can. Vet. J.* 2001 Nov; **42(11):** 865–868.

Paget, T., Jarroll, E., Manning, P., Lindmark, D. and Lloyd, D. (1989) Respiration in the cysts and trophozoites of *Giardia muris. J. Gen. Microbiol.* **135**: 145–154.

Paget, T., Manning, P. and Jarroll, E. (1993a) Oxygen uptake in cysts and trophozoites of *Giardia lamblia. J. Eukaryot. Microbiol.* **40**: 246–250.

Paget, T., Kelly, M., Jarroll, E., Lindmark, D. and Lloyd, D. (1993b) The effects of oxygen on fermentation in *Giardia lamblia. Mol. Biochem. Parasitol.* **57**: 65–71.

Paget, T., Macechko, P. and Jarroll, E. (1998) *Giardia intestinalis*: metabolic changes during cytodifferentiation. *J. Parasitol.* **84**: 222–226.

Paget, T., Maroulis, S., Mitchell, A., Edwards, M., Jarroll, E. and Lloyd, D. (2004) Menadione-generated radicals kill *Giardia intestinalis* trophozoites and cysts. *Microbiology* **150**: 1231–1236.

Peattie, D., Alonso, A., Hein, A. and Caulfield, J. (1989) Ultrastructural localization of giardins to the edges of disk microribbons of *Giardia lamblia* and the nucleotide and deduced protein sequence of alpha giardin. *J. Cell Biol.* **109**: 2323–2335.

Piva, B. and Benchimol, M. (2004) The median body of *Giardia lamblia*: an ultrastructural study. *Biol. Cell* **96**: 735–746.

Radulescu, S. and Meyer, E. (1990) In vitro cultivation of *Giardia* trophozoites. In: Giardiasis (ed. E. Meyer) Elsevier, Amsterdam, 99–110.

Regoes, A., Zourmpanou, D., Leon-Avila, G., van der Giezen, M., Tovar, J. and Hehl, A. (2005) Protein import, replication, and inheritance of a vestigial mitochondrion. *J. Biol. Chem.* **280**: 30557–30563.

Reiner, D., McCaffery, M. and Gillin, F. (1990) Sorting of cyst wall proteins to a regulated secretory pathway during differentiation of the primitive eukaryote, *Giardia lamblia. Eur. J. Cell Biol.* **53**: 142–153.

Reiner, D., McCaffery, J. and Gillin, F. (2001) Reversible interruption of *Giardia lamblia* cyst wall protein transport in a novel regulated secretory pathway. *Cell Microbiol.* **3**: 459–472.

Reiner, D.S., Douglas, H., Gillin F.D. (1989) Identification and localization of cyst-specific antigens of *Giardia lamblia. Infect Immun.* **57(3):** 963–968.

Rendtorff, R. (1954) The experimental transmission of human intestinal protozoan parasites. II. *Giardia lamblia* cysts given in capsules. *Am. J. Hyg.* **59**:209–20.

Reynoldson, J. (2002) Therapeutics and new drug targets for giardiasis. In: *Giardia*: the Cosmopolitan Parasite (eds B. Olson, M. Olson and P. Wallis). CABI, New York, NY, pp. 159–175.

Richards, T. and van der Giezen, M. (2006) Evolution of the Isd11-IscS complex reveals a single α-proteobacterial endosymbiosis for all eukaryotes. *Mol. Biol. Evol.* **23**: 1341–1344.

Russell, A., Shutt, T., Watkins, R. and Gray, M. (2005) An ancient spliceosomal intron in the ribosomal protein L7a gene (*Rpl7a*) of *Giardia lamblia. BMC Evol. Biol.* **5**: 45–56.

Sagolla, M., Dawson, S., Mancuso, M. and Cande, W. (2006) Three-dimensional analysis of mitosis and cytokinesis in the binucleate parasite *Giardia intestinalis. J. Cell Sci.* **119**: 4889–4900.

Schupp, D., Erlandsen, S., Januschka, M., Sherlock, L., Meyer, E., Bemrick, W. and Stibbs, H. (1988) Production of viable *Giardia* cysts *in vitro*: determination by flurogenic dye staining, excystation, and animal infectivity in the mouse and Mongolian gerbil. *Gastroenterology* **95**: 1–10.

Scofield, P., Costello, M., Edwards, M. and O'Sullivan, W. (1990) The arginine dihydrolase pathway is present in *Giardia intestinalis. Int. J. Parasitol.* **20(5)**: 697–699.

Scofield, P., Edwards, M., Matthews, J. and Wilson, J. (1992) The pathway of arginine catabolism in *Giardia intestinalis. Mol. Biochem. Parasitol.* **51**: 29–36.

Şener, K., Shen, Z., Newburg, D. and Jarroll, E. (2004) Amino sugar phosphate levels change during formation of the *Giardia* cyst wall. *Microbiology* **150**: 1225–1230.

Slavin, I., Saura, A., Carranza, P., Touz, M., Nores, M. and Luján, H. (2002) Dephosphorylation of cyst wall proteins by a secreted lysosomal acid phosphatase is essential for excystation of *Giardia lamblia. Mol. Biochem. Parasitol.* **122**: 95–98.

Soliman, M., Taghi-Kilani, R., Abou-Shady, A., El-Mageid, S., Handousa, A., Hegazi, M. and Belosevic, M. (1998) Comparison of serum antibody responses to *Giardia lamblia* of symptomatic and asymptomatic patients. *Am. J. Trop. Med. Hyg.* **58**: 232–239.

Soltys, B. and Gupta, R. (1994) Immunoelectron microscopy of *Giardia lamblia* cytoskeleton using antibody to acetylate alpha-tubulin. *J. Eukaryot. Microbiol.* **41**: 625–632.

Stefanic, S., Palm, D., Svärd, S. and Hehl, A. (2006) Organelle proteomics reveals cargo maturation mechanisms associated with Golgi-like encystation vesicles in the early-diverged protozoan *Giardia lamblia. J. Biol. Chem.* **281**: 7595–7604.

Steimle, P., Lindmark, D. and Jarroll, E. (1997) Purification and characterization of glucosamine 6–phosphate isomerase from encysting *Giardia. Mol. Biochem. Parasitol.* **84**: 149–153.

Sun, C-H., McCaffery, J., Reiner, D. and Gillin, F. (2003) Mining the *Giardia lamblia* genome for new cyst wall proteins. *J. Biol. Chem.* **278**: 21701–21708.

Sun, C., Su, L. and Gillin, F. (2006) Novel plant-GARP-like transcription factors in *Giardia lamblia. Mol. Biochem. Parasitol.* **146**: 45–57.

Tai, J., Chang, S., Chou, C. and Ong, S. (1996) Separation and characterization of two related giardiaviruses in the parasitic protozoan *Giardia lamblia. Virology* **216**: 124–132.

Thoden, J.B., Wohlers, T.M., Fridovich-Keil, J.L. and Holden, H.M. (2001) Human UDP-galactose 4-epimerase. Accommodation of UDP-*N*-acetylglucosamine within the active site. *J. Biol. Chem.* **276**: 15131–15136.

Thompson, R. (2002) Towards a better understanding of host specificity and the transmission of *Giardia*: the impact of molecular epidemiology. In: *Giardia*: the Cosmopolitan Parasite (eds B. Olson, M. Olson and P. Wallis). CABI, New York, NY, pp. 55–69.

Touz, M., Gottig, N., Nash, T. and Luján, H. (2002a) Identification and characterization of a novel secretory granule calcium binding protein from the early branching eukaryote *Giardia lamblia. J. Biol. Chem.* **277**: 50557–50563.

Touz, M., Nores, M., Slavin, I., Carmona, C., Conrad, J., Mowatt, M., Nash, T., Coronel, C. and Luján H.D. (2002b) The activity of a developmentally regulated cysteine proteinase is required for cyst wall formation in the primitive eukaryote *Giardia lamblia. J. Biol. Chem.* **277**: 8474–8481.

Touz, M., Conrad, J. and Nash, T. (2005) A novel palmitoyl acyl transferase controls surface protein palmitoylation and cytotoxicity in *Giardia lamblia. Mol. Microbiol.* **58**: 999–1011.

Tovar, J., Leon-Avila, G., Sanchez, L.B., Sutak, R., Tachezy, J., van der Giezen, M., Hernandez, M., Müller, M. and Lucocq, J. (2003) Mitochondrial remnant organelles of *Giardia* function in iron-sulphur protein maturation. *Nature* **426**: 127–128.

Upcroft, J., Campbell, R. and Upcroft, P. (1996) Quinacrine-resistant *Giardia duodenalis. Parasitology* **112**: 309–313.

van der Giezen, M., Tovar, J. and Clark, C. (2005) Mitochondrion-derived organelles in protists and fungi. *Int. Rev. Cytol.* **244**: 175–225.

Visvesvara, G. (1980) Axenic growth of *Giardia lamblia* in TSP-1 medium. *Trans R. Soc. Trop. Med. Hyg.* **74**: 213–215.

Wade, S., Mohammed, H., Zeigler, P. and Schaaf, S. (2002) Epidemiologic risk analysis study of *Giardia* sp. in domestic and wild animals. In: *Giardia*: the Cosmopolitan Parasite (eds B. Olson, M. Olson and P. Wallis). CABI, New York, NY, pp. 135–155.

Wang, C.C. and Aldritt, S. (1983) Pyrimidine salvage in *Giardia lamblia*. *J. Exp. Med.* **161**: 437–445.

Wang, A. and Wang, C.C. (1986) Discovery of a specific double-stranded RNA virus in *Giardia lamblia*. *Mol. Biochem. Parasitol.* **21**: 269–276.

Ward, H., Alroy, J., Lev, B., Keusch, G. and Pereira, M. (1985) Identification of chitin as a structural component of *Giardia* cysts. *Infect. Immun.* **49**: 629–634.

Ward, W., Alvarado, L., Rawlings, D., Enge, J., Franklin, C. and McKerrow, J. (1997) A primitive enzyme for a primitive cell: the protease required for excystation of *Giardia*. *Cell* **89**: 437–444.

Weisehahn, G., Jarroll, E., Lindmark, D., Meyer, E. and Hallick, L. (1984) *Giardia lamblia*: autoradiographic analysis of nuclear replication. *Exp. Parasitol.* **58**: 94–100.

Wohlers, T.M., Christacos, N.C., Harreman, M.T. and Fridovich-Keil, J.L. (1999) Identification and characterization of a mutation, in the human UDP-galactose-4-epimerase gene, associated with generalized epimerase-deficiency galactosemia. *Am. J. Hum. Gene* **64**: 462–470.

Yanke, S., Ceri, H., McAllister, T., Morck, D. and Olson, M. (1998) Serum immune response to *Giardia duodenalis* in experimentally infected lambs. *Vet. Parasitol.* **75**: 9–19.

Yu, L., Birky, W. and Adam, R. (2002) The two nuclei of *Giardia* each have complete copies of the genome and are partitioned equationally at cytokinesis. *Eukaryot. Cell* **1**: 191–199.

D2 *Trichomonas vaginalis*

Bibhuti N. Singh, John. J. Lucas
and Raina N. Fichorova

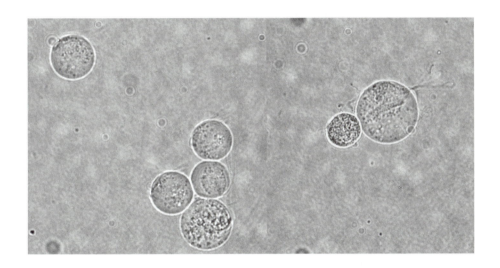

1 Introduction

Trichomonas vaginalis is a primitive, eukaryotic, parasitic, flagellated protist causing one of the most common nonviral STI, trichomoniasis in humans worldwide. It is estimated that over 180 million people are infected annually with *T. vaginalis* worldwide, including more than eight million cases reported each year in the United States alone (Petrin *et al.*, 1998; Ryu and Min, 2006; Schwebke, 2005; Schwebke and Burgess, 2004; Weinstock *et al.*, 2004). *T. vaginalis* is an extracellular pathogenic protozoan; it adheres to and damages vaginal epithelial cells (Gilbert *et al.*, 2000; Sommer *et al.*, 2005) and lives in the vagina of women and urethra of men. *T. vaginalis* is transmitted by sexual intercourse. Women can get the disease from infected men or women. Nonvenereal transmission is rare, but possible, since the organisms can survive 1–2 days in urine and a few hours in/on wet places (e.g., wet towels, sponges, swimming pools). The infection rate is much higher (>50%) in women than men. Infected women can transmit *T. vaginalis* to their female offspring during birth.

The infection is responsible for serious health consequences for women and is emerging as one of the most important cofactors in amplifying HIV-1 transmission. Vaginitis and acute inflammatory disease of the genital mucosa are often induced by this infection. In pregnant women, *T. vaginalis* infection is implicated in premature rupture of membranes and preterm birth including low-birth-weight infants (Cotch *et al.*, 1997; Sorvillo *et al.*, 2001). Infection leads to urethritis, prostatitis, epididymitis, and other genitourinary tract problems in men (Krieger, 1981; Krieger and Alderete, 1999). Infection by *T. vaginalis* can become chronic, and resistant to the most commonly used drug – metronidazole. As with many microbial infections drug resistance is on the rise and this is an emerging problem with *T. vaginalis* infection. Trichomoniasis has important social, economic, and medical complications, especially for women and children. According to a recent report from the Institute of Medicine, preterm births in the USA alone cost at least $26.2 billion in 2005, or an average of $51 600 per infant (http://www.nas.edu/morenews/20060713.html). Imagine the cost and consequences of this infection all over the world! There are several excellent reviews on various aspects of *T. vaginalis*, which are recommended for additional reading (Krieger and Alderete, 1999; Petrin *et al.*, 1998; Ryu and Min, 2006; Schwebke and Burgess, 2004).

2 Historical Perspective

A French physician, Alfred Donne, was the first to report the presence of the *T. vaginalis* parasite in female genital tract discharge in 1836. At that time it was believed that the vaginal discharge was associated with marriage and child bearing and the parasite was regarded as a harmless inhabitant of the vagina for a long time. In 1916, Hoehne came up with the idea that the protozoan is a pathogen and introduced the term 'trichomonaskolpitis', (which meant trichomonas vaginitis) to state this kinship. For many years investigators debated whether the organism is responsible for vaginal irritation. It was not until 1940 that Trussell and Plass cultured *T. vaginalis* from a vaginitis patient and demonstrated that the organism can produce abnormal vaginal discharge and irritation in women, irrespective of its association with other bacterial flora. They also showed that *T. vaginalis* could not be implanted successfully in the vaginas of cattle, sheep, goat, horses, guinea pigs, rabbits, cats or dogs. From the 1950s to 1980s the significant incidence of STIs such as gonorrhea and syphilis were commonly reported, while trichomoniasis received very little attention. Since then thousands of reports have been published regarding the prevalence and global significance of trichomoniasis. During the last two decades physicians and scientists have come to recognize the importance of this infection and several important developments have occurred in understanding this disease. Yet even today in many communities across the world, including the United Sates, *T. vaginalis* infection is not a reportable disease, perhaps due to social/cultural stigma and ignorance or lack of a public health response to the epidemic of sexually transmitted diseases.

3 Morphology and Life Cycle

T. vaginalis is a flagellated protozoan having five flagella (four anterior flagella), with the fifth flagellum being incorporated within the undulating membrane of the organism. The anterior flagella contribute to the erratic and jerking motility of the organism. The parasite is spherical or pear-shaped in culture and measures about 7–23 μm by 5–10 μm (Figure 1). However, the shape of the parasite assumes an irregular amoeboid type appearance when attached to the host vaginal epithelial cells (Arroyo *et al.*, 1993). Parasites spread over the host surface and interact by villous extensions (Nielsen and Nielsen, 1975). Also, under unfavorable or different

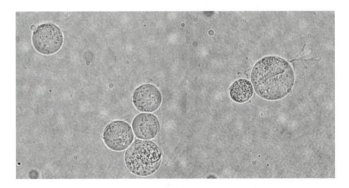

Figure 1 Microscopic observation of *T. vaginalis*. A small drop of fresh culture placed on a glass slide with 50% ethanol and visualized under the microscope (100 ×).

physiological growth conditions, *T. vaginalis* can round up and internalize the flagella.

T. vaginalis has well-developed cytoskeletal structures composed of tubulin and actin fibers (Kulda *et al.*, 1986). The Golgi is well developed, having structurally integrated with the root fibrils to form a parabasal apparatus. The nucleus is located towards the anterior portion of the organism, similar to other eukaryotes, and is surrounded by a nuclear envelope. The axostyle of *T. vaginalis* is noncontractile and runs through the cell from the anterior end to the posterior end (Nielsen and Nielsen, 1975). The parasite lacks mitochondria; instead it has numerous microbody-like particles called hydrogenosomes which produce hydrogen (Lindmark *et al.*, 1975). The hydrogenosomes are energy-producing organelles, analogous to the mitochondria present in higher eukaryotes. Muller's and Johnson's groups have extensively studied the functional significance of hydrogenosomes in trichomonads (Dyall and Johnson, 2000a, 2000b; Muller, 1988, 1990). The parasite cytoplasm also contains glycogen granules, rough endoplasmic reticulum, free ribosomes, polysomes and a variety of vacuoles. *T. vaginalis* has a simple life cycle since the parasite exists only in trophozoite form and a cyst form has never been reported. The trophozoites live in close contact with the epithelium of the urogenital tract and reproduce by longitudinal binary fission. Trophozoites are usually transmitted by sexual intercourse and continual re-infection of one sexual partner by the other is common. An infected mother can also transmit the infection to her newborn girl during passage through the birth canal (Littlewood and Kohler, 1966). However, in such cases, the infection tends to remain asymptomatic until puberty. It is of interest to note that the shape and length of trophozoites can change with changing environmental conditions, for example, from its presence in the vagina to *in vitro* culture.

4 Classification

Trichomonads make up a group of protists belonging to the phylum Parabasala (Cavalier-Smith, 1998). They are aerotolerant, anaerobic organisms that lack mitochondria and peroxisomes, but possess specialized organelles called hydrogenosomes. Hydrogenosomes are responsible for metabolic processes that extend glycolysis (Kulda, 1999; Muller, 1993). Two pathogenic organisms that infect the urogenital tract in human and cattle are *T. vaginalis* and *Tritrichomonas foetus*, respectively. These are parasitic organisms of significant medical and agricultural importance belonging to the protozoan order Trichomonadida. The following classification is based on papers by Kirby (1947), Honigberg (1963) and recent reports published by Stechmann and Cavalier-Smith (2002) and Hampl *et al.* (2006).

Kingdom   Protista
Phylum   Parabasala (amitochondrial flagellate)

Class	Trichomonadea (Parabasalea)
Order	Trichomonadida (Trichomonads; 4–6 flagella, no cyst stage)
Family	Trichomonadidae (3–5 free flagella with one on the margin of undulating membrane; axostyle protruding through the posterior of the cell)
Genus	*Trichomonas* (4 free flagella and one along the outer margin of the undulating membrane reaching only the middle of the body)
Species	*Trichomonas vaginalis* (Donne, 1836)

The other related nonpathogenic organisms that belong to the trichomonad family are *Trichomonas tenax* (found in oral gingival and tracheobronchial sites), Tetratrichomonads (found in oral cavity and respiratory tract), and *Pentatrichomonas hominis* (found in colon).

5 Clinical Symptoms

Almost half of the women with trichomoniasis are asymptomatic while the other half develops symptoms of severe inflammatory reaction. The genital inflammation caused by *T. vaginalis* infection can increase a woman's susceptibility to HIV-1 infection if she is exposed to the virus. Some reports suggest that persistent infection raises the chances of women becoming infertile and developing cervical cancer (Vikki *et al.*, 2000; Yap *et al.*, 1995). The other consequences associated with this infection are described in detail in the epidemiology section (section 7) of this review. Primarily, *T. vaginalis* infects the squamous epithelium of the urogenital tract. Once the infection is established it can persist for a long period of time in females, but only for a short time in males. Symptoms usually appear in 4 to 28 days after exposure to *T. vaginalis* in about 50% of infected individuals. The majority (~70%) of men having intercourse with infected women showed the presence of the organism within 48 hours, and 85% of the women whose male partners were infected developed trichomoniasis. The severity of infection can be classified as acute, chronic, or asymptomatic (Petrin *et al.*, 1998).

The acute infection presents with signs of diffuse vulvitis due to *copius leucorrhoea* and the discharge (50–75%) is usually frothy, yellow or green (20–30%), having a strong foul odor. The vaginal and cervical mucosa may contain small punctuate hemorrhage spots (2–5% of infected women on routine examination) and this appearance has been referred to as '*colpitis macularis*' or 'strawberry appearance' (Fouts and Kraus, 1980; Krieger and Alderete, 1999). Krieger *et al.* (1990a) reported the characteristic of *colpitis macularis* in 44% of infected women which could only be detected by colposcopy. The strawberry appearance or *colpitis macularis* is a specific clinical sign for diagnosis of *T. vaginalis* infection. These symptoms are cyclic and worsen around the time of menses.

If untreated the vaginitis may develop into a chronic infection, which is characterized by intermittent symptoms with pruritus and dyspareunia. The vaginal secretion may be very scanty, containing mucus (Petrin *et al.*, 1998). Women with these symptoms become the prime source of transmission. Asymptomatic women (50%) have a normal vaginal pH of 3.8–4.2 and a normal vaginal flora (Spiegel, 1990) and these women typically develop clinical symptoms during the following 6 months. The other symptoms associated with infection include discomfort during intercourse, frequent urination, including dysuria and some time lower abdominal pain (Markell *et al.*, 1999). In a large population urethral involvement is also common. Statistical data can vary from one clinic to another.

In men, trichomoniasis is mainly asymptomatic and they are usually considered to be the carriers of *T. vaginalis*. It is well recognized that *T. vaginalis* infection in men is an important cause of nongonococcal urethritis (68%; Krieger and Aldetrete, 1999; Petrin *et al.*, 1998; Schwebke and Hook, 2003) and is associated with prostatitis (40%; Krieger and Aldetrete, 1999; Skerk *et al.*, 2002), including epididymitis, and infertility (Gopalkrishnan *et al.*, 1990; Lloyd *et al.*, 2003; Ryu and Min, 2006). The number of nongonoccocal urethritis cases is increasing more rapidly than gonorrhoea in the United States (Sutcliffe *et al.*, 2006). A recent report by Sutcliffe *et al.* (2006) points out a history of trichomoniasis in men and its strong association with incidence of prostate cancer in a large nested case-control study. Petrin *et al.* (1998) pointed out that *T. vaginalis* infection in men can also be classified into three groups. These are: asymptomatic stage, when infected female partner is identified; acute stage, which is characterized by profuse purulent urethritis; and mild symptomatic stage, which is clinically identical to that from other causes of nongonococcal urethritis. Other symptoms may include clear to mucopurulent discharge, dysuria, and mild pruritus or burning feeling right after sexual intercourse (Krieger *et al.*, 1990a).

6 Culture and Detection of *T. vaginalis*

There are several published reports indicating that infection with *T. vaginalis* increases the risk of bacterial and viral (especially HIV-1) co-infection, cervical cancer, and adverse pregnancy outcomes (see reviews by Krieger and Alderete, 1999; Petrin *et al.*, 1998; Schwebke and Burgess, 2004). Because *T. vaginalis* infection is often asymptomatic and difficult to diagnose, accurate detection methods may play a role in reducing the incidence of the HIV epidemic. The clinical diagnosis is usually performed in STD clinics or in a physician's office only when a person comes in with complaints about vaginal discharge, itching or urethral complaints from men.

A variety of techniques are used to diagnose *T. vaginalis*. One of the most common and simple methods to detect trichomoniasis depends on the microscopic observation of the motility of this organism in vaginal or cervical discharges of patients during pelvic examinations. This microscopic method was used originally by Donne in 1836 to identify the parasite. In current medical practice physicians do not screen women for STI until and unless the women have vaginal complaints and are symptomatic. The same practice exists for men during routine physical examination. STD clinics use wet-mount preparation of vaginal discharge and male urine samples to detect *T. vaginalis* by microscopy. Although this method is simple and economical it detects only 35–75% of the cases in women and is less reliable in men. The procedure depends on the expertise of a skilled microscopist. The reliability and sensitivity of microscopic examination can be attributed to the loss of distinctive mobility of the organisms that have been removed from their natural environment of body temperature and also the amount of sample removed from the patient. Thus, this test should be performed within 10–15 min of collection of the samples in order to visualize the motility of the organisms.

The culture technique is the 'gold standard' for the diagnosis of trichomonas infection. *T. vaginalis* is cultured in Diamond's TYM (trypticase yeast maltose) media (pH 6.0; Diamond, 1968; Gilbert *et al.* 2000). It is accomplished by placing a vaginal swab collected from a suspected patient directly into the culture media. Antibiotics and antimicotics are added to the culture media at the start of culture to eliminate any bacterial or fungal contamination. This is the most accurate method for identifying *T. vaginalis* (>98% success) and the best method for research investigators. However, there are some limitations, such as the number of organisms (200–400 parasites ml^{-1}) and the incubation time of 2–7 days required for the organisms to grow. During this time infected patients may continue to transmit the infection. Also, this culture system is not widely available to clinicians. Detection of *T. vaginalis* by this method in males may be more difficult; one may have to use both urethral swabs and urine sediment for culture. It should also be noted that sometimes parasites may not grow in laboratory settings, once they are out of their natural environment.

Another commercially available InPouch culture method composed of liquid medium (perhaps similar to Diamond's) is used in many clinics for detecting *T. vaginalis* (Draper *et al.*, 1993). This method is good since it is used with both self- and clinician-collected vaginal specimens (Schwebke *et al.*, 1997, 1999). This is usually ideal for screening in the adolescent population and other patients where pelvic examination is not possible. It takes 2–5 days to obtain test results. For reliable results it is also important that the swab specimens be placed as soon as possible into the InPouch at 37°C. In some cases it takes 3–10 days before one can visualize the movement of organisms with good success (~84–92%). Levi *et al.* (1997) reported the sensitivity for InPouch culture was about 83% versus 88% for Diamond's modified medium. There are several staining methods (see Krieger and Alderete, 1999) such as acridine orange, periodic acid-Schiff,

Giemsa, Papanicolaou, and immunoperoxidase and others which have been used to identify this organism in clinics, but the reliability and sensitivity is far less accurate than direct microscopic examination. The Papanicolaou staining method has been shown to have limited success (40–70%) during routine cytological examination, but false positive-negative results (~48%) are also obtained (Wiese *et al.*, 2000). Stary *et al.* (2002) reported a modified Columbia agar (MCA) technique for routine laboratory diagnosis of *T. vaginalis* infections in asymptomatic and symptomatic patients with great success (92–98%). This culture method may be very useful in identifying *T. vaginalis* since it requires only 10–100 organisms ml^{-1} for inoculums and 24–72 h for growth. However, it should be pointed out that microscopic examination is always performed by using saline wet mount from this culture for positive identification of *T. vaginalis*.

Recently, the Federal Drug Administration (FDA; USA) approved two new tests for diagnosing trichomoniasis in women. One is the OSOM Trichomonas Rapid Test (Genzyme Diagnostic, MA), an immunochromatographic capillary flow dipstick technology, and the second is the Affirm™ VP III (Becton Dickenson, CA), a nucleic acid probe test that evaluates *T. vaginalis*. These tests are performed on vaginal secretions and have been claimed to have a sensitivity greater than 83%, and specificity greater than 97%. Results are obtained in 10–45 min. In a low prevalence population this test might give false-positive results. These tests may have potential value where microscopy is not on hand.

In men, diagnosis of this infection is rather more difficult and cumbersome since wet-mount preparation is insensitive, and culture testing of urethral swab, urine, and semen is required for optimal sensitivity. Positive identification may take up to a few weeks. That is, if they ever come to the STD clinic with their female partners, since most men are asymptomatic. PCR may provide great value for detecting this organism in the male population. PCR-based diagnosis for *T. vaginalis* is currently under active investigation in many laboratories, with good success (Schwebke and Burgess, 2004). Recently, Hobbs *et al.* (2006) reported the presence of *T. vaginalis* in 72% of the male sexual partners of women with trichomoniasis, using culture and PCR from urine, urethral swabs and semen. Others have also used PCR technology to detect more *T. vaginalis* infections in men than by using the culture method in STD clinics (Schwebke and Lawing, 2002; Wendel *et al.*, 2003). Several studies have reported the PCR-based assay sensitivity and specificity at 82% and 95%, respectively (Hobbs *et al.*, 1999; Schwebke and Lawing, 2002).

7 Epidemiology

Trichomoniasis is one of the most common cosmopolitan nonviral STIs. The percentage of population that is affected by trichomoniasis depends upon the patient population studied and the available methods used for diagnosis. There are several elements involved in reporting the prevalence of trichomoniasis, such as investigators and subjects at STD clinics; sexual behaviors; sexual partners; sexually active women/men; techniques of examination, including specimen collection; and laboratory diagnosis. Sex industry workers and mistresses appear to have higher infection rates than housewives. The highest rate of infection occurs between the ages of 16–45, the period of greatest sexual activity. However, recent data indicate an astonishingly high prevalence of *T. vaginalis* infection in 14- to 17-year-old adolescent women (Schwebke, 2005; Van Der Pol *et al.*, 2005) in certain communities. The prevalence of this STI can reach 50–85% in prison inmates and commercial sex workers (Rein, 1990).

T. vaginalis infection is observed more frequently in females attending STD clinics than in virgins and postmenopausal women. The infection is usually diagnosed when a woman visits her Obstetric/Gynecology physician, but is usually missed during a regular visit to a physician's office, unless there is some complaint from the patient. Women attending innercity STD clinics in the United States often have trichomoniasis, with rates of infection ranging from 18 to 45% or even higher in certain populations (Bachmann, *et al.*, 2000; Schwebke, 2005; Sorvillo and Kerndt, 1998). *T. vaginalis* is highly prevalent among African-Americans in major urban centers in the United States and is often the most common STI in black women. It is emerging as one of the most important cofactors in amplifying HIV-1 transmission and the pathology induced by *T. vaginalis* infection can also magnify HIV-1 shedding in the female genital tract (Draper *et al.*, 1998; Jackson *et al.*, 1997; Magnus *et al.*, 2003; Sorvillo *et al.*, 2001). The increased viral load can enhance sexual and perinatal transmission of HIV-1 (John *et al.*, 2001; Mason *et al.*, 2005; Tuomala *et al.*, 2003). Indications of lower socioeconomic status such as a lower level of education, poor personal hygiene and poverty appear to be associated with a higher rate of trichomoniasis (Catterall, 1972). A large racial disparity is another feature of prevalence in the United States, since African-Americans have an 11 times higher rate of infection than non-Hispanic white and Mexican-American women.

Around the globe infection rates have been reported as high as 67% in Mongolia (Schwebke *et al.*, 1998), more than 60% in Africa, and 40% in indigenous Australian women (Bowden *et al.*, 1999). Behavior factors, such as multiple sexual partners and unprotected sex, are also associated with an increased risk of contracting *T. vaginalis*. A few reports have suggested that the use of oral contraceptive is associated with a lower rate of infection (Birnbaum and Kraussold, 1975; Bramley and Kinghorn, 1979). This is a rather interesting observation which could open up a new line of future investigation – to see if hormonal regulation plays any direct or indirect roles in preventing/inducing trichomoniasis. That idea is also consistent with the observation that girls infected during birth do not manifest the infection until puberty. Trichomoniasis has also been linked to preterm delivery, low birth weight, especially in the African-American community (Cotch *et al.*, 1997; Sorvillo *et al.*, 2001), and to infertility (Grodstein *et al.*, 1993) and cervical cancer (Vikki *et al.*, 2000; Yap *et al.*, 1995). It has been estimated that *T. vaginalis* infection may be responsible for 20–25% of the cases of premature delivery in Africa (Bowden and Garnett, 1999).

In men, the rates of *T. vaginalis* infection are poorly understood due to lack of proper diagnostic methods, as well as their unwillingness to attend STD clinics alone or with their partners. The recent report by Hobbs *et al.* (2006) made a significant contribution in identifying this STI in men using a PCR assay. They detected *T. vaginalis* in 72% of the male sexual partners of women with trichomoniasis, using the combination of culture and PCR assays. This detection method is still in its primary stage of development and is not affordable for general use in public health settings. The prevalence of infection has been reported as 7–60% in males attending several STD clinics in the United States (Krieger and Alderete, 1999).

8 Pathobiology and Models of Host-Parasite Interactions

The host-parasite interaction is a very complex phenomenon, since a wide spectrum of symptoms, ranging from a relatively asymptomatic state to severe inflammation and irritation, are associated with trichomoniasis in women. This may suggest that the disease employs more than one pathogenic mechanism. It has been known that all clinical isolates of *T. vaginalis* are capable of initiating infection leading to trichomoniasis. Since the parasite resides in a very hostile environment, the immune evasion mechanisms employed by the parasite could be an important aspect of pathogenesis. Certain factor(s), such as phenotypic variation and release of proteases, may play roles in the chronic nature and the recurrence of infections (Corbeil *et al.*, 2003; Petrin *et al.*, 1998). However, knowledge of host and parasite factors that initiate the host inflammatory response and control the severity of the symptoms is still in its infancy (see below for discussion).

A bovine trichomoniasis model and several nonhuman and human cell culture models have been used to study the pathobiology of *T. vaginalis* infection. It has been suggested by Corbeil and her coworkers (Corbeil, 1995; Corbeil *et al.*, 2003) that bovine trichomoniasis is a useful model system for human trichomoniasis. *T. foetus* causes a sexually transmitted disease, bovine trichomoniasis, and like *T. vaginalis*, induces vaginitis, as well as abortion and sterility. Because of the similarities between the parasites and the resultant diseases, the veterinary disease provides a model system where ideas and products applicable to the human condition may be explored and tested. It has been reported that cows vaccinated with a surface glycoconjugate, lipophosphoglycan (LPG, see below; Singh *et al.*, 2001), containing the TF 17 antigen from *T. foetus*, showed faster clearance of infection when challenged with *T. foetus*, in comparison with nonvacci-nated animals (Corbeil *et al.*, 2003). This study demonstrates that a productive local immune response to trichomonads in the reproductive tract may be possible.

Most of the earlier studies involving *T. vaginalis* were performed using readily available cell lines such as the human cervical adenocarcinoma cell line HeLa or the MDCK (Madine-Darby canine kidney) cell line, but these cell types are not naturally encountered by the parasite during infection. Furthermore, both bovine and human trichomonads bind to these cells. Occasionally, crude human vaginal scrapings of unknown physiological condition were used. A consistent finding in most of these studies was that no species specificity was observed; *T. vaginalis* and *T. foetus* function equally well. To overcome these limitations, human and bovine vaginal primary epithelial cell cultures and immortalized human cell lines were developed (Fichorova *et al.*,1997; Gilbert *et al.*, 2000; Singh *et al.*, 1999). It has been shown that bovine vaginal epithelial cells (BVECs) specially bind *T. foetus* (Singh *et al.*, 1999) and human vaginal epithelial cells (HVECs) spe-cially bind *T. vaginalis* (Gilbert *et al.*, 2000). The respective parasites do not adhere to the noncognate host cells. The species specificity of host-parasite interaction extends further into known cytopathogenic effects of parasite proteases and newly discovered immunoinflammatory responses to para-site lipophosphoglycan (discussed below).

As in many other microbial infections, trichomonas infection begins with adhesion of parasites to the host target tissue, the vaginal epithelium (Alderete and Garza, 1985; Alderete *et al.*,1995 a and b; Corbeil *et al.*, 1989; Krieger *et al.*, 1985; Petrin *et al.*, 1998). Adhesion may be followed by asymptomatic colonization or may have cytopathic consequences, usually accompanied by neutrophil infiltration and vaginal discharge. Interestingly, all *T. vaginalis* isolates, whether from asymptomatic or symp-tomatic patients, are capable of inducing cytopathic effects, suggesting that the host environment may produce the factors that critically define the severity of symptoms in *T. vaginalis* infections.

Several parasite elements have been reported to play vital roles in initi-ating adhesion, apoptosis and host immunoinflammatory responses to trichomonas. These include lipids, proteins (adhesins), cysteine proteases, and glycoconjugates. Below we discuss the best characterized parasite moieties and evidence for their involvement in various aspects of host-parasite interactions.

8.1 Lipids and glycoconjugates

Trichomonads live on the surface of the epithelium of the urogenital tract where they derive many of their nutrients from the host, for example purines, pyrimidines and lipids. Over the last two decades, an understand-ing of the importance of parasite lipids and glycoconjugates in parasite survival in a hostile environment and attachment to host cell surfaces has developed (Beach *et al.*, 1990, 1991; Costello *et al.*, 1993, 2001; Holz *et al.*, 1987; Lindmark *et al.*, 1991; Singh, 1994; Singh *et al.*, 1991, 1994).

Trichomonads are fatty acid and sterol auxotrophs (Beach *et al.*, 1990, 1991) and synthesize glycolipids and glycophosphosphingolipids *de novo* with lipid components obtained from the host (Singh *et al.*, 1991, 1994). Glycerophopholipids are incorporated into most phopholipids, suggesting that turnover of lipids in the cell membrane takes place (Beach *et al.*, 1991). Parasites are unable to form cholesterol or other sterols from acetate, from mevalonic acid or from squalene. However, trichomonads incorporate free cholesterol and cholesterol esters, probably by nonspecific lipoprotein binding and endocytosis. The major phosphoglycerides are phosphatidylethanolamine and phosphatidylcholine, which account for about 50% of the extractable phospholipids. The acidic lipids – a newly identified inositol-diphosphate ceramide (TV$_1$; Figure 2B) having an ethanolamine substituent on inositol diphosphate at the 3-position of the inositol ring, inositol phosphosphingolipids (IPC; Figure 2A), phosphatidylinositol (PI), phosphatidylglycerol (PG), and a nonpolar phosphoglyceride, which has presumptively been identified as O-acylphosphatidylglycerol

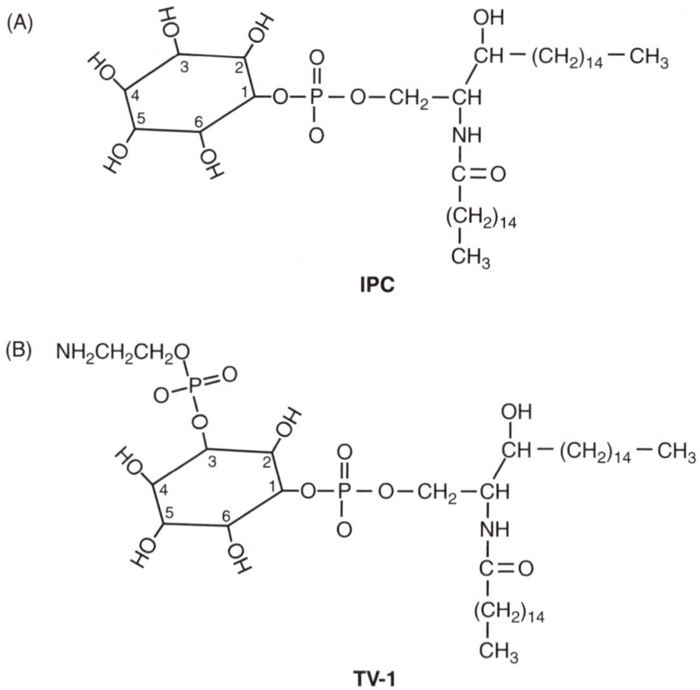

Figure 2 Structures of inositol phosphate ceramide (IPC; A) and inositol-diphosphate ceramide (TV$_1$; B) having an ethanolamine at the 3-position of the inositol ring.

(O-acyl-PG) – account for about 10–20% of the total lipids. *T. vaginalis* also contains significant amounts of sphingomyelin (ca. 15–20% of total lipid) and some *N,N*-dimethylphosphatidylethanolamine and phosphatidic acid. Trichomonads lack cardiolipin (Beach *et al.*, 1990), a lipid characteristic of mitochondria. The presence of PG and O-acyl-PG-like lipids is unusual among eukaryotes. When trichomonads were exposed to lipid precursors, [^{3}H]*myo*-inositol radioactivity was incorporated into TV$_1$, IPC and PI acid lipids; in parasites grown in the presence of the lipid precursors [^{3}H]-palmitic acid or [^{32}P]-orthophosphoric acid, radioactivity was incorporated into all five acidic lipids.

Singh and his coworkers characterized the structures of these five acidic lipids from trichomonads (Costello *et al.*, 2001, 2003). TV$_1$, may represent a metabolic intermediate for a new type of membrane anchor employed by *T. vaginalis* surface glycopeptides or glycolipids that mediate host-parasite interaction. The closely related bovine parasite, *T. foetus*, also contains an inositol di-phosphosphingilipid similar to TV$_1$, named TF$_1$. In addition, TF$_1$ contains a fucosyl residue substituted at the 4-position (Costello *et al.*, 1993). There are other reports of glycoconjugates anchored by a ceramide moiety in certain protozoans, for example *Dictyostelium discoideum* adhesion proteins (Stadler *et al.*, 1989), lipopeptidophosphoglycan from *Trypanosoma cruzi* (Previato *et al.*, 1990), and lipophosphoglycan from *Acanthamoeba castellanii* (Dearborn *et al.*, 1976).

Cell surface glycoconjugates have been implicated in many cell-cell interactions. It is known that parasitic protozoans contain a variety of complex carbohydrates on their surfaces including glycolipids, glycoproteins, and glycosylated phosphatidylinositol glycolipids (GPI). Parasite glycoconjugates play important roles in host-cell invasion and evasion of host immune responses (Ferguson, 1999; Guha-Niyogi *et al.*, 2001; Lodge and Descoteaux, 2005; Turco and Descoteaux, 1992).

Evidence began to accumulate in the mid 1980s that *T. vaginalis* surface glycoconjugates play important roles in parasite biology. Torian *et al.* (1984) and Connelly *et al.* (1985) reported that monoclonal antibodies (mAbs) prepared against *T. vaginalis* parasites reacted with periodate-sensitive cell-surface antigens (90–115 kDa), suggesting that the determinants recognized by these mAbs are carbohydrate-containing molecules. Later, Krieger *et al.* (1990b) showed that mAbs prepared against *T. vaginalis*, which inhibited the adhesion of parasites to HeLa cells, also reacted with the 90- to 115-kDa carbohydrate-containing surface antigens. The importance of *T. vaginalis* cell-surface glycoconjugates in adhesion of parasites to host cells was also shown by Mirhaghani and Warton (1998). Similar evidence has accumulated for the role of complex carbohydrates in the adhesion of the bovine parasite (Bastida-Corcuera *et al.*, 2005; Corbeil *et al.*, 1989; Hodgson *et al.*, 1990; Shaia *et al.*, 1998; Singh *et al.*, 1999, 2001).

The best charactererized surface glycoconjugate of *T. vaginalis* is *T. vaginalis* LPG. *T. vaginalis* LPG is akin to *Leishmania* LPG, which is involved in the attachment, entry, and survival of parasites in host cells (macrophages), and is capable of modulating macrophage signaling pathways (Lodge and Descoteaux, 2005; Soares *et. al.*, 2005; Turco and Descoteaux, 1992). Similarly, *T. vaginalis* LPG is involved in parasite adhesion and has been recently identified as a powerful initiator of the host inflammatory response in human vaginal epithelial cells (Fichorova *et al.*, 2006).

8.2 T. vaginalis *lipophosphoglycan*

T. vaginalis LPG has been isolated, purified and partially characterized (Fichorova *et al.*, 2006; Singh, 1993, 1994; Singh *et al.*, 1994). Cell surface radiolabeling of trichomonads by galactose oxidase indicated that the glycoconjugate is located on the parasite cell surface (Singh, 1993). There are about 3×10^6 copies of LPG per parasite and it is anchored in the membrane via an inositol-phosphate-ceramide (Singh, 1993; Singh *et al.*, 1994). LPG migrates as a broad polydisperse band (~30–80 kDa) upon SDS-PAGE (Figure 3A). It is the only PAS- (periodic acid-Schiff) positive species present in the isolated fraction when it is analyzed by SDS-PAGE. It does not stain with Coomassie blue and is not sensitive to pronase, suggesting that it does not contain a peptide component. Furthermore, mass spectral analysis showed the absence of a peptide component. Like *Leishmania*

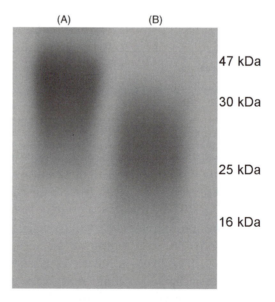

Figure 3 SDS-PAGE analysis of *T. vaginalis* LPG (A) and glycan-lipid inositol core of LPG (B) released by mild acid treatment of LPG. Gel was stained with periodic acid-Schiff (PAS) reagent.

LPGs, TV-LPG is susceptible to PI-PLC (phospholipase-C) from *Bacillus thuringiensis*, nitrous acid (HNO_2) deamination, and mild acid hydrolysis (Singh, 1994). Reductive radiomethylation of LPG provides evidence for the presence of more than one free amino group (e.g., unacetylated GlcN and/or ethanolamine) in the glycan-inositol lipid core (Singh *et al.*, 1994). LPG can be partially degraded into distinct domains by a combination of chemical degradation and enzymatic treatments. SDS-PAGE analysis of the glycan-lipid inositol core (~12–30 kDa) released by mild acid is shown in Figure 3B. Based on compositional analysis and chemical characterization (Fichorova *et al.*, 2006; Guha-Niyogi *et al.*, 2001; Singh *et al.*, 1994), a structural scheme for TV-LPG is shown below in Figure 4.

TV-LPG is analogous to LPGs from other parasites, but is clearly distinct from them. *Leishmania* LPG is a polymer of repeating Galβ-1,4,-Manα-1-PO_4 units attached to a glycan core that is embedded into the membrane via a l-O-alkyl-2-*lyso*-phosphatidyl(*myo*)inositol anchor. Unlike *Leishmania* LPG (which has promastigote and amastigote stages), trichomonad (only one stage) LPG does not undergo structural modifications during parasite growth. Additionally, unlike other GPI-anchored molecules, TV-LPG contains no Man (mannose) and recent data from Singh's group showed the presence of polylactosamine repeats (Galβ-1-4GlcNAc) in the LPG molecule (Guha-Niyogi *et al.*, 2001; Singh *et al.*, 2000a, 2000b). This is potentially very interesting, since this type of glycan structure in the LPG molecule could act

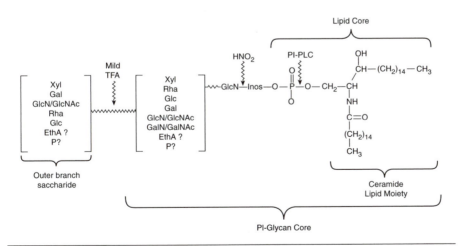

Figure 4 Structural scheme of *T. vaginalis* LPG. Domains of LPG can be released by different chemical or enzymatic treatments. The PI-glycan core (sometimes referred to as the glycan-lipid inositol core) is obtained upon mild acid (100 mM TFA, 100°C for 3 h) hydrolysis of the intact LPG molecule followed C18-SepPak (hydrophobic chromatography); the outer branch saccharide(s) elutes with aqueous solvent in the 'void' volume. The lipid core is released by treatment with HNO_2 and the ceramide moiety is released by PI-PLC.

as a ligand for galectin-1 and galectin-3, which are widely expressed in several mammalian cell types, including epithelial cells, and are known to play vital roles in host-pathogen interactions and immune recognition (Rabinovich and Gruppi, 2005; van den Berg *et al.*, 2004).

A direct binding assay was used to demonstrate the role of TV-LPG in adhesion of parasites to HVECs (Fichorova *et al.*, 2006). Another recent study, using a mutant parasite expressing LPG having a lower galactose and glucosamine content, showed reduced adherence to a human ectocervical epithelial cell line (Bastida-Corcuera *et al.*, 2005). It has also been shown that periodate treatment of *T. foetus* parasites abolishes the binding of *T. foetus* to bovine vaginal epithelial cells, once again suggesting the importance of surface glycoconjugates in a trichomonas infection (Singh *et al.*, 1999). The respective LPG molecules (*T. foetus* and *T. vaginalis*) also exhibit species-specificity in displacement binding assays. These results clearly suggest that receptor-ligand interactions are involved between trichomonads and host epithelial cells.

8.3 Other adhesion molecules

Alderete and his coworkers have long been involved in studying *T. vaginalis* host-parasite interactions (Alderete and Garza, 1985; Alderete *et al.*, 1995a, 1995b; Arroyo *et al.*, 1993; Lehker and Alderete, 2000). These investigators reported the involvement of four trichomonad surface proteins – referred to as AP65, AP51, AP33 and AP23 – in adhesion of parasites to host cells, believed to recognize specific host-cell proteins by a ligand-receptor interaction. Arroyo *et al.*, (1993) reported that polyclonal antibodies to these adhesins partially inhibited the binding of parasites to host cells. The same investigators reported that the parasites undergo transformation from the regular pear-oval shape to amoeboid form upon contact with host cells. Sequence analyses of some of these adhesins revealed structural molecular mimicry of AP65 with decarboxylating malic enzyme and the AP51 and AP33 with the β- and α-subunits of succinyl coenzyme synthetase (Engbring *et al.*, 1996), enzymes found in hydrogenesomes for oxidative decarboxylation of pyruvate (Muller, 1993). However, doubt has been raised about the identity of these four adhesins and it has been suggested that other factors and/or mechanisms are responsible for nonspecific host-parasite interactions (Addis *et al.*, 2000). Expression of these adhesins on the surface of *T. vaginalis* is induced in the presence of iron (Garcia *et al.*, 2003). Recently, Alderete's group reported that the attachment of parasites to immortalized HVECs led to increased synthesis of adhesins after contact with host cells and also led to upregulation of several HVEC genes (Kucknoor *et al.*, 2005). It was hypothesized that the increased expression of adhesins and other genes could reflect signaling events and might lead to an understanding of host-parasite interactions. However, the functional significance(s) of these findings in parasite adhesion still remains to be established.

8.4 T. vaginalis *cytopathogenic effects*

T. vaginalis adherence to host cells and damage by contact-dependent mechanisms has been studied by several investigators utilizing cell lines such as HeLa, HEp-2, MDCK epithelial cells, and Chinese hamster ovary (CHO) cells (Alderete and Garza, 1984, 1985; Alderete and Perlman, 1984; Filho-Silva and deSouza, 1988; Petrin *et al.*, 1998). Fiori *et al.* (1996, 1997) reported contact-dependent and contact-independent disruption of human erythrocytes by *T. vaginalis*. Subsequently, Singh's group used primary HVECs to explore the pathogenic effects of host-cell cytotoxicity induced by *T. vaginalis* (Gilbert *et al.*, 2000). The results showed that the cytopathogenic effect is a function of parasite density and HVEC disruption started within 2 hours; the cells are completely destroyed by 15–24 hours (see Figure 5A–D). In the absence of direct contact, there is no damage to the host-cell monolayers. Pretreatment of parasites with periodate abolishes adhesion of parasites to HVECs and the cytopathic effect (Figure 5D), suggesting involvement of carbohydrate-containing molecules in these processes (as discussed above). Metronidazole-treated *T. vaginalis* show

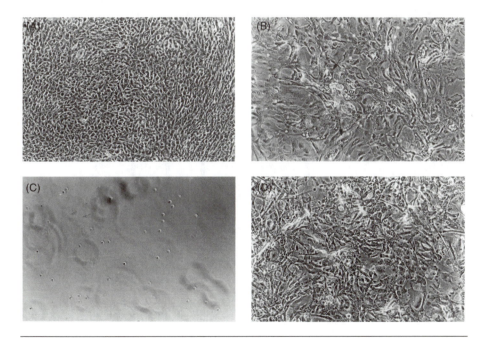

Figure 5 Microscopic observation of human vaginal epithelial cells (HVECs) co-incubated with *T. vaginalis*. (A) Normal untreated HVECs. (B) HVECs treated with *T. vaginalis* for 30 min and then washed. Note abundant adhesion of parasites to the monolayer. Large and small arrows indicate HVECs and parasites respectively. (C) HVECs co-incubated with *T. vaginalis* for 24 h. Monolayers are completely disrupted, and only dead parasites are observed. (D) HVECs exposed to periodate-treated *T. vaginalis* for 24 hours. No adherence of *T. vaginalis* and no destruction of monolayers are apparent.

some attachment of parasites to HVECs, but produce no damage, suggesting the importance of metabolic integrity in adhesion and induction of cytopathic effects of the parasites. As indicated above, all *T. vaginalis* isolates, whether from asymptomatic patients or from patients with vaginitis, are capable of damaging HVECs (Figure 6) and the levels of cytotoxicity produced by different clinical isolates may be related to different levels of cytotoxic product(s) released by the organisms in the presence of target cells.

It has been suggested that trichomonads exert their pathogenic effects on MDCK cells in culture either by direct contact or by the release of specific components (Filho-Silva and deSouza, 1988). It is possible that certain proteases and glycosidases found in trichomonad extracts play a role in modulating the interactions of trichomonads with epithelial cells. Several investigators have proposed that some types of soluble cytotoxin play a role in the pathogenic effect on host cells (Alderete and Garza, 1984; Petrin *et al.*, 1998). Garber *et al.* (1989) reported the presence of a cell-free product of *T. vaginalis*, cell detachment factor (CDF). CDF (which contained protease activity) is involved in the cytopathogenic effects in cell cultures of McCoy, HEp-2, human foreskin fibroblasts, and CHO monolayers. The detachment of cell culture monolayers is thought to be analogous to the sloughing of vaginal epithelial cells in the mucosa during acute *T. vaginalis* infection, and may be related to tissue invasion. Garber and Bowie (1990) suggested that very low pH associated with metabolically active *T. vaginalis*

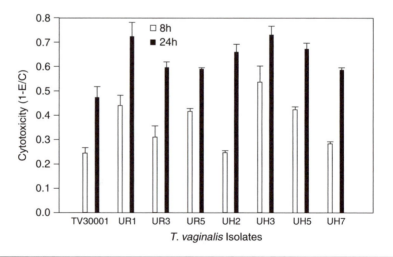

Figure 6 Cytotoxicity of HVECs by various clinical isolates of *T. vaginalis* at 8 and 24 h. Strain TV3001 has been in culture for more than 20 years, and the remainder of the isolates were obtained recently from a STD clinic. UR3, UR5 and UH7 isolates were from asymptomatic patients and the rest were from symptomatic patients. Parasites were added to each well containing HVECs. Cytotoxicity was determined by CellTiter 96 Aqueous (Promega Corp., WI, USA) assay.

may be an important factor in the contact-dependent killing of mammalian cells. Pindak *et al.* (1993) suggested that the acidic metabolites produced by *T. vaginalis* during coincubation of parasites and host cells lead to death of cultured cells. Fiori and his group have used red blood cells to study cytopathogenicity of *T. vaginalis* and suggested a protease-like (nonsecreted) molecule is responsible for lysis and damage of red blood cells (Fiori *et al.*, 1997). Recently, Vargas-Villarreal *et al.* (2005) reported the isolation of soluble and membrane-associated phospholipases A_1 and A_2 from *T. vaginalis*. It was suggested that these lipases could be responsible for the contact-dependent and contact-independent hemolytic and cytolytic activities of *T. vaginalis* as reported earlier by Fiori *et al.* (1996, 1997, 1999). These phospholipases may have biological significance in the lipid metabolism of *T. vaginalis*.

A number of microorganisms have been reported to produce extracellular components that have cytotoxic properties (Clikenbeard *et al.*, 1989; Que and Reed, 2000; Ravdin *et al.*, 1980; Young and Cohn, 1985). It is not known whether *T. vaginalis* parasites produce any cytotoxic material upon contact with the host cells. However, it is clear that cysteine proteases secreted by trichomonads are involved in triggering host-cell cytotoxicity (Singh *et al.*, 2004; Sommer *et al.*, 2005). Recent studies by Guenthner *et al.* (2005) showed that host-cell contact with *T. vaginalis* lead to cytotoxicity and disruption of epithelial cells (primary prostate and ectocervical epithelial cells), allowing HIV-1 access to underlying immune cells. This is an important observation which points to the role that *T. vaginalis* may have in the epidemiological observation that trichomoniasis is associated with the enhanced sexual transmission of HIV-1.

8.5 T. vaginalis *proteases*

Proteases have been identified in biological systems from viruses to humans in diverse physiological or pathological roles. For more than a decade there has been considerable interest in parasite proteases, especially cysteine proteases (CPs) and their roles in the infection process (Klemba and Goldberg, 2002; McKerrow *et al.*, 2006; North, 1994; Sajid and McKerrow, 2002). It is clear that proteases play vital roles in the interaction of parasites with their hosts. Parasite CPs play key roles in tissue and cellular invasion, immune evasion, degradation of hemoglobin and other blood proteins, and activation of inflammation. Parasite proteases are promising chemotherapeutic or vaccine targets.

Recently, Singh and his coworkers showed that trichomonad infection of primary vaginal epithelial cells *in vitro* results in cell detachment followed by cell destruction, the result of apoptosis (Gilbert *et al.*, 2000; Singh *et al.*, 2004; Sommer *et al.*, 2005). Apoptosis is triggered by isolated CPs, secreted by *T. vaginalis*, mimicking the infection (and apoptogenic) process observed with intact parasites (Singh *et al.*, 2004, 2005; Sommer *et al.*,

2005). In addition, trichomonad CPs have been proposed to be involved in pathogenesis when released into the host mucosa surface (Hernandez-Gutierrez *et al.*, 2004; Thomford *et al.*, 1996), or in parasite evasion of host immune responses (Bastida-Corcuera *et al.*, 2000; Draper *et al.*, 1998; Kania *et al.*, 2001; Min *et al.*, 1998), hemolysis (Fiori *et al.*, 1996, 1999), adhesion (Hernandez-Gutierrez *et al.*, 2004; Mendoza-Lopez *et al.*, 2000), and they may also serve as virulence factors (Alvarez-Sanchez *et al.*, 2000; Arroyo and Alderete, 1989, 1995; Mallinson *et al.*, 1994, 1995; Thomford *et al.*, 1996).

North and his colleagues reported the presence of a number of secreted proteases from both *T. vaginalis* and *T. foetus* (Mallinson *et al.*, 1994; North, 1994). Using a combination of SDS-PAGE analyses and synthetic substrates they showed that several different proteases are present in the secreted fraction. Four *T. vaginalis* CPs were cloned by Mallinson *et al.* (1994; GenBank accession numbers: X77218, X77219, X77220 and X77221) using active-site amino acid sequences, based on homologies within the broad family of CPs. Two of the clones appear to be full-length. The other two are partial, though near full-length. In addition, Garber *et al.* (1993) reported the cloning of a cDNA from a λGT_{11} library representing a portion of a secreted 60-kDa CP (accession no. X70823) and Leon-Sicairos *et al.* (2004) reported the cloning of an apparently intracellular CP, tvcp12 (accession nos AY371180 and AY463679). Other CPs, such as CP30, CP39 and CP65, have been reported to play some role in host-parasite interactions (Alvarez-Sanchez *et al.*, 2000; Hernandez-Gutierrez *et al.*, 2004; Mendoza-Lopez *et al.*, 2000). Multiple proteases have also been observed in parasite cell lysates and as many as 23 protease species have been detected on two-dimensional SDS-PAGE of *T. vaginalis* extracts (Neale and Alderete, 1990). It has been suggested, however, that at least some of the observed species are procedural artifacts and/or are the products of post-translational modification. For example, Garber and Lemchuk-Favel (1994) reported that a 60-kDa CP from *T. vaginalis* fragments into 23- and 43-kDa species.

Recently, Singh's group (Sommer *et al.*, 2005) reported the isolation, purification, identification and biological significance of a secreted CP fraction (CP30) from *T. vaginalis*. This is the first report to show the accurate molecular weight determination of mature CP30, its characterization, and its involvement in HVEC cytotoxicity and apoptosis. SDS-PAGE analysis of the CP30 (Figure 7) fraction showed three bands around 30 kDa. Analysis of CP30 using MALDI-TOF MS (matrix-assisted laser desorption ionization time-of-flight mass spectrometry) showed three peaks with molecular masses of about 23, 23.6 and 23.8 kDa. Extensive MS peptide sequencing of proteolytically digested CP30 resulted in matches to cDNA clones, CP2, CP3, and CP4, previously reported by Mallinson *et al.* (1994), as well as a newly identified sequence with high homology to CP4, named CPT (Sommer *et al.*, 2005). Interestingly, there is no evidence that the CP30

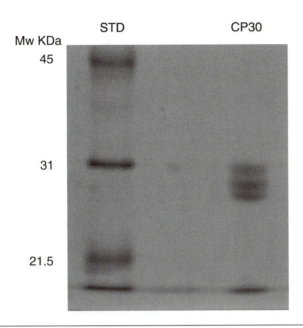

Figure 7 SDS-PAGE analysis of CP30 fraction. The gel was stained with Coomassie blue.

fraction contains CP1 as identified by Mallinson *et al.* (1994). Other evidence suggests that CP1 is lysosomal, and is not secreted by *T. vaginalis* (Dr Kirk Land, University of Pacific, CA, USA, personal communication).

CP30 is very active in induction of host cell apoptosis. Initiation of apoptosis is correlated with protease activity, because the specific CP inhibitor E64 inhibits both activities. Protease activity was assayed fluorometrically using Z-RR-AMC (*N*-carbobenzoxy-arginine-arginine-7-amido-4-methycoumarin) and activity required the presence of reducing agents (e.g. DTT). *T. vaginalis* CP30 has no effect on bovine host cells (BVECs) and conversely, *T. foetus* CP8 (Singh *et al.*, 2004), which induces apoptosis in BVECs, has no effect on HVECs, indicating that initiation of host cell apoptosis is highly species-specific. The extremely high homology amongst the four components of CP30 makes it impossible to determine which of the CPs is responsible for inducing HVEC apoptosis (it may be one or all). The four CPs, CP2/3/4 and CPT have been cloned into a number of expression vectors and the expression of individual CPs is under investigation, which may shed light on which of the CPs are apoptogenic (Singh *et al.*, unpublished observations).

Interestingly, CP30 is down-regulated by serum iron added to *T. vaginalis* culture medium and up-regulated by the addition of the iron chelator, dipyridyl (Singh *et. al.*, manuscript in preparation). This observation suggests that secretion of CPs by the parasites represents a mechanism to obtain iron. Thus, in an iron-poor environment CP30 is secreted, which leads to enhanced host-cell apoptosis, thereby releasing nutrients (especially iron) for the parasites. Others have also observed that iron affects *T. vaginalis* CP expression, especially with *T. vaginalis* CP39 and tvcp12 (Hernandez-Gutierrez *et al.*, 2004; Leon-Sicairos *et al.*, 2004).

8.6 *T.* vaginalis *and host-cell apoptosis*

Apoptosis is a significant and well-regulated form of cell death that occurs under a variety of physiological and pathological conditions (Bellamy *et al.*, 1995; Moss *et al.*, 1999; Schaumburg *et al.*, 2006). Microbes have evolved mechanisms to trigger the apoptotic signal-transduction cascade, which likely plays a role in pathogenesis (Gao and Kwaik, 2000; Moss *et al.*, 1999; Schaumburg *et al.*, 2006; Zychlinsky and Sansonetti, 1997). Apoptotic cell death has been studied in detail in response to several bacterial and viral infections and it is evident that bacterial, viral, and protozoan pathogens have evolved a variety of different strategies to modulate host-cell apoptosis (Gao and Kwaik, 2000; Schaumburg *et al.*, 2006). In some instances, microbes diminish apoptosis, and in others, they stimulate apoptosis. The signaling machinery in host-cell pathology that triggers apoptosis by microbes is still unclear (Moss *et al.*, 1999). Gavrielescu and Denkers (2003) pointed out that activation of apoptotic pathways contributes to the pathology of several protozoan infections. Knowledge and understanding of the mechanisms of apoptosis during parasitic infections may provide tools in the development of effective antiparasitic vaccines (James and Green, 2004).

Certain parasitic pathogens, such as *Acanthamoeba castellanii* (Alizadeh *et. al.*, 1994), *Plasmodium falciparum* (Balde *et al.*, 1995), *Tryapanosoma cruzi* (Lopes *et al.*, 1995), *Trypanosoma brucei* (Sites *et al.*, 2004), *Cryptosporidium parvum* (McCole *et al.*, 2000; Mele *et al.*, 2004), *Toxoplasma gondii* (Abbasi *et al.*, 2003; Wei *et al.*, 2002), and *Entamoeba histolytica* (Huston *et al.*, 2000; Que and Reed, 2000), can kill mammalian cells by an apoptotic mechanism that occurs in response to infection. The exact mechanism by which these pathogens trigger cell death in specific host cells remains to be clarified. The mechanism(s) by which extracellular pathogens such as *T. vaginalis* induce host-cell death are likely to be quite different from the mechanism(s) utilized by intracellular parasites. In another case of extracellular parasites, CPs have been reported (Huston *et al.*, 2000; Que and Reed, 2000) to be involved in *E. histolytica*-induced pathology, where they destroy host tissue.

Recently, Singh's group showed that *T. vaginalis* or its secreted CP30 induce apoptotic cell death in HVECs (Sommer *et al.*, 2005) and the bovine *T. foetus* or its CP8 induced apoptosis in BVECs (Singh *et al.*, 2004). One of the earliest intracellular events to occur following induction of apoptosis is the disruption of the mitochondrial membrane potential (Reers *et al.*, 1995). As shown by fluorescence microscopy (Figure 8), in untreated

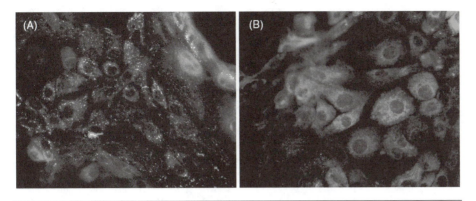

Figure 8 Fluorescence microscopy of HVECs stained with JC-1 dye. Fluorescence microscopic images of normal HVECs (A), and HVECs undergoing apoptosis (B) in the presence of CP30.

HVECs, cationic dye (JC-1) accumulates and aggregates in the mitochondria, giving off mostly bright orange-red fluorescence, indicative of high mitochondrial membrane potential. However, in HVECs treated with parasite CP30 the dye does not accumulate in the mitochondria. It remains in the cytoplasm, and fluoresces mostly green (>80%), indicative of low (altered) mitochondrial potential. Many of the events that occur during apoptosis are mediated by a family of cysteine proteases called caspases (Slee *et al.*, 1999). Anti-ACTIVE® Caspase-3 pAb (Promega Corp., USA) was employed to detect an active form of caspase-3 in apoptotic HVECs in the presence of CP30 (Figure 9). Apoptotic cells show strong antibody staining as well as cell shrinkage and nuclear condensation in contrast to nonapoptotic or untreated cells. In addition, CP30-induced apoptosis in HVECs is significantly inhibited by Z-VAD-FMK, a general caspase inhibitor, as

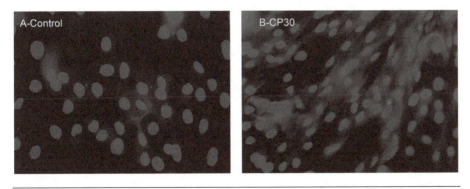

Figure 9 Activation of HVEC caspase-3 by CP30. HVECs treated with CP30, washed, formalin-fixed and labeled with anti-ACTIVE Caspase-3 pAb (Promega Corp., WI, USA), followed by fluorescence-labeled anti-rabbit secondary antibody. The cell nuclei were counterstained with DAPI (40× magnification).

shown quantitatively by the ELISA[PLUS] assay (Figure 10A). Activation of caspase-3 in HVECs in response to CP30 is also demonstrated fluorometrically using Ac-DEVD-AFC (BIO-RAD) substrate (Figure 10B). These data clearly indicate that CP30 induces activation of HVEC caspases, especially caspase-3, resulting in apoptosis. Other methods were also employed to assess cell death and showed that more than 50% of HVECs undergo apoptosis in

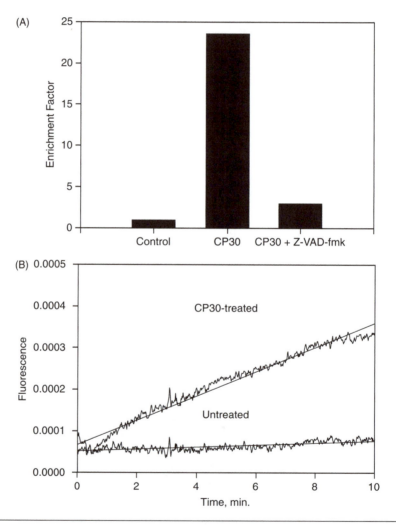

Figure 10 (A) Activation of caspase-3. HVECs were incubated in the presence and absence of CP30 and the caspase-3 activity was measured fluorometrically in cell extracts using Ac-DEVD-AFC (BIO-RAD) as described by the manufacturer. (B) Effect of caspase inhibitor. HVEC apoptosis was assayed by ELISA[PLUS] quantitation of nucleosomal DNA fragmentation in the presence of CP30 and CP30 with the addition of the caspase inhibitor Z-VAD-fmk-treated cells. Apoptosis is reflected by the enrichment of nucleosomes in the cytoplasm shown on the y-axis.

response to parasites or to CP30 (Sommer *et al.*, 2005). Although, these data showed the involvement of caspase-3 in HVECs apoptosis, the signaling pathways or mechanisms involved in initiating the activation cascade are as yet unclear. Although it seems reasonable to hypothesize that CP30 acts directly at the HVEC cell surface, perhaps by binding to one of the previously identified death receptors such as TNFR1, Fas (CD95), DR3, DR4 (TRAILR1), DR5 (TRAILR2) and DR6, or to an as yet unidentified death receptor, there may be other mechanisms involved (see Reed, 2000). Further research is needed to explain this phenomenon in detail.

Others have reported that *T. vaginalis* induced apoptotic cell death via a Bcl-xL-dependent pathway and via the phosphorylation of p38 MAPK in RAW264.7 macrophages (Chang *et al.*, 2004). Recently, Kang *et al.* (2006) reported that *T. vaginalis* increases human neutrophil apoptosis, and this contributes to the resolution of inflammation. However, it should be emphasized that the HVEC cultures described by the authors are devoid of macrophages or lymphocytes, and therefore, some apoptotic pathways may be different. For example, perforin or granzymes produced by natural killer cells or cytotoxic T lymphocyctes can also induce apoptosis – but cannot operate in HVECs, since these cell cultures do not contain lymphocytes.

8.7 T. vaginalis *lipophosphoglycan and cytokine regulation*

Shaio *et al.* (1994, 1995) reported the presence of interleukin-8 (IL-8) and leukoctriene B_4 in the vaginal discharge from patients with symptomatic trichomoniasis, which suggests they may be involved in the inflammatory response to *T. vaginalis*. *In vitro* live *T. vaginalis* trophozoites and undefined lysates induced IL-8 production by human neutrophils and by HeLa cells (Chang *et al.*, 2006; Ryu *et al.*, 2004). Similarly, undefined *T. vaginalis* membrane fragments stimulated the generation of IL-8 by monocytes (Shaio *et al.*, 1995). The molecular determinants of the parasite responsible for this stimulation remained unidentified until recently. Fichorova *et al.* (2006) demonstrated for the first time that purified *T. vaginalis* LPG is capable of initiating an inflammatory reaction in host epithelial cells. They demonstrated that *T. vaginalis*, but not *T. foetus* LPG, caused a significant upregulation of IL-8 and macrophage inflammatory protein (MIP)-3α by human vaginal, ectocervical, and endocervcial epithelial cells in the absence of cell toxicity (Fichorova *et al.*, 2006).

IL-8 and MIP-3α have well-established roles in the inflammatory process and may be part of the pathogenic mechanisms linking *T. vaginalis* with susceptibility to other infections and reproductive tract diseases. IL-8 is a CXC chemokine, which is a powerful chemoattractant for neutrophils and monocytes. Increased IL-8 levels in genital tract secretions have been correlated with upper genital tract infections, HIV-1, bacterial vaginosis and preterm labor (see review Fichorova, 2004). IL-8 cannot only attract

HIV-1 host cells to the site of inflammation, but can also activate HIV-1 replication in these cells (Narimatsu *et al.*, 2005). MIP-3α, a CC chemokine also known as CCL20, has not been studied in vaginal secretions so far, but is known to recruit Langerhans-like CD34+ progenitor dendritic cells (DC) to mucosal sites of inflammation; particularly guiding them traversing *lamina propria* on their way to the epithelial surface, which precedes their maturation (Caux *et al.*, 2002). DCs are potent antigen-presenting cells that likely play multiple roles in HIV-1 by becoming infected or by capturing and transmitting infectious virus to other CD4-positive cells in the vaginal and cervical mucosa and secondary lymphoid tissues (Lekkerkerker *et al.*, 2006). Taken together, these data suggest that *T. vaginalis*-induced IL-8 and MIP-3α may be part of the pathogenic mechanisms linking HIV-1 with trichomoniasis.

The promoter regions of both IL-8 and MIP-3α include binding sites for the major proinflammatory transcription factor nuclear factor (NF)-κB. It has been hypothesized that the production of IL-8 associated with *T. vaginalis* infection may be at least partially related to the cytopathic effects and secondary to the release of TNFα and IL-1 from damaged epithelial cells. TNFα and IL-1 activate NF-κB via the IL-1 receptor (R) I, and TNF-RI, respectively. They are increased in cell culture supernatants from *T. vaginalis*-infected cells (Chang *et al.*, 2006), as well as in the cervicovaginal secretions from women with clinical conditions commonly associated with trichomoniasis, for example, vaginitis, bacterial vaginosis, HIV-1 and preterm labor (Coombs *et al.*, 2003; Platz-Christensen *et al.*, 1993). Large intracellular stores of IL-1α and IL-1β, and less TNFα, can be released from the vaginal and cervical epithelial cells upon cell death (Fichorova *et al.*, 2001). However, these cytokines are normally counterbalanced by simultaneous release of their soluble receptors or other natural antagonsists such as the IL-1 receptor antagonist (RA). Moreover, Fichorova *et al.* (2006) demonstrated that the LPG stimulatory effects occur at comparatively low parasite burden, and in the absence of cell toxicity or a significant increase of endogenous IL-1β and TNFα levels. Moreover, while the cytoplasmic adaptor protein MyD88 is a requirement for IL-1 signaling (Adachi *et al.*, 1998), LPG-induced NF-κB activation and chemokine up-regulation is at least partially MyD88-independent (Fichorova *et al.*, 2006). Thus, a direct LPG recognition and LPG-induced signaling, rather than secondary cell death and IL-1/TNF receptor-mediated events must be taking place in the observed chemokine induction phenomenon. In addition to NF-κB, the IL-8 promoter has several binding sites for other transcription factors, for example AP-1 and C-EBP/NF-IL-6, among others, that cooperate with NF-κB for optimal transcription regulation (Brat *et al.*, 2005; Roebuck, 1999). On the other hand, both IL-8 and MIP-3α promoters contain one or more binding sites for the ETS (E26 transformation-specific sequence) transcription factors that may interact with NF-κB or act in a NF-κB-independent fashion (Hedvat *et al.*, 2004; Kwon *et al.*, 2003). IL-8 is also regulated by ERK1/2- and p38-dependent activation of ATF-1/2 transcription factors that occurs independently of NF-κB (Agelopoulos and Thanos, 2006; Marin *et al.*, 2001). Stabilization of IL-8 mRNA by the p38 mitogen-activated protein kinase pathway is an alternative pathway for increased IL-8 protein production (Hoffmann *et al.*, 2002) that may be utilized by *T. vaginalis* infection (Chang *et al.*, 2006). Fichorova *et al.* (2006) demonstrated a relatively weak activation of epithelial NF-κB by *T. vaginalis* LPG. Other transcription factors and mRNA stabilization pathways that may play roles in LPG signaling remain to be elucidated.

Although the detailed structure of LPG and its receptors on the host-cell surface are yet unknown, preliminary structural analysis suggests that it is in many ways reminiscent of pathogenic bacterial-cell surface glycoconjugates: the Gram-negative cell-wall component lipopolysaccharide (LPS) and the Gram-positive compound peptidoglycan. Those molecules involve signaling via different Toll-like receptors (TLR) in complex with adaptor proteins variously expressed on the surface of multiple host-cell types in the female genital tract (Fichorova *et al.*, 2002; Pivarcsi *et al.*, 2005; Prebeck *et al.*, 2003). It is, therefore, reasonable to hypothesize that similarly to bacterial LPS, *T. vaginalis* LPG has TLR-mediated immunoregulatory effects on the cervical and vaginal mucosa. Indeed, recent studies from Dr. Spear's laboratory (Zariffard *et al.*, 2004) provided indirect evidence suggesting that *T. vaginalis* infection induces proinflammatory factors in the female genital tract that stimulate cells through TLR-4. These investigators pointed out two possibilities: (i) *T. vaginalis* parasites produce a product or products that bind to TLR-4 or (ii) the TLR-4 stimulatory substance is a human product secreted by the genital tract either as an innate immune response to infection or as a result of tissue damage. The vaginal and cervical epithelial cells used in the LPG stimulation experiments (Fichorova *et al.*, 2006) lack TLR-4 and the adaptor protein MD2, which are both essential for signaling by the lipid A unit of bacterial LPS. Lipid A anchors LPS to the outer membrane of Gram-negative bacteria and is the most potent agonist of TLR-4 (Fichorova *et al.*, 2002). Thus, it may be hypothesized that in these cells *T. vaginalis* LPG acts via up-regulation of TLR-4 or via alternative TLR pathways, for example, TLR-2 and TLR-6 that are functionally expressed by the human vaginal and cervical epithelial cells (Fichorova *et al.*, 2002). It is also reasonable to hypothesize that *in vivo* systemic and local factors such as sex hormones, pH and bacterial microflora, may modulate the expression of TLR and related costimulatory molecules, and thus may predispose certain women with trichomoniasis to a more severe inflammatory reaction.

Further studies are needed to elucidate the role of the distinct *T. vaginalis* LPG structural domains in the regulation of the epithelial inflammatory response. A detailed understanding of the LPG/host-cell

interactions and immunoregulatory responses by the lower female reproductive tract epithelium is fundamentally important for the development
of new vaccine/drug targets and strategies to prevent and cure *T. vaginalis*
infection in women and adolescents.

9 Current Drug Strategies

9.1 Purine metabolism in T. vaginalis

T. vaginalis is incapable of *de novo* synthesis of purines and pyrimidines,
and therefore parasites rely on purine salvage pathways to nourish their
purine nucleotide pools to survive (Heyworth and Gutteridge, 1978; Wang,
1986). The purine salvage pathway in *T. vaginalis* is rather simple and distinct from *T. foetus*. It relies on the sequential actions of two key enzymes,
purine nucleoside phosphorylase (PNP) and a purine nucleoside kinase
(PNK) for salvaging purine bases. Wang and his group studied these pathways in great detail (Munagala and Wang, 2003; Zang *et al.*, 2005).
Inhibition of either enzyme can result in terminating the growth of *T. vaginalis* and could lead to therapeutic gain. *T. vaginalis* PNP is a bacterial-type
hexameric protein and does not recognize adenine as a substrate. PNK is
also a bacterial-type guanine kinase, whereas human PNK uses only
adenosine as a substrate. These two central enzymes in purine metabolism
are good potential targets for the development of therapeutic agents
because of the significant differences in their mechanisms and those of
human enzymes.

Incorporation of external adenine into the all-purine nucleotides and
guanine into guanine nucleotides only has been reported in *T. vaginalis*.
Based upon these studies, investigators have used an analogue of adenine,
formycine A, to inhibit both enzymes as well as the *in vitro* growth of
T. vaginalis (Munagala and Wang, 2003). Recently Zang *et al.* (2005) determined the crystal structure of *T. vaginalis* PNP, as well as a nontoxic
nucleoside analogue substrate, 2-fluoro-2'-deoxyadenosine, for PNP that
has potential for developing antitrichomoniasis chemotherapy. These findings have important implications for the future development of safe
drug(s) for *T. vaginalis* infection; the only effective drug currently available
is metronidazole (discussed below).

10 Treatment of Infection

Metronidazole (Flagyl; 5-nitroimidazole) has been the drug of choice to
treat trichomoniasis since 1960 and still remains effective in most cases,
with a cure rate of approximately 95% (Crowell *et al.*, 2003; Cudmore *et al.*,
2004; Dunne *et al.*, 2003). Metronidazole enters *T. vaginalis* via passive diffusion. Although the drug is inactive by itself, anaerobic reduction leads to
the formation of a cytotoxic nitro radical anion. The resulting nitro radical
is believed to bind to DNA, thereby disrupting the strands and causing the
death of the parasite. Oral administration of this drug usually cures the

infection in approximately 5 days. However, as with many microbial infections, drug resistance is on the rise (Cudmore *et al.*, 2004; Dunne *et al.*, 2003). Clinical resistance to metronidazole has been reported since 1962 (Robinson, 1962). Metronidazole-resistant *T. vaginalis* is involved in an increasing number of refractory clinical cases. It has been reported that the majority of the patients with trichomoniasis that are refractory to initial metronidazole treatment will eventually respond to increased doses of the drug. Since metronidazole is the only approved drug in the United States, recurrent or resistant *T. vaginalis* infection is treated with increasing doses of drug for a long period of time (Crowell *et al.*, 2003; Cudmore *et al.*, 2004). In addition, the drug is usually not recommended for use by women in early pregnancy or nursing mothers because of potential adverse side effects on the child, and many women (especially those over 45 years) do not tolerate the drug well. In fact, it was reported that metronidazole is correlated with increased cases of pancreatitis in patients with recurrent vaginal trichomoniasis (Feola and Thornton, 2002). Some patients with drug-resistant *T. vaginalis* remain incurable and other adverse symptoms may include nausea and hypersensitivity. The National Institute of Child Health and Human Development Maternal Fetal Medicine Units Network presented data that suggest that metronidazole treatment of symptomatic carriers of *T. vaginalis* increases the risk of preterm birth. It has been suggested that incremental doses could benefit those women who are sensitive to the drug (Cudmore *et al.*, 2004).

The Centers for Disease Control (CDC) and Prevention guidelines recommend that metronidazole be dispensed orally, with dose regimens of 250 mg three times a day for 7 days, 500 mg twice a day for 7 days, or a single 2-g dose. Many physicians favor a 2-g dose because patient compliance is better and less drug is required for achieving good results. However, there is always a slight increased risk of side effects associated with the higher dose. It is also important that a laboratory culture be obtained and tested for drug resistance. This is especially important in cases where a patient comes back with a recurring infection. Also, sexual partners should be tested and treated, if necessary, for *T. vaginalis* infection, to prevent the spread of infection.

A number of investigators have been involved in studying the mechanisms of metronidazole drug action and resistance in *T. vaginalis* (Dunne *et al.*, 2003; Kulda, 1999; Land and Johnson, 1999; Land *et al.*, 2004; Upcroft and Upcroft, 2001). Metronidazole is reduced in the hydrogenosomes of *T. vaginalis* by the enzyme pyruvate-ferredoxin oxidoreductase. Ferridoxin appears to be essential for metronidazole susceptibility and a correlation between altered expression of a ferridoxin gene and drug resistance has been well studied. Land *et al.* (2004) reported the development of a homologous gene replacement method for *T. vaginalis* to examine the effect of losing a ferridoxin gene on drug susceptibility. These studies will have

direct implications in understanding the molecular basis of drug susceptibility and resistance in *T. vaginalis*.

Another drug, trinidazole, which is a 5-nitroimidazole, has been used to
treat *T. vaginalis* infection in many countries other than the United States
for some time. This drug has recently been approved in the United States
with some success. It has a longer half-life than metronidaziole and is eliminated at a significantly lower rate. Trinidazole has been proposed as an
alternative to metronidazole for many patients, as it appears to have fewer
and milder side effects than those of metronidazole and may be better for
patients with metronidazole-resistant *T. vaginalis*. However, one of the
concerns is cross-resistance to *T. vaginalis* since the mode of action and
structure is very similar to metronidazole. In fact, studies have already
shown cross-resistance to these drugs (Crowell *et al.*, 2003; Dunne *et al.*,
2003; Upcroft *et al.*, 2006). This type of clinical resistance can become a
serious issue for many refractory patients with trichomoniasis, and it can
be extremely difficult to treat.

There are several other nitroimidazole derivatives such as ornidazole,
secnidazole, and nimorazole that have been explored for the treatment of
T. vaginalis infection (Cudmore *et al.*, 2004). Again, these drugs are not
without side effects. A number of other chemotherapeutic agents have
been developed to treat *T. vaginalis* infections. An aromatic polyene compound related to amphotericin B, called hamycin, has been shown to be
effective in killing both metronidazole-sensitive and -resistant strains of
T. vaginalis at low concentrations. Hamycin is currently in use in India as a
topical treatment for trichomoniasis, though side effects have been
reported (Cudmore *at al.*, 2004). Recently, Upcroft *et al.* (2006) examined 30
new 5-nitroimidazole synthesized drugs against metronidazole-resistant *T.
vaginalis* parasites and showed that at least three of them were very effective in inhibiting metronidazole-resistant isolates in the laboratory. The
clinical application of these newly synthesized drugs may be important to
treat cross-resistance among 5-nitroimidizole drugs. It is clear that a number of potential good drugs have been investigated, but research must
continue to evaluate their efficacy in clinical settings.

It is, however, also clear that prevention of trichomoniasis is needed, and
it has been suggested that successful elimination of *T. vaginalis* may be a
cost-effective step in decreasing the incidence and transmission of HIV-1
(Bowden and Garnet, 1999; Sorvillo *et al.*, 2001).

11 Conclusion

During the last two decades, studies of *T. vaginalis* infection have received
increased attention from a broad range of medical and scientific communities, including healthcare workers, molecular biologists, biochemists,
and social workers. This is because of its astounding prevalence around the
globe and its strong association with HIV-1 infection, low birth weight, and

other complications, such as inflammation, and their impact on women's reproductive health. The increasing number of sexually active young males and females adds to the growing burden of *T. vaginalis* infection. More studies of the host-parasite interaction at molecular and cellular levels are needed for our understanding of the pathobiology of *T. vaginalis* infections. Knowledge relevant to host cell pathology induced by *T. vaginalis* infection may reveal novel strategies for prevention and treatment of trichomoniasis in women, men and adolescents and thereby curb the spread of HIV-1 infection around the world.

12 Acknowledgments

We are very grateful to Dr Gary Hayes, Department of Biochemistry and Molecular Biology, Upstate Medical University, Syracuse, NY, for critical review and preparation of this manuscript. Supported in part by the NIH-NICHD HD 054451 grant.

13 References

Abbasi M., Kowalewska-Grochowska, K., Bahar, M.A., Kilani, R.T., Winkler-Lowen, B. and Guilbert, L.J. (2003) Infection of placental trophoblasts by *Taxoplasma gondii*. *J. Infect. Dis.* **188**: 608–616.

Adachi, O., Kawai, T., Takeda, K., Matsumoto, M., Tsutsui, H., Sakagami, M., Nakanishi, K. and Akira, S. (1998) Targeted disruption of the MyD88 gene results in loss of IL-1- and IL-18-mediated function. *Immunity* **9**: 143–150.

Addis, M.F., Rappelli, P. and Fiori, P.L. (2000) Host and tissue specificity of *Trichomonas vaginalis* is not mediated by its known adhesion proteins. *Infect. Immun.* **68**: 4358–4360.

Agelopoulos, M. and Thanos, D. (2006) Epigenetic determination of a cell-specific gene expression program by ATF-2 and the histone variant macroH2A. *Embo J.* **25**: 4843–4853.

Alderete, J.F. and Garza, G.E. (1984) Soluble *Trichomonas vaginalis* antigens in cell-free culture supernatanta. *Mol. Biochem. Parasitol.* **13**: 147–158.

Alderete, J.F. and Garza, G.E. (1985) Identification and properties of *Trichomonas vaginalis* proteins involved in cytoadherence. *Infect. Immun.* **56**: 28–33.

Alderete, J.F. and Perlman, E. (1984) Pathogenic *Trichomonas vaginalis* cytotoxicity to cell culture monolayers. *Br. J. Vener. Dis.* **60**: 99–105.

Alderete, J.F., O'Brien, J.L., Arroyo, J.A., Engbring, J.A., Musatovova, O., Lopez, O., Lauriano, C. and Nguyen, J. (1995a) Cloning and molecular characterization of two genes encoding adhesion proteins involved in *Trichomonas vaginalis. Mol. Microbiol.* **17**: 69–83.

Alderete, J.F., Arroyo, R. and Lehker, M.W. (1995b) Analysis of adhesins and specific cytoadherence of *Trichomonas vaginalis. Methods Enzymol.* **253**: 407–414.

Alizadeh, H., Pidhemey, M.S., McCulley, J.P. and Niederkorn, Y. (1994) Apoptosis is a mechanism of cytolysis of tumor cells by a pathogenic free-living amoeba. *Infect. Immun.* **62**: 1298–1303.

Alvarez-Sanchez, M.E., Avila-Gonzalez, L., Becerril-Garcia, C., Fattel-Facenda, L.V., Ortega-Lopez, J. and Arroyo, R. (2000) A novel cysteine proteinase (CP65) of *Trichomonas vaginalis* involved in cytotoxicity. *Microb. Pathol.* **28**: 193–202.

Arroyo, R. and Alderete, J.F. (1989) *Trichomonas vaginalis* surface proteinase activity is necessary for parasite adherence to epithelial cells. *Infect. Immun.* **57**:2991–2997.

Arroyo, R. and Alderete, J.F. (1995) Two *Trichomonas vaginalis* surface proteinases bind to host epithelial cells and are related to levels of cytoadhearence and cytotoxicity. *Arch. Med. Res.* **26**: 279–285.

Arroyo, R., Gonzalez-Robles, A., Martinez-Palomo, A. and Alderete, J.F. (1993) Signaling of *Trichomonas vaginalis* for amoeboid transformation and adhesion synthesis follows cytoadherence. *Mol. Microbiol.* **7**: 299–309.

Bachmann, L., Lewis, I., Allen, R., Schwebke, J.R., Leviton, L.C., Siegel, H.A. and Hooks 3rd, E.W. (2000) Risk and prevalence of treatable sexually transmitted disease at Birmingham substance abuse treatment facility. *Am. J. Publ. Health* **90**: 1615–1618.

Balde, T.A., Sarthou, J.L. and Roussilhon, C. (1995) Acute *Plasmodium falciparum* infection associated with increased percentages of apoptotic cells. *Immunol. Lett.* **46**: 196–200.

Bastida-Corcuera, F., Butler, J.E., Heyermann, H., Thomford, J.W. and Corbeil, L.B. (2000) *Tritrichomonas foetus* extracellular cysteine proteinase cleavage of bovine IgG2 allotypes. *J. Parasitol.* **86**: 328–332.

Bastida-Corcuera, F.D., Okumura, C.Y., Colocoussi, A. and Johnson, P.J. (2005) *Trichomonas vaginalis* lipophosphoglycan mutants have reduced adherence and cytotoxicity to human ectocervical cells. *Eukaryot. Cell* **4**: 1951–1958.

Beach, D.H., Holz Jr, G.G., Singh, B.N. and Lindmark, D.G. (1990) Fatty acid and sterol metabolism of cultured *Trichomonas vaginalis* and *Tritrichomonas foetus*. *Mol. Biochem. Parasitol.* **38**: 175–190.

Beach, D.H., Holz Jr, G.G., Singh, B.N. and Lindmark, D.G. (1991) Phospholipid metabolism of cultured *Trichomonas vaginalis* and *Tritrichomonas foetus*. *Mol. Biochem. Parasitol.* **44**: 97–108.

Bellamy, C.O.C., Malcomson, R.D.G., Harrison, D.J. and Wyllie, A.H. (1995) Cell death in health and disease: the biology and regulation of apoptosis. *Cancer Biol.* **6**: 3–16.

Birnbaum, H. and Kraussold, E. (1975) Incidence of blasomyces and trichomonad infections during the use of hormonal and intrauterine contraception. *Zentralbl. Gynaekol.* **97**: 1636–1640.

Bowden, F.J. and Garnett, G.P. (1999) Why is *Trichomonas vaginalis* ignored? *Sex. Transm. Infect.* **75**: 372–374.

Bowden, F.J., Paterson, B.A., Mein, J., Savage, J., Fairley, C.K., Garland, S.M. and Tabrizi, S.N. (1999) Estimating the prevalence of *Trichomonas vaginalis, Chlamydia trachomatis, Neisseria gonorrhoeae*, and human papillomavirus infection in indigenous women in northern Australia. *Sex. Transm. Infect.* **75**: 431–434.

Bramley, M. and Kinghorn, G. (1979) Do oral contraceptives inhibit *Trichomonas vaginalis? Sex. Transm. Dis.* **6**: 261–263.

Brat, D.J., Bellail, A.C. and Van Meir, E.G. (2005) The role of interleukin-8 and its receptors in gliomagenesis and tumoral angiogenesis. *Neuro-oncology* **7**: 122–133.

Catterall, R.D. (1972) *Trichomonas* infections of the genital tract. *Med. Clin. North Am.* **56**: 1203–1209.

Caux, C., Vanbervliet, B., Massacrier, C., Ait-Yahia, S., Vaure, C., Chemin, K., Die-Nosjean, M.C. and Vicari, A. (2002) Regulation of dendritic cell recruitment by chemokines. *Transplantation* **73**: S7–S11.

Cavalier-Smith, T. (1998) A revised six-kingdom system of life. *Biol. Rev. Camb. Philos. Soc.* **73**: 203–266.

Chang, J.-H., Kim, Y.-S. and Park, J.-Y. (2004) *Trichomonas vaginalis*-induced apoptosis in RAW264.7 cells is regulated through Bcl-x$_L$, but not Bcl-2. *Parasite Immun.* **26**: 141–150.

Chang, J.-H., Park, J.-Y. and Kim, S.-K. (2006) Dependence on p38 MAPK signalling in the upregulation of TLR2, TLR4 and TLR9 gene expression in *Trichomonas vaginalis*-treated HeLa cells. *Immunology* **118**: 164–170.

Clinkenbeard, K.D., Mosier, D.A. and Confer, W.W. (1989) Transmembrane pore size and role of cell swelling in cytotoxicity caused by *Pasteurella hemolytica* leukotoxin. *Infect. Immun.* **57**: 420–425.

Connelly, R.J., Torian, B.E. and Stibbs, H.H. (1985) Identification of surface antigen of *Trichomonas vaginalis. Infect. Immun.* **49**: 270–274.

Coombs, R.W., Reichelderfer, P.S. and Landay, A.L. (2003) Recent observations on HIV type-1 infection in the genital tract of men and women. *AIDS* **17**: 455–480.

Corbeil, L.B. (1995) Use of an animal model of trichomoniasis as a basis for understanding this disease in women. *Clin. Infect. Dis.* **21**(Suppl 2): S158–S161.

Corbeil, L.B., Hodgson, J.L., Jones, C.W., Corbeil, R.R., Widders, P.R. and Stephens, L.R. (1989) Adherence of *Tritrichomonas foetus* to bovine vaginal epithelial cells. *Infect. Immun.* **57**: 2158–2165.

Corbeil, L.B., Campero, C.M. Rhyan, R.C. and BonDurant, R.H. (2003) Vaccines against sexually transmitted diseases. *Reprod. Biol. Endocrinol.* **1**:118.

Costello, C.E., Glushka, J., van Halbeek and Singh, B.N. (1993) Structural characterization of novel inositol phosphosphingolipids of *Tritrichomonas foetus* and *Trichomonas vaginalis. Glycobiology* **3**: 261–269.

Costello, C.E., Beach, D.H. and Singh, B.N. (2001) Acidic glycerol lipids of *Trichomonas vaginalis* and *Tritrichomonas foetus. Biol. Chem.* **382**: 275–281.

Cotch, M.F., Pastorek 2nd, J.G., Nugent, R.P., *et al.* (1997) *Trichomonas vaginalis* associated with low birth weight and preterm delivery. The Vaginal Infections and Prematurity Study Group. *Sex. Transm. Dis.* **24**: 353–360.

Crowell, A.L., Sanders-Lewis, K.A. and Secor, W.E. (2003) In vitro metronidazole and trinidazole activities against metronidazole-resistant strains of *Trichomonas vaginalis. Antimicrob. Agents Chemother.* **47**: 1407–1409.

Cudmore, S.L., Degaty, K.L., Hayward-McClelland, S.F., Petrin, D.P. and Garber, G.E. (2004) Treatment of infections caused by metronidazole-resistant *Trichomonas vaginalis. Clin. Microbiol. Rev.* **17**: 783–793.

Dearborn, D.G., Smith, S. and Korn, E.D. (1976) Lipophosphoglycan of the plasma membrane of *Acanthamoeba castellanii. J. Biol. Chem.* **251**: 2976–2982.

Diamond, L.D. (1968) Techniques of axenic culture of *Entamoeba histolytica* Schaudin 1903 and *E. histolytica*-like amebae. *J. Parasitol.* **54**: 1047–1056.

Donne, A. (1836) Animacules observes dans les matieres purulentes et le produit des secretions des organes genitaux de l'home et da la femme. *CR Acad. Sci. Paris* **3**: 385–386.

Draper, D.O., Parker, R., Patterson, E., Jones, W., Beutz, M., French, J., Borchardt, K. and McGregor, J. (1993) Detection of *Trichomonas vaginalis* in pregnant women with the InPouch TV culture system. *J. Clin. Microbiol.* **31**: 1016–1018.

Draper, D., Donohoe, W., Mortimer, L. and Heine, R.P. (1998) Cysteine proteases of *Trichomonas vaginalis* degrade secretory leukocyte protease inhibitor. *J. Infect. Dis.* **178**: 815–819.

Dunne, R.L., Dunn, L.A. Upcroft, P., O'Donoghue, P.J. and Upcroft, J.A. (2003) Drug resistance in the sexually transmitted protozoan *Trichomonas vaginalis. Cell Res.* **13**: 239–249.

Dyall, S.D. and Johnson, P.J. (2000a) The trichomonad hydrogenosome. In: *Biology of Parasitism* (eds C. Tschudi and E.J. Pearce). Kleuwer Academic, New York, NY, pp. 169–173.

Dyall, S.D. and Johnson, P.J. (2000b) Origin of hydrogenosomes and mitochondria: evalution and organelle biogenesis. *Curr. Opin. Microbiol.* **3**: 404–411.

Engbring, J.A., O'Brien, J.L. and Alderete, J.F. (1996) *Trichomonas vaginalis* adhesin proteins display molecular mimicry to metabolic enzymes. *Adv. Exp. Med. Biol.* **408**: 207–223.

Feola, D.J. and Thornton, A.C. (2002) Metronidazole-induced pancreatitis in a patient with recurrent vaginal trichomoniasis. *Pharmacotherapy* **22**: 1508–1510.

Ferguson, M.A. (1999) The structure, biosynthesis and functions of glycosylphosphatidylinositol anchors, and the contributions of trypanosome research. *J. Cell Sci.* **112**: 2799–2809.

Fichorova, R.N. (2004) Guiding the vaginal microbicide trials with biomarkers of inflammation. *J. AIDS* **37**(Suppl 3): S184–S193.

Fichorova, R.N., Rheinwald, J.G. and Anderson, D.J. (1997) Generation of papillomavirus-immortilized cell lines from normal human ectocervical, endocervical and vaginal epithelium that maintain expression of tissue-specific differentiation proteins. *Biol. Reprod.* **57**: 847–855.

Fichorova, R.N., Desai, P.J., Gibson 3rd, F.C. and Genco, C.A. (2001) Distinct proinflammatory host responses to *Neisseria gonorrhoeae* infection in immortalized human cervical and vaginal epithelial cells. *Infect. Immun.* **69**: 5840–5848.

Fichorova, R.N., Cronin, A.O., Lien, E., Anderson, D.J. and Ingalls, R.R. (2002) Response to *Neisseria gonorrhoeae* by cervicovaginal epithelial cells occurs in the absence of toll-like receptor 4-mediated signaling. *J. Immunol.* **168**: 2424–2432.

Fichorova, R.N., Trifonova, R.T., Gilbert, R.O., Costello, C.E., Hayes, G.R., Lucas, J.J. and Singh, B.N. (2006) *Trichomonas vaginalis* lipophosphoglycan triggers a selective upregulation of cytokines by human female reproductive tract epithelial cells. *Infect. Immun.* **74**: 5773–5779.

Filho-Silva, C.S. and deSouza, W. (1988) The interaction of *Trichomonas vaginalis* and *Tritrichomonas foetus* with epithelial cell in vitro. *Cell Struct. Funct.* **13**: 301–310.

Fiori, P.L., Rappelli, P., Addis, M.F., Sechi, A. and Cappuccinelli, P. (1996) *Trichomonas vaginalis* haemolysis: pH regulates a contact-dependent mechanism based on pore-forming proteins. *Microb. Pathog.* **20**: 109–118.

Fiori, P.L., Rappelli, P., Addis, M.F., Mannu, M. and Cappuccinelli, P. (1997) Contact-dependent disruption of host cell membrane skeleton induced by *Trichomonas vaginalis. Infect. Immun.* **65**: 5142–5148.

Fiori, P.L., Rappelli, P. and Addis, M.F. (1999) The flagellated parasite *Trichomonas vaginalis*: new insights into pathogenicity mechanisms. *Microbe Infect.* **1**: 149–156.

Fouts, A.C. and Kraus, S.J. (1980) *Trichomonas vaginalis*: reevaluation of its clinical presentation and laboratory diagnosis. *J. Infect. Dis.* **141**: 137–143.

Gao, L.-Y. and Kwaik, Y.A. (2000) The modulation of host cell apoptosis by intracellular bacterial pathogens. *Trends Microbiol.* **8**: 306–313.

Garber, G.E. and Bowie, W.R. (1990) The effect of *Trichomonas vaginalis* and the role of pH on cell culture monolayer viability. *Clin. Investig. Med.* **13**: 71–76.

Garber, G.E. and Lemchunk-Favel, L.T. (1994) Analysis of the extracellular proteases of *Trichomonas vaginalis. Parasitol. Res.* **80**: 361–365.

Garber, G.E., Lemchunk-Favel, L.T. and Bowie, W.R. (1989) Isolation of a cell-detaching factor of *Trichomonas vaginalis. J. Clin. Microbiol.* **27**: 1548–1553.

Garber, G.E., Lemchunk-Favel, L.T., Meysick, K.C. and Dimock, K. (1993) A *Trichomonas vaginalis* cDNA with partial sequence homology with a *Plasmodium falsiparum* excreted protein ABRA. *Appl. Parasitol.* **34**: 245–249.

Garcia, A.F., Chang, T.-H., Benchimol, M., Klump, D.J., Lehker, M.W. and Alderete, J.F. (2003) Iron and contact with host cells induce expression of adhesins on surface of *Trichomonas vaginalis. Mol. Microbiol.* **47**: 1207–1224.

Gavrielescu, A.R. and Denkers, E.Y. (2003) Apoptosis and balance of homeostatic and pathologic responses to protozoan infections. *Infect. Immun.* **71**: 6109–6115.

Gilbert, R.O., Elia, G., Beach, D.H., Klaessig, S. and Singh, B.N. (2000) Cytopathogenic effect of *Trichomonas vaginalis* on human vaginal epithelial cells. *Infect. Immun.* **68**: 4200–4206.

Gopalkrishnan, K., Hinduja, I.N. and Kumar, T.C. (1990) Semen characteristics of asymptomatic males affected by *Trichomonas vaginalis. J. In Vitro Fertil. Embryo Transfer* 7: 165–167.

Grodstein, F., Goldman, M.B. and Cramer, D.W. (1993) Relation of tubal infertility to a history of sexually transmitted diseases. *Am. J. Epidemiol.* **137**: 577–584.

Guenthner, P.C., Secorand, W.E. and Dezzutti, C.S. (2005)*Trichomonas vaginalis*-induced epithelial monolayer disruption and human immunodeficiency virus type 1 (HIV-1) replication: implications for the sexual transmission of HIV-1. *Infect. Immun.* **73**: 4155–4160.

Guha-Niyogi, A., Sullivan, D.R. and Turco, S.J. (2001) Glycoconjugate structures of parasitic protozoa. *Glycobiology* **11**: 45R–59R.

Hampl, V., Vrlik, M., Cepicka, I., Pecka, Z., Kulda, J. and Tachezy, J. (2006) Affiliation of *Cochlosoma* to trichomonads confirmed by phylogenetic analysis of the small-subunit rRNA gene and new family concept of the order Trichomonadida. *Int. J. Syst. Evol. Microbiol.* **56**: 305–312.

Hedvat, C.V., Yao, J., Sokolic, R.A. and Nimer, S.D. (2004) Myeloid ELF1-like factor is a potent activator of interleukin-8 expression in hematopoietic cells. *J. Biol. Chem.* **279**: 6395–6400.

Hernandez-Gutierrez, R., Avila-Gonzalez, L., Ortega-Lopez, J., Cruz-Talonia, F., Gomez-Gutierrez, G. and Arroyo, R. (2004) *Trichomonas vaginalis*: characterization of a 39-kDa cysteine proteinase found in patient vaginal secretions. *Exp. Parasitol.* **107**: 125–135.

Heyworth, P.G. and Gutteridge, W.E. (1978) Further studies on the purine and pyrimidine metabolism in *Trichomonas vaginalis. J. Protozool.* **25**: 9b.

Hobbs, M.M., Kazambe, P., Reed, A.W., *et al.* (1999) *Trichomonas vaginalis* as a cure of urethritis in Malawian men. *Sex. Transm. Dis.* **26**: 381–387.

Hobbs, M.M., Lapple, D.M., Lawing, L.F., *et al.* (2006) Methods for detection of *Trichomonas vaginalis* in the male partners of infected women; implications for control of trichomoniasis. *J. Clin. Microbiol.* **44**: 3994–3999.

Hodgson, J.L., Jones, D.W., Widders, P.R. and Corbeil, L.B. (1990) Characterization of *Tritrichomonas foetus* antigens by use of monoclonal antibodies. *Infect. Immun.* **58**: 3078–3083.

Hoehne, O. (1916) Trichomonas vaginalis als haufiger Erreger einer typischen Colpitis purulenta. Zentralbl. F. Gynak. **40:** 4–15.

Hoffmann, E., Dittrich-Breiholz, O., Holtmann, H. and Kracht, M. (2002) Multiple control of interleukin-8 gene expression. *J. Leukocyte Biol.* **72**: 847–855.

Holz Jr, G.G., Lindmark, D.G., Beach, D.H., Neale, K.A. and Singh, B.N. (1987) Lipids and lipid metabolism of trichomonads. *Acta Univ. Carol. Biol.* **30**: 299–311.

Honigberg, B.M. (1963) Evolutionary and systematic relationships in the flagellate order Trichomonadida Kirby. *J. Protozool.* **10**: 20–63.

Huston, C.D., Houpt, E.R., Mann, B.J., Hahn, C.S. and Petri, W.I. (2000) Caspase 3-dependent killing of host cells by the parasite *Entamoeba histolytica. Cell Microbiol.* **2**: 617–625.

Jackson, D.J., Rakwar, J.P., Bwayo, J.J., Kreiss, J.K. and Moses, S. (1997) Urethral *Trichomonas vaginalis* infection and HIV-1 transmission. *Lancet* **350**: 1076.

James, E.R. and Green, D.R. (2004) Manipulation of apoptosis in the host-parasite interaction. *Trends Parasitol.* **20**: 280–287.

John, G.C., Nduati, R.W., Mbori-Ngacha, D.A, *et al.* (2001) Correlates of mother-to-child human immunodeficiency virus type 1 (HIV-1) transmission: association with maternal plasma HIV-1 RNA load, genital HIV-1 DNA shedding, and breast infections. *J. Infect. Dis.* **183**: 206–212.

Kang, J.H., Song, H.O., Ryu, J.S., Shin, M.H., Kim, J.M., Cho, Y.S., Alderete, J.F., Ahn, M.H. and Min, D.Y. (2006) *Trichomonas vaginalis* promotes apoptosis of human neutrophils by activating caspase-3 and reducing Mcl-1 expression. *Parasite Immun.* **28**: 439–446.

Kania, S.A., Reed, S.L., Thomford, J.W., BonDurant, R.H., Hirata, K., Corbeil, R.R., North, M.J. and Corbeil, L.B. (2001) Degradation of bovine complement C3 by trichomonad extracellular proteinase. *Vet. Immun. Immunopath.* **78**: 83–96.

Kirby, H. (1947) Flagellate and host relationships of Trichomonad-flagellates. *J. Parasitol.* **33**: 214–228.

Klemba, M. and Goldberg, D.E. (2002) Biological roles of proteases in parasite protozoa. *Annu. Rev. Biochem.* **71**: 275–305.

Krieger, J.N. (1981) Urologic aspects of trichomoniasis. *Invest. Urol.* **18**: 411–417.

Krieger, J.N. and Alderete, J.F. (1999) *Trichomonas vaginalis* and trichomoniasis. In: *Sexually Transmitted Diseases* (eds K.K. Holms, P.F. Sparling, P.A. Mardh, S.M. Lamon, W.E. Stamm, W.E. Piot and J.N. Wasserheit). McGraw/Hill, New York, NY, pp. 587–604.

Krieger, J.N., Ravdin, J.L. and Rein, M.F. (1985) Contact-dependent cytopathogenic mechanisms of *Trichomonas vaginalis. Infect. Immun.* **50**: 77–81.

Krieger, J.N., Wolner-Hanssen, P., Stevens, C. and Holmes, K.K. (1990a) Characteristic of *Trichomonas vaginalis* from women with and without colpitis macularis. *J. Infect. Dis.* **161**: 207–211.

Krieger, J.N., Torian, B.E., Hom, J. and Tam, M.R. (1990b) Inhibition of *Trichomonas vaginalis* motility by monoclonal antibodies associated with reduced adherence to HeLa cell monolayers. *Infect. Immun.* **58**: 1634–1639.

Kucknoor, A.S., Vasanthakrishna, M. and Alderete, J.F. (2005) Adherence to human vaginal epithelial cells signals for increased expression of *Trichomonas vaginalis* genes. *Infect. Immun.* **73**: 6472–6478.

Kulda, J. (1999) Trichomonads, hydrogenosomes and drug resistance. *Int. J. Parasitol.* **29**: 199–212.

Kulda, J., Nohynkva, E. and Ludvik, J. (1986) Basic structure and function of the trichomonad cell. *Acta Univ. Carol. Biol.* **30**: 181–198.

Kwon, J.H., Keates, S., Simeonidis, S., Grall, F., Libermann, T.A. and Keates, A.C. (2003) ESE-1, an enterocyte-specific Ets transcription factor, regulates MIP-3alpha gene expression in Caco-2 human colonic epithelial cells. *J. Biol. Chem.* **278**: 875–884.

Land, K.M. and Johnson, P.J. (1999) Molecular basis of metronidazole resistance in pathogenic bacteria and protozoa. *Drug Resist. Updates* **2**: 289–294.

Land, K.M., Delgadillo-Correa, M.G., Tachezy, J., Vanacova, S., Hsieh, C.L., Sutak, R. and Johnson, P.J. (2004) Targeted gene replacement of a ferredoxin gene in *Trichomonas vaginalis* does not lead to metronidazole resistance. *Mol. Microbiol.* **51**: 115–122.

Lehker, M.W. and Alderete, J.F. (2000) Biology of trichomoniasis. *Curr. Opin. Infect. Dis.* **13**: 37–45.

Lekkerkerker, A.N., van Kooyk, Y. and Geijtenbeek, T.B. (2006) Viral piracy: HIV-1 targets dendritic cells for transmission. *Curr. HIV Res.* **4**: 169–176.

Leon-Sicairos, C.R., Leon-Felix, J. and Arroyo, R. (2004) tvcp12: a novel *Trichomonas vaginalis* cathepsin l-like cysteine proteinase-encoding gene. *Microbiol.* **150**: 1131–1138.

Levi, M.H., Torres, J., Pina, C. and Klein, R.S. (1997) Comparison of the InPouch TV culture system and Diamond's modified medium for detection of *Trichomonas vaginalis*. *J. Clin. Microbiol.* **35**: 3308–3310.

Lindmark, D.G., Muller, M. and Shio, H. (1975) Hydrogenosomes in *Trichomonas vaginalis*. *J. Parasitol.* **61**: 552–554.

Lindmark, D.G., Beach, D.H., Singh, B.H. and Holz Jr, G.G. (1991) Lipids and lipid metabolism of trichomonads (*Tritrichomonas foetus* and *Trichomonas vaginalis*). In: *Biochemical Protozoology* (eds G. Coomb and M. North). Taylor and Francis, Abingdon, pp. 329–335.

Littlewood, J.M. and Kohler, H.G. (1966) Urinary tract infection of *Trichomonas vaginalis* in a new born baby. *Arch. Dis. Child.* **41**: 693–695.

Lloyd, G.L., Case, J.R., De Frias, D. and Braninigan, R.E. (2003) *Trichomonas vaginalis* orchitis with associated severe oligoasthenoteratospemia and hypogonadism. *J. Urol.* **170**: 924.

Lodge, R. and Descoteaux, A. (2005) Modulation of phagolysosome biogenesis by lipophosphoglycan of *Leishmania*. *Clin. Immun.* **114**: 256–265.

Lopes, M.F., da Veiga, V.F., Santos, A.R., Fonseca, M.E. and DosReis, G.A. (1995) Activation-induced CD4+ T cell death by apoptosis in experimental Chagas' disease. *J. Immunol.* **154**: 744–752.

Magnus, M., Clark, R., Myers, L., Farley, T. and Kissinger, P.J. (2003) *Trichomonas vaginalis* among HIV-infected women: are immune status or protease inhibitor use associated with subsequent *T. vaginalis* positivity? *Sex. Transm. Dis.* **30**: 839–843.

Mallinson, D.J., Lockwood, B.C., Coombs, G.H. and North, M.J. (1994) Identification and molecular cloning of four cysteine proteinase genes from the pathogenic protozoan *Trichomonas vaginalis*. *Microbiology* **140**: 2725–2735.

Mallinson, D.J., Livingstone, J., Appleton, K.M., Lees, S.J., Coombs, G.H. and North, M.J. (1995) Multiple cysteine proteinases of the pathogenic protozoon *Tritrichomonas foetus*: identification of seven diverse and differentially expressed genes. *Microbiology* **141**: 3077–3085.

Marin, V., Farnarier, C., Gres, S., Kaplanski, S., Su, M.S.-S., Dinarello, C.A. and Kaplanski, G. (2001) The p38 mitogen-activated protein kinase pathway plays a critical role in thrombin-induced endothelial chemokine production and leukocyte recruitment. *Blood* **98**: 667–673.

Markell, E.K., John, D.T. and Krotoski, W.A. (1999) *Markell and Voge's Medical Parasitology*, 8th Edn. WB Saunders, Philadelphia, MD, pp. 64–68.

Mason, P.R., Fiori, P.L., Cappuecinelli, P., Rappelli, P. and Greson, S. (2005) Seroepidemiology of *Trichomonas vaginalis* in rural women in Zimbabwe and patterns of association with HIV infection. *Epidemiol. Infect.* **133**: 315–323.

McCole, D.F., Eckmann, L., Laurent, F. and Kagnoff, M.F. (2000) Intestinal epithelial cell apoptosis following *Cryptospridium parvum* infection. *Infect. Immun.* **68**: 1710–1713.

McKerrow, J.H., Caffrey, C., Kelly, B., Loke, B. and Sajid, M. (2006) Proteases in parasitic diseases. *Ann. Rev. Pathol.* **1**: 497–536.

Mele, R., Gomez Morales, M.A., Tosini, F. and Pozia, E. (2004) *Cryptospridium parvum* at different developmental stages modulates host cell apoptosis in vitro. *Infect. Immun.* **72**: 6061–6067.

Mendoza-Lopez, M.R., Becerril-Garcia, C., Fattel-Facenda, L.V., Avila-Gonzalez, L., Ruiz-Tachiquin, M.E, Ortega-Lopez, J. and Arroyo, R. (2000) CP30, a cysteine proteinase involved in *Trichomonas vaginalis* cytoadherence. *Infect. Immun.* **68**: 4907–4912.

Min, D.Y., Hyun, K.H., Ryu, J.S., Ahn, M.H. and Cho, M.H. (1998) Degradations of human immunoglobins and hemoglobin by a 60 kDa cysteine proteinase of *Trichomonas vaginalis*. *Korean J. Parasitol.* **36**: 261–268.

Mirhaghani, A. and Warton, A. (1998) Involvement of *Trichomonas vaginalis* surface-associated glycoconjugates in parasite/target cell interaction. A quantitative electron microscopy study. *Parasitol. Res.* **84**: 374–378.

Moss, J.E., Aliprantiz, A.O. and Zychlinsky, A. (1999) The regulation of apoptosis by microbial pathogens. *Int. Rev. Cytol.* **187**: 203–259.

Muller, M. (1988) Energy metabolism of protozoa without mitochondria. *Ann. Rev. Microbiol.* **42**: 465–488.

Muller, M. (1990) Biochemistry of *Trichomonas vaginalis*. In: *Trichomonads Parasitic in Humans* (ed. B.M. Hongberg). Springer-Verlag, New York, NY, pp. 53–83.

Muller, M. (1993) The hydrogenosome. *J. Gen. Microbiol.* **139**: 2879–2889.

Munagala, N.R. and Wang, C.C. (2003) Adenine is the primary precursor of all purine nucleotides in *Trichomonas vaginalis. Mol. Biochem. Parasitol.* **127**: 143–149.

Narimatsu, R., Wolday, D. and Patterson, B.K. (2005) IL-8 increases transmission of HIV type 1 in cervical explant tissue. *AIDS Res. Hum. Retrovirus* **21**: 228–233.

Neale, K.A. and Alderete, J.E. (1990) Analysis of proteinases of representative *Trichomonas vaginalis* isolates. *Infect. Immun.* **58**: 157–162.

Nielsen, M.H. and Nielsen, R. (1975) Electron microscopy of *Trichomonas vaginalis* Donne'; interaction with vaginal epithelium in human trichomoniasis. *Acta Pathol. Microbiol. Scand.* **83**: 305–320.

North, M.J. (1994) Cysteine endopeptidases of parasite protozoa. *Method Enzymol.* **244**: 523–539.

Petrin, D., Delgaty, K., Bhatt, R. and Garber, G. (1998) Clinical and microbiological aspects of *Trichomonas vaginalis. Clin. Microbiol. Rev.* **11**: 300–317.

Pindak, F.F., de Pindak, M.M. and Gardner, J. (1993) Contact-dependent cytotoxicity of *Trichomonas vaginalis. Genitourin. Med.* **59**: 35–40.

Pivarcsi, A., Nagy, I., Koreck, A., Kis, K., Kenderessy-Szabo, A., Szell, M., Dobozy, A. and Kemeny, L. (2005) Microbial compounds induce the expression of pro-inflammatory cytokines, chemokines and human beta-defensin-2 in vaginal epithelial cells. *Microbes Infect.* **7**: 1117–1127.

Platz-Christensen, J.J., Mattsby-Baltzer, I., Thomsen, P. and Wiqvist, N. (1993) Endotoxin and interleukin-1 alpha in the cervical mucus and vaginal fluid of pregnant women with bacterial vaginosis. *Am. J. Obstet. Gynecol.* **169**: 1161–1166.

Prebeck, S., Brade, H., Kirschning, C.J., da Costa, C.P., Durr, S., Wagner, H. and Miethke, T. (2003) The Gram-negative bacterium *Chlamydia trachomatis* L2 stimulates tumor necrosis factor secretion by innate immune cells independently of its endotoxin. *Microbes Infect.* **5**: 463–470.

Previato, J.O., Gorin, P.A.G., Mazurek, M., Xavier, M.T., Fournet, B., Wieruszeski, L.M. and Mendanca-Previato, L. (1990) Primary structure of oligosaccharide chain of lipopeptidophosphoglycan of epimastigote form of *Trypanosoma cruzi. J. Biol. Chem.* **265**: 2518–2526.

Que, X. and Reed, S.L. (2000) Cysteine proteinases and the pathogenesis of amoebiosis. *Clin. Microbiol. Rev.* **13**: 196–206.

Rabinovich, G.A. and Gruppi, A. (2005) Galectins as immunoregulators during infectious processes: from microbial invasion to the resolution of the disease. *Parasite Immunol.* **27**: 103–114.

Ravdin, J.E., Croft, B.Y. and Guerrant, R.L. (1980) Cytopathic mechanisms of *Entamoeba histolytica. J. Exp. Med.* **152**: 377–390.

Reed, J.C. (2000) Mechanisms of apoptosis. *Am. J. Pathol.* **157**: 1415–1430.

Reers, M,, Smiley, S.T., Mottola-Hartshorn, C., Chen, A., Lin, M. and Chen, L.B. (1995) Mitochondrial membrane potential monitored by JC-1dye. *Methods Enzymol.* **260**: 406–417.

Rein, M.F. (1990) Clinical manifestations of urogenital trichomoniasis in women. In: *Trichomonads Parasitic in Humans* (ed. B.M. Honigberg), Springer-Verlag, New York, NY, pp. 225–234.

Robinson, S.C. (1962) Trichmonol vaginitis resistant to metronidazole. *Can. Med. Assoc. J.* **86**: 665.

Roebuck, K.A. (1999) Regulation of interleukin-8 gene expression. *J. Interferon Cytokine Res.* **19**: 429–438.

Ryu, J.S. and Min, D.Y. (2006) *Trichomonas vaginalis* and trichomoniasis in the Republic of Korea. *Korean J. Parasitol.* **44**: 101–116.

Ryu, J.S., Kang, J.H., Jung, S.Y., Shin, M.H., Kim, J.M., Park, H. and Min, D.Y. (2004) Production of interleukin-8 by human neutrophils stimulated with *Trichomonas vaginalis. Infect. Immun.* **72**: 1326–1332.

Sajid, M. and McKerrow, J.H. (2002) Cysteine proteases of parasite organisms. *Mol. Biochem. Parasitol.* **120**: 1–21.

Schaumburg, F., Hippe, D., Vutovo, P. and Luder, C.G.K. (2006) Pro- and anti-apoptotic activities of protozoan parasites. *Parasitology* **132**: S69–S85.

Schwebke, J.R. (2005) Trichomoniasis in adolescents: a marker for the lack of a public health response to the epidemic of sexually transmitted diseases in the United States. *J. Infect. Dis.* **192**: 2036–2038.

Schwebke, J.R. and Burgess, D. (2004) Trichomoniasis. *Clin. Microbiol. Rev.* **17**: 794–803.

Schwebke, J.R. and Hook III, E.W. (2003) High rates of *Trichomonas vaginalis* among men attending a sexually transmitted diseases clinic: implications for screening and urethritis management. *J. Infect. Dis.* **188**: 465–468.

Schwebke, J.R. and Lawing, L.F. (2002) Improved detection by DNA amplification of *Trichomonas vaginalis* in males. *J. Clin. Microbiol.* **40**: 3681–3683.

Schwebke, J.R., Morgan, S.C. and Pinson, G.B. (1997) Validity of self-obtained vaginal specimen for diagnosis of trichomoniasis. *J. Clin. Microbiol.* **35**: 1618–1619.

Schwebke, J.R., Aira, T., Jordan, N., Jolly, P.E. and Vermund, S.H. (1998) Sexually transmitted diseases in Ulaanbaatar, Mongolia. *Int. J. STD AIDS* **9**: 354–358.

Schwebke, J.R., Venglarik, M.F. and Morgan, S.C. (1999) Delayed versus immediate bedside inoculation of culture method for diagnosis of vaginal trichomoniasis. *J. Clin. Microbiol.* **37**: 2369–2370.

Shaia, C.I., Voyich, J., Gillis, S.J., Singh, B.N. and Burgess, D.E. (1998) Purification and expression of the Tf190 adhesin in *Tritrichomonas foetus. Infect. Immun.* **66**: 1100–1105.

Shaio, M.F., Lin, P.R., Liu, J.Y. and Tang, T.D. (1994) Monocyte-derived interleukin-8 involved in the recruitment of neutrophils induced by *Trichomonas vaginalis* infection. *J. Infect. Dis.* **170**: 1638–1640.

Shaio, M.F., Lin, P.R., Liu, J.Y. and Yang, K.D. (1995) Generation of interleukin-8 from human monocytes in response to *Trichomonas vaginalis* stimulation. *Infect. Immun.* **63**: 3864–3870.

Singh, B.N. (1993) Lipophosphoglycan-like glycoconjugate of *Trichomonas foetus* and *Trichomonas vaginalis. Mol. Biochem. Parasitol.* **57**: 281–294.

Singh, B.N. (1994) The existence of lipophosphoglycanlike molecules in Trichomonads. *Parasitol. Today* **10**: 152–154.

Singh, B.N., Costello, C.E. and Beach, D.H. (1991) Structures of glycophosphosphingolipids of *Tritrichmomonas foetus. Arch. Biochem. Biophys.* **286**: 409–418.

Singh, B.N., Beach, D.H., Lindmark, D.G. and Costello, C.E. (1994) Identification of the lipid moiety and further characterization of the novel lipophosphoglycan-like glycoconjugates of *Trichomonas vaginalis* and *Trichomonas foetus. Arch. Biochem. Biophys.* **309**: 273–280.

Singh, B.N., Lucas, J.J., Beach, D.H., Shin, S.T. and Gilbert, R.O. (1999) Adhesion of *Tritrichomonas foetus* to bovine vaginal epithelial cells. *Infect. Immun.* **67**: 3847–3854.

Singh, B.N., Hayes, G.R., Lucas, J.J., Levery, S.B., Mirgorodskaya, E. and Costello, C.E. (2000a) Novel lipophosphoglycans of trichomonad parasites (abstract). *Glycobiology* **10**: 1127.

Singh, B.N., Lucas, J.J., Hayes, G.R., Costello, C.E., Viseux, N and Levery, S. (2000b) Characterization of novel trichomonad glycoconjugates (abstract). *FASEB J.* **14**: 1339.

Singh, B.N., BonDurant, R.H., Campero, C.M. and Corbeil, L.B. (2001) Immunological and biochemical analysis of glycosylated surface antigens and lipophosphoglycan of *Tritrichomonas foetus. J. Parasitol.* **87**: 770–777.

Singh, B.N., Lucas, J.J., Hayes, G.R., Kumar, I., Beach, D.H., Frajblat, M., Gilbert, R.O., Sommer, U. and Costello, C.E. (2004) *Trichomonas foetus* induces apoptotic cell death in bovine vaginal epithelial cells. *Infect. Immun.* **72**: 4151–4158.

Singh, B.N., Hayes, G.R., Lucas, J.J., Beach, D.H. and Gilbert, R.O. (2005) *In vitro* cytopathic effects of a cysteine protease of *Tritrichomonas foetus* on cultured bovine uterine epithelial cells. *Am. J. Vet. Res.* **66**: 1181–1186.

Sites, J.K., Whittaker, J., Srfo, B.Y., Thompson, W.E., Powel, M.D. and Bond, V.C. (2004) Trypanosome apoptotic factor mediates apoptosis in human brain vascular endothelial cells. *Mol. Biochem. Parasitol.* **133**: 229–240.

Skerk, V., Schonwald, J., Granic, J., Krhen, I., Barsic, B., Markekovic, I., Roglic, S., Desnica, B. and Zeljko, Z. (2002) Chronic prostatitis caused by *Trichomonas vaginalis* – diagnosis and treatment. *J. Chemother.* **14**: 537–538.

Slee, E.A., Adrain, C.E. and Martin, S.J. (1999) Serial killers: ordering caspase activation events in apoptosis. *Cell Death Differ.* **6:** 167–174.

Soares, R.P.P., Cardoso, T.L., Barron, T., Araijo, M.S.S., Pimenta, P.F.P. and Turco, S.J. (2005) *Leishmania braziliensis*: a novel mechanism in the lipophosphoglycan regulation during metacyclogenesis. *Int. J. Parasitol.* **35**: 245–253.

Sommer,U., Costello, C.E., Hayes, G.R., Beach, D.H., Gilbert, R.O., Lucas, J.J. and Singh, B.N. (2005) Identification of *Trichomonas vaginalis* cysteine proteases that induce apoptosis in human vaginal epithelial cells. *J. Biol. Chem.* **280**: 23853–23860.

Sorvillo, F.L. and Kerndt, P. (1998) *Trichomonas vaginalis* and amplification of HIV-1 transmission. *Lancet* **351**: 213–214.

Sorvillo, F., Smith, L., Kerndt, P. and Ash, L. (2001) *Trichomonas vaginalis*, HIV, and African-Americans. *Emerg. Infect. Dis.* **7**: 927–932.

Spiegel, C.A. (1990) Microflora associated with *Trichomonas vaginalis* and vaccination against vaginal trichomoniasis. In: *Trichomonads Parasitic in Humans* (ed. B.M. Honigberg). Springer-Verlag, New York, NY, pp. 213–224.

Stadler, J., Keenan, T.W., Bauer, G. and Gerisch, G. (1989) The contact site: a glycoprotein of *Dictyostelium discoideum* carries a phospholipid anchor of a novel type. *EMBO J.* **8**: 371–377.

Stary, A., Kuchinka-Koch, A. and Teodorowicz, L. (2002) Detection of *Trichomonas vaginalis* on modified Columbia agar in the routine laboratory. *J. Clin. Microbiol.* **40**: 3277–3280.

Stechmann, A. and Cavalier-Smith, T. (2002) Rooting the eukaryote tree by using a derived gene fusion. *Science* **297**: 89–91.

Sutcliffe, S., Giovannuci, E., Alderete, J.F., Chang, T.-H., Gaydos, C.A., Zenilman, J.M., DeMarzo, A.M., Willett, W.C. and Platz, E.A. (2006) Plasma antibodies against *Trichomonas vaginalis* and subsequent risk of prostate cancer. *Cancer Epidemiol. Biomarkers Prev.* **15**: 939–945.

Thomford, J.W., Talbot, J.A., Ikeda, J.S. and Corbeil, L.B. (1996) Characterization of extracellular proteinases of *Tritrichomonas foetus*. *J. Parasitol.* **82**: 112–117.

Torian, B.E., Connelly, R.J., Stephens, R.S. and Stibbs, H.H. (1984) Specific and common antigens of *Trichomonas vaginalis* detected by monoclonal antibodies. *Infect. Immun.* **43**: 270–273.

Trussell, A.B. and Plass, E.D. (1940) The pathogenicity and physiology of a pure culture of *Trichomonas vaginalis*. *Am. J. Obstet. Gynecol.* **40**: 883–890.

Tuomala, R.E., O'Driscoll, P.T., Bremer, J.W., *et al.* (2003) Cell-associated genital tract virus and vertical transmission of human immunodeficiency virus type 1 in antiretroviral experienced women. *J. Infect. Dis.* **187**: 375–384.

Turco, S.J. and Descoteaux, A. (1992) The lipophosphoglycan of *Leishmania* parasites. *Annu. Rev. Microbiol.* **46**: 65–94.

Upcroft, P. and Upcroft, J.A. (2001) Drug targets and mechanisms of resistance in anaerobic protozoa. *Clin. Microbiol. Rev.* **14**: 150–164.

Upcroft, J.A., Dunn, L.A., Write, J.A., Benakli, K., Upcroft, P. and Vanelle, P. (2006) 5-Nitroimidazole drugs effective against metronidazole-resistant *Trichomonas vaginalis* and *Giardia duodenalis. Antimicrob. Agents Chemother.* **50**: 344–347.

van den Berg, T.K., Honing, H., Franke, N., *et al.* (2004) LacdiNAc-glycans constitute a parasite pattern for galectin-3-mediated immune recognition. *J. Immunol.* **173**: 1902–1907.

Van Der Pol, B., Williams, J.A., Orr, D.P., Batteiger, B.E. and Fortenberry, J.D. (2005) Prevalence, incidence, natural history, and response to treatment of *Trichomonas vaginalis* infection among adolescent women. *J. Infect. Dis.* **192**: 2039–2044.

Vargas-Villarreal, J., Mata-Cardenas, B.D., Palacios-Corona, R., Gonzalez-Salzar, F., Cortes-Gutierrez, E.I., Martinez-Rodriguez, H.G. and Said-Fernandez, S. (2005) *Trichomonas vaginalis*: identification of soluble and membrane-associated phospholipase A_1 and A_2 activities with direct and indirect hemolytic effects. *J. Parasitol.* **91**: 5–11.

Vikki, M., Pukkala, E., Nieminen, P. and Hakama, M. (2000) Gynecological infections as risk determinants of cervical neoplasia. *Acta Oncol.* **39**: 71–75.

Wang, C.C. (1986) Nucleic acid metabolism in trichomonads. *Acta Univ. Carol. Biol.* **30**: 267–280.

Wei, S., Marches, F., Borvak, J., Zou, W., Channon, J., White, M., Radke, J., Cesbron-Delauw, M.-F. and Curiel, T.J. (2002) *Toxoplasma gondii*-infected human myloid dendritic cells induce T-lymphocyte dysfunction and contact-dependent apoptosis. *Infect. Immun.* **70**: 1750–1760.

Weinstock, H., Berman, S. and Cates Jr, W. (2004) Sexually transmitted diseases among American youth: incidence and prevalence estimates, 2000. *Perspect. Sex. Reprod. Health* **36**: 6–10.

Wendel, K.A., Erbelding, E.J., Gaydos, C.A. and Rompalo, A.M. (2003) Use of urine polymerase chain reaction to define the prevalence and clinical presentation of *Trichomonas vaginalis* in men attending STD clinic. *Sex. Transm. Infect.* **79**: 151–153.

Wiese, W., Patel, S.R., Patel, S.C., Ohl, C.A. and Astrada, C.A. (2000) A meta-analysis of the Papanicolaou smear and wet mount for the diagnosis of vaginal trichomoniasis. *Am. J. Med.* **108**: 301–308.

Yap, E.H., Ho, T.H., Chan, Y.C., Thong, T.W., Ng, G.C., Ho, L.C. and Singh, M. (1995) Serum antibodies to *Trichomonas vaginalis* in invasive cervical cancer patients. *Genitourin. Med.* **71**: 402–404.

Young, J.D. and Cohn, Z.A. (1985) Molecular mechanisms of cytotoxicity modified by *Entamoeba histolytica*: characterization of pore-forming protein (PFP). *J. Cell Biochem.* **29**: 299–308.

Zang, Y., Wang, W.-H., Wu, S.-W., Ealick, S.E. and Wang, C.C. (2005) Identification of a subversive substrate of *Trichomonas vaginalis* purine nucleoside substrate and the crystal structure of the enzyme-substrate complex. *J. Biol. Chem.* **280**: 22318–22325.

Zariffard, M.R., Harwani, S., Novak, R.M., Graham, P.J., Ji, X. and Spear, G.T. (2004) *Trichomonas vaginalis* infection activates cells through toll-like receptor 4. *Clin. Immunol.* **111**: 103–107.

Zychlinsky, A. and Sansonetti, P.J. (1997) Host/pathogen interactions apoptosis in bacterial pathogenesis. *J. Clin. Invest.* **100**: 493–496.

E

Protozoan Pathogens of Major Medical Importance

Naveed Ahmed Khan

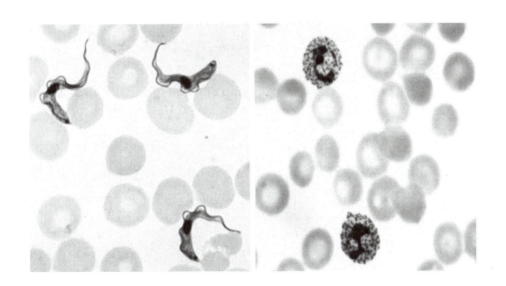

1 *Entamoeba histolytica*

Biology: Trophozoites are about 20–30 μm long, contain a single nucleus, feed on nutrients obtained from the host, reproduce by binary fission and form cysts for transmission.

Disease: Diarrhea or amoebic dysentery and liver abscesses. Symptoms include intestinal lesions developed in the gut and with increasing number of parasites the mucosal destruction becomes extensive with abdominal pain, diarrhea (bloody), cramps, vomiting, malaise, weight loss with extensive scarring of the intestinal wall. Death can occur with gut perforation, exhaustion and liver abscesses.

Diagnosis: Diagnosis is made by direct demonstration of parasites in stool samples using serology-based assays.

Transmission: Oral uptake of *Entamoeba* cysts from human feces in contaminated food or drinking water.

Treatment: Treatment includes oral application of metronidazole or diiodohydroxyquin and iodoquinol.

Occurrence: Worldwide (most prominent in warm countries).

2 *Plasmodium* spp.

Biology: Parasite differentiates into different forms in the vertebrate host and depending on the stage; the size varies from 2.5 to 15 μm; reproduce both by asexual and sexual reproduction.

Disease: Malaria: infects liver endothelial cells and red blood cells. Symptoms include abdominal pain, headache, and typical intermittent fever-chills and anemia associated with destruction of the red blood cells (periods vary depending on species). *Plasmodium falciparum* may produce continuous fever and results in death.

Diagnosis: Diagnosis is made by direct demonstration of parasites in stool samples and serology-based assays.

Transmission: Disease transmission is via insect bite, that is, the *Anopheles* mosquito (vector), as parasites are present in the salivary glands.

Treatment: Use of chloroquine, primaquine and qinghaosu in early disease is effective.

Occurrence: Warm countries (mostly tropical and subtropical countries).

3 *Trypanosoma brucei gambiense/Trypanosoma brucei rhodesiense*

Biology: Trophozoites are approximately 18–29 μm long, contain a single nucleus, feed on nutrients obtained from the host and reproduce by binary fission but do not form cysts.

Disease:	Sleeping sickness or African trypanosomiasis. Symptoms observed within weeks and include fever, headache, swollen lymph nodes, weight loss, and heart involvement (usually with *T. b. rhodesiense*). It is important to note that *T. b. rhodesiense* causes acute disease and the host often dies before the disease can develop fully. In contrast, *T. b. gambiense* produces chronic disease and the parasite invades the CNS and produces typical symptoms such as tremors, sleepiness, paralysis and finally death within months. Of interest, *T. b. brucei* causes fatal infections in nonhuman mammals: horses, sheep, and so on.
Diagnosis:	Based on symptoms and demonstration of parasites by microscopy.
Transmission:	Disease transmission occurs via insect bite, that is, the tsetse fly (vector), where parasites are present in the salivary glands.
Treatment:	Use of suramin, pentamidine, Berenil and difluoromethylornithine in early disease may have beneficial effects and Melarsoprol in late stages.
Occurrence:	Africa.

4 *Trypanosoma cruzi*

Biology:	Trophozoites are approximately 20 μm long, contain a single nucleus, feed on nutrients obtained from the host and reproduce by binary fission but do not form cysts.
Disease:	Chagas' disease or American trypanosomiasis. Following invasion, parasites infect susceptible tissues, in particular the heart. Parasites penetrate the myocardial fibers, multiplying for several days and producing a cavity in the invaded tissue, then escape into the bloodstream and invade other susceptible tissues including liver, spleen, muscles, and intestinal mucosa resulting in organ failure and finally death within months to years.
Diagnosis:	Based on symptoms and demonstration of parasites by microscopy.
Transmission:	Disease transmits via insects, that is, the triatomid bugs (vector). While feeding, these bugs defecate on the host, and parasites in the fecal material gain entry into the human body through the bite wound.
Treatment:	No effective treatment but use of nifurtimox, ketoconazole and benznidazole may be effective.
Occurrence:	South and Central America.

5 *Leishmania tropica/Leishmania major*

Biology:	Trophozoites are approximately 2–5 μm long, contain a single nucleus, feed on nutrients obtained from the host and reproduce by binary fission but do not form cysts.
Disease:	Cutaneous leishmaniasis. Following invasion, parasites multiply within reticuloendothelial and lymphoid cells. Symptoms include cutaneous lesions within days to months at the site of insect bite.
Diagnosis:	Microscopic demonstration of parasites.
Transmission:	Disease transmits via insects, that is, the sandfly. While feeding, these bugs excrete material onto the host, which contains parasites. These parasites gain entry into the human body through the site of the bite.
Treatment:	Injections of Pentostam and Glucantime provide effective treatment. In addition, sodium stibogluconate may be useful.
Occurrence:	Africa, Middle East, Asia.

6 *Leishmania donovani*

Biology:	Trophozoites are approximately 2–5 μm long, contain a single nucleus, feed on nutrients obtained from the host and reproduce by binary fission but do not form cysts.
Disease:	Kala-azar or visceral leishmaniasis. Following invasion, parasites are taken up by macrophages and multiply, eventually killing the macrophages and severely affecting the host defenses. Symptoms appear within a few weeks to months and include fever, malaise, anemia, enlarged liver and spleen and finally death. Of interest, *L. donovani* taken up by neutrophils are killed but this does not have a major effect on the outcome of the disease.
Diagnosis:	Microscopic demonstration of parasites.
Transmission:	Disease transmits via insects, that is, the sandfly.
Treatment:	Injections of Pentostam and Glucantime provide effective treatment. In addition, sodium stibogluconate may be useful.
Occurrence:	Americas, Africa, Asia, Mediterranean.

F

Protozoan Biology

Naveed Ahmed Khan

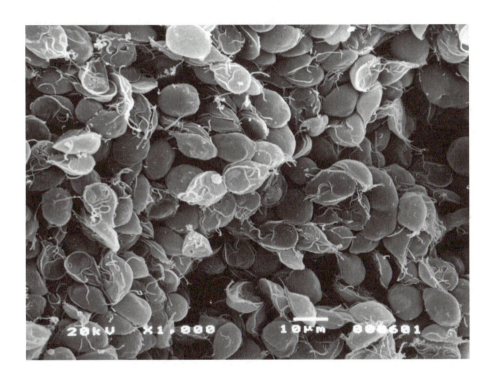

1 Introduction

The term protozoan is derived from 'proto' meaning 'first' and 'zoa' meaning 'animal'. Protozoa are 'first animals' which generally describes their animal-like nutrition. Protozoa are the largest group of single-celled, microscopic organisms with more than 20 000 species that are found in all aspects of life. Protozoa are widely distributed in various environments from favorable rainforests to sandy beaches to the bottoms of oceans to snow-covered mountains. With the availability of improved diagnostic methods, protozoa are being discovered from diverse habitats. However, the abundance and diversity of protozoa in ecosystems is dependent on abiotic factors such as water, temperature, pH, salinity, osmolarity, and biotic factors including the availability of food particles. Protozoa include the causative agents of some of the most notorious and deadly diseases. For example, malaria alone causes between one to two million deaths worldwide, annually. Other protozoa play important roles in the food chain, maintaining a balanced ecosystem, or act as commensal organisms (not harmful) of nearly all humans. Some of the protozoan pathogens have only recently been identified as a major threat to human health. For example, *Cryptosporidium* was originally described in the 19th century, but has recently been associated with serious human infections in AIDS patients. With the increasing number of AIDS patients during the last few decades, many of the mild protozoan pathogens have become a major problem for human health.

2 Cellular Properties

Protozoa are the largest single-cell nonphotosynthetic animals that lack cell walls. The study of protozoa, invisible to the naked eye, was initiated with the discovery of the microscope in the 1600s by Antonio van Leeuwenhoek (1632–1723). The majority of protozoa pathogens are less than 150 μm in size, with the smallest one between 1 and 10 μm. The plasma membrane forms the outer boundary of most protozoa and may possess locomotory organelles, such as pseudopodia, flagella and cilia. Although there is diverse structural dissimilarity, the majority of the intracellular organelles are similar to other eukaryotic cells. The DNA is arranged in chromosomes that are contained in one or more nuclei. Protozoa are mostly aerobic and contain mitochondria to generate energy, although many intestinal protozoa are capable of anaerobic growth. The anaerobes lack recognizable mitochondria but may contain hydrogenosomes, mitosomes or glycosomes instead. The cytoplasm, particularly in free-living protozoa, contains contractile vacuoles. Their function is to regulate osmotic pressure by expelling water. Some protozoa feed by transporting food across the plasma membrane by pinocytosis (engulfing liquids/particles by invagination of the plasma membrane) and/or phagocytosis (engulfing large particles, which may require specific interactions). Some require specialized structures to take up food. For example, ciliates take in food by waving cilia towards a mouth-like opening known as a cyto-some. The feeding and/or growing stage is known as the trophozoite. However, under harsh conditions, some protozoa transform into a protective cyst form. Cysts can survive lack of food, extremes in pH or temperature, and resist toxic chemicals or chemotherapeutic compounds. These properties allow some protozoa to find new hosts, and thus helps in their transmission. Protozoa reproduce asexually by binary fission (parent cell mitotically divides into two daughter cells), multiple fission, also known as schizogony (parent cell divides into several daughter cells), budding, and spore formation, or sexually by conjugation (two cells join, exchange nuclei and produce progeny by budding or fission). Some protozoa produce gametes (gametocytes, i.e., haploid sex cells), which fuse to form a diploid zygote. Protozoa are among the five major classes of pathogens: intracellular parasites (viruses), prokaryotes, fungi, protozoa and multicellular pathogens. From more than 20 000 species of protozoa, only a handful cause human disease. However, these few are a major burden on human health and have a severe economic impact. For example, malaria is the fourth leading cause of death, worldwide. To produce disease, protozoa access their hosts via direct transmission through the oral cavity, the respiratory tract, the genitourinary tract and the skin, or by indirect transmission through insects, rodents, as well as by inanimate objects such as towels, contact lenses and surgical instruments. Once the host tissue is invaded, protozoa multiply to establish themselves in the host, and this may be followed by physical damage to the host tissue or depriving it of nutrients, and/or by the induction of an excessive host immune response resulting in disease.

3 Classification

The development of molecular methods in recent years divided all living organisms into three domains based on ribosomal RNA sequences: (i) the ancient 'Archaea', containing prokaryotes that inhabit the extreme environments, (ii) the Bacteria, prokaryotes that comprise an extensive large group of microorganisms, and (iii) the Eukarya, including all remaining nucleated uni- and multicellular organisms. Within eukaryotes, protists were first to evolve, emerging approximately 1.5 billion years ago (diplomonads, e.g., *Giardia* are among the ancient single-cell eukaryotes that possess a nucleus but lack mitochondria), leading to the emergence of the remaining single-cell organisms such as amoebae and higher animals. Before the availability of molecular tools, taxonomists divided the protozoa into four groups, based on the organisms' mode of locomotion (Figure 1A) as follows:

3.1 Phylum Mastigophora

Protozoa placed in this phylum were characterized by the presence of a flagellum, at least during some phases of their life cycles. These organisms

Figure 1 (A) The old classification scheme of protozoa, based on mode of locomotion. (B) The new classification scheme of protists, based on nucleotide sequencing.

are parasitic or free-living in anoxic environments and lack mitochondria and Golgi apparatus. Examples include *Giardia*, *Trypanosome*, *Leishmania* and *Trichomonas*.

3.2 Phylum Ciliophora

The members of this phylum were ciliated during at least one stage of their life cycle, moving typically by the beating of the cilia. They exhibit both asexual and sexual reproduction, that is, asexual reproduction by budding, binary and multiple fission (schizogony), as well as sexual reproduction by conjugation, autogamy or cytogamy. They are parasitic, commensal, or free-living. Examples include *Balantidium*, *Isotricha* and *Sonderia*.

3.3 Phylum Sarcodina

The majority of protozoa placed in this phylum exhibited movement using characteristic pseudopodia (moving of the protoplasm in a certain direction). They are typically uninucleate and possess mitochondria. Reproduction is by asexual fission and they may be parasitic or free-living. Examples include *Entamoeba, Acanthamoeba, Naegleria, Balamuthia* and *Blastocystis*.

3.4 Phylum Apicomplexa

Organisms in this group were all parasitic and characterized by the presence of an apical complex, located at one end of the organisms. They exhibited both sexual and asexual reproduction. Examples include *Plasmodium, Toxoplasma, Babesia, Isospora* and *Cryptosporidium*.

However, this scheme is based on the organisms' locomotion and did not reflect any genetic relatedness. Based on nucleotide sequencing, protozoa are now classified into the following taxa (Figure 1B):

3.5 Parabasala

Protozoa placed in this group lack mitochondria, contain a single nucleus and a parabasal body, which is a Golgi-body-like structure. Examples include *Trychonympha* and *Trichomonas*.

3.6 Cercozoa

Cercozoa is a group of amoebae with thread-like pseudopodia. These include foraminifera containing a porous shell, composed of calcium carbonate. Pseudopodia extend through holes in the shell. These may be microscopic or several centimeters in diameter. Commonly, foraminifera live on the ocean floor.

3.7 Radiolaria

Radiolaria is a group of amoebae that also have thread-like pseudopodia. The organisms have ornate shells composed of silica and live in marine waters as part of plankton. The pseudopodia of radiolarians radiate from the central body like spokes of a spherical wheel.

3.8 Amoebozoa

This group presents the third taxon of amoebae that can be distinguished from the other two taxa by having lobe-shaped pseudopodia and no shells. Examples include *Acanthamoeba, Balamuthia, Naegleria* and *Entamoeba*. In addition, slime molds are also included in this group (previously thought to be fungi), based on lobe-shaped pseudopodia, no cell wall (cell wall is present in fungi), and feeding. Slime molds can be further divided into cellular slime molds such as *Dictyostelium* and acellular slime molds (also known as Plasmodial slime molds) that are characterized by filaments of cytoplasm that creep as amoebae and may contain millions of nuclei.

Table 1 Representative protozoa pathogens of human importance

Taxon		Human Pathogens	Disease	Source of Infection
Parabasala		*Trichomonas vaginalis*	Urethritis, vaginitis	Contact with vaginal-urethral discharge
Alveolata	Ciliates	*Balantidium coli*	Dysentery	Fecal contamination of drinking water
	Apicomplexan	*Plasmodium*	Malaria	Mosquito bite
		Cryptosporidium	Diarrhea	Humans
		Toxoplasma	Toxoplasmosis	Cats, beef, congenital
		Babesia	Babesiosis	Domestic animals, ticks
		Isospora	Coccidiosis	Domestic animals
Amoebozoa		*Acanthamoeba*	Keratitis, encephalitis	Soil/water
		Balamuthia	Encephalitis	Soil/water
		Naegleria	Encephalitis	Water
		Entamoeba	Dysentery	Fecal contamination of drinking water
Diplomonadida		*Giardia*	Giardial enteritis	Fecal contamination of drinking water
Euglenozoa		*Trypanosoma cruzi*	Chagas' disease	Triatoma (kissing bug) bite
		Trypanosoma brucei	African trypanosomiasis	Bite of tsetse fly
		Leishmania	Leishmaniasis	Bite of sand fly
Stramenopila		*Blastocystis hominis*	Diarrhea	Contaminated food/water

3.9 Alveolata

Protozoa in this taxon contain small membrane-bound cavities known as alveoli beneath their cell surface, although the function of these structures remains unclear. This group is further divided into three sub-groups: (i) ciliates such as *Balantidium coli*, (ii) Apicomplexans such as *Plasmodium*, *Cryptosporidium parvum*, *Toxoplasma gondii*, *Babesia microti* and *Isospora belli*, and (iii) dinoflagellates, which are phototrophic, such as *Gymnodinium* and *Gonyaulax*.

3.10 Diplomonadida

Protozoa in this taxon lack mitochondria, Golgi bodies and peroxisomes. The organisms have two equal-sized nuclei and multiple flagella, for example, *Giardia*.

3.11 Euglenozoa

This group is further subdivided into two groups. (i) The Euglenids are photoautotrophic unicellular microbes with chloroplasts containing pigments (historically thought to be plants). However, they possess flagella, lack cell wall and are chemoheterotrophic phagocytes (in the dark), for example, *Euglena*. (ii) The Kinetoplastids have a single large mitochondrion that contains a unique region of mitochondrial DNA, called a kinetoplast. Kinetoplastids live inside animals and some are pathogenic, for example, *Trypanosoma* and *Leishmania*.

3.12 Stramenopila

This group is a complex assemblage of 'botanical' protists with both heterotrophic and photosynthetic representatives. The evolutionary history of this group is unclear. Generally, the organisms included in this group are slime nets, water molds and brown algae, and are characterized by possessing flagella. Recent molecular phylogenetic studies revealed that *Blastocystis* belongs to this group.

It is important to indicate that no single classification scheme has gained universal support and future studies will almost certainly dictate changes in the above scheme. The representative protozoa pathogens that are covered in this book are indicated in Table 1.

4 Locomotion

Motility in protozoa is usually mediated by cilia, flagella, or cellular appendages adapted for propulsion, or amoeboid movement. Other modes of protozoa locomotion involve gliding movements in which no changes in body shape are observed. The various protozoa have evolved to exhibit distinct movements depending on where they normally live. For example, protozoans with amoeboid movements using pseudopodia are normally present in environments with abundant organic matter or in flowing water with plant life. Cilia or flagella are used to travel longer distances *per se* that maximizes the possibility of encountering food particles.

4.1 Pseudopodia

Pseudopodia are not permanent structures present on the surface of organisms but are formed upon a stimulus. Pseudopodia are observed in amoeboid movement. They are characterized as a flow of cytoplasm in a particular direction. The cytoplasmic membrane temporarily attaches to the substratum and cytoplasm is drawn into the new attachment protruding as a foot-like structure, hence it is called pseudopodia. These extensions may exhibit the following distinct phenotypes: (i) broad, round-tipped pseudopodia are known as lobopodia; (ii) extensively branched, forming a net-like structure are rhizopodia; (iii) sharp, pointed projections are filopodia; and (iv) axopodia are similar to filopodia but contain slender filaments. Pseudopodia play important roles in locomotion as well as in food uptake. Examples include *Acanthamoeba* spp. and *Entamoeba histolytica*.

4.2 Cilia and flagella

In contrast to pseudopodia, flagella and cilia are permanent microtubular organelles that are anchored within the plasma membrane of certain protozoa, and project from the cell surface. Flagella are long slender structures (50–200 μm), usually one to a few on a cell, with whip-like movements starting at the tip or the base of the cell. This results in forward, backward or spiral movements. Cilia are similar to flagella but smaller in size (5–20 μm) and usually more numerous. Cilia move with a back-and-forth stroke. Their thickness is approximately 0.2 μm. Both cilia and flagella exhibit bending motion resulting in fluid propulsion and cellular movements. However, due to the extended length of the flagellum, bends are propagated along the flagellum pushing the surrounding water symmetrically on both sides of the cell. In contrast, with the shorter length and large numbers, cilia beat in co-ordination with one another.

4.3 Gliding movements

Another form of locomotion in protozoa is gliding. For example, several flagellates glide over surfaces. In such cases, flagella do not exhibit whip-like movement but make contact with the surface and slide over it with the aid of microtubules. Other examples include the sporozoite stage during the life cycle of *Plasmodium* spp.

4.4 Locomotory proteins

The major proteins involved in locomotion are: (i) microtubules, cylindrical fibrils formed of tubulin molecules around 25 nm in diameter, and (ii) microfilaments, also known as actin filaments that are composed of actin molecules and are about 7nm thick. The polymerization and depolymerization of the tubulin molecules in microtubules and their ability to interact with dynein adenosine triphosphatase proteins cause movement in protozoa. In contrast, actin is polymerized and forms microfilaments. These microfilaments are pushed along by interacting with myosin ATPase molecules. Other proteins important in protozoa locomotion are myonemes of centrin, which form filaments of about 10μm thickness. Depending on calcium concentration, centrin exist in two states: filaments shorten in the presence of calcium (centrin binds to calcium) and extend when calcium is withdrawn. These contractions result in cellular movements, for example, *Vorticella* and *Stentor*.

5 Feeding

The Protozoa consists of photosynthetic organisms (autotrophs that synthesize their own food), nonphotosynthetic organisms (heterotrophs that obtain food such as organic molecules or particulate matter from the environment) and organisms that use both modes, called mixotrophs. Photosynthetic protozoa usually occur among the flagellates. Some protozoa, that is, certain ciliates, temporarily acquire chloroplasts from ingested algae and use them to synthesize food. In other cases, protozoa maintain a symbiotic relationship with another organism for their nutrition needs. The partner in the symbiotic relationship may involve a virus, a prokaryote, a protist or a multicellular eukaryote. The resultant outcome of this relationship may be commensalism, that is, a condition in which one organism benefits. Usually the smaller organism feeds on the food that is unusable or unwanted by the host. For example, *Entamoeba coli* feeds on bacteria in the large intestine of humans. Other relationships may be mutualisms, in which both partners are nutritionally dependent on each other. For example, termites are unable to digest cellulose and thus cannot live without the cellulolytic bacterial flagellates in their digestive tract, which obtain nutritional benefits in return. And lastly parasitism, a condition in which one organism lives at the expense of the other by obtaining essential nutrients. It is widely accepted that the 'true parasites' do not intend to kill their host and maintain a balanced relation with the host. However, the overwhelming hosts' immune response and/or the burden of the parasite could result in disease and even host death.

Bacteria form the major food source for the majority of protozoa including ciliates, flagellates, and amoebae. The mode of nutrient uptake is dependent on their nature. For example, particles are taken up by phagocytosis. Phagocytosis is a process in which particles are bound to the cell surface or captured by pseudopodia, followed by invagination of the plasma membrane, thus engulfing the food particle in an intracellular vacuole. These vacuoles are fused with lysosomes containing digestive enzymes which degrade the food particles. In contrast, fluid or dissolved nutrients are taken up by pinocytosis or transported across the plasma membrane by diffusion or active transport. These processes are facilitated by cytoskeletal rearrangements involving microtubules and actin filaments. Some protozoa, for example, *Amoeba proteus* or *Balamuthia*, are capable of attacking and ingesting other protozoa.

5.1 Metabolism

The nonphotosynthetic (heterotrophic) protozoa engulf smaller organisms or organic matter and digest them to generate energy to perform their routine functions. Alternatively, protozoa use waste organic particles produced by the metabolism of other organisms. Protozoa use inorganic solutes including potassium, chloride, essential metals, nitrates, ammonium and amino acids, inorganic phosphates as well as organic phosphate compounds, ethanol, other short-chain alcohols, and organic acids such as acetate, pyruvate and lactate as carbon sources for their cellular metabolism. However, the photosynthetic (autotrophic) protozoa can metabolize organic particles or use inorganic solutes to synthesize their food. Many catabolic (pathways involved in degradation) and anabolic (synthetic) pathways are similar to those found in other eukaryotes.

Glycolysis

The majority of protozoa metabolize carbohydrates as energy sources and exhibit minimal catabolism of fatty acids and amino acids. They maintain intracellular stores of carbohydrates in the form of glycogen (or trehalose). Usually the bloodstream parasites such as *Plasmodium* and *Trypanosoma* do not store polysaccharides as they have ready access to glucose. Glucose or glycogen is broken down by glycolysis to give pyruvate. In *Trypanosoma* most of the glycolytic pathway is sequestered in an organelle, the glycosome, and its link with a mitochondrion is shown in Figure 2. This adaptation allows glycolysis to function optimally. In some protozoa that are unable to utilize glucose, organic acids or short-chain alcohols are used as alternative carbon sources to glucose.

Mitochondrial metabolism

In aerobic protozoa, mitochondria have an electron transport system exhibiting mammalian-like respiration sensitive to cyanide and having a cytochrome aa_3 as terminal oxidase. The glycolytic products enter the mitochondria and generate adenosine triphosphate (ATP) via the oxygen dependent cycle, similar to the mammalian electron transport system. Usually protozoa do not fully degrade the end products of metabolism to carbon dioxide and water as observed in the mammalian system. These differences may have potential implications in the identification of novel pathways to inhibit parasite energy metabolism, which may prove fatal for parasites and offer targets for therapy.

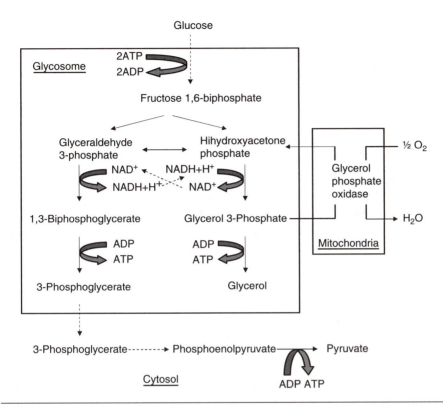

Figure 2 Aerobic glycolysis in bloodstream forms of *Trypanosoma*, showing compartmentalization of much of the pathway in the glycosome. *Dotted lines* indicate several reaction steps.

Hydrogenosomal metabolism

Protozoa that inhabit oxygen-poor environments (anaerobes), such as *Trichomonas*, do not contain mitochondria but possess organelles called hydrogenosomes. In such organisms, pyruvate or malate formed by glycolytic pathways enter the hydrogenosomes and are converted by fermentation reactions into a range of incompletely oxidized compounds such as acetate with the production of ATP. In addition, during this process, molecular hydrogen is generated by the action of an anaerobic electron-transport pathway (Figure 3).

Nitrogen metabolism

Free-living protozoa can utilize inorganic as well as organic sources of nitrogen when available. Of the inorganic sources, ammonium is preferentially used. However, many species have nitrate reductase (converts nitrate to nitrite) and nitrite reductase (converts nitrite to ammonium) suggesting that these organisms can use many available sources of nitrogen. Most parasitic protozoa and those involved in symbiotic relationships will exclusively use amino acids as sources of nitrogen. Parasites secrete

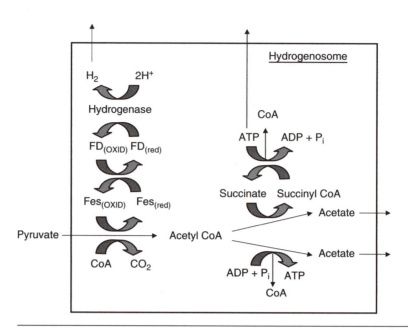

Figure 3 Hydrogenosomal metabolism in the Trichomonads. Fes, iron-sulfur proteins; FD, ferridoxins; OXID, oxidized state; red, reduced state.

protease enzymes that degrade proteins releasing small peptides and amino acids. Amino acids may also be end products of metabolism and may be excreted; storage and excretion of amino acids may play a role in osmoregulation in protozoans lacking contractile vacuoles.

Overall, protozoa employ diverse mechanisms for adaptation of their energy metabolism. A complete understanding of the biochemical pathways associated with protozoa energy metabolism is crucial for the rationale of development of antiparasitic drugs.

6 Reproduction

To ensure survival, a species must reproduce in large numbers when conditions are favorable. The reproduction in protozoa can be sexual, that is, formation and fusion of gametes producing new offspring, or asexual, that is, mitotic division of a parent cell into two or more identical offspring. Many protozoa only use asexual, while others use both during their life cycles.

6.1 Asexual reproduction

Asexual reproduction is the primary mode of reproduction in many protozoa. Asexual reproduction occurs by binary fission, multiple fission (schizogony) and budding. The process of cell division is divided into an S phase (DNA synthesis phase) and an M phase (division of nucleus that involves DNA condensation and organized distribution of chromosomes). Both phases are separated by gap phases, that is, G1 and G2. In addition, cytoplasmic organelles are duplicated, followed by cytokinesis. The result-ing daughter cells are identical to their parent cell and are produced in sufficient numbers for successful transmission. Asexual reproduction occurs in the majority of amoebae, ciliates, and flagellates.

Binary fission

In this form, a single cell divides into two daughter cells and it is the most common type of asexual reproduction, lasting from 6 to 24 h. This results in very large numbers of identical parasite populations within days. The DNA and cytoplasmic organelles are duplicated followed by nuclear division and cytokinesis, and finally a constriction ring bisects the cell producing two daughter cells. Examples include *Toxoplasma* and *Acanthamoeba.*

Multiple fission (schizogony)

In this form of reproduction, mitotic nuclear division occurs several times. This results in several nuclei within the cytoplasm. Each nucleus together with a layer of cytoplasm gives rise to independent daughter cells that are released. In parasitic protozoa, it results in the rapid production of large protozoan populations, which overwhelms the host immune system. This occurs in various protozoa including *Pelomyxa palustris, Volvox* and *Gonium.*

Budding

In this form of reproduction, nuclei divide and the daughter nuclei migrate into a cytoplasmic bud. This is followed by cytoplasmic fission and release of the cell, which develops into a mature reproductive organism. The resulting daughter cells may differ from the parent cells. Such reproductive schemes are limited to several protozoa such as *Trichophrya* and *Ephelota.*

6.2 Sexual reproduction

Many parasitic protozoa have complex life cycles and reproduce both sex-ually and asexually. Sexual reproduction involves formation of haploid gametes by meiosis that fuse to form a diploid zygote generating a new organism.

Gamete formation

Gametes with haploid chromosomes are formed in the vegetative stage, which fuse with one another to produce diploid zygotes. The zygote under-goes meiosis, where numbers of chromosomes are halved, followed by asexual reproduction to produce large populations of the organisms. Gamete formation directly from the original population followed by their fusion into a zygote is called 'hologamy'. Examples include *Chlamydomonas, Dunaliella* and *Polytoma.* If gametes are identical (at least morphologically, but there may be minor genetic or physiological differences), the term 'isogamy' is used. Examples include *Chlorogonium*, as well as many foraminiferan sarcodinids and some sporozoa. When there are clear differences between male and female gametes, typically known as micro- and macrogametes (size differences, presence of flagella, physiological or biochemical properties), the gamete formation and their fusion into a zygote is called 'anisogamy'. This is a common form of gamete formation in protozoa. Examples include *Plasmodium* spp. The factors that induce gamete formation are not clearly known but may involve environmental conditions such as salinity, pH, temperatures, nutrients, and so on.

If both gametes arise from the same clone, the species is called monoecious but if they arise from different clones, the species is called dioecious. However, if self-fertilization occurs, the species is called hermaphrodite. If the new population of an organism develops from unfused gametes without fertilization, it is known as parthenogenesis. Examples of this include the genus *Volvox* and *Eucoccidium dinophili*.

Gametic nuclei: conjugation

Some ciliated protozoa are unable to produce gametes. Instead they possess dual nuclei and, during their sexual reproduction, the nuclei from two organisms (instead of gametes) fuse together yielding a zygotic nucleus. This is followed by asexual fissions producing large populations. Again, diverse factors are responsible for this process, termed conjugation, including temperature, salinity, pH, and so on. Examples include *Paramecium* and *Tetrahymena*.

7 Life Cycle

The protozoa that are major human pathogens can be broadly subdivided into three main categories: those which are blood-borne that are usually transmitted via insects (e.g., *Plasmodium*), those which are transmitted via contaminated food/water (e.g., *Giardia*) and those which are sexually transmitted (e.g., *Trichomonas*). The life cycles of blood-borne protozoa generally involve humans and insects. The life cycle in insects may be necessary for parasite development or the insects may merely act as vectors to transmit it to a new host. In contrast, other protozoa pathogens are acquired by a new host through exposure to contaminated water and/or food, or through sexual intercourse.

7.1 Plasmodium *spp.*

Although the precise life cycle of *Plasmodium* varies between species, it can be divided broadly into two hosts as follows:

Life cycle in the vertebrate host

When an infected mosquito takes a blood meal, it injects saliva containing sporozoites (approx. 10–15 μm long and 1 μm in diameter). The sporozoites penetrate hepatocytes (liver cells) and undergo asexual reproduction, a process known as the pre-erythrocytic (PE) cycle or exoerythrocytic cycle (EE). Within the hepatic cell, the parasite transforms into a trophozoite stage that feeds on the host-cell cytoplasm. Within a few days, the trophozoite matures and produces daughter nuclei and at this stage is called a schizont. A single schizont undergoes cytokinesis and produces many daughter cells called merozoites (2.5 μm long and 1.5 μm in diameter). The merozoites are released and infect new hepatocytes or enter the erythrocytic cycle. Upon entry into an erythrocyte, a merozoite again transforms into a trophozoite and feeds on the host-cell cytoplasm, forming a large food vacuole giving the characteristic ring appearance. The trophozoite becomes a schizont again and produces many merozoites which infect new erythrocytes. After several generations, some merozoites infect erythrocytes and become macrogametocytes and microgametocytes. These are taken up by the mosquito, where the remaining life cycle continues (Figure 4).

Life cycle in the invertebrate host

The gametocytes are taken up by the mosquito during their blood meal. If gametocytes are taken up by a susceptible mosquito (*Anopheles* spp. in the

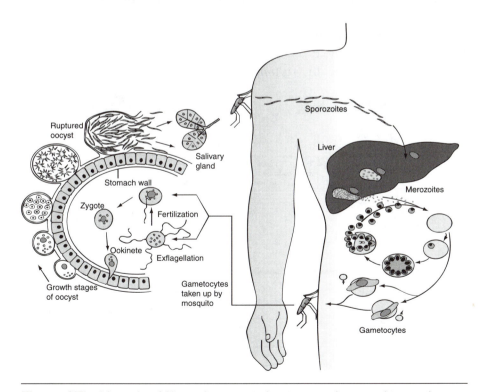

Figure 4 The life cycle of *Plasmodium* spp. indicating vertebrate and invertebrate hosts.

case of human *Plasmodium*), gametocytes develop into gametes, that is, micro- and macrogametes. The nucleus of a microgametocyte divides to produce 6–8 nuclei and exflagellates. The microgamete fuses with a macrogamete to form a diploid zygote. The zygote elongates to become a motile ookinete that is 10–15 μm. The ookinete penetrates the gut wall of the mosquito and develops into an oocyst. The oocyst undergoes meiosis and produces sporozoites, which penetrate salivary glands and are injected into a new host at the next blood meal, thus completing the cycle (Figure 4).

7.2 Trypanosoma brucei

As for *Plasmodium*, the life cycle of *Trypanosoma brucei* can be divided broadly into two hosts as follows:

Life cycle in the vertebrate host

When an infected tsetse fly (*Glossina* spp.) takes a blood meal, it injects trypomastigotes into the host. The trypomastigotes then travel through the lymphatic and circulatory systems to other sites, where they reproduce by binary fission. Eventually, some trypomastigotes enter the central nervous system and the cerebrospinal fluid, while others continue to circulate to be picked up by the tsetse fly, where the remaining life cycle continues (Figure 5).

Life cycle in the invertebrate host

The trypomastigotes are taken up by the tsetse fly during their blood meal. In the mid-gut of the fly, trypomastigotes multiply by binary fission,

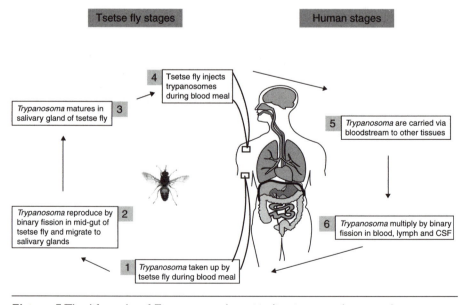

Figure 5 The life cycle of *Trypanosoma brucei* indicating vertebrate and invertebrate hosts.

producing immature epimastigotes that migrate to the salivary glands. Here, they mature and become trypomastigotes and are injected into a new host at the next blood meal, thus completing the cycle (Figure 5).

7.3 Trypanosoma cruzi
Life cycle in the vertebrate host

Trypomastigotes are shed in the feces of the kissing bug (*Triatoma* spp.), while the bug feeds on the mammalian host. When the host scratches the itchy wound, trypomastigotes enter the wound and then travel throughout the body and penetrate certain cells, especially macrophages and heart muscle cells, where they transform into nonflagellated amastigotes. The amastigotes divide by binary fission and eventually rupture the host cells. The released amastigotes either infect new cells or transform into trypo-mastigotes that circulate in the bloodstream to be picked up by the kissing bug, where the remaining life cycle continues (Figure 6).

Life cycle in the invertebrate host

The circulating trypomastigotes are taken up by the kissing bug during their blood meal. In the mid-gut of the bug, trypomastigotes multiply by binary fission, producing immature epimastigotes that migrate to the hindgut of the kissing bug. Here, they mature and become trypomastigotes and are injected into a new host at the next blood meal, thus completing the cycle (Figure 6).

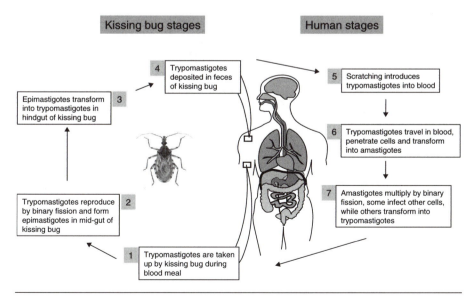

Figure 6 The life cycle of *Trypanosoma cruzi* indicating vertebrate and invertebrate hosts.

7.4 Leishmania *spp.*

Life cycle in the vertebrate host

When an infected sandfly (*Phlebotomus* spp. or *Lutzomyia* spp.) takes a blood meal, it injects promastigotes into the host. Macrophages near the bite site phagocytize the promastigotes, which then transform into amastigotes. The amastigotes reproduce by binary fission until the macrophage ruptures. The released amastigotes infect other macrophages, as well as circulating in the bloodstream to be picked up by the sandfly, where the remaining life cycle continues (Figure 7).

Life cycle in the invertebrate host

During a blood meal, the sandfly takes up phagocytes containing amastigotes. In the mid-gut of the fly, amastigotes are released from the phagocytes and transform into promastigotes. Promastigotes rapidly divide by binary fission, filling the fly's digestive tract and migrate to the proboscis, and are injected into a new host at the next blood meal, thus completing the cycle (Figure 7).

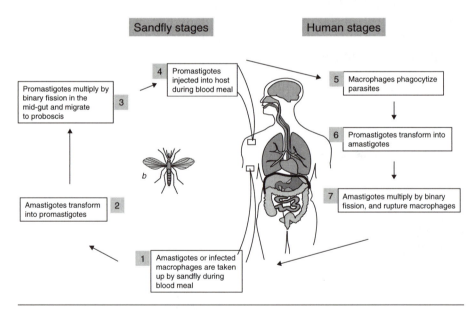

Figure 7 The life cycle of *Leishmania* spp. indicating vertebrate and invertebrate hosts.

G Host Response

Naveed Ahmed Khan

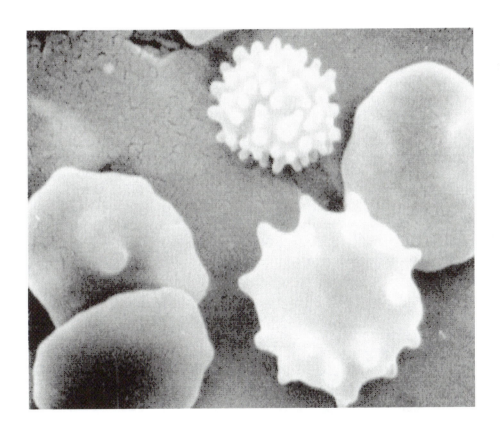

I Human Defense Mechanisms

Once protozoan pathogens access into the human body, they encounter highly professional defense systems. Traditionally, the defense system has been divided into two components:

1. Nonspecific/constitutive/innate immune response that includes: skin that acts as a physical barrier, neutrophils that police the bloodstream and attack any foreign invaders, complement and cytokines that direct the activities of neutrophils, and natural killer cells, which kill infected cells.
2. Specific/inducible/acquired immune response that includes: antibodies (produced by B-lymphocytes) and T-lymphocytes.

These defense systems do not operate independently but communicate with each other to build an effective defense against the invading organism (Figure 1).

1.1 Nonspecific immune responses

Defense of surfaces

Skin is a dry, thick layer of dead, keratinized cells and provides a physical barrier to pathogens. Natural openings in the skin such as pores, hair follicles, and sweat glands are protected by secretion of toxic chemicals such as antimicrobial peptides called dermicidins, lysozyme (an enzyme that digests cell walls of pathogens), and sebum, which contains fatty acids that lower the pH of the skin's surface to about pH 5, which is inhibitory to many microbes. Generally speaking, pathogens can only breach this barrier through wounds. However, underneath the skin, is skin-associated

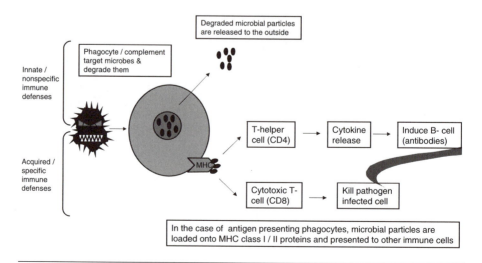

Figure I A simplified view of the host immune response to microbes.

lymphoid tissue (SALT). In SALT, the resident Langerhans cells (similar to macrophages) attack pathogens and initiate specific immune defenses.

Inside the body, intestinal tract, respiratory tract, vaginal tract and bladder are covered in mucosal membranes. Cells in these membranes are protected by a thick layer of mucus/mucin. The mucus (an extremely sticky substance secreted by goblet cells) consists of proteins and polysaccharides that trap pathogens and prevent them from reaching the epithelial cells. Mucin is constantly shed and replaced, together with any trapped microbes. In addition, low pH (~2) in the stomach, cilia movement over trachea, flushing of urinary tract and competition from the normal flora prevents colonization of the tissue by pathogenic protozoa. Similar to the skin, mucosal membranes are associated with the specific immune defense by mucosa-associated lymphoid tissue (MALT). For intestinal tract, it is called gastrointestinal-associated lymphoid tissue (GALT). In GALT, the resident macrophages (similar function to Langerhans cells of SALT) initiate specific defenses.

In addition to the above, the skin and mucus membranes of the body are home to a variety of virus, bacteria, protozoa and fungi. This normal flora plays an important role in protecting the body by competing with potential pathogens, a situation known as microbial antagonism. The normal flora achieves this by: (i) consuming the available nutrients, making them unavailable to pathogens; (ii) stimulating the body's defenses; and (iii) occupying binding sites required by pathogens.

The violation of initial defenses may lead to inflammation. Inflammation is the tissue reaction to infections or injury characterized by redness (increased blood flow), swelling (increased extravascular fluid and phagocyte infiltration), heat (increased blood flow and pyrogens, fever-inducing agents) and pain (local tissue destruction and irritation of sensory nerve). Inflammation leads to recruitment of neutrophils or eosinophils to the infection site, which leads to ingestion and destruction of pathogens (Figure 2). Neutrophils are also known as polymorphonuclear leukocytes (PMNs) indicating that the nucleus is a multilobed structure that may appear to be multinuclei, when viewed in cross-section. Both neutrophils and eosinophils are primarily found in the bloodstream, but can exit the blood to attack invading microbes in the tissues by squeezing between cells lining capillaries. Basophils can also leave the blood but are nonphagocytic; instead, they release inflammatory chemicals. The steps of ingestion and killing by phagocytic cells involve the following:

1. Attachment of the pathogen – formation of pseudopodia.
2. Uptake of the pathogen in a vacuole – 'phagosome'.
3. Fusion of lysosome (containing chemicals and enzymes) with phagosome. The lysosomes contain myeloperoxidase enzyme that is normally inactive but following fusion with a phagosome this enzyme becomes

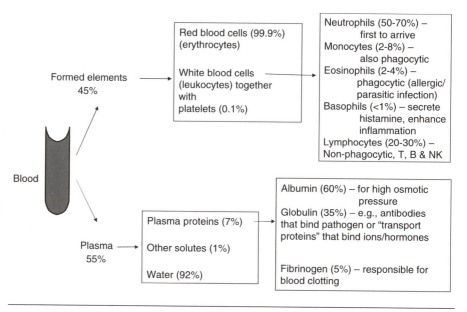

Figure 2 Components of blood.

activated and produces superoxide and hypochlorite. Superoxide is a toxic form of oxygen that can oxidize and inactivate proteins and other molecules on the microbial surface. When mixed with chlorine, it forms hypochlorus acid, highly toxic to invading organisms. In addition, toxic nitric oxide (NO) is produced in response to invading microbes.

4. Overall result is killing and digestion of the ingested pathogen.
5. Release of digested products to the outside.

Other phagocytic cells are initially monocytes, which leave the blood and mature into phagocytic macrophages. Macrophages are named for their location. For example, wandering macrophages leave the blood by squeezing through cells lining the capillaries and target microbes, as for neutrophils/eosinophils. Other macrophages do not wander. These include alveolar macrophages of the lungs, microglia of the central nervous system and Kupffer cells of the liver. Generally, fixed macrophages phagocytoze microbes within a specific organ. Another group of phagocytes are dendritic cells, which are multibranched and are scattered throughout the body, particularly skin and mucous membranes, where they await microbial invaders. In addition, plasma (acellular part of blood) also contains: (i) lactoferrin, iron binding protein that deprives pathogens of iron; (ii) lysozyme, an enzyme that digests cell walls of pathogens; and (iii) defensins, small proteins that interfere with pathogen intracellular signaling pathways and metabolism, and may produce holes in the pathogen membranes.

Complement

In addition to neutrophils, serum contains a potent complement system. Complement is a set of proteins produced by the liver that circulate in blood and are activated with proteolytic cleavage, a process called complement activation. They recognize invading pathogens and directly cause damage to the pathogen cell through a process known as membrane attack complex (MAC) or help phagocytes (macrophages and neutrophils) to eliminate the pathogen from the body (Figure 3). There are

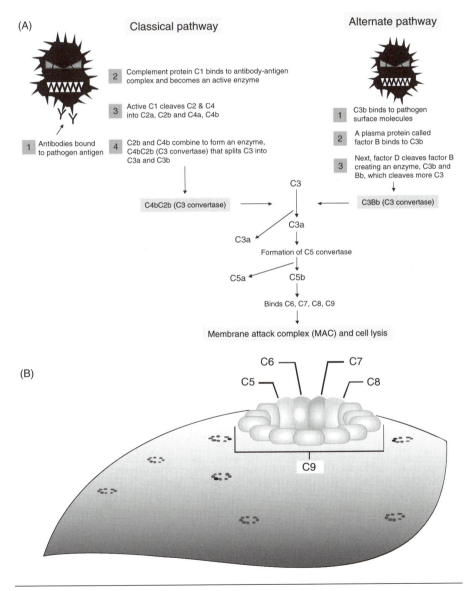

Figure 3 (A) Two pathways by which complement is activated. (B) Membrane attack complex (MAC).

nine complement proteins, C1–C9. There are two pathways of complement activation:

1. the classical pathway which is initiated by antibody-antigen complex, and
2. the alternative pathway which is triggered by microbial surface molecules such as cell membranes.

The activation causes complement proteins to be proteolytically cleaved. The cleaved products are indicated by small fragment 'a' or large fragment 'b'. In the classical pathway, antibodies specifically bind to microbial antigens. A complement protein, C1, binds to the antibody-antigen complex and becomes an active enzyme. Active C1 splits several molecules of C2 and C4. The fragments of C2 and C4 combine to form an enzyme, C4bC2b (C3 convertase), that splits C3 into C3a and C3b. C3b combines with the remaining fragments of C2 and C4 to form an enzyme that cleaves C5 into C5a and C5b. C5b combines with C6, C7, C8 and several molecules of C9 to form a membrane attack complex (MAC). A MAC drills a circular hole in the pathogen's membrane leading to hypotonic lysis of the cell. In contrast, the alternate pathway occurs independently of antibodies. It begins with cleavage of C3 into C3a and C3b. This occurs naturally in the plasma but at a slow rate and C3b is further degraded. However, in the presence of a microbe, C3b binds to the microbial surface, to a protein called factor B. Another plasma protein, factor D, cleaves factor B creating an enzyme composed of C3b and Bb, which continues with the complement pathway and MAC. The ultimate effect of complement is damage of the microbe by creating pores (i.e., MAC) or their opsonization (pathogen uptake by phagocytes).

Cytokines

Cytokines are produced primarily by endothelial cells (which form blood vessels) and phagocytes (such as macrophages and neutrophils), when they are attacked by pathogens. This leads to the release of cytokines. The function of cytokines is like smoke-signals, as they attract phagocytes to the site of infection and link nonspecific and specific immune systems. Figure 4 provides a summary of innate/nonspecific immune responses.

1.2 Specific immune response

This system involves cells that respond to the invading pathogen in a specific fashion. These include cytotoxic T-cells (CD8 cells), T helper cells (CD4 cells), B cells (produce antibodies) and active macrophages (Table 1). Macrophages are phagocytes similar to neutrophils that are mostly found in the tissues. Macrophages are first produced by stem cells as monocytes, which circulate in the bloodstream. On attraction to a site of action or entry

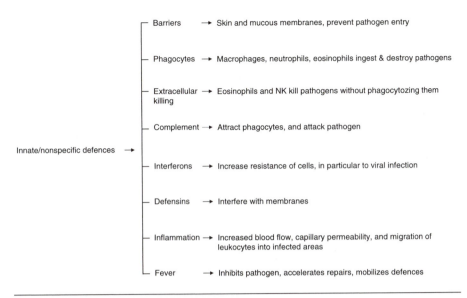

Figure 4 Summary of components of the innate/nonspecific defences.

into a tissue they differentiate into macrophages. The invading pathogens are taken up by macrophages and killed (as with neutrophils) but instead of releasing the digested products to the outside, some microbial products are loaded onto a complex of proteins called the major histocompatibility

Table 1 Cells involved in immune defence against protozoa

Eosinophils and Neutrophils (PMN)	Phagocytic Cells
Basophils	Nonphagocytic cells, that release inflammatory chemicals
Monocyte	Immature macrophage – limited killing but produce cytokine
Macrophage	Phagocytic cell that kills, and present antigens to other immune cells
Dendritic cell	Phagocytic cell, active in presenting antigens to other immune cell
NK cell	Kills cells infected with cytoplasmic pathogens
B cells	When activated, produce antibodies
CD8 T cells bacteria	Cytotoxic T cells, kills cells containing cytosolic
CD4 T cells	Helpes cell, for B cells and activates macrophages
Fibroblasts	Act in repair after infection
Platelets	Produce clotting, which acts as barrier to pathogen dissemination and blood loss

complex class II proteins (MHC II) and presented on the cell surface. Hence, these cells are called antigen-presenting cells (APCs). Why certain microbial products are selected to be expressed on APCs (macrophages, dendritic, B cells) and not other microbial products is not known. Of interest, macrophages in the brain are called microglial cells, while macrophages in the liver are called Kupffer cells. The B cells also act like APCs in taking up and processing foreign antigens and loading them onto the MHC II complex. Finally, the antigen-MHC II complex is recognized by T helper cells.

1. In the case of T helper-macrophage interactions, T helper cells become activated and produce cytokines, such as interferon-gamma (IFN-γ) and interleukin-2 (IL-2), which activate macrophages and stimulate B cells to produce antibodies.
2. In the case of T helper-B cell interactions, T helper cells become activated as above, which in turn stimulates B cells to differentiate into antibody-producing cells, that is, plasma cells.

This approach is used in response to extracellular microbial pathogens. In contrast, if the microbial pathogen is intracellular, their products are loaded onto the MHC I complex. The MHC I displays proteins on almost all cells. Finally, the antigen-MHC I complex is recognized by cytotoxic T-cells. Cytotoxic T-cells kill the host cells displaying antigen-MHC I complex, thus killing any intracellular microbe. This is particularly useful against viral infections.

Antibodies
Antibodies are proteins called immunoglobulins (Igs). They are made of two heavy and two light chains, which create two functionally important areas, as shown in Figure 5. The region that recognizes and binds antigen is called the Fab region, while the Fc region binds to phagocytes (phagocytes have an Fc receptor). The result of an antibody binding to a specific antigen is:

- Enhanced phagocytosis by providing a binding site (Fc) for phagocytes
- Microbes trapped in mucin
- Complement activation
- Neutralization of toxins
- Inactivation of proteins
- Inhibition of binding of toxins/microbes to the host cells.

This ultimately results in the neutralization and removal of microbes and/or their toxins from the host. The antibodies present in the serum are IgG (70–75% of total antibodies) and IgM (10% of total antibodies), while

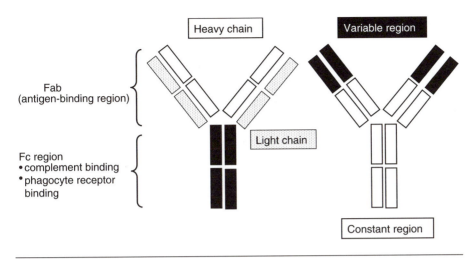

Figure 5 Structure of an antibody.

IgA (15% of total antibodies) are present in secretions such as milk, tears, saliva and mucosal secretions, and are called secretory IgA (sIgA). In addition, IgE and IgD are present in the serum and are involved in allergic reactions.

2 Parasite Strategies of Immune Evasion

Upon invasion of the human body, some pathogenic protozoa are recognized as foreign and have to cope with the highly professional host immune system. The involvement of the immune system in protozoan infections is further indicated with the finding that the majority of parasitic infections are limited to babies/children or individuals with less developed (impaired) immune systems and these infections may be self-limiting in adults with a fully developed immune system. To this end, parasite-specific antibodies in the presence of phagocytes provide a highly effective defense system. For example, the role of antibodies in malaria is demonstrated with the finding that passive transfer of antibodies from immune individuals to those suffering from acute malaria results in the reduction of parasitemia. In addition, B cell deficiency severely impairs the host's ability to clear malaria. Antibodies specifically bind to antigens on the microbial surface with their Fab region, and bind to phagocytes (neutrophils/macrophages) using their Fc region. Thus, phagocytes in the presence of specific antibodies can effectively clear parasites from the human body.

Therefore, parasites must develop mechanisms to evade the host immune system. This is essential for some parasites to buy sufficient time to complete development (reproduction) and in particular for parasites that cause chronic infections, lasting months to years. There are several strategies used by pathogenic protozoa to overcome the host's immune response, as described below (Figure 6).

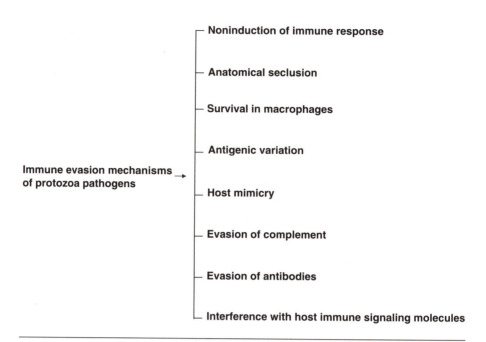

Figure 6 Mechanisms of immune evasion in protozoa pathogens.

2.1 Noninduction of immune response

The pathogens may target organs and/or tissues that have a limited immune response. For example, *Acanthamoeba* produces infection of avascular cornea and *Naegleria* produces infection of the brain tissues, both of which have limited immune responses.

2.2 Anatomical seclusion

The parasites can invade nonimmune cells, replicate intracellularly and hide from immune cells, thus avoiding an overwhelming immune response. For example, some stages of *Plasmodium* spp. live inside liver cells, while other stages live inside erythrocytes, which do not have a nucleus and thus are not recognized by cytotoxic T-cells or natural killer cells. Other benefits of anatomical seclusion are to avoid being exposed to antimicrobials.

2.3 Mechanisms of survival in macrophages

Even if taken up by immune cells, some parasites have the ability to survive the onslaught. For example, *Leishmania* spp. are taken up by macrophages but evade macrophage proteolytic processes by: (i) induction of cytokines, such as TGF-β, that inhibit or deactivate macrophages; (ii) by inhibiting phagolysosome formation; (iii) by producing antioxidases to counter products of the macrophage's oxidative burst; or (iv) by inhibiting inducible nitric oxide synthase (iNOS) expression or activity.

2.4 Antigenic variation

Parasites can evade the immune response by changing the antigens that are expressed on their surfaces, every few generations. This is achieved by random activation or rearrangement of genes that code for surface antigens. Thus antibodies produced against the former antigens become ineffective. This is best studied in the African trypanosomes. *Trypanosoma brucei* continuously changes its VSG coat to keep one step ahead of the immune system that mounts an antibody response to that particular VSG. The host's antibodies facilitate neutralization and killing of approximately 99% of the original *Trypanosoma* population. However, during this time a few of the *Trypanosoma* have shed their coat, that is, switched VSG, and covered themselves with a new antigenically distinct VSG coat. These distinct *Trypanosoma* give rise to a new population expressing the new VSG coat. The immune system again responds to this proliferated population by producing a new set of antibodies that successfully kill 99% of the *Trypanosoma* population. But once again, VSG switching among a small portion of *Trypanosoma* renders them undetectable, and they successfully evade the host immune response, and this cycle continues. Hundreds of genes encode VSGs in each trypanosome's genome. These genes lie either in the interior of the chromosome or near the telomeres. The transcribed or expressed VSG gene always lies near the telomere and, although there are multiple telomere-proximal VSGs, expression is mono-allelic so each individual cell expresses a single VSG at a time. VSG switching occurs by at least two mechanisms: (i) by exchanging or altering active gene transcription within a single chromosome – an interior VSG copy is duplicated onto cassette, translocated to the expression site at the telomere and becomes activated: and (ii) by changing the transcriptionally active telomere. Some VSG genes are expressed without being duplicated and translocated. In these cases, an active telomeric site on one chromosome is silenced and a telomeric site on another chromosome is activated. There are about 20 different expression sites. In the case of *Entamoeba histolytica*, shedding of the surface antigens provides immune evasion. This may be achieved even when the antibody-antigen complex is formed and directs the immune response away from the parasite. Some parasites secrete protease that breaks down the antigen-antibody complex or cleaves an antibody before it binds to the parasite surface. In the case of *Plasmodium*, development of an effective immune response is hampered by the fact that stage-specific proteins tend to be highly polymorphic or antigenically variable.

2.5 Host mimicry

Some parasites use molecular mimicry to escape host immunosurveillance. This can be achieved by expressing proteins/glycoproteins on the parasite surface that mimic host proteins/glycoproteins or by adsorbing host components on their surfaces. The ultimate purpose is to be recognized as host cells and thus avoid an overwhelming immune response.

2.6 Evasion of complement

Complement is a set of proteins that circulate in the blood and are activated through an antibody binding to microbial antigens or, directly, by the bacterial surface, leading to MAC attack (forming pores in microbial membranes) or opsonization of the pathogen (helps pathogen uptake by phagocytes). In the case of *Trypanosoma*, conversion of the parasites from epimastigotes to infectious trypomastigotes coincides with acquisition of resistance to lysis by the alternative complement pathway. In addition, human serum also contains *Trypanosome* lysis factors (TLF): however pathogenic species of *Trypanosoma* (*T. brucei rhodesiense*/*T. brucei gambiense*) are resistant to TLF-mediated lysis. *Leishmania* spp. resists complement by shedding the MAC complex and by inactivation of some MAC components by phosphorylation (C3, C5, C9).

2.7 Evasion of antibodies

At the mucosal surfaces, protozoa have to cope with potent secretory immunoglobulins (sIgA) and then again, in the bloodstream, IgG and other antibodies are highly effective in trapping microbes and neutralizing toxins. Pathogenic protozoa have been shown to evade antibodies by antigenic variation (indicated above), and by inactivating antibodies. For example, *Entamoeba histolytica* is known to evade humoral immunity by degradation of antibodies by secreted proteases.

2.8 Interference with host immune signaling molecules

Some protozoan pathogens interfere with the host immune signaling molecules thus affecting host ability to build an effective immune response. For example, *Leishmania* spp. reduces the production of interferon-gamma by the host, which is an important cytokine in the activation of macrophages, immune cells important in clearing protozoa from the host as well as building an effective immune response. In addition, *Leishmania* spp. suppresses the transcription of the IL-12 gene by interfering in the protein kinase C, protein tyrosine kinase and Janus family tyrosine kinase (JAK)/Stat pathway, thus blocking the helper cell 1, Th1, response. *Entamoeba histolytica* induces IL-4 and IL-10 and modulates the Th1 response. *Trypanosoma brucei* increases the levels of IFN-γ and decreases IL-2 and IL-2R, which affects T-cell responses. *Trypanosoma cruzi* blocks IL-12 and TNF-α by interfering with MAPK re-phosphorylation. *Toxoplasma* interferes with NF-κB translocation, Stat-1 translocation and MAPK re-phosphorylation resulting in blockage of the effector molecules IL-12, TNF-α and MHC-II.

3 Strategies Against Protozoa Pathogens

Despite our efforts to control and/or eradicate protozoa pathogens, they have continued to cause millions of deaths per year and remain a major contributing factor to human misery with both economical and social implications.

The importance of sanitation, good housing, nutrition and clean water supplies is paramount in reducing the number of protozoa infections. Other than improved public health measures, vaccines and antimicrobial chemotherapy are the major weapons in our fight against protozoa diseases. To maximize the chances of controlling protozoan infections, a variety of approaches must be employed and continued research should identify novel strategies to prevent (prophylaxis), eradicate and/or control these infections. Some of these strategies are indicated below.

3.1 Vaccines

Vaccination is one of the most powerful means to save lives and to increase the level of health in mankind. Vaccination has proved to be a useful tool for protecting individuals and the population and can result in disease eradication. Despite this great potential, vaccines have had little impact on human parasitic infections. A prime example is *Plasmodium*, one of the most evasive protozoan pathogens for more than 5,000 years. Even today, malaria remains a major killer, responsible for more than a million annual deaths, worldwide, reminding us of the challenges associated with effective vaccine development against protozoa pathogens. There may be many reasons for this: protozoa are eukaryotes that are biologically and genetically complex organisms, some with elaborate life cycles; our understanding of the mechanisms of immune control of many parasitic infections and how to induce high-level immunological memory is poorly understood. Despite the setbacks, vaccines remain the real hope in controlling parasitic infections. In particular, with the genome sequencing projects of many parasites underway, there is real optimism of identifying novel targets for vaccine discovery.

A vaccine is a material originating from a microorganism that is introduced into individuals in a controlled way leading to the stimulation of the immune responses without the symptoms of full-blown disease. The material may be produced artificially or directly obtained from the pathogen. Ultimately, this results in the production of memory cells within the host. If the host encounters the pathogen again, the immune system can generate a rapid antibody response thus preventing infection. However, many of the potent parasitic protozoa exhibit antigenic variation, thus the previously generated antibodies/memory cells become obsolete leading to their inability to prevent infection. Generally, vaccines are produced with the following objectives:

1. Vaccine promotes an effective resistance to disease
2. Sustained protection – lasts several years
3. Induce humoral response (antibodies)
4. Induce protective T-cell responses (cellular immunity)
5. Vaccine is safe – minimal side effects

6. Vaccine is stable and will remain so during transportation
7. Vaccine is reasonably cheap and easily administered.

However, not all vaccines fit these criteria and trials have to be conducted to determine whether the beneficial effects outweigh the potential side effects. Briefly, there are the following types of vaccines:

Whole cell vaccines

- Killed pathogen (ruptured or formalin-preserved): these vaccines should stimulate maximum protective immunity but may have severe side effects.
- Live attenuated or low virulence vaccines: these can be produced by knocking-out the virulent genes from the pathogen but may have side effects.

Protein subunit vaccines (single/multiple components)

Produced from purified proteins from the pathogen, these may induce maximum protection but may be toxic. Also, protein purification is expensive. In contrast, recombinant protein antigens are cheap and consistent. Other problems may arise from the fact that pathogen populations might be polymorphic. The vaccine may be ineffective against some strains. A simple recombinant might not stimulate all the required immune components, for example, immunogenic carbohydrate wrongly processed in production.

DNA vaccines (nucleic acid vaccines)

For these, DNA (gene) encoding one or multiple vaccine protein(s) is cloned into a plasmid-containing host expression promoter. Individuals are vaccinated with this plasmid. The plasmid infects the host cell and expresses a foreign gene or genes and produces the protein inside the cell. The antigen will be expressed on the cell's MHC class I molecules, due to the antigen being endogenous. This stimulates the activation of cytotoxic T-cells, which are important for clearance of pathogen-infected cells.

Traditional vaccines enter the MHC class II pathway, due to the fact that the antigens encountered are exogenous. This primarily stimulates antibody responses which are not effective at clearing viruses which are protected by the cells they reside in. However with DNA vaccines, once the cytotoxic T-cells have been activated, they may lyse infected cells and release the antigen allowing the antibody response to also become stimulated. These are cheaper and easier to store than purifying proteins (very stable) and may also be immunogenic but side effects remain unclear. The transient production of protein for about a month is enough to evoke a robust immune response. If any of the above vaccines are used, this process is called active immunization. In contrast, passive immunization is injection of purified antibody to produce rapid but temporary protection. Passive immunization is used to prevent a disease after known exposure, improve the symptoms of an ongoing disease, protect immunosuppressed patients, or block the action of parasite toxins and prevent disease.

With the emergence of parasitic strains that are resistant to antimicrobial compounds, there is renewed effort in the development of effective vaccines against protozoan pathogens. To this end, researchers at the National Institutes of Health have developed an experimental vaccine that could, theoretically, eliminate malaria from entire geographic regions, by eradicating the malaria parasite from an area's mosquitoes. The vaccine, so far tested only in mice, would prompt the immune system of a person who receives it to eliminate the parasite from the digestive tract of a malaria-carrying mosquito. However, the vaccine would not prevent or limit malarial disease in the person who received it. The vaccine is based on a protein, *Plasmodium falciparum* surface protein 25 (Pfs25), that is found only on the surface of the ookinette, a stage of the parasite living in the mosquito gut, and does not appear on any other stage of the parasite. However, when injected alone into humans, Pfs25 fails to generate a sufficient level of antibodies to target the parasite. Using conjugate technology, researchers chemically joined, or conjugated, Pfs25 to ovalbumin, a protein found in egg whites. The conjugate produced high levels of antibodies in mice. When the immune blood containing the antibodies was fed to mosquitoes carrying *Plasmodium falciparum*, it completely eliminated the ookinettes from the mosquito digestive tract. These findings resulted in great optimism, in that the conjugate technology could be adapted to make effective vaccines against protozoan pathogens.

3.2 Chemotherapy

Chemotherapeutic approaches remain the most common form of treatment for parasitic diseases (Table 2). These include antimicrobial compounds that are derived from other microbes, chemically synthesized or, most commonly, derived from natural products that inhibit the growth of, or kill protozoa and are called antiprotozoal agents. These provide the most direct and cheapest way of controlling these infections. Many of the currently available antiparasitic drugs were identified by screening large numbers of compounds. A limited number of drugs against some parasitic diseases are available as chemoprophylaxis, which can be used before, during and after parasite exposure (especially for individuals who are not immune and are visiting endemic areas). A major problem with drug therapy is the ability of a subpopulation of parasites to develop drug resistance. Multiple-resistant protozoa are therefore becoming a problem and the fear is that it will not be long before our limited arsenal of the antiprotozoal

Table 2 The spectrum of activity and mode(s) of action of the major antiprotozoal drugs

Drug	Mode of Action	Spectrum of Activity
Sulfonamides (a group of antibiotics such as sulfadiazine)	Inhibition of folic acid synthesis, which is crucial for DNA synthesis	*Plasmodium, Toxoplasma, Acanthamoeba*
Proguanil and Mefloquine	Inhibition of folic acid synthesis	*Plasmodium*
Suramin	Binds to enzymes in nucleus and inhibits DNA synthesis	*Trypanosoma*
Chloroquine	Inhibition of DNA synthesis	*Plasmodium*
Pentamidine	Inhibition of DNA replication and translation	*Trypanosoma, Leishmania*
Amphotericin B	Disrupts membrane function	*Leishmania, Acanthamoeba*
Eflornithine	Inhibition of protein synthesis	*Trypanosoma*
Tetracycline	Inhibition of protein synthesis	*Plasmodium, E. histolytica*
Albendazole	Inhibition of microtubule assembly	*G. intestinalis, Microsporidia*
Buparvaquone	Inhibition of energy production	*Plasmodium*
Megumine antimonate	Inhibition of energy production	*Leishmania*
Metronidazole	Inhibition of energy production	*G. intestinalis, E. histolytica*
Primaquine	Inhibition of energy production	*Plasmodium*
Difluoromethylornithine	Inhibition of polyamine synthesis, inhibiting cellular growth	*Trypanosoma*
Berenil	Inhibition of DNA synthesis	*Trypanosoma*
Nifurtimox and Benznidazole	Increases oxidant stress on intracellular parasites	*Trypanosoma*
Ketoconazole, Fluconazole and Itraconazole	Inhibits biosynthetic pathways	*Trypanosoma*
Pentostam	Inhibits macromolecular (DNA, RNA, protein) synthesis	*Leishmania*

(Continued)

Table 2 (Continued)

Drug	Mode of Action	Spectrum of Activity
Glucantime	Inhibits protein synthesis	*Leishmania*
Diiodohydroxyquin	Similar to metronidazole but works for longer periods	*E. histolytica*
Chlorhexidine and Polyhexamethylene biguanide	Disrupts membrane function	*Acanthamoeba*
Propamidine isethionate	Inhibits DNA synthesis	*Acanthamoeba*
Azithromycin	Inhibits protein synthesis	*Acanthamoeba*
Alkylphosphocholine	Induces apoptosis	*Acanthamoeba*
Qinghaosu (Artemisinin)	Produces oxidative effects	*Plasmodium, Naegleria*
Trimethoprim and Sulfamethoxazole (combination is called Co-trimoxazole)	Inhibits folic acid synthesis	*Pneumocystis*
Pyrimethamine	Inhibits DNA synthesis	*Toxoplasma*
Nitazoxanide	Inhibits energy metabolism	*Cryptosporidium*

compounds become obsolete. Potential mechanisms of drug resistance may include:

- conversion of the drug to an inactive form by an enzyme or failure to activate the drug,
- modification of a drug-sensitive site,
- permeability changes leading to decreased influx or increased efflux,
- alternative pathways to bypass inhibited reactions or decreased requirement for product of inhibited reaction,
- increased production of drug-sensitive enzymes.

Many of these mechanisms have been observed in parasites, such as alteration in cell permeability, modifications of drug-sensitive sites and increased quantities of the target enzymes. These modifications tend to arise in a population of parasites by various mechanisms, such as:

- physiological adaptations,
- differential selection of resistant individuals from a mixed population of susceptible and resistant strains,
- spontaneous mutations, followed by selection, and
- changes in gene expression.

The appearance of resistance within a population has been observed to occur within 5–50 generations. One factor that contributes to rapid acquisition of resistance is the use of a suboptimal concentration of drug, resulting in more survivors. In addition, there are no effective and/or recommended antimicrobial compounds for some of the protozoan infections, such as *Balamuthia* granulomatous encephalitis. Thus, there is an urgent need to identify novel targets in protozoa pathogens. A complete understanding of the parasite metabolism and their life cycle should help identify novel targets for the rational development of antiprotozoal drugs. However, with the growing evidence of drug resistance in protozoa pathogens, other control measures should be explored in conjunction with chemotherapeutic approaches.

3.3 Control measures
Sanitary measures
The maintenance of good sanitary measures such as draining swamps, building sewage systems and, most importantly, providing clean water supplies (in particular for waterborne protozoa), although expensive, are crucial for controlling parasitic infections.

Vector control
- Reduce human-vector contact
- Use of repellent-impregnated bed nets
- Use of vector repellents and protective clothing
- Screens, house spraying with insecticides.

Reduce vector capacity
- Environmental modifications – vector habitat alterations – measures that reduce the risks of flooding, which may provide breeding sites for vectors
- Larvacides/insecticides used during various stages of insect life cycle
- Biological control – introducing genetically modified vector in the endemic area that cannot act as host for parasites.

Reduce parasite reservoir
- Diagnosis and treatment
- Chemoprophylaxis.

Index